PROCEEDINGS OF THE FIRST SOUTHERN AFRICAN GEOTECHNICAL CONFERENCE

PROCEEDINGS OF THE FIRST SOUTHERN AFRICAN GEOTECHNICAL CONFERENCE, 5–6 MAY 2016, SUN CITY, SOUTH AFRICA

Proceedings of the first Southern African Geotechnical Conference

Editor

S.W. Jacobsz

Department of Civil Engineering, University of Pretoria, South Africa

CRC Press
Taylor & Francis Group
Boca Raton London New York

CRC Press is an imprint of the
Taylor & Francis Group, an **informa** business

A BALKEMA BOOK

Published by:
CRC Press/Balkema
P.O. Box 447, 2300 AK Leiden, The Netherlands
e-mail: Pub.NL@taylorandfrancis.com
www.crcpress.com – www.taylorandfrancis.com

First issued in paperback 2020

Typeset by V Publishing Solutions Pvt Ltd., Chennai, India

ISBN 13: 978-0-367-73712-2 (pbk)
ISBN 13: 978-1-138-02971-2 (hbk)

Visit the Taylor & Francis Web site at
http://www.taylorandfrancis.com

and the CRC Press Web site at
http://www.crcpress.com

Proceedings of the first Southern African Geotechnical Conference – Jacobsz (Ed.)
© 2016 Taylor & Francis Group, London, ISBN 978-1-138-02971-2

Table of contents

Soil reinforcement and slopes

Proceedings of the first Southern African Geotechnical Conference – Jacobsz (Ed.)
© *2016 Taylor & Francis Group, London, ISBN 978-1-138-02971-2*

Preface

The First Southern African Geotechnical Conference was organised by the Geotechnical Division of the South African Institution of Civil Engineering (SAICE) under the auspices of the International Society of Soil Mechanics and Geotechnical Engineering (ISSMGE) and took place at Sun City, South Africa on 5 and 6 May 2016. It has long been the opinion of members of the SAICE Geotechnical Division that a need existed for a local forum where geotechnical practitioners and academics from the Southern African region can present their work and interact. This is because it is often difficult for local engineers and engineering geologists to attend the African Regional Conferences on Soil Mechanics and Geotechnical Engineering due to the distances and costs associated with travelling in Africa. The First Southern African Geotechnical Conference was organised to take place approximately one year after the 16th African Regional Conference so as to not detract from this conference, while presenting a regional forum for geotechnical engineers and engineering geologists.

More than 60 papers were received form authors in South Africa, Botswana, Kenya, Tanzania, Uganda, Algeria, Austria, France, Germany, Switzerland and the United Kingdom. They represent consulting engineers and engineering geologists, contractors, academics and product specialists. The papers were grouped into the following themes: Foundations, Mining and Tailings, Modelling and Design, Site investigation, Soil Properties and Soil Reinforcement and Slopes. The wide range of topics is considered to be thoroughly representative of the current activities of the geotechnical industry in the Southern African Region. The conference, hosted over two days, was divided into two parallel sessions to allow all papers to be presented.

The conference took place approximately two years after the passing of Prof. Geoffrey Blight, Professor of Soil Mechanics at the University of the Witwatersrand and world renowned expert on unsaturated soils. The organising committee thought it appropriate to organise a memorial lecture in honour of Prof. Blight and was delighted that Prof. Andy Fourie, long-time colleague of Prof. Blight and now Head of the Department of Civil Engineering at the University of Western Australia in Perth, accepted the invitation to present the Geoffrey Blight Memorial Lecture.

The conference was organised by a small committee under the chairmanship of Prof. Gerhard Heymann of the University of Pretoria. He was assisted by Prof. S.W. Jacobsz, Chair of the Scientific Committee, Heather Davis of Verdicon as Treasurer, Trevor Green of Verdicon and current chair of the SAICE Geotechnical Division and Dr. Peter Day of Jones & Wagener, also affiliated to the University of Stellenbosch. The committee wishes to express their thanks to RCA Conference Organisers an Selah Productions for assistance with the conference arrangements and to the CRC Press/Balkema—Taylor & Francis Group who published the conference proceedings.

The organising committee is grateful to all authors who submitted technical papers and to the members of the Scientific Committee who conducted the review process. All papers were peer reviewed by two independent reviewers. The committee also wishes to thank all sponsors who contributed financially to ensure the success of the conference.

S.W. Jacobsz & Gerhard Heymann
Pretoria, South Africa

Proceedings of the first Southern African Geotechnical Conference – Jacobsz (Ed.)
© 2016 Taylor & Francis Group, London, ISBN 978-1-138-02971-2

Committees

ORGANISING COMMITTEE

Gerhard Heymann, *Chair, University of Pretoria, Pretoria*
S.W. Jacobsz, *Chair of Scientific Committee, University of Pretoria, Pretoria*
Heather Davis, *Treasurer, Verdicon, Midrand*
Trevor Green, *Verdicon, Midrand*
Peter Day, *Jones & Wagener, Johannesburg*

SCIENTIFIC COMMITTEE AND PAPER REVIEWERS

S.W. Jacobsz, *Chair, University of Pretoria, Pretoria*
Alan Parrock, *ARQ, Pretoria*
Charles Macrobert, *University of the Witwatersrand, Johannesburg*
Danie Labuschagne, *University of Pretoria, Pretoria*
Dawie Marx, *University of Pretoria, Pretoria*
Denis Kalumba, *University of Cape Town, Cape Town*
Dimiter Alexiew, *Huesker Synthetic GmbH, Gescher, Germany*
Eduard Vorster, *Aurecon, Pretoria*
Florian Hörtkorn, *SRK, Johannesburg*
Gary Davis, *Aurecon, Pretoria*
Luis Torres-Cruz, *University of the Witwatersrand, Johannesburg*
Gerhard Heymann, *University of Pretoria, Pretoria*
Heather Davis, *Verdicon, Midrand*
Irvin Luker, *University of the Witwatersrand, Johannesburg*
Jacobus Breyl, *Jones & Wagener, Johannesburg*
Jean Potgieter, *University of Pretoria, Pretoria*
Nico Vermeulen, *Jones & Wagener, Johannesburg*
Nicol Chang, *Franki Africa, Johannesburg*
Louis Geldenhuys, *University of Pretoria, Pretoria*
Peter Day, *Jones & Wagener, Johannesburg*
Phil Paige-Green, *Tshwane University of Technology, Pretoria*
Tiago Gaspar, *University of Pretoria, Pretoria*
Trevor Green, *Verdicon, Midrand*

Foundations

Proceedings of the first Southern African Geotechnical Conference – Jacobsz (Ed.)
© 2016 Taylor & Francis Group, London, ISBN 978-1-138-02971-2

Capacity and performance of Franki enlarged-base piles

J.H. Engelbrecht & H. Kgole
Franki Africa, Johannesburg, South Africa

ABSTRACT: The Franki driven cast in situ enlarged-base pile has been used extensively throughout Southern Africa for the past 68 years and remains one of the more effective pile types. This self-testing, procedure reliant pile exhibits excellent load/deflection characteristics and capacity, primarily accredited to the soil compaction and pre-loading when forming the enlarged bulbous base. This paper presents the results of 63 static load tests performed on Franki enlarged-base piles. The results indicate higher than expected bearing capacities achieved across all founding conditions. An empirical correlation was also developed between the ultimate bearing pressure of the Franki enlarged-base piles and the unconfined compressive strength of the founding weathered rock, q_a.

1 INTRODUCTION

The Franki driven cast in situ enlarged-base pile remains one of the most effective and economical pile types, generating exceptional capacity at a shallower founding depth, when compared to other pile types, without compromising performance. This is largely due to the densification of insitu material developed during the dynamically driven driving tube and formation of the bulbous base process (Tchepak, 1986). The capacity and performance of the Franki enlarged-base pile is inherently attributed to this densification process.

This paper presents the results of 63 statically load tested Franki enlarged-base piles from 27 test locations throughout Southern Africa. These test piles were generally pre-drilled and installed in various conditions, generalized to piles founded in cohesive (Silt/Clay), non-cohesive (Sand) and on weathered rock (Rock). The test data is used to provide insight into the bearing capacity performance of the Franki enlarged-base piles and compared to industry design parameters.

1.1 *Construction of the Franki enlarged-base pile*

The installation process of a Franki enlarged-base pile, or simply enlarged-base pile, consists of a bottom-driven plugged driving tube and the formation of the bulbous base and pile shaft.

The driving tube is advanced by bottom driving on a zero slump concrete or gravel plug, formed at the bottom of the tube (Nordlund, 1982). This driving process can either be carried out from piling platform level or from a predrilled depth, with the latter improving production rates. The appropriate founding depth is established by measuring

Figure 1. Pile test locations.

the penetration rate of the tube for ten blows and compared to the design requirement. This is also referred to as the set criteria. Once this founding depth is achieved, the driving tube is held in position to prevent further penetration and the plug expelled (Byrne & Berry, 2008).

The formation of the enlarged bulbous base is followed by compacting a zero slump concrete, which is introduced through the tube, with the compaction blow counts and volume of expelled concrete recorded. This process is repeated until the enlarged base has reached the desired energy levels (Byrne & Berry, 2008).

The pile reinforcement is lowered into the tube, and driven into the base when tension capacity is required, and the tube filled with high slump self-compacting concrete whilst the tube is slowly extracted. This installation sequence is illustrated in Figure 2.

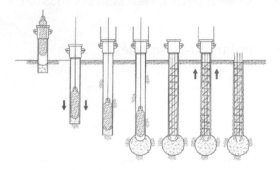

Figure 2. Installation sequence of an enlarged-base pile (Byrne & Berry, 2008).

Enlarged-base piles are predominantly designed in end-bearing with pile shaft diameters ranging from 250 mm to 610 mm for conventional enlarged-base piles with pre-drilled enlarged-base piles seldom exceeding 1200 mm. These less common larger diameter enlarged-base piles are installed as a pre-drilled piles, utilizing a mini Dynamic Compaction (DC) pounder to create the enlarged-base.

There are numerous advantages of this self-testing procedure reliant pile type mainly attributed to its driving and compaction process.

The main advantage is the excellent bearing capacity and load/deflection characteristics generated by the enlarged-base, developing lower base deflections in both cohesive and non-cohesive material. This results in a shallower founding level for a more economical pile type in comparison to conventional bored piles. This exceptional capacity is also achieved on weathered rock where discontinuities within the rock can also be improved to some extent (Tchepak, 1986).

Confidence is improved with a guaranteed bearing capacity of the enlarged bulbous base due to the self-testing installation process using the wave and cavity expansion equations by Smith (1960) and Nordlund (1982) respectively.

Large tension loads can also be resisted due to the enlarged base forming an anchor effect at the base of the pile (Byrne & Berry, 2008). The specifically designed tension pile reinforcement is constructed and properly driven into the enlarged bulbous base to ensure that tension failure does not occur and the pile shaft/base interface.

2 METHOD OF INVESTIGATION

2.1 Geology

Southern Africa is defined by eleven countries, namely, South Africa, Swaziland, Lesotho, Namibia, Botswana, Zimbabwe, Mozambique, Tanzania, Malawi, Zambia and Angola (Figure 1). The geology in this area is highly variable and complex. The rocks range from silicic to mafic igneous rocks; from clastic to organic sedimentary rocks; and from contact to regionally metamorphosed metamorphic rocks. The soils are either of transported (i.e.: Alluvial, Hillwash, Colluvial, etc.) or residual material and to a lesser extent, soils of human origin. For the purpose of this article we shall look at the regional geology of the piles throughout portions of South Africa, Botswana, Mozambique and Angola.

The sites on the southwestern portion of South Africa are underlain by Quartzites, Shales and Tillites of the Table Mountain Group; Syenitic rocks, Granites, Quartz porphyry of the Cape Granite Suite; Quartzites, Arkose, Shales, Phylite, Tillites, Lavas and Tuffs of the Nama Group; and Unconsolidated Superficial Deposits, Conglomerates, Limestones, Sandstones, Marls and Gravels of Quaternary to Tertiary Deposits (Geological Survey, 1970).

The sites on the eastern portion of South Africa are underlain by Sands of the Berea formation; variable soils of the Mkwelane Formation; Siltstones and Sandstones of the St Lucia Formation (Brink, 1979).

The site on the northeastern Portion of South Africa is underlain by Basalts of the Sibasa formation; and Gneiss, Granites and Migmatites of the Sand River Gneiss (Geological Survey, 1970).

The sites on the north-northeastern Portion of South Africa are underlain by Shales of the Silverton Formation; Shales and occasional Quartzite of the Government Subgroup; and Norites and Anorthorsites of the Rusternburg Layered Suite and Quartzites of the Pretoria Group (Geological Survey, 1970).

The site on the north-northwestern portion of South Africa is underlain by Manganese Bearing Deposits and Unconsolidated Superficial Deposits of the Quaternary to Tertiary Era (Geological Survey, 1970).

The sites in Swaziland are underlain by Ngwane and Mahambe Gneisses and Granite Greenstone Terrane with intruded plutons such as, Usushwana Gabbros and Pyroxinites complexes; Diabase and Dolerite Dyke intrusions (van Straaten, 2002).

The site in Botswana are underlain by Alluvial Sands and Granites of the Basement Gneiss Complex (Geological Survey Department Lobatse, 1984).

The site in Angola are underlain by Marine Sands of Quaternary and Tertiary Deposits (International Geological Mapping Bureau, 1985–1970).

2.2 Pile testing procedure

The pile tests were carried out under static condition conforming to SABS 1200F and BS 8004:1986 and/or specific contract and building code requirements with test loads ranging from 150% to 300% of working load. Pile testing commenced at least 7 days after casting or when shaft concrete had reached its required strength.

A recently calibrated cylindrical hydraulic jack was used to develop the reaction load with four enlarged-base piles generally used as reaction piles. Strand anchors and micro piles may also be used to provide reaction where access is restricted.

The testing load was applied according to the loading procedure with the pile head deflection recorded by four dial gauges mounted on an independent reference beam, also monitored by two separate dial gauges. The results of the tests is graphically presented on a load/deflection plot including, in most cases, a time/deflection and time/load plot as per BS 8004:1986.

2.3 Geotechnical design

Enlarged-base piles are predominantly designed in end bearing with only the end bearing capacity for the tested piles analyzed as part of this paper.

2.3.1 Piles founded in cohesive and non-cohesive material

Pile design in southern Africa is largely reliant on empirical correlations, for ultimate shaft and end bearing pile capacities, using insitu tests such as Standard Penetration Tests (SPT). This insitu correlation using SPT N-values was used for piles founded in non-cohesive material and cohesive material.

Typically used design correlations and guidelines for piles founded in cohesive and non-cohesive

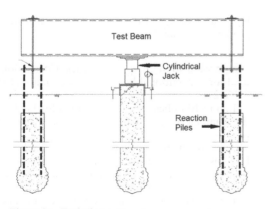

Figure 3. Typical test pile setup.

material as well as piles founded on rock are given in Table 1.

2.3.2 Piles founded on rock

Pile capacity design for piles founded in rock is based on the allowable bearing pressure of the rock based on the unconfined compressive strength of the intact rock, q_a. These typical ultimate base capacities are indicated in Table 2.

3 RESULTS AND DISCUSSIONS

The collated test data, depicted in the figure below, indicate a distinction between the various founding strata. These distinctions are separated and analyzed individually to establish the typical pile capacities of the founding strata.

Ultimate pile capacities are back analyzed using the hyperbolic load transfer function developed by Fleming (1992). The failure criteria for the ultimate pile capacity back analysis for piles not tested to failure is, for the purpose of this study, assumed at 10% of the pile shaft diameter.

An interesting observation was made in comparing the predicted to measured pile settlement and ultimate capacities as depicted in Figure 5. The data indicates conservative settlement predictions. The ultimate pile capacity predictions for piles founded in sand and silt/clay are optimistic in comparison to the conservative predictions made for piles founded on rock.

Table 1. Ultimate bearing capacity correlations for enlarged-base pile using insitu test data.

	Cohesive		Non-cohesive	
Source	q_b	$q_{b\,max}$	q_b	$q_{b\,max}$
	MPa	MPa	MPa	MPa
Byrne & Berry (2008)	0.06 N	6.0	0.5 N	15.0
Meyerhof (1956)			0.4 N	

Table 2. Range of ultimate base capacities for driven and bored piles founded in rock (Byrne & Berry, 2008).

	Q_b
Description	MPa
Code of practice	$0.5q_a$
General range	$4q_a$ to $11q_a$
Reasonable correlation for rock with minimal jointing effects	$5q_a$
Socket length to pile diameter ratio >2	$20q_a$

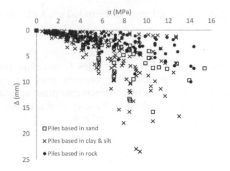

Figure 4. Pile head settlement (Δ) vs applied shaft stress (σ) plot.

3.1 *Test piles founded in non-cohesive sand*

Test piles founded in material with a PI < 10 are defined as piles founded in non-cohesive sand with a designed ultimate base capacity limited to 15 MPa (Byrne & Berry, 2008).

The Figure 7 indicates a good correlation for the piles founded in medium dense to dense material with q_b = 0.43 N and a better than expected correlation of q_b = 0.76 N for piles founded in dense to very dense material. The latter does have a large degree of scatter with a lower bound SPT N-value correlation of q_b = 0.59 N.

3.2 *Test piles founded in cohesive silt/clay*

Test piles founded in a sandy silts and clays with a plasticity index greater than 10 and unified soil classification of ML, CL and OL are, for the purpose of this paper, classified as cohesive material. No piles were founded in pure silt or clay, with high PI and liquid limit, with a cohesive classification in Byrne & Berry (2008).

Figure 8 indicates the capacity performance according to the consistency of the founding material also indicating a distinction between the consistencies of the founding material.

Figure 9 indicates the exceptional performance of the enlarged-base piles founded in sandy silt/clay with similar performance, as expected, to enlarged-base piles founded in sand. The average correlation for bearing capacity vs SPT N-Value is calculated to q_b = 0.45 N, 0.57 N and 0.41 N for piles founded in firm becoming stiff, stiff becoming very stiff and very stiff soil consistencies. A much larger degree of scatter is observed, especially for piles founded in stiff material, with a respective lower bound correlation of q_b = 0.37 N, 0.38 N and 0.2 N.

3.3 *Test piles founded on rock*

The bearing capacity design for enlarged-base piles founded on rock is limited by the unconfined

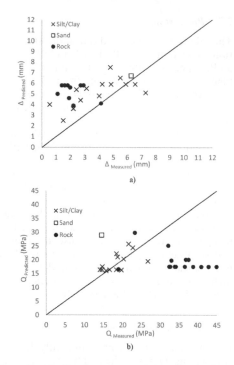

Figure 5. Predicted vs measured data for a) Pile head settlement and b) Pile capacity.

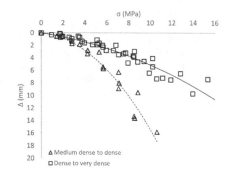

Figure 6. Pile head settlement vs applied shaft stress plot for piles founded in sand.

compression strength, q_c, of the rock. Correlations used for conventional bored piles are indicated in Table 2.

Figure 10 indicates the exceptional load/deflection characteristics for enlarged-base piles founded on rock.

The ultimate bearing capacity, q_b, was back analyzed using Fleming (1992) and plotted in relation to the unconfined compression strength of the intact founding rock, q_a as indicated in Figure 11.

Figure 11 indicates the test result for piles founded on very soft rock Shale and Norite and

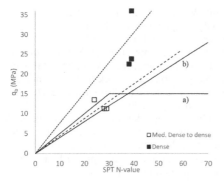

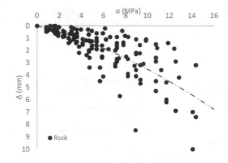

Figure 7. Back analyzed bearing capacities for piles founded in sand correlated to SPT N-values and plotted against to the design capacity correlations; a) 0.5 N limited to 15 MPa and b) 0.4 N as per Table 1.

Figure 10. Pile head settlement vs applied shaft stress plot for piles founded on rock.

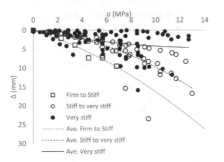

Figure 8. Pile head settlement vs applied shaft stress plot for piles founded in silt/clay.

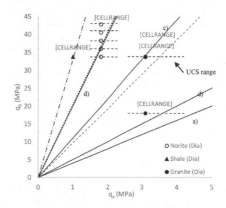

Figure 11. Ultimate bearing capacity, q_b, plotted against unconfined compression strength, q_a including typical design correlations, q_b; a) $4q_a$, b) $5q_a$, c) $11q_a$ and d) $20q_a$ as per Table 2.

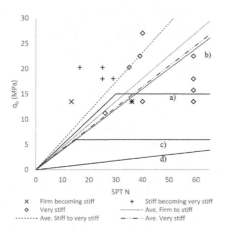

Figure 9. Back analyzed ultimate bearing capacities for piles founded in silt/clay correlated to SPT N-values plotted against the design correlations; a) 0.5 N limited to 15 MPa, b) 0.4 N, c) 0.5 N limited to 6 MPa and d) 0.06 N as per Table 1.

very soft to soft rock Granite with an ultimate bearing capacity correlation, $q_b = 33.0q_c$, $20.3q_a$ and $9.6q_c$ when using the average q_c of 1.0 MPa, 1.8 MPa and 3.1 MPa respectively. The average q_c is based on the available information from founding depth to a depth of $3 \times$ pile diameters below with the range of indicated in Figure 11.

The results indicate that the enlarged-base piles performed exceptionally well compared to Code of Practice recommendation and generally accepted correlations of $0.5q_c$ and $5q_c$ respectively (Byrne & Berry, 2008; Williams et al., 1980). The measured performance relates closer to rock socketed bored piles with a socket length to pile diameter ratio exceeding 2.0 (Byrne & Berry, 2008).

An interesting observation was made in comparing q_b to the pile diameter indicating a reduction in q_b with an increasing pile diameter. Although speculative, and highly dependent on the jointing and weathering of the founding rock, this reduction can be attributed to the basing inefficiency as noted in section 1.1 and also possibly due to scale effects by Meyerhof (1983).

Table 3. Summarized recommendations for ultimate bearing capacities correlated to SPT N-values according to the consistency of the founding material.

Founding material category	Consistency/Range	q_b MPa	$q_{b\,max}$ MPa
Cohesive sand	Medium dense to dense	0.4 N	15
	Dense to very dense	0.5 N	0.7 N, 15
Non-cohesive sandy silt/clay	Lower bound	0.2 N	15
	Average	0.4 N	15

Table 4. Summarized ultimate bearing capacities correlations for enlarged-base piles founded on weak rock.

Rock type	Rock strength classification	Tested q_a range MPa	q_b MPa
Shale	Very soft rock	1	$33q_a$
Norite	Very soft rock	1.5–2.3	$20q_a$
Granite	Very soft rock to Soft rock	2.6–4.2	$9q_a$

4 CONCLUSION

63 Driven cast in situ enlarged-base pile tests were carried out across southern Africa and back analyzed using the hyperbolic load transfer function by Fleming (1992). The back analyzed ultimate bearing capacities could be correlated to SPT N-values and unconfined compression strengths, q_c and compared to documented design correlations by Byrne & Berry (2008), Meyerhof (1956) and Williams et al. (1980). The following conclusions can be made following the analysis of the data.

The ultimate bearing capacity correlations for enlarged-base piles founded in non-cohesive sand and cohesive sandy silt/clay, of medium to low plasticity, was founded to be similar to current design correlations for non-cohesive material by Byrne & Berry (2008) and Meyerhof (1956). Recommendations can be made for piles founded in a material of specific consistency as summarized in Table 3 with additional maximum capacities indicated limited to 15 MPa (Byrne & Berry, 2008). These values are given as a guideline as consolidation settlement over time was not evaluated and should form part of the detailed geotechnical design.

An initial ultimate bearing capacity (q_b) correlation to unconfined compression strength of the intact founding rock (q_a) could be established for enlarged-base piles founded on weak rock. There are currently no such correlation specific to enlarged-base piles with design correlations for driven and bored piles by Byrne & Berry (2008) and Williams et al. (1980) currently adopted. The analyzed pile test results indicate a much higher correlation similar to bored piles socketed in weak rock by Williams et al. (1980) as summarized in Table 4. It is therefore recommended that enlarged-based piles based on rock to be designed according to Williams et al. (1980) with q_b correlation ranging from $5q_a$ to 20qa.

Design consideration must be made for discontinuities of the founding rock, scale effects by Meyerhof (1983) and base compaction inefficiency of large diameter pre-drilled enlarged-base piles > 610 mm in diameter.

REFERENCES

Brink, A.B.A. 1979. *Engineering Geology of Southern Africa. Post-Gondwana Deposits*. Vol. four.

British Standard Institute. 1986. *British Standard Code of Practice, Foundations*, BS 8004:1986, 98–99, 105–108.

Byrne, G.P.B & Berry, A. 2008. *A practical guide to geotechnical engineering in Southern Africa*. Johannesburg: Franki Africa.

Fleming, W.G.K. 1992. A new method for single pile settlement prediction and analysis. *Geotechnique* 42(3): 411–425.

Geological Survey. 1970. *Geological Map of the Republic of Southern Africa and the Kingdoms of Lesotho and Swaziland*. 1:1 000 000.

Geological Survey Department Lobatse. 1984. *Geological Map of the Republic Of Botswana*. 1:1 000 000.

International Geological Mapping Bureau. 1985–1970. *International Geological Map of Africa*. 1:5 000 000.

Meyerhof, G.G. 1956. Penetration Tests and Bearing Capacity of Piles. *Journal of the Soil Mechanics and Foundation Division*, ASCE, Vol. 82(1): 1–19.

Meyerhof, G.G. 1983. Scale Effects of Ultimate Pile Capacity. *J. Geotech. Engrg*, 109:6(797): 797–806.

Nordlund, R.L. 1982. Dynamic formula for pressure injected footings. *Journal of the Geotechnical Engineering Division*, ASCE, 108(GT3): 419–435.

Smith, E.A.K. 1960. Pile driving analysis by the wave equation, *Journal of the Geotechnical and Foundation Engineering Division*, ASCE, Vol. 86(4): 35–61.

South African Bureau of Standards, 1983. *Standardized Specification for Civil Engineering Construction, F:Piling*. SABS 1200F-1983. 12–14.

Tchepak, S. 1986. Design and construction aspects of enlarged base Frankipiles. *Proceedings of the Speciality Geomechanics Symposium, Adelaide, 18–19 August*: 160–165.

van Straaten, P. 2002. *Rocks for crops: Agrominerals of sub-Saharan Africa. Swaziland*, pp. 275–277. van Straaten, P. [online]. Available from: http://www.uoguelph.ca/~geology/rocks_for_crops/51swaziland. PDF/ [Accessed 28th December 2015].

Williams, A.F., Johnston, I.W. & Donald, I.B. 1980. The design of socketed piles in weak rock. *International Conference on Structural Foundations on Rock, Sydney*. Vol. one.

Proceedings of the first Southern African Geotechnical Conference – Jacobsz (Ed.)
© 2016 Taylor & Francis Group, London, ISBN 978-1-138-02971-2

Recent use of the internally-jacked pile test in South Africa

I. Luker
University of the Witwatersrand, Johannesburg, South Africa

ABSTRACT: The use of the internally-jacked load test of foundation piles is described. Two recent examples in South Africa are given, one done by an overseas contractor and one done by a South African contractor.

1 INTRODUCTION

In an internally-jacked pile test, load on the pile is obtained by incorporating into the shaft of the pile a (disposable) hydraulic jack, as shown in Figure 1. When activated, the jack pushes up and down with equal force, loading the shaft length above it and the shaft length plus base below it. Separate measurements of movement of the lengths above and below the jack are made and their behaviours are added to give the equivalent head load vs. head displacement graph.

The main advantage of internal jacking is that an external reaction (from kentledge or anchors) is not needed, and sufficient testing force can be generated for any size of pile. This makes internally-jacked tests cheaper than head load tests in many cases, particularly for piles bigger than approximately 600 mm, and sites where only a small number of tests will be done. A further advantage is that most of the preparation for the test is done offsite, so that the time taken on the site construction programme is less than needed for a conventional externally jacked test.

The disadvantage of the internally-jacked test is that after one of the shaft lengths reaches its maximum capacity, no greater force can be generated to see the maximum capacity of the other length. This means that the choice of the position of the jack in the length of the shaft must be done carefully, taking into account the strengths of the strata along the shaft and under the base, so that the maximum resistances of the pile lengths above and below the jack are as equal as possible.

The internally-jacked pile test was first done by Professor Osterberg at north-Western university and a jack device was patented by him in the USA in 1986, (Osterberg, 1998). Another patent for an improved device was granted in the USA in 1996. The patents have also been lodged in other countries, but not South Africa. Internally-jacked tests have been vigorously promoted by osterberg and his associates, so that the name given to his internal jack, "osterberg cell" has become, to many people, synonymous with the test technique.

Compared to many other countries, few internally—jacked tests have been done in South Africa. The earliest was at the university of the Witwatersrand in 1986, by undergraduate students supervised by the present author, and sponsored by the Franki Pile company. Unfortunately few readings were obtained in the loading of the pile and so the results were not published. Since then, to the author's knowledge, only 12 more internally-jacked tests have been done: nine commercially and three as research/development exercises.

2 INTERNALLY-JACKED PILE TESTS AT MOUNT EDGECOMBE INTERCHANGE, DURBAN

In 2013 a UK company, which is licensed in the uK to use the name osterberg cell, was the nominated sub-contractor for internally-jacked tests on three bored piles of 900 mm diameter and approximately 30m length at the Mt. Edgecombe interchange. The soil strata were coarse and fine soils of great depth, and the toe of the pile did not reach rock. Most of the lengths of the piles were below the water table so the piles were constructed using the screwed-in cased auger pile technique.

The position of the jack in each pile was approximately 2/3 of the length of the pile from the top, and in all cases the length below the jack reached its maximum capacity first. This meant that the maximum capacity of the upper 2/3 of the shaft was not seen and its force vs. displacement behaviour only seen up to approximately 3 mm. To get the equivalent head load vs. head displacement graph up to a desired mount of displacement, it was therefore necessary to extrapolate the force vs. displacement behaviour of the upper 2/3 of the shaft from approximately 3 mm to 20 mm.

3 INTERNALLY-JACKED PILE TEST AT KING SHAKA AIRPORT, DURBAN

3.1 General description

For a deep basement excavation at King Shaka Airport, 900 mm diameter piles were constructed round its perimeter by the continuous flight auger method. Soil strata were mostly coarse sand with some silt, both transported (including fill) and residual, underlain by sandstone of the Karoo geological group. All piles were socketed into material of rock consistency.

The pile tested was a dedicated test pile, shown in Figure 1. The "strain rods" mentioned on the figure are 1m lengths of y12 with electrical resistance strain gauges on them. The reinforcement cage was chosen to be just sufficient for the jack and the strain rods to be attached to it and lowered into the augered hole. Because the cage and jack can be accurately centralised in the hole, very little bending of the pile occurs during an internally-jacked pile test, and the (heavier) working pile reinforcement makes a negligible contribution to the axially loaded behaviour so is unnecessary in a dedicated internally-jacked test. Figure 2 shows

Figure 2. Jack ready to be lowered into the pile hole.

the jack attached to the longitudinal reinforcement just prior to its installation in the hole.

Values of force provided by an internal jack during the test are obtained from measurement of the hydraulic pressure applied, multiplied by its internal area. This is checked by calibration before the jack is brought to site. Displacements in this test were measured using tell-tale rods from the bottom and top of the jack, through tubes, to the top of the pile.

Displacement of the top of the pile was also easured.

3.2 Test procedure

This was specified by the supervising geotechnical engineer as the "Quick load test method" of AStM d1143 "Standard test Methods for deep Foundations under Static Axial compressive Load". The Quick method description (clause 8.2.2) can be abbreviated as follows.

> Apply the test load in increments of 5% of the anticipated failure load. uring each load interval, keep the load constant for not less than 4 minutes and not more than 15 minutes.
> Remove the load in five to ten equal decrements, keeping it constant as for load increasing.

This test method allows the test to be completed in a few hours, compared to the present South African standard (SAnS 1200 Part F 1983) which requires many days, even in freely draining soils.

Figure 3 shows the graph of jack force vs. time that actually occurred in this test. Planned increments were 250 kN, but the first one was accidentally 400 kn. When applying the increment to 3500 kn, the electric pump was initially unable to hold the required pressure, which fell, but then rose to 4200 kn. This was because a leak had developed somewhere below ground, the pump's reservoir

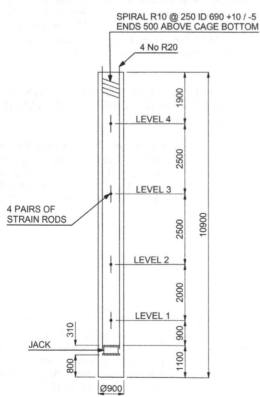

SPIRAL R10 @ 250 ID 690 +10 / -5
ENDS 500 ABOVE CAGE BOTTOM

4 No R20

1900

LEVEL 4

2500

LEVEL 3

4 PAIRS OF STRAIN RODS

2500

10900

LEVEL 2

2000

LEVEL 1

900

310

1100

JACK

800

Ø900

Figure 1. Section through test pile at King Shaka Airport.

nearly emptied and the supply of fluid to the pump had become intermittent. While the load was being held at 4200 kn, large movements were seen at the top of the pile and on the tell-tale from the top of the jack, indicating that the maximum capacity of the length of the pile above the jack had been reached.

Because of the leak and the irregular behaviour of the pump caused by it intermittently sucking air, the unloading of the pile was not as smooth as was specified, as can be seen in Figure 3.

3.3 Test results

Figure 4 shows the force applied by the jack plotted against: (i) displacement of the top of the pile; (ii) displacement of the top of the jack; (iii) displacement of the bottom of the jack. no serious irregularities in the graphs were caused by the less than optimum load vs. time record.

3.4 Interpretation of measurements into the equivalent head load vs. head displacement behaviour

When a pile is loaded at its top, and if the elastic compression of the pile is ignored, the top and bottom of the pile will displace by the same amount. Hence

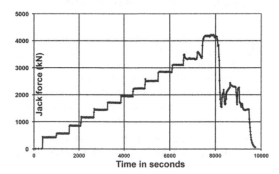

Figure 3. Jack force vs. time.

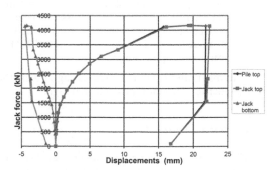

Figure 4. Jack force vs. displacements at three places.

the basic procedure (ignoring pile compression) for interpreting the internally jacked test is:

i. choose intervals of displacement of the part of the pile that displaced the lesser;
ii. for each displacement value, see from Figure 4 what forces on the pile parts above and below the jack produced the same displacement value;
iii. add the forces together to get a point on the head load vs. head displacement graph.

This has been done and is shown on Figure 5. the maximum force seen on this graph is 6900 kn, which is less than the maximum total applied force in the test of $2 \times 4160 = 8320$ kn seen on Figure 3. This is because the maximum displacement of the lower part of the pile is 4.5 mm, and at this displacement the upper part of the pile has reached a force of only 2740 kn.

To synthesise a longer head force vs. head displacement graph, the force displacement behaviour of the lower part of the pile (in this test) must be extrapolated to a displacement closer to that for which the behaviour of the upper part of the pile is known. However, because the toe of the pile is socketed into rock, this is difficult to do. In this case, because 6900 kn was a satisfactory test value for this pile, such an extrapolation was not done. For a "floating" pile, founded above rock, such as those at Mt edgecombe, the extrapolation can be easily done, because the side shear stress does not change much in the vicinity of the jack.

Note that a further refinement of the procedure to synthesise the head load vs. head displacement graph is to include the effect of elastic shortening of the pile shaft. The effect of this shortening can be seen on Figure 4, where the measured displacements at the top of the jack are slightly greater than those at the top of the pile. However the differences are small and so this refinement was not done.

3.5 Interpretation of the strain rod measurements

The mean of the two measured strains at each strain rod level is plotted along the length of the pile,

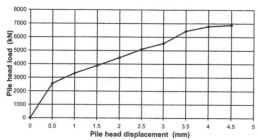

Figure 5. Equivalent head load vs. head displacement.

for chosen intervals of the jack load increments, as shown in Figure 6. the strains that occurred at the level of the bottom of the jack (where the jack opened) are extrapolated from the strains at rod level 1 using the same slope as from level 1 to level 2. This is shown on Figure 6 by dotted lines. In fact the ground will probably be stiffer from level 1 to the jack, meaning that the extrapolation lines ought to be flatter, but it is difficult to estimate this increase in stiffness.

Figure 7 shows the jack force plotted against the extrapolated strains. The slope of this graph is the conversion factor from strain to force in the shaft.

The shear stress values in Figure 8 are the means over the lengths between strain rod levels. They are obtained from the difference in the forces in the pile at sequential strain rod levels, divided by the shaft area between those levels.

In Figure 9, graphs of shaft shear stress vs. shaft displacement are plotted using the mean of the measured values of displacement of the top of the jack and the top of the pile. As already mentioned, there is a negligible difference between these two.

The strain rod positions are usually chosen to be at the boundaries between the distinguishable soil strata. then Figure 9 would give the shear stress v. displacement for each of the strata. However in the

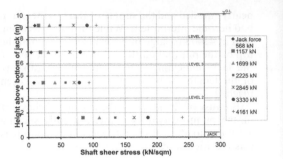

Figure 8. Mean shaft shear stresses for 7 stages of loading.

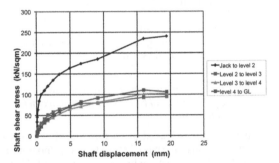

Figure 9. Shaft shear stress vs. shaft displacement.

present case, the depth to rock was found to be 30% less than expected, causing the pile cage to protrude a long way above ground and consequently the strain Rod positions did not correspond to strata boundaries as well as could be wished.

The advantages from strain rod measurements are as follow.

i. The distribution of shear stress along the pile can be seen and compared to what the pile designer predicted.
ii. The proportions of load carried in end bearing and shaft side shear can be seen.
iii. The graphs of shaft shear stress vs. shaft displacement and end bearing stress vs. displacement can be used with the load transfer method of modelling pile mechanics (Everett, 1991) to optimise the pile design.

Items (i) and (ii) may be reassuring or alert the designer to a need for a change in design.

4 CONCLUSIONS

The comparative economy of the tests can be seen from the fact that the cost to the piling contractor of using the UK testing contractor at Mt

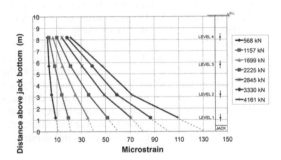

Figure 6. Mean strains at strain rod levels for 7 stages of loading.

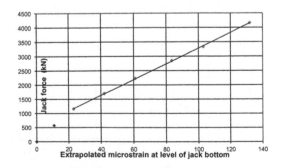

Figure 7. Jack force vs. extrapolated strain at level of jack bottom.

edgecombe was more than r600 000 per test, and the cost for the King Shaka test was r78 000. The latter could have been less, but the piling contractor Wepex generously paid what had already been allowed for a conventional test, to help finance further research into other methods of pile testing.

Although no changes to the pile design were done at King Shaka Airport as a result of this particular test, the potential for improvements in pile design has been shown. this potential can be realised providing: i. strain rods are incorporated; (ii) the pile test is done early in the project programme.

ACKNOWLEDGEMENTS

The author gratefully acknowledges the assistance of the following companies in the preparation of this paper: Franki Africa, SMec, tambew, Wepex.

REFERENCES

Everett, J.P. 1991. Load transfer functions and pile performance modeling. *Geotechnics in the African Environment*. 229–234. Balkema.
Osterberg, J.O. 1998. The osterberg load test method for drilled shafts and driven piles—the first ten years. *Proc 7th intern. conf. on piling and deep foundations, 1–17*. Vienna, Austria.

Proceedings of the first Southern African Geotechnical Conference – Jacobsz (Ed.)
© 2016 Taylor & Francis Group, London, ISBN 978-1-138-02971-2

The use of strain measurements along the shaft of a pile being load tested

I. Luker
University of the Witwatersrand, Johannesburg, South Africa

ABSTRACT: Two methods of measurement of strain in pile tests are described in terms of cost and reliability. The implications of their reliability are discussed.

1 INTRODUCTION

By measuring the longitudinal strain at points along the shaft of a pile while it is loaded, the way in which it transfers load into the ground can be deduced. Forces in the shaft at the strain measurement points are found using a conversion factor, then, by extrapolation, the force transferred by bearing stress under the toe is found. The difference in forces at two sequential points in the shaft is the load transferred by shear stress between the shaft and the soil. The mean shear stress between the points is therefore the difference in the forces divided by the shaft area between them.

This technique has been used since the invention of the electrical resistance strain gauge in 1938 and much literature has been generated describing its application. For example, a good review is given by Hayes & Simmonds (2002).

2 METHODS OF STRAIN MEASUREMENT

Three methods have been used: (i) Electrical Resistance (ER) strain gauges; (ii) Vibrating Wire (VW) strain gauges; (iii) fibre optic strain measurement.

Electrical resistance strain gauges: These are tiny metal grids which are usually stuck onto reinforcing steel bars in the pile. Strain in the steel causes proportional change in the grid's electrical resistance, causing a small change in a voltage from an electrical circuit arrangement of the grids called a Wheatstone bridge. The voltage changes are very small, in the range up to a few hundred microvolts, therefore the system is very vulnerable to the ingress of any moisture to the soldered contacts between the gauge grids and their interconnecting wires.

Vibrating wire strain gauges: These use the fact that the frequency of vibration of a plucked taut wire is proportional to the tension in the wire. If strain occurs over the length between the end anchors of the wire, changing its tension, the frequency change gives a measure of the strain. They use higher voltages than ER gauges so are much less vulnerable to moisture and therefore more reliable, (Hayes & Simmonds 2002). However, they are much more expensive than ER gauges.

Fibre-optic strain measurement: This is presently even more expensive than VW strain gauges, both for the devices cast into the concrete and for the instrumentation. However, it is better able to measure strain at multiple points than ER and VW strain gauges, so has good potential if prices reduce.

3 COMPARATIVE COSTS OF ELECTRICAL RESISTANCE AND VIBRATING WIRE STRAIN GAUGES

For the purpose of comparison, an example pile is assumed to have two diametrically opposite gauge points at each of five levels and the mean length of cable per gauge is 20 m. It is further assumed that the instrumentation for monitoring the gauges needs to be capable of driving 20 gauges and automatically recording the measurements.

A choice must be made of the manner in which the gauges are used. Consider first the ER gauges. These can be stuck on the reinforcing bars in the pile's cage on site, but this is difficult. An easier and cheaper method is to stick them, under workshop conditions, onto a separate piece of steel. This is a length (approximately 1 m) of concrete reinforcing steel that is then fastened into the pile's cage, as shown in Figure 1. Such a bar has been called a "sister bar" in the overseas literature, but has acquired the name "strain rod" in South Africa.

Consider now the VW gauges. The end blocks of the gauge can be easily welded on site onto the pile's longitudinal steel, as shown in Figure 2. However, two gauges per strain point (four at each level in the pile) will then be needed to compensate for local bending of the steel bar and overall bending of the pile section. Furthermore, the protrusion of the VW gauges from the surface of the steel bar makes them vulnerable to damage during installation of the steel and pouring of the concrete. Manufacturers of VW gauges also make strain rods that contain a centrally placed VW gauge. These are more expensive than two separate, surface mounted, VW gauges, but the work on site is easier and they are much less prone to damage, so this option is chosen for this comparison.

In Table 1, purchase costs of strain rods made with ER and VW gauges are compared. Both figures are the means of three recent quotations from commercial companies. Also in Table 1 are comparative costs of the instrumentation needed. For the ER instrumentation the cost is again the mean of three quotations. However, only two suitable monitoring and recording systems could be found for VW gauges. Of course, the instrumentation may already be available to whomever will do the monitoring of the strains during the pile test, but its capital cost must still be taken into account.

Figure 1. Strain rods in reinforcing steel cage.

Figure 2. Strain rods in reinforcing steel cage. (Photo by courtesy of Geokon.).

It is assumed that the time of competent personnel to install the strain rods and monitor strains during a test is the same for both types of gauge. From the experience of the author, the times are as in Table 2. No estimate of Rand costs is made because a well-trained and supervised young apprentice may competently do the work while earning little compared to his/her supervisor.

4 RELIABILITY OF ELECTRICAL RESISTANCE GAUGE STRAIN RODS

Although ER gauges and the instruments needed for them are much cheaper than VW gauges, their reliability must also be considered and the consequences of any failures amongst them.

From records of the author, a total of 150 strain gauged points have been installed under his supervision on reinforcing bars and strain rods in piles and anchors. Of these, 8 have failed to work when the test was done, an average of 5.3%. Most failures were probably because of moisture ingress, but possibly also because of physical damage to gauges or leadwires. The latter risk also affects VW gauges.

The 8 failures of the ER gauged points were not randomly distributed over the 18 piles or anchors

Table 1. Comparative costs of strain gauge types.

	Total costs in Rands, including VAT	
Item	Electrical resistance strain gauges	Vibrating wire strain gauges
10 No. strain rods, each with 20 m leadwire	21 000	73 900
Monitoring and recording instruments	44 000	89 500

Table 2. Times of on-site operations for strain measurement during a pile test.

Task	Man-hours on-site
Fix 10 No. strain rods into pile reinforcement cage	3
Monitoring strain measurement:	
• Set up equipment under cover provided by others.	3
• Initiate measuring of strains when pile test begun.	1
• Return to stop measuring and remove equipment.	2

in which they were used. All occurred when new assistants were beginning to work on ER strain gauging. Despite the supervisor stressing the importance of careful work, it is an unfortunate fact that experience to build skill and attention to detail in a manual task cannot be taught, it can only be acquired.

5 CONSEQUENCES OF A LOSS OF A STRAIN GAUGE POINT

A pile test will typically contain ten strain rods, so there is a significant chance of losing one or more of them. The consequences of the loss depend on the circumstances.

Two strain measurement points are usually put diametrically opposite each other at the chosen levels in the pile, so that the mean of the two will compensate for any bending in the pile during the loading. If the pile is loaded close to axially, the shape of the strain distribution down one side of the pile will be a good guide to the distribution down the other side where a gauge is missing, enabling the missing strain values to be interpolated.

However, if enough bending has occurred, then such interpolation may not be possible and no strain readings at the level in the pile of the failed gauge are available. For example, consider the strain gauge positions in Figure 3, which is an internally jacked pile test, and imagine that gauge loss at level 3 means that no reliable strain readings are available for that level. Without the loss (which was the actual case for this pile) two separate graphs for the shear stress vs. displacement relationships from level 2 to 3 and from 3 to 4 can be drawn, (Luker, 2016), but with the loss, only one graph, for the length from levels 2 to 4 can be drawn. Other levels in the pile would be unaffected.

The reduced number of shear stress vs. displacement graphs affects the optimisation of the pile design using the Load Transfer method of modelling pile behaviour, (Everett, 2002, Abdrabbo & Gaaver, 2012). Having one fewer element in the subdivision of the pile when using this method will have only a small effect on the accuracy with which the method can be used to optimise the pile design.

In the author's pile testing experience there have been two piles where two gauge points were lost and one where three were lost. In these instances the effect on the value of the test results depended on where the gauges were lost and whether the amount of bending allowed interpolation of strain values to be accurately done. However, in all three cases, sufficient information about the mechanics of the piles was obtained for the purposes of the tests.

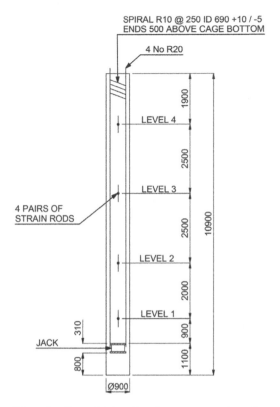

Figure 3. Example of strain rods arrangement in a pile.

6 CONCLUSIONS

The cost of using highly reliable vibrating wire gauges can be prohibitive when added to the cost of the pile test itself, causing potential savings in optimising the pile designs for a whole site to be not realised. With sufficient care in usage, electrical resistance strain gauged strain rods can be adequately reliable and their cost an acceptable investment to gain better efficiency of the working piles on a site.

REFERENCES

Abdrabbo, F.M. & Gaaver, K.E. 2012. Application of the observational method in deep foundations. *Alexandria University Engineering Journal, http://dx.doi.org/10,1016/j. aej.2012.10.004http://dx.doi.org/10.1016/j.aej.2012.10.004.*

Everett, J. 2002. Load transfer functions and pile performance modelling. *Geotechnics in the African Environment, Blight et al. (eds).* 229–234. Balkema.

Hayes, J. & Simmonds, A. 2002. Interpreting strain measurements from load tests in bored piles. *Proc. 9th Int. Conf. on Piling and Deep Foundations, Nice, France.*

Luker, I. 2016. Recent use of the internally-jacked pile test in South Africa. *1st Southern African Geotechnical Conference.* Balkema.

Proceedings of the first Southern African Geotechnical Conference – Jacobsz (Ed.)
© *2016 Taylor & Francis Group, London, ISBN 978-1-138-02971-2*

Execution and engineering principles of Control Modulus Column (CMC)

K. Coghlan, C. Plomteux & J. Racinais
Menard SAS, Nozay, France

ABSTRACT: CMC is a soil improvement technique consisting of a mortar semi rigid column which was first developed by Menard-France in the mid-nineties. The concept was initially conceived as an alternative to stone columns. The development of this new technology resulted in Menard carrying out the first CMC project in 1996 in the north of France. The CMC foundation system is now used across Europe, North America, Asia and Australia. CMCs are effective in all soil types, but particularly in soft cohesive clayey or silty soils in order to prevent excessive or adverse settlements, but also provide stability and adequate structural bearing capacity. It is based on a load sharing system between the soil and the column. It is commonly used beneath structures with uniformly distributed loads such as warehouses, caisson quaywalls, embankments, wind turbines and storage tanks. This paper will provide basic execution and design principles of the CMC.

1 HISTORY OF CMC TECHNIQUE

1.1 *Brief history*

The concept of deep foundation systems or reinforcing underlying soils has existed for many centuries. Examples can be found across the globe, detailing wooden piling systems in 15th century Venice, Italy to the use of oak trees in the 16th century to consolidate soils beneath the chateau of Chambord, Loire in France.

For a modern structure requiring a deep foundation system due to the presence of soft soils, there are various types of soil reinforcement techniques available. Not unlike the science of geotechnics, the idea of soil reinforcement is still a relatively new concept which has developed through various innovative techniques since the early nineteen sixties. Soil reinforcement techniques as a concept, make use of what capacity is within the soil and compensate as necessary using stiffer materials to achieve the required bearing capacity or settlement reduction. This is in contrast to piling which transfers the full structural load to a deeper substrata, and effectively replaces any founding capacity in surface layers.

The invention of techniques such as Dynamic Compaction in 1969 by Louis Menard, Hamidi, Nikraz & Varaksin, (2009), allowed geotechnical engineers to improve in-situ soil parameters of loose granular soils rather than installing traditional piled solutions.

The CMC is an unreinforced concrete column which was initially developed to fill the gap between potentially more expensive pile solution and classical soil improvement techniques such as vertical drains or stone columns. The combination of very soft soils and a more stringent settlement criteria, would have traditionally ruled out the use of soil reinforcement, therefore requiring a more expensive piling solution. Like all soil reinforcement techniques, the principle of 'controlled modulus column' is not necessarily to stop settlement but to control it to within acceptable limits.

The CMC technique was first developed and executed in 1996 for the Amien Football Stadium project in Amien, in the north of France.

Today, CMC is a widely accepted Rigid Inclusion technique within the geotechnical industry and has been executed in various regions around the world since 1996.

Recommendations for the design and execution of CMCs, now exist within the first ever guideline for soil reinforcement techniques, ASIRI (2013) ground improvement guideline.

2 CONSTRUCTION OF CMC

Execution of CMC can be implemented using a variety of techniques including soil extraction or soil displacement methods.

2.1 *Bored CMC with soil extraction*

2.1.1 *Simple bored*
The process of simply bored CMCs is carried out in soils which are self-supporting and generally this means cohesive soils.

2.1.2 Continuous Flight Auger method (CFA)

The CFA method employs the use of a hollow flighted auger similar to that used for standard piling methods. The principle is the same as that of CFA in that concrete/mortar is injected under pressure as the tool is raised up to platform level. One particular example of this technique is the Nigh Son Petrochemical Project in Vietnam. The project comprised of 32 no steel tanks varying in diameter from 25 m to 70 m. CMCs were designed up to 20 m in depth in soft to stiff clays with sand lenses. The specification required consisted of differential settlement—tank center to edge < R/300; Circumferential Settlement—13 mm per 10 m; Tilt Settlement < Dia/200.

2.1.3 Bored CMC with displacement auger method

The CMC displacement method requires a rig of high hydraulic capacity and specifically designed auger with reverse pitch to laterally displace material. Concrete is then pumped through the hollow stem of the auger under pressure as the auger is raised from the required depth to the working platform level. This technique ensures minimal spoil over the platform.

2.1.4 Cast in situ Vibro Concrete Column (VCC)

This execution method consists of lowering a tube with a valve or clamp at the lower end, into the soil to the required depth. The vibrated void is then filled with concrete. Penetration is either carried out using vibrator at the end of the tube or a hydraulic or diesel powered vibrator attached to the top of the tube.

2.2 Load transfer platform

It is required to place a Load Transfer Platform (LTP) over the CMCs in order to uniformly diffuse the load amongst the inclusions. This ensures that the CMC/soil combination acts as a composite layer. The LTP can consist of various different materials.

2.2.1 Granular mattress

The most common type of LTP is a layer of granular material compacted layer by layer. This generally is a well graded sand or gravel with less than 10% fines with a thickness ranging between 0.4 and 0.8 m in thickness however this depends on the conditions on site and the geometry of the foundation.

2.2.2 Geotextile reinforcement

Should the required thickness of LTP not be possible, a single or numerous geotextile membranes can also be incorporated into the platform.

2.2.3 Steel reinforcement

Where calculation shows that horizontal or lateral loads are beyond the capacity of a CMC section, a steel reinforcement mesh can be incorporated into the transfer platform. A typical example of this is beneath high embankments (>8 m on very soft soils) where lateral loads can become large. This was used successfully on the LGV High Speed Train project from Paris to Bordeaux as seen below in figure 2.

2.2.4 CMC cap

Where CMCs are installed in very soft soils to support an embankment height of 2.5 m or less, this may be susceptible to an undulating effect on the platform. The load for such a small embankment is generally quite low, resulting in a wide grid of CMCs. This type of arrangement can be susceptible to settlement between columns. To counter this effect, a system of caps can be constructed on the head of the CMCs. The dimensioning of the CMC and the Caps should satisfy the condition in equation 1, ASIRI (2103) and figure 3:

Figure 1. Vibro concrete column.

Figure 2. Steel reinforcement cage under embankment—LGV project Paris—Bordeaux.

$$H_M > 1.5(s - a), \qquad (1)$$

3 DESIGN OF CMC

Generally CMCs are designed to achieve specifications such as minimum total settlement, minimum differential settlement, bearing capacity; minimum factor of safety for stability.

3.1 Principles of settlement—deformation analysis

The analysis of deformation of composite soils reinforced with CMCs can be divided into a number of steps.

Step 1: Application steps of load at surface level. This will in turn induce a deformation within the CMC, δ_{cmc} and within the surrounding soil, δ_{soil}.

Step 2: The deformation within the soil is greater than the deformation within the CMC, $\delta_{soil} > \delta_{cmc}$. This will cause negative skin friction along the surface of the CMC above the Neutral Axis (NA). Therefore, this subsequently transfers the stress from the surrounding soil to the CMC.

Step 3: At the Neutral Axis, $\delta_{soil} = \delta_{cmc}$ therefore this is point of maximum stress in the CMC. Below the neutral axis, $\delta_{cmc} > \delta_{soil}$ creating a mobilisation of positive skin friction and end bearing resistance.

Step 4: Finally once sufficient capacity within the founding layer has been mobilised, a state of stress equilibrium is reached.

In order to understand the localised stress distribution between the CMC and the natural soil, generally an analysis is carried out using FEM software in two steps.

Localised axi symmetric model is performed considering the grid, column diameter, length of CMC column, characteristics of concrete/mortar etc to understand the skin friction and end bearing interaction.

Once localised parameters have been calculated and validated, these are then used to compile a global model incorporating all external factors.

3.2 Global bearing capacity

As a primary check, and as the CMC concept is reliant on a load sharing system between column and soil, based on basic CMC design of grid and column diameter, it is required to verify the following formula 2, ASIRI, (2013):

$$q_{ref}.S \leq q_{soil}.(S - A_{cmc}) / \gamma_{soil} + S . q_{cmc} / \gamma_{cmc} \qquad (2)$$

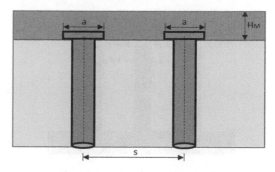

Figure 3.   CMC cap with minimum LTP.

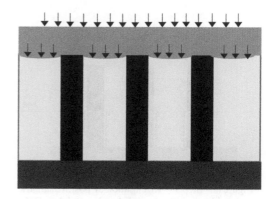

Figure 4.   Step 1—Application of load at surface with deformation of soil between CMCs.

where q_{ref} = average stress of structure on the soil; q_{soil} = average stress over area of the soil; q_{cmc} = average stress over area of the CMC Column; A_{cmc} = Area of CMC Column; S = Area of influence for one CMC.

3.3 Design checks

3.3.1 Allowable stress at CMC head

It is required to verify that the maximum allowable stress value at the head of the inclusion is compatible with the LTP characteristics (material and geometry).

Using equations of the failure mechanism (Prandtl's diagram), of the load equilibrium and of the load bearing capacities of the soil and the inclusion material, it is possible to determine the allowable domain of the limit stress in the LTP at the inclusion head.

In figure 10, q_s+ and q_p+ are the stresses applied on the supporting soil and at the inclusion head respectively. The allowable domain is reduced to the intersection of the segment (4) with the shaded surface.

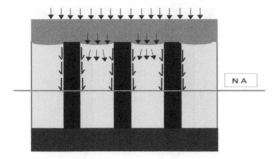

Figure 5. Step 2—Induced negative skin, above the Neutral Axis (NA) along CMC due to deformation of soil.

Figure 6. Step 3—Mobilisation of positive skin friction and end bearing resistance below Neutral Axis (NA).

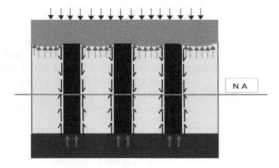

Figure 7. Step 4—State of equilibrium.

4 ENGINEERING APPLICATIONS

The following provides some previous applications of CMC and how the technique has been used in order to achieve the specification required for each structural application.

4.1 Tanks

Tank structures concentrate a high load over a relatively small surface area. The critical criteria

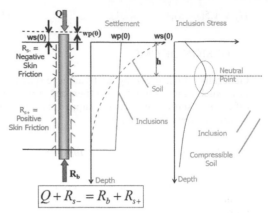

$$Q + R_{s-} = R_b + R_{s+}$$

Figure 8. Example of stress distribution in CMC/soil interface.

for any steel or concrete tank being the control of differential settlement.

One recent example of tanks founded on a CMC system, is the Marrero Magellan Project in Louisiana, USA. The project was carried out in 2014 and detailed five no 40 m dia tanks with reinforcement of soft silty clays to 25–32 m with column diameters ranging from 320 mm and 420 mm. An innovative ring beam using reinforced earth was also implemented as a part of this project in order to reduce differential settlements between the edge and the centre of the tank.

4.2 Embankments

As previously mentioned, CMC is an effective system as a foundation solution to control total and differential settlements as well as providing slope stability against shear failure below embankments.

The ideal load case for this technique is where the CMC is almost fully in compression, however due to the inclined nature of an embankment, and the shear forces induced within the soil due a potential slip circle, the CMCs can be susceptible to lateral loading, subsequently inducing a moment within the concrete column itself. The verification of the inclusion integrity in terms of axial force and bending moment in the column is carried out in accordance with the half-moon method shown in formula 3 below and derived from the Eurocode 2, section 12:

$$N_{Ed} \leq N_{Rd} = A_{ref} f_{cd} \qquad (1)$$

Where N_{Ed} = design value of the applied axial force; f_{cd} = design compressive strength of the CMC grout; A_{ref} = compressive area of the CMC section under vertical load and bending moment.

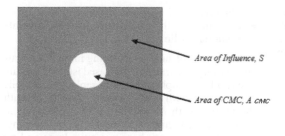

Figure 9.   Area of influence of CMC.

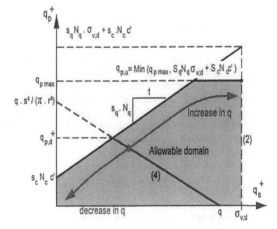

Figure 10.   Stress interaction between CMC & soil, ASIRI (2013).

As long as $N_{Ed} \leq N_{Rd}$, no additional measures are required. Should $N_{Ed} \geq N_{Rd}$, some potential solutions are as follows:

- Increased the thickness of LTP
- Reinforcement of LTP using techniques previously described in section 2 above (steel or polymers—Reinforced Earth)
- Reinforcement of column as necessary
- Include further shear reinforcement between CMCs detailing Soil Mixing.

One example of a project with CMC supporting an embankment was the Forth Replacement Crossing (FRC) project in Scotland (2015). The project detailed an embankment of 6 m height which was built on a soft soils up to 17 m in thickness, reinforced with CMC of diameter 360 mm.

The plane-strain finite element calculations showed an expected horizontal displacement of around 14 cm at the edge of the embankment.

Implementation of a soil mixing trench reinforcement between CMCs allowed columns to remain at 360 mm in diameter while also improving the stability factor of safety against failure.

4.3   Commercial structures and warehouses

The nature of the CMC technique increases the global stiffness of the native soil, therefore it allows structural engineers to reduce the thickness and subsequently the reinforcement required within slabs on grade.

Generally, slabs are designed using a subgrade reaction coefficient, kv, considering a homogenous soil. The influence of rigid inclusions induces a slight additional moment within the slab. However, a method of analysis of the moments transferred to structural slabs considering rigid inclusions and slab joints has been developed and is detailed within the ASIRI Recommendations for Design of Rigid Inclusions (2013). This method as per Racinais and Plomteux (2011), consists of separating the sum of moment into three separate parts.

The load cases are as follows:

- 'ma'—calculation of a slab on equivalent homogenised soil
- 'mb'—influence of rigid inclusions on a continuous slab without joints
- 'mc'—Interaction between rigid inclusions and joints.

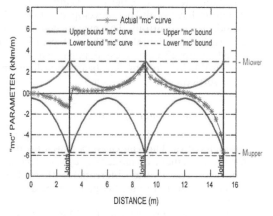

Figure 11.   Bending moment envelope for slab with joints, ASIRI (2013).

Figure 12.   CMC Execution; Porto Di Vado.

23

This method is now commonly used in tandem with rigid inclusions with one such example being the recently constructed FM Logistics platform in 2015 in Moscow, Russia. The use of this design approach allowed for the reduction of the slab to 20 cm with a reduced steel reinforcement on a soil profile that was often of a peaty nature with a depth profile in the range of 7 m-14 m of reinforcement.

4.4 Wind turbines

CMC foundation solutions has seen an exponential growth in the Wind Farm industry in tandem with an increased requirement for governments to supply a minimum power source from renewable energies.

Turbine bases in terms of foundation design require the following specifications:

- Bearing Capacity
- Min Differential Settlement across the base
- Max Total Settlement
- Min Dynamic Rotational Stiffness, k_{dyn}.

The largest combined Wind Farm in Europe, the Fantanele and Cogealac Wind Farms in Romania have both been developed using CMC foundations solutions. The project consisted of 250 turbines in total and the soil profile consisted of up 27 m of loess deposits which were susceptible to collapse or bearing failure and large settlements.

4.5 Quaywalls and port structures

Soils in port locations can often have very poor characteristics. This can cause potential problems for the placement of modern quaywall structures which are sensitive to differential settlement due to more often than not, the presence of a gantry crane.

A successful installation of a CMC system was the recent Porto Di Vado Trial Project in Italy in 2014. Execution was carried out in offshore conditions in 20–30 m of soft soils in the location of the proposed container terminal. A granular mattress was placed over the head of the CMCs. The caissons where then floated into place and positioned on top of the CMCs.

4.6 Limitations

As previously described, CMC is an ideal solution under Uniformly Distributed Loads (UDL) such as embankments or warehouses however CMC may not be suitable for projects with highly concentrated loads such as:

- structures of G + 10 and above
- foundations with very high overturning moments or lateral loading.

In the case of high overturning moments or lateral loading, generally an increased isolated footing size can be a simple solution to replace a deep pile and pile cap with CMCs placed directly beneath the footing with or without a LTP. However, in some cases, for higher moments or overturning, the footing necessary becomes quite large and no longer would be considered feasible in a practical or financial sense.

5 CONCLUSIONS

The design, application and execution of the CMC Rigid Inclusion technique has been developed through intense research, development and design over the past 20 years. It has been shown to be effective across a wide range of diverse industries. By using the existing capacity within the soil the CMC has proved to be an efficient and economical alternative to traditional piling.

REFERENCES

AFPS-CFMS. 2012. *Procédés d'amélioration et de renforcement des sols sous actions sismiques—Guide technique*, Presses des Ponts, Paris.

ASIRI National Project. 2013. Recommendations for the Design Construction and Control of Rigid Inclusion Ground Improvements, Presses des Ponts, Paris.

BSI. 2004. *BS EN 1992-1-1:2004 Eurocode 2: Design of concrete structures Part 1-1: General rules and rules for buildings*.

Hamidi, B., Nikraz, H. & Varaksin, S. 2009. A Review on Impact Oriented Ground Improvement Techniques, *Australian Geomechanics Journal*: 44(2), 17–24.

Plomteux, C. & Ciortan, R. 2010. Integrated Ground improvement solution for the largest wind farm project in Europe. *Proc 14th Danube-European Conference on Geotechnical Engineering* 'From Research to Design in European Practice, 2010. Bratislava, Slovakia.

Racinais, J. & Plomteux, C. 2011. Design of slab-on-grades supported with soil reinforced by rigid inclusions. *EYGEC*; Rotterdam: September 2011.

Racinais, J., Varaksin, S. & Yee, K. 2013. *New approach to determine ultimate stress mobilization in the interface "rigid inclusion head and load transfer platform"*.

Yee, K, Setiawan, R.A., & Bechet, O. 2012. Controlled Modulus Columns (CMC): A New Trend in Ground Improvement and Potential Applications to Indonesian Soils. *ISSMGE—TC211—International Symposium on Ground Improvement*; Brussels: 31 May & 1 June 2012.

Proceedings of the first Southern African Geotechnical Conference – Jacobsz (Ed.)
© 2016 Taylor & Francis Group, London, ISBN 978-1-138-02971-2

Ageing of driven piles in silica sands

S. Rimoy

University of Dar es Salaam, Dar es Salaam, Tanzania

ABSTRACT: Pile load testing has shown that axial capacities of piles driven in silica sands can increase dramatically with age after driving. This paper reports an extended database that has been used to assess; the effects of prior loading history, the distinct contributions of base and shaft loads, the potential influence of axial loading direction (compression or tension), the influence of groundwater type as well as the effects of pile material and diameter on the ageing characteristics. The analysis is reported in parallel with highly instrumented model tests that were designed to investigate the fundamental mechanisms of driven pile ageing at scaled and controlled laboratory environment conditions. The presented findings have important implications in the design and assessment of pile foundations in projects with similar settings.

1 BACKGROUND

Axial shaft capacities of piles driven in silica sands have been reported to increase (set-up) with age well beyond the end of primary consolidation, while base capacities remain relatively unaffected. Empirical capacity gains with time relationships have been proposed for the set-up from generally widely scattered databases; see Skov & Denver (1988) and others. Jardine et al. (2006) reported first-time axial static tension tests to failure present clearer and higher shaft capacity trends than multiple re-tests.

The beneficial pile ageing trends have remained hard to model quantitatively. Pile designers are likely to remain cautious in application of ageing benefits until the fundamental mechanisms of the ageing processes are clearly understood. Existing hypotheses for the ageing processes have been posed by amongst others Axelsson (2000). We summarise these as:

a. *Radial stress re–distribution:* Pile driving generates extreme stress distributions, with initially elevated circumferential (hoop) effective stresses (σ'_θ) and reduced radial effective stresses (σ'_r) existing close to the shaft, and both developing their maxima at some radial distance (r) away from the shaft. Creep processes then promote redistribution over time, with the extreme $\partial\sigma'_\theta/\partial r$ gradients reducing as the arching protection provided to the shaft from the higher, more distant radial stresses gradually weakens. Radial stresses on the shaft increase as $\partial\sigma'_r/\partial r$ gradients flatten in response and this continues until stable distributions are established.

b. *Enhanced shaft dilation:* Potential ageing benefits by this mechanism are associated with growth of the contribution made to shaft resistance through the kinematically constrained dilatant radial effective stress changes, $\Delta\sigma'_{rd} = 2G\Delta r/R$ (Lehane et al. 1993) that develop at the soil-pile interface under axial loading. Micro-structural re-arrangements and grain interlocking might increase either sand shear stiffness G or the radially outward interface movement Δr required to allow interface slip, leading to larger $\Delta\sigma'_{rd}$ and shaft capacity for a pile of radius R.

c. *Physiochemical processes:* Involving interactions between sand particles and/or the shaft material leading to bonding, increased interface roughness and/or sand stiffness that either push the ultimate cylindrical shearing surface out into the surrounding soil or raise the stationary radial stresses. These processes are also likely to enhance dilation during pile loading.

Chow et al. (1998) resolved that installation related processes are critical to pile ageing. Driving and jacking might lead to different ageing trends and set-up is not generally reported with replacement (bored or CFA) piles. Bowman and Soga (2005) speculated that dilative creep observed in triaxial apparatus tests enhances σ'_r growth around piles over time and can be accelerated by cyclic perturbations. Further, Jardine & Standing (2012) reported that low-level axial cycling enhances shaft capacity for piles driven in sand while high level cycling led to marked degradation.

Axelsson (2000) interpreted enhanced shaft dilation as a key set-up mechanism in field and laboratory tests though Jardine et al. (2006) showed that initial axial load–displacement stiffness is largely unaffected by time. Chow et al. (1998) and White & Zhao (2006) concluded that physiochemical processes apply to oxidisable steel piles in the presence of air and water. Concrete and timber driven piles also have reported set-up in sands, indicating that chemical processes involving the pile material may not be the primary cause of shaft set-up.

The current study contributes to the discussion through both database studies and model tests. Rimoy (2013) assembled an extended set of fifteen case studies of static and dynamic tests on aged piles driven in silica sands adding to Chow et al. (1998) and Jardine et al. (2006) to help test the stated hypotheses. While Rimoy et al. (2015) present the complete analysis of the database and model pile studies, this paper focuses only on the trends established from the full scale field tests.

2 ANALYSIS OF THE FIELD PILE TESTS DATABASE

Rimoy (2013) and Rimoy et al. (2015) assembled a large database from 24 separate studies involving static and dynamic tests on 160 steel and concrete piles driven in sands. Pile diameters D varied from around 200 to 1220 mm and lengths L from 3 to 57 m. Rimoy (2013) presents a global analysis showing:

i. Saline groundwater is not a necessary condition for shaft capacity growth.
ii. Closed and open-ended piles follow broadly similar trends, as do concrete and steel piles.
iii. Static pile test capacity measurements exhibit more consistent ageing trends than dynamic re-strikes.
iv. First-time static tension testing provides the clearest indicator of shaft capacity growth.

The dataset comprised 20 unambiguous first-time static tension tests to failure on steel (340 mm < D < 508 mm) piles from four sites. Figure 1 shows the first-time tension test capacities from Blessington by Gavin et al. (2013) and Larvik by Karlsrud et al. (2014) normalised by the ICP-05 (Jardine et al. 2005) estimated pile capacities as per approach recommended by Jardine et al. (2006). The test data were analysed to find a best fitting tri-linear Intact Ageing Characteristic (IAC) trend similar to that interpreted by Jardine et al. (2006). The 'day 1' tension shaft capacities for Dunkirk and Larvik were determined by instrumented driving stress-wave matches from several piles and that for Blessington from a static day 1 test. Tension re-tests showed a greater scatter, reflecting the dominant impact of prior failures on ageing.

Site specific Cone Penetration Tests (CPT) information is required for ICP-05 normalisation. In cases where this was not possible, Rimoy et al. (2015) normalised the time dependent capacities by interpreted End of Driving (EoD) or day 1 static

test capacities. This process leads to greater scatter as shown in Figure 2 and ICP capacity normalisation is clearly preferable. Despite these difficulties, the Ryggkollen tests by Karlsrud et al. (2014) confirm very strong growth in first-time tension shaft capacity over time.

The compression base capacity measurements collected by Rimoy et al. (2015) from re-tested piles scattered around a shallow mean gain of ~2.5% per log time, Figure 3; no measurements were available to assess first-time base capacity trends. Figure 4 shows the static first-time compression tests on 305 mm equivalent diameter piles. The broad scatter limits attempts for tri-linear 'IAC' fit, but the broad trend of 44% gain per log-time over 60 days falls, as expected, well below that applying to pile shafts over the mid-range IAC shown in Figure 1.

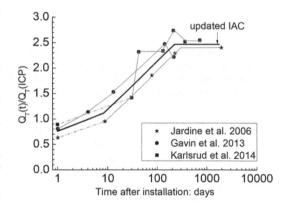

Figure 1. Static tension capacity-time trends from first-time tests on steel pipe piles driven at three sand sites, normalised by ICP-05 tension capacities (Rimoy et al. 2015).

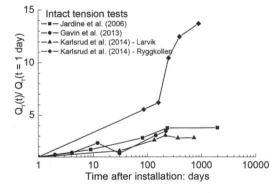

Figure 2. Static tension capacity-time trends from first-time tests on steel pipe piles driven at four sand sites, normalised by field EoD tension capacities.

The total compression capacity trends established from first dynamic re-strikes and plotted in Figure 5 scatter around a relatively gentle 41% gain per log-time cycle mean trend over several hundred days. Rimoy (2013) found a similar result from static compression tests normalised by EoD values derived from pile driving monitoring. Steeper gains (~67% per log cycle) were noted for shaft resistances interpreted from the subset of static and dynamic compression re-tests whose base resistances could be isolated via strain gauging or dynamic monitoring.

Rimoy et al. (2015) investigated possible diameter effects by plotting field and laboratory model pile tests' $\Delta Q_C/Q_{C(t=1\,\text{day})}$ ratios after normalising for the field databases' mean total compression capacity trend with time. The mean time trend gauged from Figures 4 and 5 was:

$$\Delta Q_C/Q_{C\,(t=1\,\text{day})} = f(t) = 0.425\log(t) \qquad (1)$$

The dataset comprised tests conducted up to 400 days after driving on piles whose diameters ranged between 200 m up to 2134 mm. Adding 36 mm diameter laboratory jacked and driven model pile tests described by Rimoy (2013) led to the composite dataset plotted in Figure 6. While the data scatter is widespread, capacity growth appears to increase markedly with pile diameter D over the D < 400 mm range. Tentative median, upper and low bound relationships are shown on Figure 6 that trend to zero with very small piles. The median curve indicates an approximately neutral (typical field) ageing trend at the mid-range of the pile test dataset (D ≈ 0.6 m) and rises above unity with larger diameter piles.

3 LABORATORY MODEL INVESTIGATIONS OF AGEING BEHAVIOUR

Field tests provide the best means of confirming the behaviour of construction piles. However, they do not allow for a systematic isolation of effects

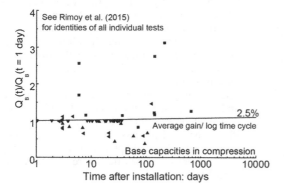

Figure 3. Normalised static axial compression base capacity-time trends of re-tested aged piles.

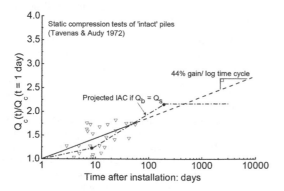

Figure 4. Normalised static axial compression total capacity-time trends: first–time tests on aged piles.

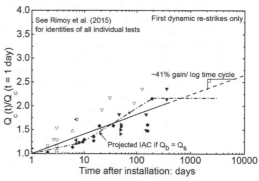

Figure 5. Normalised axial compression total capacity-time trends from dynamic re-tests.

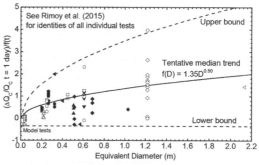

Figure 6. Examination of re-strike database from Fig. 5 for potential influence of pile diameter on growth of total compressive capacity with time: $\Delta QC/QC_1$ values normalised by 0.425log(t), where t is time after driving.

of factors such as grain size, water conditions or applied stresses and are limited by the considerable difficulties of measuring local stresses in the sand and on the pile shafts. Large piles tests are expensive to monitor over extended periods and sites with uniform sand deposits are difficult to acquire and maintain.

Laboratory model testing can offer advantages over field tests, provided they adequately represent full scale behaviour. Various model pile instrumentation schemes and test arrangements have been reported. However, model studies have generally been indicating far less impressive set-up trends than field tests. Potential causes for this paradox include mismatches with field stress conditions, physiochemical effects, multiple retesting, scale effects including potentially low chamber-to-pile diameter ($D_{chamber}/D$) and pile-to-grain diameter (D/d_{50}) ratios (Baxter & Mitchell 2004) and interactions with neighbouring test piles. Most models offer limited control over environmental conditions and lack instruments to track local stress changes over installation, ageing and load testing.

This study attempted to overcome these drawbacks by employing a relatively large and fully pressurised calibration chamber test facility. Piles were installed into the Laboratoire 3S-R at Grenoble Tech 1.20 m diameter, 1.50 m deep, calibration chamber (Figure 7) which was repeatedly filled with fresh batches of silica NE34 or GA39 Fontainebleau sands (d_{50} = 0.210 and 0.127 mm respectively) within which multiple miniature stress sensors were placed, whose cell action effects were recognized and calibrated (Zhu et al. 2009). Rimoy et al. (2015) describe how a series of factors that might affect ageing were considered systematically in the model pile study.

4 THE PARADOX BETWEEN MODEL AND FIELD CAPACITY-TIME TRENDS

The field and the laboratory model piles described were shown by Rimoy et al. (2015) to share some common trends. However, a central paradox emerged that was persistent throughout the extended model pile testing campaign, despite multiple variations being made in test procedure and equipment, the model tests showed far weaker shaft capacity–time trends. Reducing pile diameter, then sand grain size, testing with stainless then mild steel piles, employing water saturated sand, varying testing procedures and installation jack strokes, changing boundary conditions, imposing environmental stress cycles and shifting from cyclic jacking to driving all had little effect on the model pile ageing trends.

Rimoy et al. (2015) offer potential explanations for this central paradox that include:

a. Scale effects relating to the annular shaft shear zone. The model pile testing programme increased the pile to grain diameter ratio D/d_{50} from 57 to 283 by changing the pile diameter (to 12 from 36 mm) in one case and the mean grain size (to 0.12 from 0.21 mm) in other experiments. While these changes had little influence, far higher D/d_{50} ratios apply in the field. It is not possible to match the field ratios in feasible scale model tests.
b. The results may be more sensitive to pile end and dynamic installation details than had been appreciated in the laboratory programme.
c. Physiochemical processes may be more important in the field, perhaps due to water reacting with minerals or fines that were absent in the model tests.

5 CONCLUSIONS

The following four main conclusions follow from the studies outlined in this paper:

1. Open and closed-ended piles made from steel, concrete or timber develop marked set-up after driving in silica sands, under offshore and onshore conditions, with salty or fresh groundwater. The ageing benefits apply principally to shaft resistance.
2. First-time field tension tests, analysed within the ICP-05 framework to allow for pile and ground conditions, provide the clearest way of characterising shaft capacity growth. Re-tested piles show staggered trends that are dominated by their testing histories. Normalising by dynamic EoD or static EoI capacities leads to greater scatter.

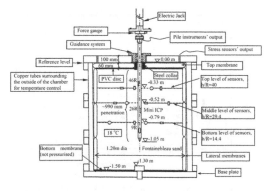

Figure 7. Vertical section of fully instrumented Calibration Chamber and Mini–ICP after installation; Rimoy (2013).

3. Laboratory model tests show far lower rates of shaft capacity growth than field piles. Of the multiple parameters varied in the model tests, only calibration chamber boundary conditions had a significant influence. Reducing the pile or grain size, adding free water, altering the number of jack strokes and cycling the boundary (radial and axial) stresses had little influence under the conditions explored.
4. Factors that may explain the central paradox between the laboratory and field capacity ageing trends include scale effects related to pile diameter (especially over the D < 400 mm range) and the fractured shear zone around the pile shaft. The D/d_{50} ratios applying in the field cannot be matched at model scale. Pile end conditions and physiochemical action may also be more important in the field, perhaps due to trace minerals, fines or pile material type. Other, as yet unidentified, processes may also operate.

ACKNOWLEDGEMENTS

The author gratefully acknowledges all colleagues he worked with during this study at Imperial College and Grenoble INP in particular Prof. R. Jardine, Prof. P. Foray, Prof. Z. Yang, Dr. C. Tsuha, Dr. M. Silva, Mr. S. Ackerley, Mr. A. Bolsher, Mr. C. Dalton, Dr. M. Emerson, Mr. F. La Malfa and Mr. J.B. Toni. Financial support provided by Atkins, the Commonwealth Scholarship Commission, Engineering and Physical Sciences Research Council (EPSRC), Centre National de la Recherche Scientifique (CNRS), the Health and Safety Executive (HSE), Shell (UK) and Total France is gratefully acknowledged.

REFERENCES

Axelsson, G. 2000. *Long-term set-up of driven piles in sand*. PhD thesis, Royal Institute of Technology, Stockholm, Sweden.

Baxter, C.D.P. & Mitchell, J.K. 2004. Experimental study on the aging of sands, *J. of Geotech. and Geoenviron. Eng,* ASCE, 130(10): 1051–1062.

Bowman, E.T. & Soga, K. 2005. Mechanisms of set-up of displacement piles in sand: Laboratory creep tests. *Can. Geotech. J.,* 42(5): 1391–1407.

Chow, F.C., Jardine, R.J., Brucy, F. & Nauroy, J.F. 1998. Effects of time on capacity of pipe piles in dense marine sands, *J. of Geotech. and Geoenviron. Eng,* ASCE, 124(3): 254–264.

Gavin, K., Igoe, D. & Kirwan, L. 2013. The effect of ageing on the axial capacity of piles in sand. *Proc. of the ICE Geotech. Eng,* 166(2): 122–130.

Jardine, R.J., Chow, F.C., Overy, R.F. & Standing, J.R. 2005. *ICP design methods for driven piles in sands and clays*. Thomas Telford, London.

Jardine, R.J., Standing, J.R. & Chow, F.C. 2006. Some observations of the effects of time on the capacity of piles driven in sand. *Géotechnique,* 56(4): 227–244.

Jardine, R.J., Zhu, B., Foray, P. & Dalton, C.P. 2009. Experimental arrangements for the investigation of soil stresses developed around a displacement pile. *Soils and Founds.* 49(5): 661–673.

Jardine, R.J. & Standing, J.R. 2012. Field axial cyclic loading experiments on piles driven in sand. *Soils and Founds.* 52(4): 723–736.

Jardine, R.J, Zhu, B.T., Foray, P. & Yang, Z.X. 2013. Interpretation of stress measurements made around closed-ended displacement piles in sand. *Geotechnique,* Vol. 63, No. 8, pp. 613–628.

Jardine, R.J., Thilsted, C.L., Thomsen, N.V. & Mygindt, M. 2015. Axial capacity design practice for North European Wind-turbine projects. *Proc. Int. Symp. On Frontiers in Offshore Geotechnics (ISFOG)*. Oslo: 581–586.

Karlsrud, K., Jensen, T.G., Wensaas Lied, E.K., Nowacki, F. & Simonsen A.S. 2014. Significant ageing effects for axially loaded piles in sand and clay verified by new field load tests. *Offshore Technology Conf.,* Houston, doi:10.4043/25197-MS

Rimoy, S. 2013. *Ageing and axial cyclic loading studies of displacement piles in sands*. PhD thesis, Imperial College London.

Rimoy, S.P. & Jardine, R.J. 2011. On strain accumulation in a silica sand under creep and low level cyclic loading. *Proc. 5th Int. Symp. on Deformation Characteristics of Geomaterials (IS-Seoul)*. Seoul, Korea: 463–470.

Rimoy, S., Silva, M., Jardine, R.J., Yang, Z.X., Zhu, B.T. & Tsuha, C.H.C. 2015. Field and model investigations into the influence of age on axial capacity of displacement piles in silica sands. *Géotechnique,* 65(7): 576–589.

Tavenas, F. & Audy R. 1972. Limitations of the driving formulas for predicting the bearing capacities of piles in sand. *Can. Geotech. J,* 9(1): 47–62.

White, D.J. & Zhao, Y. 2006. A model-scale investigation into 'set-up' of displacement piles in sand. *6th International Conference on Physical Modelling in Geotechnics,* Hong Kong, China.

Zhu, B.T., Jardine, R.J. & Foray, P. 2009. The use of miniature soil stress measuring sensors in applications involving stress reversals. *Soils and Foundations,* 49(5): 675–688.

Proceedings of the first Southern African Geotechnical Conference – Jacobsz (Ed.)
© *2016 Taylor & Francis Group, London, ISBN 978-1-138-02971-2*

Stiffness considerations of local wind turbine gravity foundations

B.W. Mawer
Jeffares & Green, Cape Town, South Africa

D. Kalumba
University of Cape Town, Cape Town, South Africa

ABSTRACT: One of the critical, and often governing design considerations for wind turbine foundations is the inherent in-situ soil stiffness and its effect on the structure's resistance to resonance and dynamic vibrations. Wind turbines fall in a unique class of structure, in the sense that, they exist as structural systems exposed to dynamic loads, which in turn, create additional new dynamic loads for which they must be designed. This paper introduces the concept of soil stiffness as it pertains to wind turbine foundation design and presented procedures currently adopted throughout the world in assessing in-situ soil stiffness. The effect of this on structural dynamic considerations, in particular those of 1st order natural frequency, are discussed with reference to three local design examples, each based on typical soil conditions from each of the three main wind development corridors of the RSA.

1 INTRODUCTION

1.1 *Wind energy in South Africa*

Wind turbine structures are a relatively new addition to a South African energy market that has, for the majority of its industrialised history, relied upon its abundant non-renewable energy resources such as coal, to meet its power and energy demands. With the shift in global awareness towards climate change and the effects of global warming, sustainable energy production has moved to the forefront of South Africa's plan to reach a sustainable development path. This is evidenced in the updated Integrated Resource Plan (IRP) of 2013 which outlined the countries commitment to 8400 MW of new wind energy infrastructure, equal to 20% of the countries existing generation capacity, over the next 20 years.

The country already boasts an installed wind generation capacity of 2,000 MW by the end of 2015, consisting of approximately 20 major wind farms, each of which were constructed within three main development corridors, concentrated along the south west and south east coasts of the country as well as in the arid southern interior, often referred to as the Karoo.

2 FOUNDATION STIFFNESS

2.1 *Introduction*

Wind turbine structures are dynamic systems that generate and experience cyclic loading. This is due to the variability of the wind driving the turbine combined with the rotational nature of the rotor and generators. The dynamic nature of these forces can cause dynamic effects which can amplify static stresses and strains experienced within the body of the structure. This is based on the idea of resonance. Resonance, in a structural sense, is the phenomenon in which a structural element tends to oscillate or vibrate at high amplitude when subjected to vibration or cyclic loads at a specific frequency. This specific frequency is called the natural frequency of the structure.

In planning for dynamic systems, designers can prevent the effects of resonance by two methods. These methods are based in either dissipating resonant energy that accumulates, or designing the system to not experience vibrations at a frequency close to the resonant frequency.

In terms of dissipating resonant energy, the structural system can be designed to allow for damping—a property of a material or system that tends to drain resonant energy from accumulating. This can include using materials such as rubber, which inherently possess high damping properties, or including cable supports to large towers that allow a path for resonant energy to dissipate quickly. The result in both cases is that the system is not allowed to vibrate to the point where stresses and strains become dangerous to the structure's stability. The problem for wind turbines is the fact that they are made primarily from steel and concrete, and do not allow for additions of further materials such as cables, and therefore this method is not always possible.

Therefore it is often favoured in turbine design, to limit resonant frequencies of the turbine to fall outside the range of frequencies at which the structure will operate. The resonant, or natural frequency of the structure is, very simply, a product of two main elements—the structures mass as well as its stiffness:

$$f_n = \frac{1}{2\pi}\sqrt{\frac{k}{m}} \qquad (1)$$

where f_n = 1st natural frequency of the system, k = the combined stiffness of soil-structure system, and m = the total mass of the structure.

Using the assumption of a simple response, three main parameters are required in order to assess the dynamic stability of the turbine. Firstly, the mass of the system, which is simply obtained by addition of turbine structural weight combined with mass of the gravity footing calculated by the designer. The two additional parameters of system stiffness and the range of turbine operating frequencies, require more in depth consideration.

2.2 Operating frequencies of turbines

Wind turbines possess two main working frequencies that must be avoided when accounting for the resonant frequency of the system. These are often referred to in literature as blade passing frequencies and are effectively frequencies at which the blades rotate during normal operation. Common to both a two-blade and three-blade turbines, is the first rotational frequency (1P) which is the frequency at which a turbine rotor completes one full revolution. In addition to this, the blade passing frequency—the number of occasions that a turbine blade passes a certain point—must also be avoided. Intuitively, for a two bladed turbine, this will occur at double the rotational frequency (2P) and similarly, three times the rotational frequency (3P) for a three bladed turbine. The result, is that for three bladed turbines that are most commonly found commercially, the 1P and 3P frequencies must be avoided. As turbine operating speeds can vary throughout operation, this produces a range of frequencies that must be avoided in order to ensure that the system does not resonate at any point in its lifetime.

To avoid these frequencies, manufacturers of turbines, design their structural components (blades, nacelle, and tower) with an overall combined stiffness that produces a natural frequency that will not overlap the range of operating frequencies. As a result, one of three design philosophies are followed based on the operating frequency range to which the system was designed. These are called the soft-soft, soft-stiff and stiff-stiff ranges, each with their own advantages.

A soft-soft tower is designed so that the natural frequency of the turbine structure occurs well below that of the 1P limiting value, and in this way avoids any resonant effects. The problem with this, is that the low stiffnesses required to fall in this range, can often lead to the system not having sufficient stiffness to resist the static structural loads. The stiff-stiff range allows for a natural frequency well above that required but this is often costly, as high stiffness materials are more expensive to procure than those usually found in production. The soft-stiff range, where the natural frequency falls in a safe area between the 1P and 3P values, is therefore the most ideal choice as it is the most economical of all the ranges, and still provides for the static structural resistance to loading. As a result, most turbines are designed to fall within this range. However, the safe range between 1P and 3P frequencies is often very narrow and minor changes in operating speeds outside of the expected design speed, such as in extreme weather conditions, can result in resonance. Figure 1 shows this relationship graphically with the 1P and 3P ranges plotted for varying operating speeds of the turbine. This includes a positive and negative 10% allowable range for an 80% confidence interval as well as an indication of the range of operating speeds for a particular 3MW turbine (as shown by large light grey block). The goal is to produce a natural frequency for the system that falls consistently between the 1P and 3P range over the full range of operating speeds of the turbine, identified by the narrow dark grey block on the figure. This provides a good indication of the narrow margin of allowable natural frequencies for which the system must be designed.

As structural designs do not allow for amplified stresses and strains due to dynamic effects, and considering that the mass of the system is predominately fixed, the stiffness of the system must therefore be carefully controlled to prevent the natural frequency occurring within these ranges.

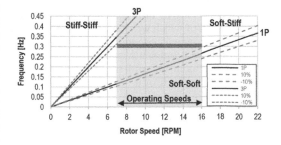

Figure 1. Campbell diagram for natural frequency design range (van der Haar 2014).

2.3 System stiffness

The problem with this type of approach of dealing with dynamic actions for foundation designers is that the manufacturers, during the modelling of resonance effects, often assume the foundation is infinitely rigid and stiff. This is not the case in practice as soils are not rigid bodies and they are not infinitely stiff. For this reason, manufacturers insist that a certain foundation stiffness is achieved in order for the above assumptions to be valid. This value differs from manufacturer to manufacturer as well as by turbine class and model. The foundation designer is then required to analyze in-situ soil stiffness and validate whether it meets the minimum requirements as stipulated by the supplier.

To calculate this stiffness value, it is common to model the soil mass, as a set of springs with an assumed elastic spring stiffness. These springs consist of stiffness components in three directions typically the z, y and θ directions as shown in Figure 2. As soils and rock bodies actually behave in a nonlinear fashion, it is important for the stiffness value to reflect stiffness of the body at strains experienced under operating conditions. According to Warren-Codrington (2014), DNV/Risφ (2002) and Bonnett (2005), typical strains in the founding material for a wind turbine structure is within the range of 10^{-3} to 10^{-2}. Equations for calculating spring stiffness values under these assumptions often relate stiffness to the shear modulus of the soil (G) which can be calculated by applying stiffness reduction to the small strain shear modulus (G_0) obtained from geophysical testing such as Continuous Surface Wave tests that are popular in RSA.

DNV/Risφ (2002) present a number of equations that can be used to calculate soil stiffness in each of the z, y and θ directions based on the soil properties as well as dimensions and embedment conditions of the foundation. This is based on theory covered in some description by authors such as Bowles (1997) and Das (2011). The z, y and θ directions are commonly referred to as vertical, lateral and rocking directions and are presented below for a foundation founded directly onto a uniform and continuous soil/rock material:

$$KV = \frac{4GR}{1-v} \tag{2}$$

$$K_H = \frac{8GR}{2-v} \tag{3}$$

$$K_\theta = \frac{8GR^3}{3(1-v)} \tag{4}$$

Where k = soil stiffness (MN/m), G = strain reduced shear modulus (MN/m²), R = radius of gravity footing (m³), v = Poisson Ratio of soil.

2.4 Stiffness of South African soil conditions

Three specific local examples were chosen within three major wind turbine development corridors in South Africa in order to assess the natural frequency effect. General location and brief description of the soils are listed below:

1. EASTERN CAPE: From inception of wind energy projects in South Africa, initial wind farms were focused in the corridor between border of the Western Cape and the city of Port Elisabeth. This corridor currently houses five of the largest wind farms currently in SA. Soil conditions in the area often comprise of varying degrees of silty sands with inclusions of a calcrete pedogenic horizons occurring in various forms and depths.
2. WESTERN CAPE: The south west coast of South Africa housed some of the first pilot projects for wind energy in South Africa including the Klipheuwel and Darling installations. This area has recently begun to be developed into one of the major wind energy corridors in the country. Soil profiles close to the town of Vredenburg, where the largest wind farm in the Western Cape exists, was chosen for this study. These conditions included varying degrees of sandy and clayey silts as well as some gravelly clays all typical of residual granite material. This was underlain by hard granite bedrock at a depth of approximately 30 m.

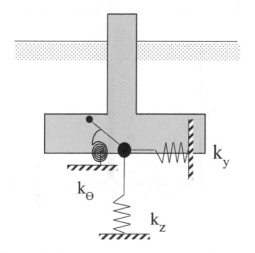

Figure 2. Campbell diagram for natural frequency design range (Mawer 2015).

3. KAROO: The Karoo area of South Africa consistently records the the highest wind speeds in the country and therefore is prime for the development of wind energy projects. The area forms part of the escarpment between the coast and the higher plateau of Gauteng. Typical soil conditions in the area are those of medium to hard rock very close to the surface including sandstone, mudstone and siltstone and occasional dolerite intrusions.

Based on soil profiles and test data taken from wind project geotechnical investigations in each of the three corridors, the following properties were assumed for analysis:

Stiffness values for each example are calculated in line with the theory presented. These are then compared with the limits on stiffness put in place by the manufacturer in order to avoid natural frequency effects. For the Vestas V112 3MW turbine used as a control model, these stiffness limits are 5000 MN/m in both vertical and lateral directions and 60 GNM/rad in rocking. Minimum foundation radius is then solved for, in order to meet stiffness requirements (see Table 2). This resulted in a minimum foundation radius of 4 m, 11 m and 4 m for the Eastern Cape, Western Cape and Karoo wind farm respectively. In the Eastern Cape and Karoo examples, stiffness was not the governing design criteria, and therefore larger bases had to be adapted resulting in higher stiffness values to be used as shown in Table 3. The Western Cape example however, was limited by stiffness criteria, which allows for a good comparison of natural frequency effects in the next section.

3 NATURAL FREQUENCY ASSUMPTION

To recapture the purpose of this study, the problem related to designing wind turbine foundations accounting for resonance, is encompassed in avoiding the operating frequency range of the turbine. As the natural frequency of the system is directly linked to the global system stiffness, the effect of the turbine manufacturer's assumption of an infinite soil stiffness, and the effect of this assumption on the turbines expected behaviour, must be assessed. This assessment is based in comparing two main assumptions, namely infinite soil stiffness assumptions that turbine manufacturers assume, and finite stiffness assumptions that soils do not possess infinite stiffness in reality. DNV\Risφ (2002) accounts for this by stating that differences in natural frequency for this assumption should be within the range of 0–5% in most cases, which will therefore fall within the +−10% range as highlighted in Figure 1. This effect, in most cases, will be negligible.

3.1 System model

In order to assess this claim specific to SA conditions, a basic dynamic model was developed by

Table 3. Adopted soil stiffness values based on final design dimensions.

	K_V	K_H	K_φ	Radius
	MN/m	MN/m	GNm/rad	m
Eastern cape	26339	17069	1347	10
Western cape	11377	6000	762	11.5
Karoo	16597	14546	5535	9

Table 1. Soil properties foundation dimensions assumed for natural frequency effect analysis.

	G		D
Example	MPa	ν	m
Eastern cape	285	0.3	21
Western cape	63	0.3	23
Karoo	1079	0.22	18

Table 2. Minimum allowable footing dimensions based on soil stiffness limits.

	K_V	K_H	K_φ	Radius
	MN/m	MN/m	GNm/rad	m
Eastern cape	7862	5804	71	4
Western cape	7978	5091	600	11
Karoo	7376	6465	486	4

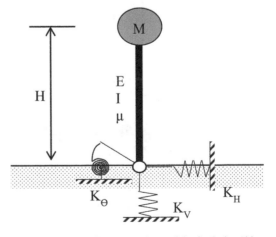

Figure 3. Simplified dynamic model of wind turbine system. Adapted from: (Byrne 2011).

modelling the Vestas 3MW V112 turbine, founded in each of the soil environments presented. Based on these systems, a natural frequency equation was formed for both infinite and finite soil stiffness assumptions and the reduction in natural frequency assessed.

Using dynamic theory, a simplified discrete representation of a wind turbine system can be created as per the suggestions of Byrne (2011). This system, as shown in Figure 3, consists of a lumped mass M, supported by a weightless cantilevered column with stiffness E, moment of inertia I and length H. This system is then supported by a set of equivalent linear springs, each with their own distinct stiffness based on strain-dependent stiffness values presented in Table 3. This model can then be used in order to derive equations required to calculate the natural frequency of the system, from first principles. This was also investigated by Molenaar & Van der Tempel (2002) using a similar system with the exception of an additional distributed mass to account for the tower weight.

Using these two systems, natural frequency using the infinite stiffness assumption is calculated as shown in Equation 5 and 6 for the Byrne (2011) and Molenaar & Van Der Tempel (2002) respectively:

$$f_{ni} = \frac{1}{2\pi}\sqrt{\frac{3EI}{MH^3}} \tag{5}$$

$$f_{ni} = \frac{1}{2\pi}\sqrt{\frac{3EI}{(m+0.227\mu H)H^3}} \tag{6}$$

where EI = flexural rigidity of the structure in N/m^2, M = total concentrated mass of the system in kg, m = mass of system excluding tower mass in kg, μ = mass per meter of the tower in kg/m H = height from foundation level to the center of the turbine hub in m.

Making use of the assumption that soil springs are attached to the tower in series, stiffness of the system including effects of soil stiffness, can be calculated using Equation 7:

$$k_{eff} = \frac{1}{\dfrac{1}{k_{tower}} + \dfrac{1}{k_\phi}} \tag{7}$$

Combining this with Equation 1, yields a natural frequency equation which accounts for soil stiffness as shown in Equation 8 & 9:

$$f_{ni} = \frac{1}{2\pi}\sqrt{\frac{1}{M(\dfrac{H^3}{3EI}+\dfrac{H^2}{k_\phi})}} \tag{8}$$

$$f_{ni} = \frac{1}{2\pi}\sqrt{\frac{1}{(m+0.227\mu H)[\dfrac{H^3}{3EI}+\dfrac{H^2}{k_\phi}]}} \tag{9}$$

Using a number of physical properties provided by the turbine manufacturer such as mass and flexural properties of the system, a basic natural frequency under both an infinite and finite stiffness assumption can be calculated.

3.2 Results

Using both natural frequencies of the finite and infinite assumptions, a percentage reduction between the two values are calculated to assess the extent to which the infinite assumption effects design. Results for each of the three design examples are highlighted in Table 4. For the Eastern Cape and Karoo Wind Farm sites, that were found to not be governed by stiffness requirements, design radi of 10 and 9 m respectively, are within the expected 0–5% range predicted by DNV/Risφ (2002). In this sense, the infinite stiffness assumption of the manufacturers is valid and can safely be assumed. For the Western Cape example, where stiffness considerations governed, the natural frequency reduction is between 5 and 8%, coming very close to the 10% allowable.

Although the reduction still falls within the 10% window, changes in natural frequency can still have a significant effect on strains that are experienced in the structure, compared to what would be expected under the infinite assumption. Considering the dynamic amplification curve in Figure 4, for the Western Cape example and focusing on a β value (ratio of experienced frequency to natural frequency) of 0.9, the curve shows a significant shift due to the reduction calculated. This translates into a change in dynamic amplification, assuming an undamped system, of the strains from 4 times up to 8.5 times the static strains, which if not taken into account can seriously affect the expected structural response of the system.

This illustrates the importance of taking into account the effects of the infinite stiffness assumption in design, in particular when minimum stiffness requirements are the governing factor. Ultimately, this supports an argument that a limit should be placed on the applicability of the infinite soil stiffness assumption to cases where stiffness

Table 4. % reduction in natural frequency for local examples due to finite stiffness assumptions.

Radius	EC Byrne	EC Tempel	Karoo Byrne	Karoo Tempel	WC Byrne	WC Tempel
m	% Red	% Red	% Red	% Red	% Red	% Red
3	19.2	19.8	45.0	45.7	355.5	357.7
4	8.5	9.0	20.9	21.5	129.9	131.0
5	4.4	4.9	11.1	11.6	63.7	64.5
6.25	2.3	2.8	5.8	6.3	35.8	36.4
7.5	1.3	1.8	3.4	3.8	21.9	22.5
9	0.8	1.3	1.9	2.4	13.0	13.5
10	0.6	1.0	1.4	1.9	8.3	8.8
12.5	0.3	0.8	0.7	1.2	5.2	5.7

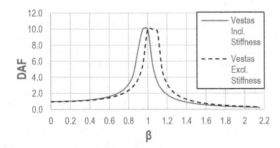

Figure 4. Dynamic amplification of Vestas V112 system including and excluding effect of soil stiffness in design (Mawer 2015).

requirements are not governing. Checks should then become standard on the global system stiffness should the soil stiffness be the governing factor in design and the effect this will have on the structural design should be analyzed.

4 CONCLUSIONS

The infinite soil stiffness assumption made by turbine manufacturers can have significant implications to structural design and stability of the turbine system if not fully understood or planned for. Soils have a finite stiffness that should be considered in design as the resulting reduction in natural frequency can have a significant effect on dynamic strains when the stiffness criteria govern design, for which wind turbine design guidelines such as the DNV/Risφ (2002) do not account. Local examples such as the fine residual granite soils of the Western Cape that have low inherent lateral and rocking stiffness, provide evidence for the need for these checks by South African designers.

REFERENCES

Bonnett, D. 2005. Wind Turbine Foundations—loading, dynamics and design. *The Structural Engineer*. ICE Publishing, February 2005, pp. 41–45.
Bowles, J.E. 1997. Foundation analysis and design, 5th edition. *McGraw-Hill Companies, Inc.*, Illinois, USA.
Byrne, B. 2011. Foundation Design for Offshore Wind Turbines, Géotechnique Lecture *Géotechnique ICE*, London.
Das, B. 2011. Principles of Foundation Engineering. 7th Edition. *Cengage Learning*, Stamford, USA.
Department of Energy. 2013. *Integrated Resource Plan For Electricity (IRP) 2010–2030—Update Report*. Prepared for the Presidency of South Africa. Available at: http://www.energy.gov.za/IRP/irp%20files/IRP2010_2030_Final_Report_20110325.pdf.
Det Norske Veritas & Risφ National Laboratory. 2002. Guidelines for the design of wind turbines. *Det Norske Veritas*. Copenhagen, Denmark.
Mawer, B.W. 2015. An Introduction to the Geotechnical Design of South African Wind Turbine Gravity Foundations. MSc Dissertation. University of Cape Town, South Africa.
Molenaar, D.P., Van der Tempel, J. 2002. Wind Turbine Structural Dynamics—A Review of the Principles for Modern Power Generation, Onshore and Offshore. *Wind Engineering*. Volume 26, No. 4 2002 pp. 211–220
Van der Haar, C. 2014. Design Aspects of Concrete Towers for Wind Turbines. *Proceedings of the International Seminar on design of Wind Turbine Support Structures*. 3 September 2014. University of Stellenbosch, Cape Town. 37–46.
Vestas Report. 2013. Foundation Loads—V112 3MW 1540rpm HH 94 IEC2 A. Document No. 0024–5696. V03. 2013.03.12
Warren-Codrington, C. (2013). Geotechnical Considerations for Onshore Wind Turbines. MSc Dissertation. Department of Civil Engineering. University of Cape Town, South Africa.

Proceedings of the first Southern African Geotechnical Conference – Jacobsz (Ed.)
© 2016 Taylor & Francis Group, London, ISBN 978-1-138-02971-2

Laboratory investigation of the performance of rammed sand columns

L. Sobhee-Beetul & D. Kalumba
University of Cape Town, Cape Town, South Africa

ABSTRACT: Rammed sand column is one of the many approaches to ground improvement which has been used successfully in several countries. It is often preferred since it can be relatively economical when treating ground for the construction of low to moderate capacity structures. Despite its recognition worldwide, South Africa is still new to the technology. Therefore, this study was pursued to investigate the possibility of improving a South African clay. Laboratory tests were conducted on clay reinforced with singular columns at different moisture contents, under a compressive force. The results were analysed in terms of stress concentration and settlement reduction ratios. Generally, the inclusion of sand columns increase the stiffness. More specifically, the well graded sand showed better improvement in vertical bearing stresses when analysed at a maximum displacement of 50 mm. On the other hand, the uniformly graded material performed better in settlement reduction at any particular vertical bearing stress.

1 INTRODUCTION

Rapid infrastructure growth has caused scarcity of competent land for development purposes. To meet the land demand, numerous ground improvement methods are being continuously investigated in order to establish the most appropriate technique to treat specific problematic sites. Granular column installation, rammed or vibrated, is regarded as one of the most versatile and cost-effective ground improvement approach. In recent times, these granular columns have been constructed from materials such as sand, gravel, quarry dust and crushed aggregates (Andreou et al. 2008, Ambily & Gandhi 2007, Isaac & Madhavan 2009). Irrespective of the particle size and nature of the column material, a certain degree of improvement in load bearing capacity has been recorded with column inclusions. Despite its proven effectiveness in Europe, Asia and the United States over the last few decades (McKelvey et al. 2004), the technology is still new and unfamiliar to many in South Africa. In conjunction, the proportion of problematic land (for the construction industry) in South Africa is approximated as 50% of the total soil coverage of the country (AGIS 2011). Therefore, to promote the use of this environmentally friendly technique locally, this study was conducted on a typical local clay to evaluate the improvement due to inclusions of granular columns.

2 EXPERIMENTAL MATERIALS AND PROCEDURE

2.1 Materials

2.1.1 Base material
The properties of the yellowish-brown Cape Town clay, excavated from a construction site in the Central Business District of Cape Town, are summarised in Table 1. Its liquid limit and plastic limit have been determined as 34.7% and 21.8% respectively.

2.1.2 Column material
Two types of locally sourced sands, namely Klipheuwel (well graded) and Cape Flats (uniformly graded), were utilised. They were chosen since they are readily available, clean, consistent and easy to work with. In addition, since the particle size and shape of both materials differed, it was possible to observe their effect on the performance of the columns. Tested properties of the sands are provided in Table 2.

2.2 Experimental procedure

For the purpose of this investigation, a rectangular box of dimensions 1000 mm × 150 mm × 450 mm was used whereby 400 mm long singular sand columns were manually installed for each test. Tests were conducted on wet clay beds at Optimum Moisture Content (OMC), liquid limit (LL) and 1.2 times Liquid Limit (1.2 LL) for both types of sand

Table 1. Mechanical properties of the base material.

Property	Unit	Cape Town clay
Specific gravity, G_s	Mg/m³	2.74
Optimum moisture Content (OMC)	%	17
Maximum dry density	kg/m³	1800
Angle of friction (Peak)	°	39[*1], 0[*2], 0[*3]
Cohesion (Peak)	kN/m²	37[*1], 3[*2], 3[*3]

*1—Optimum Moisture Content (OMC), *2—Liquid Limit (LL), *3–1.2 times Liquid Limit (1.2 LL).

Table 2. Mechanical properties of the column materials.

Property	Unit	Cape Flats	Klipheuwel
Specific gravity, G_s	Mg/m³	2.68	2.68
Average densest dry density[*1]	kg/m³	1735	1762
Average loosest dry density	kg/m³	1684	1578
Optimum Moisture Content (OMC)	%	14.2	9.8
Maximum dry density	kg/m³	1740	1840
Particle range	mm	1.18–0.075	0.075–2.36
Mean grain size, D_{50}	mm	0.65	0.90
Coefficient of uniformity, C_u	–	3.0	6.5
Coefficient of curvature, C_v	–	1.1	1.0
Angle of friction, φ (Peak)	°	38	37
Cohesion, c (Peak)	kN/m²	5	8

*1—obtained by compacting oven dried sand in a proctor mould.

columns. High moisture contents were used to gain an understanding of the extent to which improvement could be achieved when sand columns were subjected to lower lateral support. Three diameters of columns were investigated namely 50 mm, 70 mm and 100 mm with corresponding loading plates of widths 150 mm and lengths 100 mm, 140 mm and 200 mm respectively. The performance resulting from the column materials, for each of these varying parameters, was later analyzed.

Figure 1 depicts a few stages involved in the column installation and testing. The clay bed was prepared by initially mixing water at OMC, LL or 1.2 LL and subsequently filling the box with the mixture in 8 equal layers to form a bed of 400 mm thick. Each layer was subjected to compaction by a 2 kg steel hammer, falling through 180 mm twelve times.

A greased steel cylinder of the desired column diameter was then manually pushed through until

(a) Steel cylinder pushed in clay bed - ready for the cutting process

(b) Installed sand column in clay bed

(c) Improved clay bed subjected to compressive force under Zwick machine

Figure 1. Sample preparation and testing.

the base was touched. A manual cutter was then used to empty the metal cylinder. After cutting and cleaning the inside of the cylinder, a predetermined mass was poured. This was rammed 12 times by a 2 kg weight dropping through 180 mm. The process was repeated 8 times until the sand column was level with the clay bed. A 25 mm thick rectangular loading plate of corresponding dimensions (150 mm × twice the diameter of the column) was placed centrally on the column and subsequently loaded through the computerised Zwick Universal Testing Machine at a speed of 1.2 mm/min. Tests were run up to a vertical displacement of 50 mm and results were electronically recorded. Full details of the methodology and justification are provided in Sobhee-Beetul (2012).

3 RESULTS AND DISCUSSIONS

Electronic experimental data were obtained in terms of the vertical stress-settlement relationship for each tested condition. Full details of these results are provided in Sobhee-Beetul (2012) which facilitated the determination of the stress concentration ratios, n, and the settlement reduction ratios, SRR, which are important in computing settlement. Stress concentration ratio was calculated as the ratio of the vertical stress exerted on a reinforced clay to that on an unimproved one at a maximum settlement of 50 mm, implying an improvement with ratios higher than 1. This paper presents the relationship between each of these ratios and the diameter of both Klipheuwel and Cape Flats sand columns, at the tested moisture content.

3.1 Relationship between stress concentration ratio and column diameter

Generally, Klipheuwel sand produced higher n-values, ranging between 1.33 and 3.29, than Cape Flats sand. This behavior can be explained in terms of the grading of the Klipheuwel sand. The well graded material allows for the finer particles to fit in the voids between the larger ones, thereby producing much stiffer columns than those from Cape Flats sand (uniformly graded). The angular shape of Klipheuwel sand particles additionally favours the higher stiffness through an enhanced interlocking of particles.

In terms of moisture content, small columns (50 mm and 70 mm) produced relatively similar results at a low Moisture Content (OMC). However, an increase in moisture content of the clay bed indicated higher n-values for both materials. This observation can be attributed to the reduction in strength of the clay bed as its moisture content increases. Furthermore, larger columns produce higher improvements due to the augmented proportion of the strong replacement material.

3.2 Relationship between settlement reduction ratio and column diameter

In this study, the settlement reduction ratio is defined as the ratio of settlement in improved ground to that in unimproved ground, with a lower SRR denoting a higher improvement.

The SRR-diameter relationships are presented in Figure 3 which indicates a general decrease in settlement reduction ratio with large Cape Flats and columns (100 mm) installed in clays at all the moisture contents. In fact, Cape Flats sand shows higher improvement in settlement than the other sand which demonstrated negligible change in SRR-values with variation in diameters (except in clays of very high moisture content). In base clays of LL and 1.2 LL, settlement reduction ratio of large Klipheuwel columns normalise to a constant

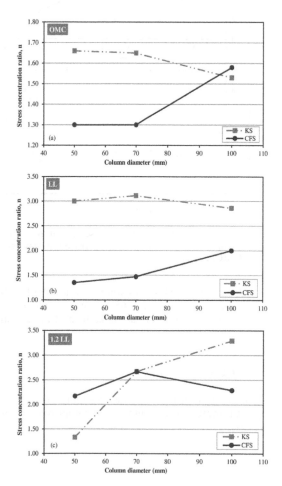

Figure 2. Shows the relationship between stress concentration ratio and column diameter in wet clay beds at OMC, LL and 1.2 LL.

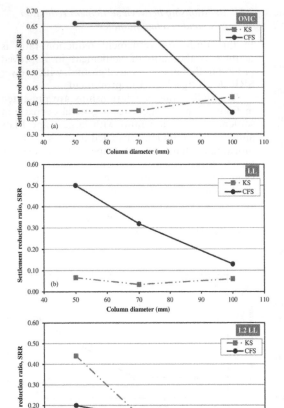

Figure 3. Relationship between settlement reduction ratio and column diameter for the two sands in clay beds at OMC, LL and 1.2 LL.

value of about 0.06. This study produced SRR values between 0.05 and 0.65, a range which shows good agreement with Zahmatkesh and Choobbasti (2010).

4 CONCLUSIONS

In an attempt to study the possibility of employing rammed sand columns in the improvement of a South African clay, laboratory tests were conducted on 400 mm long singular columns (diameters of 50 mm, 70 mm and 100 mm) installed in a local clay prepared at three different moisture contents. Two types of column materials were used and their effect on stress concentration and settlement reduction ratios were observed. With respect to vertical applied stress, Klipheuwel sand (well graded) showed better improvement with values ranging between 1.33 and 3.29. However, Cape Flats sand exhibited higher enhancement in settlement.

ACKNOWLEDGEMENTS

The authors thank the Geotechnical Engineering Research Group at the University of Cape Town for the financial support, and Esorfranki in Cape Town for allowing access to their District Six construction site to collect the Cape Town clay.

REFERENCES

AGIS. 2011. Natural resources atlas. *Soil classes:* www.agis.agri.za.

Ambily, A.P. & Gandhi, S.R. 2007. Behaviour of Stone Columns Based on Experimental and FEM Analysis. *Journal of Geotechnical and Geoenvironmental Engineering, ASCE* 133(4): 405–415.

Andreou, P., Frikha, W., Frank, R., Canou, J., Papadopoulos, V. & Dupla, J.C. 2008. Experimental study on sand and gravel columns in clay. *Proc. of the Institution of Civil Engineers, Ground Improvement* 161(GI4): 189–198.

Isaac, D.S. & Madhavan, S.G. 2009. Suitability of Different Materials for Stone Columns Construction. *Engineering Journal of Geotechnical Engineering* 14: 1–12.

McKelvey, D., Sivakumar, V., Bell, A. & Graham, J. 2004. Modelling vibrated stone columns in soft clay. *Proc. of the Institution of Civil Engineers, Geotechnical Engineering* 157(GE3): 137–149.

Sobhee-Beetul, L. 2012. An investigation into using rammed stone columns for the improvement of a South African silty clay. *MSc Thesis, Dept. of Civil Engineering, University of Cape Town, South Africa.*

Zahmatkesh, A. & Choobbasti, A.J. 2010. Investigation of bearing capacity and settlement of strip footing on clay reinforced with stone columns. *Australian Journal of Basic and Applied Sciences* 4(8): 3658–3668.

Proceedings of the first Southern African Geotechnical Conference – Jacobsz (Ed.)
© 2016 Taylor & Francis Group, London, ISBN 978-1-138-02971-2

The movement of soil moisture under a government subsidy house

D.M. Bester, P.R. Stott & E. Theron
Central University of Technology, Bloemfontein, South Africa

ABSTRACT: The South African government's attempts to provide affordable, subsidised housing for the very poor has suffered from a large number of structural failures, many due to heaving foundations. These houses are particularly susceptible to damage by heaving clay because they are exceptionally light and clay can lift them very easily. Rational design requires knowledge of the pattern of heave which will occur under the foundation. The pattern of heave depends on the pattern of moisture movement. Currently available methods of rational design rely on assumptions about the shape of the mound which will develop due to moisture movement under the foundation. The shape assumed is largely guided by measurements made on test foundations. Instrumentation has been installed under a Government Subsidy house in the Free State and moisture movement is being monitored. The actual pattern of moisture movement observed is substantially different to what is normally assumed and could point to more reliable estimates of the heave which needs to be designed for.

1 INTRODUCTION

Since 1994 the South African government has built more than 2.68 million subsidised houses throughout the country (Government Communication and Information System 2015).

The construction of raft foundations for Government Subsidy housing has become common in areas affected by expansive soils as is the case in much of the Free State and Northern Cape. The intention of raft foundations on active soils is to limit the differential movements of the underlying soils to a level which can be tolerated by the superstructure (Day 1991). The large number of failures suggests that the problem of providing sufficient stiffness to the foundation has not yet been solved. This impression is reinforced by the observation that many houses are being built on "stock design" rafts which bear only general correspondence with the likely heave potential of the soils involved. There is the impression that there may be inadequate understanding of the actions which need to be designed for. In many cases, when structures have become structurally unsound due to heaving foundations, it is more economical to demolish than to attempt repair. The seriousness of this problem therefore merits a search for a solution.

Current raft design relies on assumptions about the shape of the mound or dome which will develop due to moisture movement underneath the foundation. The shape which is assumed is often based on heave measurements on simulated foundations (Pidgeon 1987), (Pidgeon and Pellisier 1987) or foundations simulated by sheet covers (Fityus et al. 2004), (Miller et al. 1995), (de Bruijn 1973). Such test foundations do not take account of two important factors concerning the influence of a building constructed on the foundation—the influence of the building on the temperature regime under the slab and the influence of the building on solar radiation reaching the soil surrounding the raft. Both of these factors can have a profound effect on moisture movement under the foundation. (Fityus et al. 2004) found that over a period of measurement lasting seven years, temperature appeared to have a greater effect than rainfall on moisture movement and consequent heave of clayey soils at their test site.

2 SITE DESCRIPTION AND INSTRUMENTATION

The location of the study area is in Botshabelo section K, which forms part of the Mangaung Metropolitan area and is situated 45 km east of Bloemfontein. Each year many government subsidy houses are built throughout Botshabelo. A significant number of them experience structural distress well short of their design lifetime.

Botshabelo is underlain by mudstones, shales and sandstones of the Beaufort Group, with frequent intrusions of dolerite. All of these rocks frequently produce expansive clays when weathered in the semi-arid conditions of the central Free State. The area is therefore likely to be very suitable for the study being undertaken. A Government

Subsidy House (GSH) was selected based on the soil conditions and the fact that it has a raft foundation of a very common "stock" design. Continuous Logging Soil Moisture (CLSM) probes were installed to measure water content at various depths under the house. Measurements are being taken automatically at hourly intervals. The installation layout is from east to west and north to south in direct alignment with the house.

The CLSM probes allow measurement of temperature and water content at depths of 150 mm, 300 mm, 450 mm, 600 mm, 800 mm and 1000 mm. The soil profile underneath the house has a thin layer of dark brown clayey sand with a thickness of 150 mm. This is underlain by a layer of black transported clay and a layer of olive residual clay. Both clays were assessed by Van der Merwe's method (Van der Merwe 1964) as having medium expansiveness. Rock is found at a depth of approximately 1.1 meter. The first clay layer is from 150–900 mm, the second clay layer is from 900–1100 mm. Figures 2 and 3 illustrate the layout of the CLSM probes inside and below the house.

3 RESULTS AND DISCUSSION

3.1 Moisture content seasonal change models

Results of hourly measurement of water content taken over a period of two years were analysed. The following figures illustrate moisture content values measured in summer and winter at each depth recorded by the probes. These graphical rep-

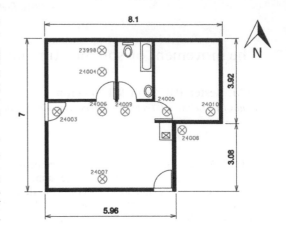

Figure 3. CLSM probe layout in raft foundation: dimensions m.

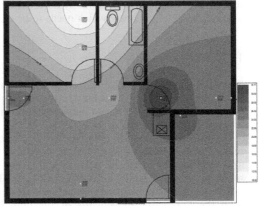

Figure 4. Summer: 150 mm w = 10%–32%.

resentations of the data were produced using the programme "3D field". The scale was selected as a best fit for each specific model. Areas of dark gray/black have higher moisture content. A grey scale is shown next to each figure.

From the figures it can be observed that the north side consistently shows the lowest moisture content compared to the rest of the house. This is almost certainly due to the fact that solar energy reaching the ground on the northern side is more intense than that reaching the other sides of the building. When rainfall occurs the north side is prone to rapid moisture change. This is probably due to the fact that the dry, cracked soil allows immediate access to rainwater. At the north side water contents range from less than 5% to 25%. In contrast, water content near the south east re-entrant corner remains within the range 31%

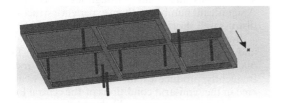

Figure 1. Botshabelo section K.

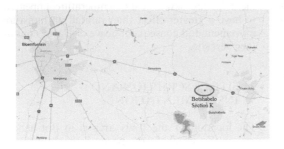

Figure 2. 3D layout of CLSM probes.

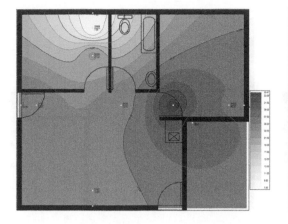

Figure 5.   Summer: 300 mm w = 7%–33%.

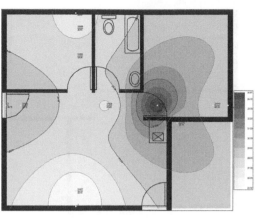

Figure 8.   Summer: 800 mm w = 25%–35%.

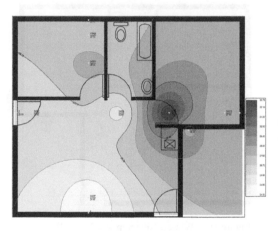

Figure 6.   Summer: 450 mm w = 24%–33%.

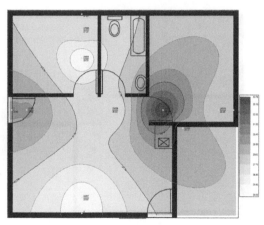

Figure 9.   Summer 1000 mm w = 25%–34%.

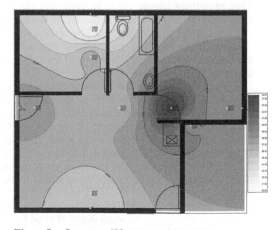

Figure 7.   Summer: 600 mm w = 20%–35%.

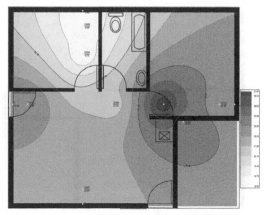

Figure 10.   Winter: 150 mm w = 15%–31%.

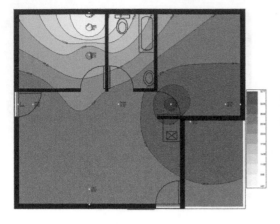

Figure 11.   Winter: 300 mm w = 5%–33%.

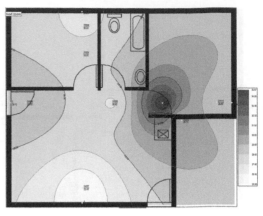

Figure 14.   Winter: 800 mm w = 25%–35%.

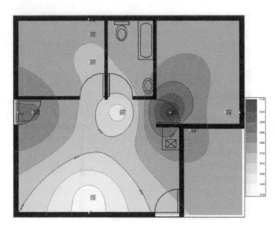

Figure 12.   Winter: 450 mm w = 24%–31%.

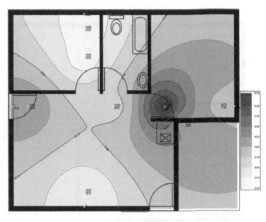

Figure 15.   Winter: 1000 mm w = 24%–33%.

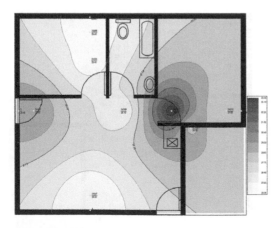

Figure 13.   Winter: 600 mm 25%–33%.

to 34% at all depths in both summer and winter. This re-entrant corner never receives any significant solar radiation. The architecture of the building guarantees that it is practically always in the shadow of the building. Its moisture content did not change by more than 3% in any of the different layers of soil.

Figure 16 illustrates the seasonal changes from March 2014 to February 2015 of the north side (CLSM probe 23998) that consistently has the lowest moisture content compared to the rest of the house. While Figure 17 shows the south east side (CLSM probe 24005) seasonal moisture changes for the same time that has the most consistently high moisture content compared to the rest of the house.

The observed moisture variations suggest that an approximately symmetrical dome-shaped heave pattern, centred roughly on the centre of the

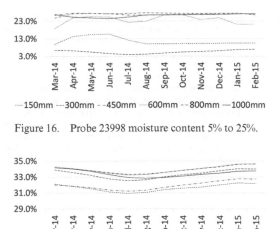

Figure 16. Probe 23998 moisture content 5% to 25%.

Figure 17. Probe 24005 moisture content 31% to 34%.

foundation, as found by Pidgeon (1987), Pidgeon and Pellissier (1987), Fityus et al. (2004) etc. in simulated foundation tests is not likely to develop. Simulated foundation tests take no account of the influence of shadows cast by the building. It appears that the south east re-entrant corner will remain in a high water-content, highly expanded state (varying by only about 3%) whereas the northern wall will assume a lower water-content, less expanded state with far more variability. This tends to confirm the findings of Fityus et al. (2004) that temperature can have a more pronounced effect than rainfall on swelling patterns in active soils.

3.2 Further considerations

It has been noted (Fityus et al. 2004) that heave predictions commonly ignore the effect of the loads applied by the building, though this may not be true in all cases. The load applied to the soil directly under a raft foundation by a completed GSH house is approximately 10 kPa. The three layers of material under the instrumented foundation were tested to assess their ability to heave against constraining pressures. The clayey sand of the thin upper layer was found to be able to expand against a pressure of 11 kPa up to a water content of slightly above 28%. The transported clay, of the second layer, was found to be able to expand against a pressure of 14 kPa up to a water content of 35% and the residual clay of the third layer could expand against a pressure of 14 kPa up to a water content exceeding 45%. These pressures were measured by apparatus under development by the

CUT Soil Mechanics Research Group. At lower water content far higher pressures can develop. Assuming a unit weight of 2 000 kg/m³ for all layers, the maximum pressure exerted at the base of the lowest layer would be approximately 21 kPa. It would therefore appear that the loads applied by overburden and structure might cause a restriction to heave in the thin upper layer in the wettest area only (around the south-east re-entrant corner). The lowest layer of clay would probably experience no restriction to heave, and the lower part of the middle layer would probably be restrained in the area of maximum water content.

4 CONCLUSIONS

The measurements of moisture content under a very common type of Government subsidy house built on clayey soil suggest that currently accepted patterns of heave are unlikely to provide good guidance for foundation design. The instrumentation used in this investigation has proved itself convenient and reliable. It is hoped that it will be possible to instrument several other light structures in order to work towards a general modeling procedure. This should enable reliable predictions of the moisture conditions which need to be designed for in the general case. This in turn should allow reliable and economic design of a wide range of raft foundations with the prospect of fewer failures. In the meanwhile, by applying the findings of this investigation it should be possible to at least obtain better estimates of moisture pattern development and to estimate more realistic heave patterns than are currently being employed.

ACKNOWLEDGEMENTS

The authors wish to express their thanks to Prof. L. van Rensburg of the Soil Science Department of Free State University for his inputs in planning, instrumentation and general help at all stages of the venture and to the NRF for financial support of the project.

REFERENCES

Day, P.W. 1991. Design of Raft foundations for houses on expansive clay using Lytton's method. *Course on design of foundation to suit various soil conditions. SAICE Structural Division 1991.*
De Bruijn, C.M.A. 1974. Moisture redistribution in Southern African soils. *Proceedings of the 8th International Conference on Soil Mechanics and Foundation Engineering. Moscow. Vol2.2.1973, pp. 37–44. (NBRI Reprint No R/Bou 442,1974).*

Fityus, S.G., Smith, D.W. & Allman, M.A. 2004. Expansive soil test site near Newcastle. *Journal of Geotechnical and Geoenvironmental Engineering. ASCE. July 2004 pp. 686–695.*

Government Communication and Information System. 2015. *Housing delivery in South Africa—How have we fared?* Government Communication and Information System (http://www.sanews.gov.za/south-africa/housing-delivery-sa-how-have-we-fared).

Miller, D.J., Durkee, D.B., Chao, K.C. & Nelson, J.D. 1995. Simplified heave prediction for expansive soils. *Unsaturated soils/sols Non Satures, Alonso & Delage (eds).* Roterdam: Balkema.

Pidgeon, J.T. 1987. The results of a large scale field experiment aimed at studying the interaction of raft foundations and expansive soils. *International Conference on soil structure Interactions. Paris. May 1987.* Paris: ENPC Press.

Pidgeon, J.T. & Pellissier, J.P. 1987. The behaviour of an L-shaped raft subjected to non-uniform support conditions. *International Conference on Soil Structure Interactions. Paris. May 1987.* Paris: ENPC Press.

Van der Merwe, D.H. 1964. The prediction of heave from the Plasticity Index and the percentage clay fraction. *The Civil Engineer in South Africa Vol.6.No.6 June 1964.*

Proceedings of the first Southern African Geotechnical Conference – Jacobsz (Ed.)
© 2016 Taylor & Francis Group, London, ISBN 978-1-138-02971-2

The design and construction of drilled and flush-grouted micropiles as covered by standard in the European Union (EU)

A.Y. Kayser
Friedr. Ischebeck GmbH, Ennepetal, Germany

A. Bartkowiak-Higgo
Titan Geotechnical Ltd., Gauteng, South Africa

ABSTRACT: Deep foundations commonly stabilizes structures such as buildings, bridges, telecommunications masts, wind turbines, etc. include weak local ground conditions or earthquake-prone regions. A new technology is the micropile, which was developed in the early 1980s. After 30 years of harmonizing standards in Europe, standard EN 14199 "Execution of special geotechnical works—micropiles" was introduced for micropile systems in 2005. According to EN 14199 micropiles have a diameter < 300 mm. The self-drilling injection system with hollow bar is predestined for building sites with restricted access or difficult conditions for large plant and—compared with large-diameter bored piles—also better suited to stabilizing structures in seismic areas. This paper focuses on self-drilled and flush-grouted micropiles (hollow bars). An overview will be given about latest developments regarding European standards, installation methods, corrosion protection (head details and hollow bar) and applications. Furthermore, various projects will be presented to illustrate and introduce solutions to different engineering/building challenges using micropiles according to European standards.

1 INTRODUCTION

Deep foundations are often designed and built to stabilize different kinds of structure in difficult ground conditions. They are generally in the form of driven or drilled piles. Common materials for driven piles are timber, precast concrete or steel, which are delivered to the building site and driven into the ground.

Drilled piles are in situ deep foundations. Micropiles are categorized as drilled piles, with the diameter being determined and limited to max. 300 mm according to EN 14199. Drilled and flush-grouted micropiles use a special installation method that offers not only speed advantages. The components needed to install a micropile are shown in the following. The installation methods are introduced and explained in a separate section. Basic corrosion influences such as soil aggressivity, cracks in concrete and head details in general are discussed in detail as well.

Engineering structures such as bridges, which can be seen as permanent structures, have to comply with Eurocode 0 (EC 0)—EN 1990. For a permanent structure, EC 0 requires the service life to exceed 100 years for engineering structures. How can a service life >100 years be achieved according to European standards? How can bare steel be

protected in soil, which could also be an aggressive environment to steel? Commonly, every country has its own methods of corrosion protection. If there are no standards, guidelines or recommendations available for certain topics, standards from other countries can be taken into consideration, analysed and evaluated in order to have something to work with.

This paper concludes with examples of different applications based on the requirements according to the European standards and a final summary.

2 SELF-DRILLED AND FLUSH-GROUTED MICROPILES—INSTALLATION METHOD

The origin of the idea to secure a structure with many small piles instead of a few large-diameter piles started with Dr F. Lizzi around 1952. His vision was based on a large tree, which has big roots but also many smaller roots working together in a network. The idea of the so-called root pile (*paliradice*), or nowadays micropile, had been born and since then has been further developed. Many small piles < 300 mm dia. work as a group to transfer tension and/or compression loads from a structure into the ground.

Conventional piles, either small or large, require several set-up steps: 1) drill a hole, 2) pull out drilling tool, 3) insert reinforcement (cage, solid bar, etc.) and 4) fill up the borehole with cement grout or concrete.

Drilling plant (drill rigs) for micropiles is much smaller than the equipment for conventional piles. This is especially advantageous in restricted areas, e.g. retrofitting in existing buildings or difficult access near bridges, harbour areas, steep slopes, etc. With micropiles, these different, challenging applications can also be managed.

Drilled and flush-grouted micropiles have only one step for each pile. With the hollow bar connected to a sacrificial drill bit, the drilling procedure starts with rotary percussive drilling into the ground. While drilling, cement suspension (water/cement ratio = 0.5–0.7) is pumped in to flush out the borehole. The cement suspension is transferred from the drill rig (e.g. flushing head) via the hollow core of the bar and exits through the flushing holes of the drill bit at the bottom of the borehole. The length of a micropile varies depending on the ground conditions and form of construction, but with coupling nuts, the hollow bars, 3 m long, for instance, can be coupled together to provide the required design depth of up to 60 m. After reaching the final depth, the injected thin cement suspension is replaced by a thick one (water/cement ratio = 0.4–0.5). As the thick cement suspension is stiffer, it displaces the thin cement suspension inside the borehole. The hollow bar is uncoupled from the drill rig and left in the borehole as reinforcement.

One micropile (either short or long) is finished and the drill rig can move on to the next drilling point. The micropile head (head detail depends on application) will be prepared similar to other systems to connect to the superstructure.

This method shows the advantage of a fast installation process, but as the hollow bar with a drill bit is used as a drilling tool and left in the ground as reinforcement, it requires enhanced steel qualities. Depending on the ground conditions, the drilling procedure is either rotation only or rotary percussive to drill through rock, for instance.

Flushing with a thin cement suspension stabilizes the borehole while transporting drilling debris out of the borehole, and the drilling tool (drill bit and hollow bar) is also cooled constantly. Furthermore, the thin cement suspension as the initial injection material has a soil improvement effect. The final grout, the thick cement suspension, achieves an even a better bond with the cement-saturated ground. Direct flushing from the beginning also ensures a constant grout body over the full length of the micropile to provide corrosion protection of the reinforcement (hollow bar and coupling nuts).

Hollow bars have a typical outer diameter range of 30–127 mm. The micropile consisting of the hollow bar surrounded by cement grout has a diameter equal to the drill bit diameter plus, theoretically, min. 20 mm, depending on soil type. Pressure grouting will result in a bigger diameter than the estimated fixed enlargement of 20 mm.

Drill bit types will be chosen subject to soil type. There are drill bits suitable for different ground conditions such as clay, peat, sand, a mixture of different soil strata, etc. The drill bit size matches the hollow bar/coupling nut outer diameter to ensure a minimum grout cover thickness for corrosion protection according to EN 14199.

In addition to drill bit, hollow bar and coupling nut, further accessories are available. For instance, centralizers keep the hollow bar centred in the borehole. End plates are needed to connect the superstructure to the micropile. Spherical collar nuts hold the plates in position, but also transfer the loads from the structure to the micropile.

As shown in Figure 1, an assembled micropile consists of a hollow bar in the centre surrounded by thick grout and, finally, thin grout that permeated the soil—resulting in a composite pile.

Besides the classic applications previously described, a micropile can also be used as a soil nail for slope stabilization or as a tie-back in the form of a passive anchor. The definition of soil nail is basically a short pile installed close to other nails (requirements can be found in EN 14490).

3 STANDARDS AND LATEST DEVELOPMENTS CONCERNING MICROPILES

Originating from a primary economics-oriented organization, a major topic of the European

Figure 1.   Section through micropile.

Union is to standardize the economies and politics among their numerous member countries. This includes harmonization of standards in the field of civil engineering. The self-drilling system with hollow bars still can be seen as a new technology, although Ernst-Friedrich Ischebeck developed this system in Germany approximately 30 years ago. There are recommendations prepared by working groups, such as the German Geotechnical Society, referring to design and execution based on existing standards (Eurocodes) in "Recommendations on Piling" or "Recommendations of the Committee for Waterfront Structures Harbours and Waterways". Through the harmonization of standards in Europe, the micropile, as a new technology, received the opportunity to be introduced and accepted within its own standard labelled EN 14199 "Execution of special geotechnical works—micropiles" in 2005.

Standards are reviewed every five years under the aspect of technical updates. Some of the latest updates to EN 14199 from July 2015 are provided here as background information with a focus on topics such as material requirements and corrosion protection.

EN 14199 provides all different kinds of requirements necessary for setting up micropiles.

Chapter 6, for instance, describes details of materials and products which are essentially important to know and need to be considered when choosing hollow bars for micropiles to achieve the advantages that come with this solution. Hollow bars ("hollow sections") have to comply with EN 10210 "Hot finished structural steel hollow sections" or EN 10219 "Cold formed welded structural steel hollow sections". Due to the installation process (described in section 2), special characteristics are required for threaded seamless tubes, the hollow bar. Although there are many steel types on the market (and therefore further standards related to other steel types), no changes to this strict requirement have been made in the latest update of EN 14199.

EN 10210 and EN 10219 provides information on chemical composition and general mechanical properties. These values are important with regard to the long-term load-carrying characteristics after being used as a drilling tool and finally in use as a loadbearing element to secure a structure. The explanation of the installation method for self-drilling and flush-grouted micropiles visualizes the forces during the drilling procedure. The whole system (hollow bar and coupling nut) has to withstand these forces without losing its characteristics after installation. Detailed explanations of steel characteristics and consequences can be found in section 4.4.

4 CORROSION INFLUENCES AND CORROSION PROTECTION

It is commonly known that carbon dioxide in combination with moisture initiates carbonation in concrete. This depassivation of concrete leads to corrosion of the steel in the end. In Europe, corrosion protection in the micropile standard is based on requirements for structural design. The micropile, with a hollow bar functioning as reinforcement in a cementitious grout body, is a composite element of steel and concrete just like reinforced concrete. Below, different corrosion influences will be introduced together with what needs to be done to delay or even prevent the onset of the corrosion process.

4.1 *Air access—head detail*

Corrosion influences above ground begin with the head detail of an application such as a micropile, soil nail or anchor. This is one of the most important issues, and is often not taken seriously in the building industry. Unprotected steel starts to corrode when an electrolyte (humidity/water) is present. Head details are connected to building structures to transfer loads through the hollow bar down to loadbearing soil strata. It is extremely important to protect the head detail against corrosion. Air and humidity must not be allowed to reach the heads of micropiles, anchors or soil nails. Apart from the head detail, the "pile neck" has to be protected as well. Some examples of head details and protection options for the pile neck area are given later. A rusty pile head without any protective cap or coating can be used as a temporary installation, but that would not be appropriate for a permanent project.

4.2 *Soil classification and aggressivity*

Soil can be aggressive to reinforced concrete if certain conditions are present. As the geographical landscape differs for each country in Europe, there are national standards for characterizing the soil. As an example, German standard DIN 50929 will be explained shortly to show a soil is evaluated and categorized. Characteristics and parameters are considered for soil evaluation, and with the estimation of these, soil can be classified as shown in DIN 50929, Table 1. In terms of prestressed rock and soil anchors, it has to be added that aggressive atmospheric conditions have to be considered in addition to water chemistry and air permeability of the ground. An estimation of corrosion likelihood in water is also explained in national standard DIN 50929, but not discussed in this paper.

Each parameter will be estimated and assigned a negative or positive evaluation number. All numbers are summed up and classified. Some parameters have a positive influence and others are negative, which means although negative influences exist, there may be positive soil conditions that reduce the negative effect. The soil is evaluated in its complete mixture with all complex interactions. Values for parameters in DIN 50929 are not given here, as the evaluation is divided into several intervals for each parameter and that would exceed the scope of this paper. Further details and the complete soil evaluation can be found in DIN 50929-3, Table 1.

After classification of the soil, the engineer is able to decide whether or not additional corrosion protection is required, but minimum requirements according to national approvals have to be taken into account as well. In European national approvals for hollow bar systems, they require only the use of cement grout in a certain range of soil classification, e.g. "…can be used for cohesive and non-cohesive soils…". These requirements are based on detailed tests and calculations.

Based on the result of a soil investigation, the type of corrosion protection will be chosen by the engineer with recommendations by the manufacturer as a supporting source. Protection methods are given later for each application.

4.3 Difference between cracks

Concrete will crack during the curing process due to various parameters such as water/cement ratio, grain size of aggregate, weather conditions, post-treatment, etc. In addition, steel will only be activated and bear tension loads when the concrete has cracked. This is one of the biggest advantages regarding why steel and concrete work with each other perfectly when it comes to tension and compression. Cracks in concrete can be divided into two groups: very small non-visible and visible cracks. Non-visible cracks are harmless with regard to long-term durability. Visible cracks can be deep and allow air or fluid to reach the steel element and accelerate the depassivation process. As concrete is an alkaline environment for the steel and therefore protective (passivation), the smaller the cracks are, the later the depassivation process is completed.

Crack width limitation is mentioned in EN 1992, also known as Eurocode 2 (EC 2) "Design of concrete structures", and EN 14490.

EN 14490 "Execution of special geotechnical works—Soil nailing" states: "Research has shown that crack widths controlled to less than 0.1 mm can be considered to be self-healing. Therefore, cement grout is considered acceptable as an impermeable protective encapsulation, provided that the crack width within the grout can be demonstrated not to exceed 0.1 mm."

A board of experts (DIBt, Deutsches Institut für Bautechnik—German Institute of Building Technology) for national approvals in Germany also agreed to this requirement, which means different suppliers can have their systems approved, as this requirement can be fulfilled through detailed laboratory testing and calculations.

Carbon dioxide, which is part of atmospheric air, with sufficient humidity initiates carbonation. After depassivation of the steel/concrete interface is complete, the steel will start to corrode when air and humidity or even aggressive elements have access. Minimizing crack width with reasonable grout thickness and proper water/cement ratios is a common solution for protecting steel from corrosion in Europe.

4.4 Steel quality, thread shape and grout body

For reinforced concrete, and the hollow bar system is seen as such in Europe, steel has to comply with EN 10080 "Steel for the reinforcement of concrete" and also with EN 10210/10219 according to EN 14199 and EN 14490. The yield strength of the steel has to be limited to a max. concrete strain of 0.2% for Germany according to EN 1992 (EC 2) in order to achieve the bond between steel and concrete.

Calculation:

$$E_{steel} \text{ (Young's modulus of steel)} = 200\,000 \text{ MPa}$$
$$f_{0.2\%} = E_{steel} \times 0.002$$
$$= 200\,000 \times 0.002$$
$$= 400 \text{ MPa}$$

According to EN 10080, rebar ribs must comply with certain dimensions for height, spacing and angle (each has min. and max. values). Different thread shapes are available on the market: special trapezoidal thread and rope thread (R-thread). Special trapezoidal thread has a sharper angle (min. 45° to the longitudinal axis to comply with EN 10080) and is similar to rebar rib shape so that it is able to provide the necessary shear bond between concrete and steel. The R-thread has a shallow angle (<17°) and a very smooth (round) thread surface, which permits easy and fast splicing of accessories to drill rods, but it does not comply with EN 10080.

In addition, for hollow sections, EN 14490 requires a global elongation A_{gt} of at least 5% at failure. With this appropriate steel quality, the behaviour of the element will be ductile and match concrete.

EN 10210 and 10219 gives the requirements for the mechanical characteristics and chemical

Figure 2. Transmission tower, Germany.

Figure 3. Base of transmission tower, Germany.

Figure 4. Seismic retrofitting, New Zealand.

A combination of steel quality, thread shape and grout body is needed for 100 year design life. Besides meeting the requirements regarding crack width and steel quality, grout cover is also a key issue. Grout cover around the steel member is needed to preserve the passivation of the steel surface and ensure that the diffusion of harmful substances such as chlorides is delayed. Due to passivation, the grout has a corrosion protection function for steel (see "Difference between cracks" above). A minimum thickness is needed to fulfil these requirements according to EC 2, EN 14199 and EN 14490.

5 APPLICATION TYPES AND DETAILS

Based on the explanation of the installation method and introduction of relevant standards, some examples of application types can be presented.

Figure 5. K-frame—seismic retrofitting, New Zealand.

Figure 6. Head detail for embedment in concrete.

composition of fine-grain steel such as yield strength, ultimate strength, Charpy value and a carbon equivalent value (CEV).

The impact toughness for fine-grain steel is specified in EN 10210 (Table B.3; EN 10219 Table B.4) as 40 J in an unfavourable test environment condition of −20°C. CVN (Charpy V-notch) toughness requirements ensure that the system take the working loads even after a rotary percussive drilling process.

6 SUMMARY AND CONCLUSION

Generally speaking, a micropile is a pile with a diameter < 300 mm. The benefit of this solution comes with the flexibility of small product components, use of the same components for different application types and easy handling.

European engineering structures such as bridges are designed for a minimum lifetime of 100 years according to EN 1990. National approvals, e.g. for Germany and Austria, are based on many test results and calculations based on European standards that prove the function of the system. Tests need to be administered constantly for every new batch to ensure a high level of quality.

The self-drilling system with hollow bar is quite new, although the hollow bar system is used for various applications in deep foundations and approved by European standards such as EN 14199 or EN 14490 and comply with material standards EN 10080 and EN 10210 or EN 10219.

Owing to the installation process, self-drilling systems with hollow bars provide a continuous grout body around the steel tendon which function as corrosion protection. Various additional corrosion protection systems can be chosen to enhance the hollow bar system.

All European requirements such as steel stress limit up to concrete strain, steel quality (Charpy value for ductility or chemical composition) and thread shape on steel bar focus on providing compatibility between steel and concrete. By doing so, bond is maximized, crack width is limited and therefore long-lasting corrosion protection is accomplished exclusively by cement grout in more than 90% of all cases. The crack width limitation of 0.1 mm is a strict requirement. If the supplier of the hollow bar meets all the requirements regarding steel stress limits, steel quality and thread shape, the only protection required is the grout cover. Hollow bars that do not meet the above requirements can have additional corrosion protection added or be accounted for with sacrificial losses.

The single grout protection is sufficient for complying with requirements for limiting crack widths. Corrosion protection in soil is very important, but the pile head must also be protected.

Various systems exist for different application types such as micropiles, soil nails, etc. These systems can be used differently depending on the application, although they consist of the same main components. Permanent corrosion protection is only guaranteed if steel shape/geometry comply with EN 10080, which is not fulfilled by hollow bars with rope threads. Further, no guarantee of permanent corrosion protection can be given if the hollow bar is not made of steel according to EN 10210 or EN 10219.

REFERENCES

DIBt, Z-34.14–209. 2013. National Technical Approval of Titan micropiles.

DIN 50929. 1985. Corrosion of metals; probability of corrosion of metallic materials when subject to corrosion from the outside; buried and underwater pipelines and structural components.

DIN 55633. 2009. Paints and varnishes—Corrosion protection of steel structures by powder coating systems—Assessment of powder coating systems and execution of coating.

DIN SPEC 18539. 2012. Supplementary provisions to DIN EN 14199.

EA-Pfähle. 2013. Recommendations on Piling, Ernst & Sohn.

EAU 2012. 2015. Recommendations of the Committee for Waterfront Structures Harbours and Waterways, Ernst & Sohn.

EN 1537. 2014. Execution of special geotechnical works—Ground anchors.

EN 1990. 2010. Eurocode: Basis of structural design.

EN 1992. 2004. Eurocode 2: Design of concrete structures—Part 1-1: General rules and rules for buildings.

EN 1993. 2010. Eurocode 3: Design of steel structures—Part 1-1: General rules and rules for buildings.

EN 10080. 2005. Steel for the reinforcement of concrete—Weldable reinforcing steel.

EN 10210–1. 2006. Hot finished structural hollow sections of non-alloy and fine grain steels.

EN 10219. 2016. Cold formed welded structural steel hollow sections.

EN 14199. 2016. Execution of special geotechnical works—Micropiles.

EN 14490. 2010. Execution of special geotechnical works—Soil nailing.

EN ISO 1461. 2009. Hot dip galvanized coatings on fabricated iron and steel articles—Specifications and test methods.

EN ISO 12944. 1998. Paints and varnishes Corrosion protection of steel structures by protective paint systems.

Proceedings of the first Southern African Geotechnical Conference – Jacobsz (Ed.)
© 2016 Taylor & Francis Group, London, ISBN 978-1-138-02971-2

Vibration associated with construction of DCIS enlarged-base piles

H.N. Chang & B.D. Markides
Franki Africa, Johannesburg, South Africa

ABSTRACT: Driven Cast In-Situ (DCIS) enlarged-base piles have been used since the early 1900s and remains an effective pile type with excellent capacity and load-deflection characteristics, particularly in non-cohesive soil profiles. As with all driven piles, vibrations may be a concern, particularly when working close to existing structures and services. This paper presents the results of vibration monitoring during various stages of pile construction for two projects in Johannesburg. Higher vibration intensities were measured in dense granular soils compared to soft cohesive soils and in larger diameter piles. The results also indicate higher ground vibrations during pile driving. Based on the results, minimum distances between piling and existing structures were recommended.

1 INTRODUCTION

1.1 *Background*

Construction of DCIS enlarged-base piles involves bottom driving of piling tubes with a drop hammer. As with other driven piling systems, ground vibration is generated during construction of DCIS enlarged-base piles and occasionally raises concerns as to possible damage to adjacent structures or services.

Ground vibration due to impact pile driving has been investigated by many authors (e.g. Wiss 1961; Attewell & Farmer 1973; Massarsch & Fellenius 2008). These publications however, deal with prefabricated steel or concrete piles, where driving energy is imparted on the pile head and may have different vibration mechanisms/intensities to DCIS enlarged-base piles where the tube is advanced by driving on the gravel plug at the base of the tube. In addition, construction of DCIS enlarged-base piles require the gravel plug to be expelled before construction of the enlarged-base, both of which may generate different ground vibrations.

This paper aims to provide insight into ground vibrations generated during construction of DCIS enlarged-base piles. Ground vibration was monitored on two sites in Johannesburg during various stages of pile construction and for various pile sizes to assess the impact of these factors on the vibration generated. A vibration attenuation curve was established using vibration data measured at various distances from piling and results were compared to published threshold values for structural damage and human perception to make recommendations for future piling works.

1.2 *The DCIS Franki enlarged-base pile*

The original Franki driven cast in-situ pile was introduced by Edgard Frankignoul in 1909 and involves a tube driven into the ground by impact of a drop hammer on a plug of compacted granular material inside the toe of the tube (Nordlund 1982). Once the tube has reached an appropriate founding depth, the tube is held in position and the compacted granular plug expelled. Zero slump concrete is then rammed out at the end of the tube to form the enlarged-base. The bearing capacity of the enlarged-base is determined during construction through the use equations formulated by Nordlund (1982) based on cavity expansion theory.

The pile exhibits excellent capacity and load-deflection characteristics compared to conventional bored piles due to the compaction/preloading of the soil surrounding the DCIS enlarged-base (Tchepak 1986; Jaksa *et al.* 2002). The construction sequence of a typical DCIS Franki enlarged-base pile is illustrated in Figure 1. During construction of Franki piles, ground vibration is generated from impact of the drop hammer during driving, expelling of the gravel plug and forming of the enlarged-base.

1.3 *Ground vibration and propagation*

Ground vibrations consists of four main types of waves: compression and shear waves which propagate through the body of the soil mass; and Rayleigh and Love waves which travel along the upper ground surface (Richart *et al.* 1970; Barkan 1992). Rayleigh waves are particularly important to geotechnical and structural engineers as it contains over two thirds of the total vibration energy

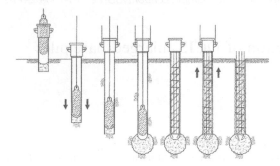

Figure 1. Construction sequence of DCIS Franki enlarged-base pile (Byrne & Berry 2008).

(Svinkin 2004; Heymann 2007). Furthermore, building foundations are generally positioned on or near the ground surface and may be susceptible to damage arising from ground vibrations.

The intensity of ground vibration can be measured in various ways, namely peak particle displacement, velocity or acceleration. Peak Particle Velocity (PPV) is frequently used to quantify vibration intensity, and when used in conjunction with frequency and duration, is the most appropriate and accurate indicator of potential structural damage (Rosenthal & Morlock 1987).

Attewell & Farmer (1973) analyzed vibrations generated by driving various types of piles in different soil conditions. They proposed that the peak particle velocity measured at a distance from the energy source can be estimated as:

$$v = k \frac{\sqrt{E}}{D} \quad (1)$$

where v = peak particle velocity (mm/s), E = theoretical energy input at source (J), k = empirical vibration factor (m²/s√J) and D = distance from energy source (m).

Brenner & Viranuvut (1975) used Equation 1 to compare vibration measurement from pile driving and suggested that values of the empirical factor k to be between 0.3 and 4.0 with an average of 0.75. Typical k-values for various soil types are summarized in Table 1.

Wiss (1981) published a state-of-the-art paper on construction vibrations and proposed an equation similar to Equation. 1:

$$v = K \left(\frac{D}{\sqrt{E}} \right)^{-n} \quad (2)$$

where v = peak particle velocity (mm/s). E = theoretical energy input at source (J), D = distance from

energy source (m), K = intercept value of vibration amplitude (mm/s) at scaled-distance $(D/\sqrt{E})$, and n = slope or attenuation rate.

Neither Attewell & Farmer (1973) nor Wiss (1981) provides guidance on how the distance should be chosen when a pile penetrates into the ground to a certain depth. Dowding (1996) commented that even when the actual vibration source is below the ground surface (as with pile driving), Rayleigh waves are formed within a few meters of a point on the surface directly above the vibration source and the propagation can be modeled in terms of Rayleigh waves. Svinkin (1999) also commented that pile driving from depths of between 4 and 10 m will generate Rayleigh waves within 0.4 and 3 m of the pile. This suggests that Rayleigh waves remain the dominant source of vibration regardless of the depth of the energy source, at least for practical depths of pile installation. The distance from the vibration source can therefore be taken as the plan (surface) distance between the piling tube and the monitoring point.

The value of attenuation rate, n, generally lies between 1.0 and 2.0 with a commonly assumed value of 1.5 (Massarsch & Fellenius 2008) whilst K can range between 0.05 and 0.3 (Heckman & Hagerty 1978). Woods & Jedele (1985) found the value of n follows closely to a value of 1.525 for silts,

Table 1. Summary of published n- and K-values.

Reference	Empirical vibration factor, K and attenuation rate, $n*$
Attewell & Farmer (1973)	1.5
Whyley & Sarsby (1992)	0.25 (soft or loose soils) 0.75 (stiff or medium dense soils) 1.5 (Stiff or dense soils)
Attewell et al. (1992)	0.76 ($n = 0.87$)
Head & Jardine (1992)	1.5 (for r > 0.5 m) 0.2 ($n = 1.54$) (at base of foundation)
BSI (1992)	0.75
Hiller & Crabb (1998)	3 (stiff or medium dense soils)
CEN (1998)	0.5 (soft cohesive soils) 0.75 (stiff cohesive soils) 1.0 (very stiff cohesive soils)
ArcelorMittal (2008)	0.5 (soft cohesive, loose granular media, loose fill and organic soils) 0.75 (stiff cohesive soils, medium dense granular media, compact fill) 1.0 (very stiff cohesive soils, dense granular media, rock, fill with large obstructions)

*Attenuation rate is only given for values other than 1.0.

sandy silts, sands, clayey sands and gravels with $5 < N_{SPT} < 15$ and 1.108 for dense compacted sands and dry consolidated clays with $15 < N_{SPT} < 50$. Note that the value of K (Equation 2) is only equivalent to k (Equation 1) if $n = 1.0$. Other published n- and K-values are summarized in Table 1.

1.4 Threshold values for ground vibration

Although considerable data has been published on ground vibration generated by construction activities, including pile driving, most refer to recommendations developed by the blasting and mining industry for guidance on threshold values for structural damage. Crandell (1949) proposed an energy based approach to assess potential damage to structures and recommended a threshold value of 84 mm/s for structure damage. Other threshold values have been proposed for various structure types and conditions as summarized in Table 2 compiled by Amick & Gendreau (2000).

Various studies have concluded that the human threshold to vibration is lower than that which has a potential to cause structural damage (Whiffen & Leonard 1971; Wiss 1981; Nugent & Amick 1992). Research carried out by the Federal Highway Administration (2012) also reported that most homeowners believed that if vibration can be felt then it must be causing damage to their buildings. A summary of human perception to vibration levels is given in Table 3.

2 FIELDWORK

Ground vibration measurements were taken on two project sites in the residual soils of Johannesburg. Brief descriptions of the ground profile at the two test sites are given below.

Table 2. Typical threshold vibration values for building damage (Amick & Gendreau 2000).

Category	Source	PPV (mm/s)
Industrial Buildings	Wiss (1981)	100
Building of substantial construction	Chae (1978)	100
Residential	Wiss (1981)	50
Residential, new construction	Chae (1978)	50
Residential, poor condition	Chae (1978)	25
Residential, very poor condition	Chae (1978)	12.5
Buildings visibly damaged	DIN 4150	4
Historic buildings	SNV (1992)	3
Historic and ancient buildings	DIN 4150	2

Table 3. Approximate vibration levels associated with human perception (Wiss 1981; Selby 1991).

Vibration level PPV (mm/s)	Degree of human perception
0.10	Not felt
0.15	Threshold of perception
0.35	Barely noticeable
1.0	Noticeable
2.2	Easily noticeable
6.0	Strongly noticeable

Site 1: Residual or completely weathered granite of very dense to very soft rock consistency. Material generally described as silty sand. Depth of bedrock in excess of 30 m, ground water not encountered. Profile described as *dense sandy* in discussions.

Site 2: Residual andesite soft to firm consistency, generally described as sandy clayey silt or silty clay. Rock generally in excess of 15 m depth, ground water not encountered. Profile described as *soft clayey* in discussions.

On both sites piles were installed as predrilled enlarged-base piles with diameters ranging from 410 to 610 mm and final length of between 8 and 11 m. Pile construction sequence can be described as follows:

- A slightly oversized hole was drilled with a conventional auger flight, generally to depths of between 5 and 10 m,
- The plugged tube was lowered into the predrilled hole and driven to set,
- The gravel plug was expelled,
- The enlarged-base was constructed,

Ground vibrations were monitored on the ground surface at plan distances of between 1 and 35 m from the piling tube. Measurements were taken using a commercially available seismograph Instantel Blastmate III with a measurement accuracy of 3 percent. The seismograph is housed in a tripod which was firmly planted in hard and compact ground for monitoring. Distinction was made between vibration readings during driving, expelling of the gravel plug and basing.

3 ANALYSIS AND DISCUSSIONS

The measured ground vibration in terms of maximum PPV is plotted against plan distance from the piling tube in Figure 2 for both sites investigated as well as data published by Tchepak (1986) and BSI (2009). Maximum PPV generally follows a similar attenuation trend with values reducing with increasing plan (surface) distance from piling works according to some non-linear function.

Wiss (1981) published data comparing ground vibration generated by typical construction activities against plan distance from the vibration source. Wiss' data is overlaid with the vibration data from this research and given in Figure 3. The vibrations generated from construction of DCIS enlarged-base piles are generally similar to that generated by a diesel or vibratory pile driving, particularly at close distances to the vibration source.

The vibration data can also be normalized against input energy by plotting maximum PPV against inverse of scaled distance ($\sqrt{E/D}$) as shown in Figure 4.

Equation 1 is preferred over Equation 2 as a means of comparison and is used throughout the discussions in this paper. Although Equation 2 provides more flexibility, it contains two mutually

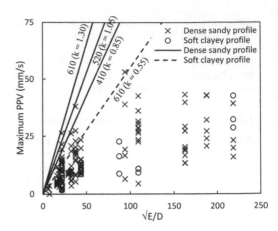

Figure 4. Plot of maximum PPV vs inverse of scaled distance.

dependent and unknown constants, K and n. For this reason, attenuation will be assumed to follow the linear function of Equation 1 with one single constant, k. The function however results in over-estimation of PPV values at distances within 2 m of the piling works by as much as one order of magnitude as PPV values tend toward infinity as the distance tends to zero. For practical purposes where piling is generally carried out some distance away from existing structures, Equation 1 gives sensible estimations.

Comparison of maximum PPV values for Site 1 indicates that the empirical vibration factor k increases with the diameter of the pile. Since the vibration data is normalized against input energy (which is higher for larger diameter piles), the increase in k can possibly be related to the volume expansion associated with pile construction and is roughly proportional to the ratio of the pile diameters.

The results also indicate distinctly lower k-value of 0.55 for a clayey profile (Site 2) when compared to a sandy profile (Site 1) where k-values ranged between 0.85 and 1.30. These values are similar to the published values summarized in Table 1 and indicate higher damping (i.e. lower k-value) observed in soft clayey soils when compared to dense sandy soils. Due to a lack of detailed profile characterization, these findings should be viewed qualitatively. Nevertheless, ground vibrations measured in generally soft clayey soils may be as low as 50 percent of that measured in dense sandy soils.

Differentiation can distinctly be made between vibration intensities during pile driving, expelling of the gravel plug and basing. Measured k-values are summarized in Table 4, with maximum PPV

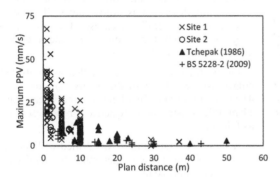

Figure 2. Plot of maximum PPV vs plan distance.

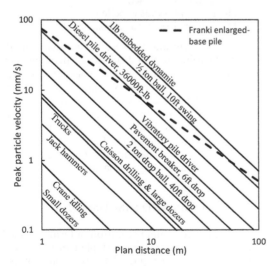

Figure 3. Comparison of ground vibrations from construction of enlarged-base Franki piles with typical construction activities (data after Wiss 1981).

readings monitored during driving of the tube. Ground vibrations may be reduced by up to 20 percent by reducing the amount of, or completely eliminating the driving component in pile construction.

A vibration attenuation curve following the general form of Equation 1 is proposed for DCIS enlarged-base piles (shown in Figure 5). The curve is capped at a PPV value of 75 mm/s for distances less than 2 m as values become unrealistic and tend to infinity as distance reduces to zero. The curve also appears to slightly overestimate PPV values for distances greater than 20 m but gives good estimate for short distances which are generally of concern.

Combining the proposed attenuation curve and the typical threshold values given in Tables 1, recommendation regarding the minimum distance between piling and existing buildings can be made. These values are summarized in Table 5 and graphically illustrated in Figure 5.

It should be noted that these recommended values are based on the building damage resulting from dynamic strains generated by ground vibration, and does not include damage resulting from secondary effects such as soil settlement, heave or liquefaction.

Table 4. Summary of k-values for various stages of pile construction.

Site	Pile diameter (mm)	Driving	k-value Expelling	Basing
1	410	0.85	0.65	0.60
	520	1.05	0.75	0.65
	610	1.00	0.80	1.30
2	610	0.55	0.50	0.50

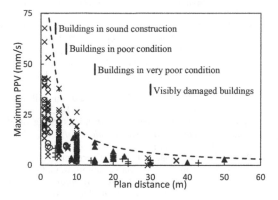

Figure 5. Vibration attenuation curve for DCIS enlarged-base piles.

Table 5. Recommended minimum distance from structure for building damage.

Structure description	Min. plan distance (m)	Limiting PPV (mm/s)
Industrial buildings or buildings or substantial construction	–	100
Buildings in sound construction	4	50
Buildings in poor condition	7	25
Buildings in very poor condition	15	12
Visibly damaged buildings	30	5

4 CONCLUSION AND RECOMMENDATION

Ground vibration was monitored during construction of DCIS enlarged-base piles to assess the vibration intensity generated. The results are assessed in terms of vibration factor k which is directly proportional to the peak particle velocity (i.e. higher k-value implies higher vibration intensity). From the analysis of the data, the following conclusions can be drawn:

The vibration generated during the construction of a DCIS enlarged-base pile is similar to that generated during driving of prefabricated piles.

The k-value is dependent on the consistency of the soil profile. Ground vibrations measured in a *soft clayey* profile may be as low as 50 percent of that measured in a *dense sandy* profile.

The k-value is dependent on the diameter of the pile, with the increase roughly equal to the proportion of the pile diameters.

Highest ground vibrations are generated during pile driving. Reducing the amount of, or completely removing the component of driving may reduce vibrations by up to 20 percent.

Ground vibration attenuates with the plan distance from the source piling work according to the power function given in Equation 1, but capped at 75 mm/s for distances less than 2 m. Based on the vibration attenuation curve, recommendations can be made for the minimum distance from existing structure to the piling work.

These values are given as a guideline only as building damage is a function of numerous factors such as vibration frequency and duration, and the natural frequency of the structure. If the risk of building damage is high, it is recommended that vibration intensity be verified by pre-construction and site specific monitoring. It is also recommended that a detailed dilapidation survey be conducted before and after pile installation to verify the extent of damage (if any) to the structure directly resulting from the piling works. Since the threshold of human perception to ground vibration is significantly lower than that which may potentially cause structural damage, communication may be important to assure clients and neighboring property owners. It has been reported that the tolerated vibration level is higher if the cause of the vibration is known and no damage is expected (Stille & Hall 1995).

REFERENCES

Amick, H. & Gendreau, M. 2000. Construction vibration and their impact on vibration sensitive facilities. *Proc. 6th ASCE Construction Congress, Orlando, Florida, 22 Feb, 2000.*

ArcelorMittal. (2008). *Piling Handbook, 8th Edition.* ArcelorMittal Commercial RPS.

Attewell, P.B. & Farmer, I.W. 1973. Attenuation of ground vibrations from pile driving. *Ground Engineering,* 3(7): 26–29.

Attewell, P.B., Selby, A.R. & O'Donnell, L. 1992. Estimation of ground vibration from driven piling based on statistical analyses of recorded data. *Geotechnical and Geological Engineering,* 10: 41–59.

Barkan, D.D. 1962. *Dynamics of Bases and Foundations.* New York: McGraw Hill Co.

Brenner, R.P. & Viranuvut, S. 1971. Measurement and prediction of vibrations generated by drop hammer piling in Bangkok subsoils. *Proc. of the 5th Southeast Asian Conference on Soil Engineering, Bangkok,* 109–112.

British Standard Institute. 2009. BS 5228.2 *Code of practice for noise and vibration control on construction and open sites—Part 2: Vibrations.* London: BSI.

Byrne, G.P.B & Berry, A. 2008. *A practical guide to geotechnical engineering in Southern Africa.* Johannesburg: Franki Africa.

Chae, Y.S. 1978. Design of excavation blasts to prevent damage. *Civil Engineering, ASCE,* 48(4), 77–79.

Crandell, F.J. 1949. Ground vibration due to blasting and its effect upon structures. *Journal of Boston Society of Civil Engineers,* April, 222–245.

Dowding, C. 1996. *Construction Vibrations.* New Jersey: Prentice Hall.

European Committee for Standardization. 1998. *Eurocode 3: Design of steel structures—Part 5: Piling.* Brussels: CEN.

Federal Highway Administration. 2012. *Ground Vibration Emanating from Construction Equipment.* New Hampshire: FHWA.

German Institute for Standardization. 1970. *DIN 4150 Vibration in Building Construction.* Pforzheim: DIN.

Head, J.M. & Jardine, F.M. 1992. Ground-borne vibrations arising from piling. *CIRIA Technical Note 142.* London: CIRIA.

Heckman, W.S. & Hagerty, D.J. 1978. Vibrations associated with pile driving. *Journal of the Construction Division, ASCE,* 104(C04): 385–394.

Heymann, G. 2007. Ground stiffness measurement by the continuous surface wave test. *Journal of the South African Institution of Civil Engineering, SAICE,* 49(1): 25–31.

Hiller, D.M. & Crabb, G.I. 2000. Groundborne vibrations from mechanized construction work. *Transport Research Laboratory Report 429.* Crowthorne: Transport Research Laboratory.

Jaksa, M.B., Griffith, M.C. & Grounds, R.W. 2002. Ground vibrations associated with installation of enlarged base driven cast-in-sit piles. *Australian Geomechanics, March, 2002:* 67–73.

Massarch, K.M. & Fellenius, B.H. 2008. Ground vibrations induced by impact pile driving. In S Prakash (eds), *Proceedings of the sixth international Conference on Case Histories in Geotechnical Engineering, Virginia, 12–16 August 2008.* Virginia: Missouri University of Science and Technology.

Nordlund, R.L. 1982. Dynamic formula for pressure injected footings. *Journal of the Geotechnical Engineering Division, ASCE,* 108(GT3): 419–435.

Nugent, R.E. & Amick, H. 1992. Environmental Monitor Vibration Consideration in Land Use Ordinances. *Proc. International Society of Optical Engineering (SPIE),* 1619.

Richart, F.E., Woods, R.D. & Hall, J.R. 1970. *Vibrations of soils and Foundations.* New Jersey: Prentice-Hall Inc.

Rosenthal, M.F. & Morlock, G.L. 1987. *Blasting Guidance* Manual. Washington DC: OSMRE.

Selby, A.R. 1991. Ground vibrations caused by pile installation. *Proceedings of the 4th International Conference on Piling and Deep Foundations, Stresa, Italy, 7–12 April,* 497–502.

Stille, H. & Hall, L. 1995. *Vibrationer genererade av Byggnadsverksamheter—natur och krav.* Institutionen för Jordoch Bergmekanik, Kungliga Tekniska Högskolan.

Svinkin, M.R. 1999. A novel approach for estimating natural frequencies of foundation vibrations. *Proc. of the 17th International Model Analysis Conference, SEM, Kissimmee, Florida.* 1633–1639.

Svinkin, M.R. 2004. Some uncertainties in high-strain dynamic pile testing. *Geotechnical Engineering for Transportation Projects, Geotechnical Special Publication, ASCE,* 126(1): 705–714.

Swiss Standards Association. 1992. *SN 604 312a Effects of Vibration on Construction.* Switzerland: SNV.

Tchepak, S. 1986. Design and construction aspects of enlarged base Frankipiles. *Proceedings of the Speciality Geomechanics Symposium, Adelaide, 18–19 August:* 160–165.

Whiffen, A.C. & Leonard, D.R. 1971. *A Survey of Traffic-Induced Vibrations.* Research Report LR418, Road Research Laboratory, Department of Transport, UK.

Whyley, P.J. & Sarsby, R.W. 1992. Ground borne vibrations from piling. *Ground Engineering,* May 1992: 32–37.

Wiss, J.F. 1967. Damage effects of pile driving vibration. *Highway Research Board Record 155:* 14–20.

Wiss, J.F. 1981. Construction vibrations: State of the art. *Journal of the Geotechnical Division, ASCE,* 107(GT2): 167–181.

Woods, R.D. & Jedele, L.P. 1985. Energy attenuation relationship from construction vibrations. In G. Gazetas & E.T. Selig (eds), *Proc. of ASCE Symposium on Vibration Problems in Geotechnical Engineering, Detroit, Michigan,* 229–246.

Proceedings of the first Southern African Geotechnical Conference – Jacobsz (Ed.)
© 2016 Taylor & Francis Group, London, ISBN 978-1-138-02971-2

The influence of foundation stiffness on the load distribution below strip foundations

H.E. Lemmen
EDS Engineering, Pretoria, South Africa

E.P. Kearsley & S.W. Jacobsz
University of Pretoria, Pretoria, South Africa

ABSTRACT: A foundation system consists of a footing which distributes an applied load over an area of soil large enough to form a stable system and the supporting soil. Foundation systems are complex interactive systems due to the variable material properties of the different constituents of the system. Neither the size nor the stiffness of the footing is taken into account during the structural design procedure, which could lead to the overdesign of footings. In this paper the extent to which increasing footing thickness influences the performance in terms of the settlement, deflection and stress distribution beneath a footing is investigated to determine whether current design procedures are valid and, if so, whether they are over conservative. The effect of footing stiffness on the settlement of and stress distribution beneath a strip footing on dense cohesionless sand was established.

1 BACKGROUND

A foundation system consists of not only a footing which distributes an applied load over an area of soil large enough to form a stable system, but also the supporting soil. Foundation systems are therefore interactive systems. However, the behaviour of a foundation system is complex due to the variable material properties of the different constituents of the system. Structural engineers design foundations against both ultimate bearing capacity failure and serviceability limit states (with the latter mostly the dominant mechanism) (Aiban & Zndarčic 1995). Due to the complexity of the failure mechanisms, various simplifications have been used to design foundation systems, all with varying degrees of limitations.

The simplest and most common method used is the Winkler hypothesis where the soil is modelled as a simple linear spring system. This method delivers fairly accurate settlement and deflection results at low soil strain conditions (Conniff & Kiousis 2007). However, the method is limited to one directional movement and does not take the nonlinear behaviour of soil during loading into account. The Winkler hypothesis is a basic simplification of the behaviour of soil under loading (Morfidis & Avramidis 2002).

Neither the size nor the stiffness of the footing is taken into account during the structural design procedure, which could lead to the overdesign of footings. As a foundation system consists of various constituents interacting with one another, an optimal footing thickness for a specific soil stiffness needs to be obtained to ensure the most economical design of the footing. The analysis and design of foundation systems needs to correctly model actual foundation behaviour, while remaining simple and economical enough for practical applications (Morfidis & Avramidis 2002). The basis of current footing design procedures was pioneered by Westergaard (1925) through his study of concrete pavements. He proposed a mathematical method to determine the stresses in a concrete slab by assuming that the soil is an isotropic, homogeneous, elastic solid with perfectly vertical reactions proportional to the size of the slab. He assumed that the soil is an elastic medium where the stiffness thereof was considered as the force which would cause unit deflection when applied over a unit area (Beckett 1995). Although his work has been refined over time, it remains the basis for current design procedures.

A foundation system is interactive and the movement of the soil and the footing directly influence one another (Hallak 2012). One of the most important aspects of this interaction is the contact stress distribution beneath the footing (Conniff & Kiousis 2007). The most common contact stress distributions beneath a footing assumed by engineers are illustrated in Figure 1. The uniform pressure distribution is representative of the Winkler hypothesis used by structural engineers during the design procedure. However, this distribution would only be valid for an infinitely stiff footing placed on a perfectly elastic soil consisting of springs (Mosley & Bungey 1987).

Leshchinsky & Marcozzi (1990) hypothesised (based on experimental evidence) that an increase in the flexibility of a footing would result in a more uniform pressure distribution beneath a footing, illustrated in Figure 2. They concluded that a rigid footing exhibited large local pressures at the edge of the footing, which would lead to local failure of the soil, which in turn would lead to a lower bearing capacity compared to flexible footings.

The current design procedure for reinforced concrete footings is based on an iterative process, where the initial footing size is based on the educated guess of an engineer, checked against relevant code requirements and altered until both ultimate and serviceability limit states are satisfied (Pisnaty & Gellert 1972). It is assumed that sufficient depth of the concrete footing would result in adequate shear, bending and diagonal tension capacity. With the limitations of the design procedure described, the majority of footings are overdesigned, redundant systems (Baumann & Weisgerber 1983). Due to the interactive nature of a foundation system,

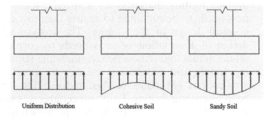

Figure 1.   Stress distribution beneath a footing on various soil types (Mosley & Bungey, 1987).

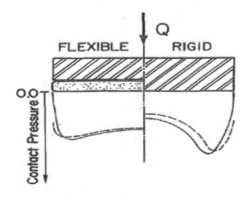

Figure 2.   Contact pressure distribution beneath a rigid and flexible footing (Leshchinsky & Marcozzi 1990).

it is important to consider the entire system as a whole, instead of considering certain elements in isolation during the design process (Arnold et al. 2010). In order to determine whether a foundation system is either flexible or rigid, a framework which combines the stiffness of the footing with the Winkler spring model was developed to obtain an equivalent system stiffness (Canadian Foundation Engineering Manual 2006). Arnold et al. (2010) used Equation 1 to determine whether a foundation system can be classified as either stiff or flexible.

$$K_s = \left(\frac{1}{12} \right) \left(\frac{E_b}{E_s} \right) \left(\frac{d}{1} \right)^{\frac{1}{3}} \tag{1}$$

where K_s = system stiffness, E_b = Young's modulus of the footing, E_s = secant stiffness modulus of soil, d = depth of the foundation and l = length of the foundation. Based on the values obtained for K_s, Arnold et al. (2010) proposed criteria to classify the stiffness of the foundation system as indicated in Table 1.

In this paper the extent to which increasing footing thickness influences the performance in terms of the settlement, deflection and stress distribution beneath a footing is investigated to determine whether current design procedures are valid and, if so, whether they are over conservative. The effect of footing stiffness on the settlement of and stress distribution beneath a strip footing on dense cohesionless sand was established.

To determine the effect of the footing stiffness on the soil-structure interaction, physical modelling of footings on dry, cohesionless sand was conducted in a geotechnical centrifuge. Due to the large size of the prototype footings, full scale testing was not a feasible option and a geotechnical centrifuge was required to produce equivalent stress conditions comparable to full scale testing (Haigh et al. 2010). The experiment was conducted in the centrifuge at the University of Pretoria (Jacobsz et al. 2013).

It is important to realistically model stiffness and strength within physical scale models. Knappett et al. (2010) determined that concrete exhibits considerable scale effects, resulting in an increase in

Table 1.   Classification of the stiffness of a foundation system (Arnold et al. 2010).

K_s Value	System Stiffness
0	Absolutely flexible
0–0.01	Semi-flexible
0.01–0.1	Semi-stiff
0.1–Infinity	Stiff

the strength of the concrete with an increase in the scale (which should be considered in the concrete mix design), while the properties of steel remain constant regardless of the scale. They showed that reinforced concrete can be used in scale models for centrifuge testing. The displacement of the both the footing and the soil beneath the footing was determined with the Particle Image Velocimetry (PIV) technique developed by White et al. (2003).

2 RESEARCH METHODOLOGY

To evaluate the effect of the footing stiffness on the settlement, deflection and contact stress distribution thereof, seven different aluminium footings were used in the experiments. All the tests were conducted on dry cohesionless sand, classified as a poorly graded, slightly silty sand (Archer 2014), with a relative density of approximately 70%. Aluminium footings were used during testing because it has a density similar to reinforced concrete. Large strip footings were modelled. The tests were conducted at a centrifugal acceleration of 30 G (model scale of 1:30). The sizes of the footings were chosen to achieve a balance between the model footing size (scaling of the materials) and the geometrical confinements of the strongbox. To ensure that a wide range of system stiffnesses were achieved, the thickness of the footings was varied from very thin to thick, as indicated in Table 2.

The centrifuge strongbox width was reduced to slightly larger than the model footings and simulated the foundations in plane-strain conditions during loading. The displacement of the footing during loading was measured with the PIV system, as well as inductive displacement transducers. The contact stress distribution at the soil-structure interface was determined with the use of a Tekscan™ system. A gradually increasing load to a maximum of 30 kN was applied to the footings, resulting in a maximum mean contact stress of 1333 kPa which significantly exceeds typical design loads for surface foundations on soils.

Table 2. Geometry of model footings.

Number	Length mm	Width mm	Thickness mm	E.I. N.m²
1	150	150	3	23.9
2	150	150	6.7	266.2
3	150	150	10	885.0
4	150	150	16	3 625.0
5	150	150	25	13 828.1
6	150	150	35	37 944.4
7	150	150	50	110 625.0

3 EXPERIMENTAL RESULTS

As a result of the stress level and sand packing density changing during the course of the experiment, the density as well as the accompanying average secant Young's modulus of the soil were calculated after completion of each experiment. As a result of the symmetrical nature of the experiment, the stresses and strains were calculated along the centre of the footing for different layers throughout the depth the soil mass. Uniformly spaced vertical displacement contours were evident throughout the soil mass, indicating relatively constant strain along the centreline. The average strain was determined by dividing the settlement of the footing by the depth of the soil layer. The mean stress at a particular depth was calculated with the use of Boussinesq equations. The secant Young's modulus was determined with Equation 2.

$$\sigma_y = \frac{E_s}{(1+v)(1-2v)} \left[\varepsilon_x v + \varepsilon_y (1-v) \right] \qquad (2)$$

where σ_y = average vertical stress in soil, E_s = secant stiffness modulus of soil, v = soil's Poisson's ratio, ε_x = horizontal strain in sand and ε_y = vertical strain in sand.

The equation is based on a uniform pressure distribution under the footing. Therefore, the results of the stiff footings were used (discussed further below) to calculate the secant Young's modulus. The results were verified by back calculation using a linear elastic finite element analysis of the experimental model which indicated a good correlation with the behaviour of the footings tested. The measured and calculated test parameters for the footings tested are listed in Table 3.

Due to the low sand density in the experiment on the 10 mm aluminium footing, the results thereof might differ slightly from the other footings, especially the settlement during loading. The settlement of the footing is influenced by factors such as the load applied to the footing, the geometry thereof, as well as the material properties of both the footing and the soil. Utilising Equation 1, the relative stiffness of each experimental model was determined as indicated in Table 3. The normalised settlement at the edge of the column on the footing is presented in Figure 3. This is the location where the maximum moment occurs. The settlement data has been normalised by the width of the footing. It is evident in the figure that the settlement results fall in three distinct groups, correlating well with the three different classifications of foundation stiffness described in Table 3. The settlement of both the stiff and semi-stiff footings appear to increase linearly with an increase in the

Table 3. Test parameters.

Thickness mm	Density kg/m³	Relative density %	Young's modulus MPa	Bearing capacity kPa	K_s	Stiffness classification
3	1578	67	44.4	1686	0.001	Semi-flex
6.7	1587	70	45.4	1901	0.012	Semi-stiff
10	1560	61	39.0	1387	0.045	Semi-stiff
16	1587	70	45.3	1883	0.158	Stiff
25	1587	70	45.9	1892	0.596	Stiff
35	1577	67	44.2	1664	1.700	Stiff
50	1587	71	45.1	1901	4.850	Stiff

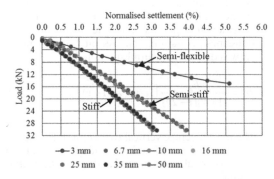

Figure 3. Normalised settlement at column edge.

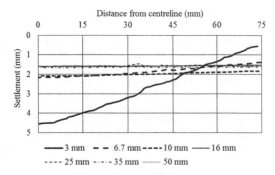

Figure 4. Deflected shapes of aluminium footings.

load applied, with the stiff footings settling at a lower rate than that of the semi-stiff footings. The settlement pattern of the semi-flexible footing differs considerably from the other footings. The rate of settlement increase appears linear at smaller applied loads, where after the settlement increase appears to increase at larger loads. The 3 mm footing could not be loaded beyond 15 kN, as failure of the footing occurred at this load. Upon closer inspection of the stiff footings, it appears that these footings did not exhibit any deflection during loading, apart from the 16 mm footing. Although this footing is also classified as stiff, examining the settlement increase along the width of the footing, different rates of settlement were observed, indicating that deflection of the footing did occur. Due to this deflection of the footing, a refinement of the classification criteria used by Arnold et al. (2010) could be justified, as this footing behaved similar to that of the semi-stiff footings. The deflected shapes of the footings clearly indicate the difference in the behaviour of the three different classifications, as illustrated in Figure 4. The deflected shapes are plotted as the actual settlement which occurred along one half of the footing (due to the symmetrical nature of the problem) at an applied load of 10 kN, as this represents a relatively high, but realistic uniform contact pressure of 444 kPa.

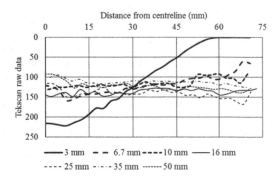

Figure 5. Contact pressure distribution beneath footings.

The figure clearly illustrates the excessive deformation of the semi-flexible footing, compared to that of the other footings.

The contact pressure distribution was determined with the use of the Tekscan™ system. The contact pressure distribution of the aluminium footings at 10 kN applied load is presented in Figure 5. The variable nature of all the distributions is due to the granular nature of the sand beneath the footings, which causes stress concentrations at particular points. The trend of contact pressure distribution is however clearly evident, with the stiff footings indicating a predominantly uniform pressure distribution (slight increase in pressure towards the edge of the footings). A decrease in the system stiffness causes the pressure distribution to become non-linear, mirroring the deflected shape of the footings during loading. Based on the pressure distribution and deflected shape of the footings, the results appear to contradict the work of Leshchinsky & Marcozzi (1990), as an increase in the flexibility of the footing did not increase the bearing capacity thereof, nor did it have a uniform contact stress distribution.

4 INTERPRETATION OF RESULTS

To determine whether the settlement and deflection of the footings could be expressed as a function of the stiffness of the foundation system, the results for both of these were plotted as a function of the foundation system stiffness (3 mm footing was only plotted up to 15 kN). Figure 6 illustrates this relationship for the data at the edge of the column. Data along the width of the footing illustrates that a trend is visible based on the stiffness of the foundation system. From Figure 6 it is clear that the normalized settlement becomes constant when the stiffness of the foundation system is classified as stiff. A sharp increase in the rate of settlement is also visible for the semi-flexible footing, with the amount of settlement between each 5 kN interval increasing with an increase in load. The results indicate that by considering the foundation stiffness at realistic loading values, the settlement of the footing can be predicted based on the relative foundation stiffness and the load applied. The contact stress distribution as a function of the foundation stiffness is represented in Figure 7. The results in the figure do not give an absolute value for the contact stress distribution, but rather the relationship between the pressures beneath the centre of the footing compared to the pressures at the edge of the footing.

It is evident that a reduction in the foundation stiffness leads to a non-uniform pressure distribution, with the pressure at the centre of the footing much larger than that at the edge of the footings. The figure demonstrates that the uniform pressure distribution assumption used by structural engineers is only valid for stiff foundation systems and careful consideration is required when designing more flexible systems.

Based on both Figure 6 and 7, three distinct pressure distribution shapes could be identified beneath the various footings. The shapes are illustrated in Figure 8.

Utilising the contact stress distributions illustrated in the figure, the deflection of the footings based on the stiffness and the pressure beneath the footings were calculated, to determine which contact stress profile is best suited for each footing. For the non-uniform pressure distributions, the pressure at the edge of the footing compared to the centre of the footing was determined with the use of Figure 5.

The deflected shapes were calculated with the use of Equation 3.

$$\frac{d^2z}{dx^2} = \frac{M}{EI} \tag{3}$$

where z = vertical deflection of footing, x = distance along width of footing, M = bending moment at specific point along footing, E = Young's modulus of footing and I = inertia around neutral axis of footing.

The results for an applied load of 10 kN are illustrated in Figure 9. The results of the stiff footings are not shown, as they did not exhibit any noticeable deflection during loading. The pressure distributions used vary linearly for the 16 mm, 10 mm and 6,7 mm footing, with a trapezoidal pressure

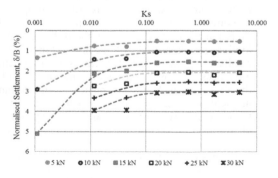

Figure 6. Settlement of footings at the edge of the column.

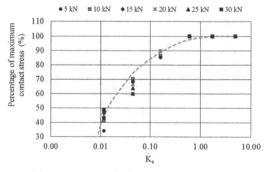

Figure 7. Contact stress distribution of footings.

distribution being utilised for the 3 mm footing. The pressure beneath 3 mm footing was calculated with the use of a triangular distribution (dotted line), as well as a triangular distribution which did not extend to the end of the footing (dashed line), based on the pressure distribution examined at this particular load in the model. This is however impractical as it is difficult to determine to what stage the edge of the footing would not experience any contact pressure anymore. It is clear that the use of the alternative pressure distribution shapes delivers fairly accurate results for the deflected shape of the footing and correlates well with the results illustrated in Figure 9.

Based on all the experimental evidence, it is recommended that the classification criteria used by Arnold et al. (2010) can be refined in the case of dense sands, with the interval for a semi-stiff footing extending to a K_s value of 0,2. It is further recommended that the pressure distribution used during the design process be altered from a uniform pressure distribution to the distributions

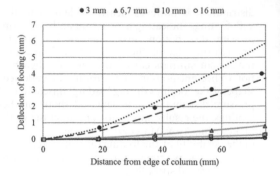

Figure 9. Calculated and measured deflected shapes of footings.

Table 4. Pressure distribution used to determine deflected shape based on foundation system stiffness.

K_s Range	Contact stress distribution shape
0.01	Triangular
0.01–0.2	Linear variable (Trapezoidal when considering symmetry)
> 0.2	Uniform

illustrated in Table 4 based on the system stiffness of the foundation.

5 CONCLUSIONS AND RECOMMENDATIONS

Based on the experimental evidence obtained on the behaviour of strip foundation on dense sand, it is evident that the use of the Winkler hypothesis (uniform pressure distribution) to design footings is only valid for a stiff foundation system. The stiffness of the system has a considerable influence on the settlement and deflection of a footing, as well as the contact stress distribution beneath a footing during loading. The use of the proposed pressure distributions to determine the deflected shapes of the footings based on the stiffness of the footing, correlates well with the physical modelling results. It is evident that the stiffness of the system influences the contact stress distribution beneath the footing, which directly relates to the design moment which engineers use in their designs. It is therefore important to use an appropriate pressure distribution to ensure that the optimum footing design can be achieved.

Finally, the testing of scaled reinforced concrete footings would result in a more accurate indication of the behaviour of actual foundation systems during loading, as it would account for variable

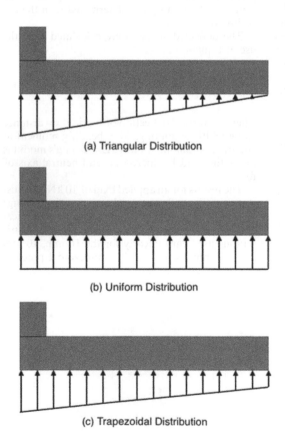

Figure 8. Contact stress distribution shapes beneath footing.

stiffness due to cracking of the footing during loading.

REFERENCES

Aiban, S.A. and Znidarčic, D. 1995. Centrifugal modelling of bearing capacity of shallow foundations on sands. *Journal of Geotechnical Engineering* 121(10): 704–712.

Archer, A. 2014. Using small-strain stiffness to predict the settlement of shallow foundations in sand. Masters Dissertation. University of Pretoria.

Arnold, A., Laue, J., Espinosa, T. and Springman, S.M. 2010. Centrifuge modelling of the behaviour of flexible raft foundations on clay and sand. In Springman, S.M., Laue, J. and Seward, L. (ed.) *Physical Modelling in Geotechnics*: 679–684. London: Taylor and Francis.

Baumann, R.A. and Weisgerber, F.E. 1983. Yield-line analysis of slabs-on-grade. *Structural Engineering Journal* 109(11): 1553–1567.

Beckett, D. 1995. Thickness design of concrete industrial ground floors. *Concrete* 29(4): 21–23.

Canadian Geotechnical Society. 2006. *Canadian Foundation Engineering Manual*. 4th ed. Richmond.

Conniff, D.E. and Kiousis, P.D. 2007. Elastoplastic medium for foundation settlements and monotonic soil-structure interaction under combined loadings. *International Journal for Numerical and Analytical Methods in Geomechanics* 31(1): 789–807.

Haigh, S.K., Houghton, N.E., Lam, S.Y., Li, Z. and Wallbridge, P.J. 2010. Development of a 2D servo-actuator for novel centrifuge modelling. In: Springman, S.M., Laue, J. and Seward, L (ed.) *Physical Modelling in Geotechnics*: 239–244. London: Taylor and Francis.

Hallak, A. 2012. Soil-Structure Interaction and Foundation Vibrations. [Online] Available from: http://www.slideshare.net/ahmadhallak1973/soil-structure-interaction-amec-presentationfinal?qid = eea2ecf3–23b5–442e-84ed-5791caeed555&v = default&b = &from_search = 1. [Accessed: 22th November 2014].

Jacobsz, S.W., Kearsley, E.P. and Kock, J.H.L. 2014. The geotechnical centrifuge facility at the University of Pretoria. In: Springman, S.M., Laue, J. and Seward, L. (ed.) *Physical Modelling in Geotechnics*: 169–174. London: Taylor and Francis.

Knappett, J.A., O'Reilly, K., Gilhooley, P., Reid, C. and Skeffington, K. 2010. Modelling precast concrete piling for use in the geotechnical centrifuge. In: Springman, S.M., Laue, J. and Seward, L (ed.) *Physical Modelling in Geotechnics*: 141–146. London: Taylor and Francis.

Leshchinsky, D. and Marcozzi, G.F. 1990. Bearing capacity of shallow foundations: Rigid versus flexible models. *Journal of Geotechnical Engineering* 116(11): 1750–1756.

Morfidis, K. and Avramidis, I.E. 2002. Formulation of a generalised beam element on a two-parameter elastic foundation with semi-rigid connections and rigid offsets. *Computers and Structures* 80(1): 1919–1934.

Mosley, W.H. and Bungey, J.H. 1987. *Reinforced Concrete Design*. 3rd ed. 270–291. Houndmills: Macmillan Education LTD.

Pisanty, A. and Gellert, M. 1972. Automatic design of sloped spread footings. *Building Science* 7(1): 53–59.

Westergaard, H.M. 1937. What is known of stresses. *Engineering news record* 118(1): 26–29.

White, D.J., Take, W.A. and Bolton, D. 2003. Soil deformation measurement using Particle Image Velocimetry (PIV) and photogrammetry. *Géotechnique* 53(7): 619–631.

Proceedings of the first Southern African Geotechnical Conference – Jacobsz (Ed.)
© *2016 Taylor & Francis Group, London, ISBN 978-1-138-02971-2*

Bored piles—cased- and uncased piles, or CFA-piles, their features and limitations

W. Schwarz
BAUER-Maschinen, Schrobenhausen, Germany

ABSTRACT: The different features of executing bored piles are shown. In doing so, the various options of available equipment are considered, the present soil conditions, the function of the piles and lasty the effort of execution. This is based on the European Standards and the German "Recommendations on Piling" (EA-Pfähle).

1 INTRODUCTION

Pile systems are commonly distinguished between Bored Piles, Displacement Piles and Micropiles. Bored piles may be executed with or without casing, unsupported or supported with fluids or soil.

In Figure 1, a systematic of bored pile options according to European usage is presented.

The application of the methods depends on the site and soil conditions as well as on the function of the pile construction. In the following the several characteristics of the bored piles systems are presented and discussed.

2 PRINCIPLES AND APPLICATION RANGE OF BORED PILES

Bored piles are characterised by ground being loosened and transported during installation (EN 1536:2015). This is in contrast to displacement piles which are installed in the ground without excavation or removal of material from the ground (EN 12699:2015). Bored piles are applicable in all soils and rock, under dry conditions and under groundwater table, down to all depth which can be reached with the drilling tools. Those piles may be used for foundations (foundation piles) as well as for retaining constructions (pile wall and secant pile wall).

For excavation of the soil material rope- or hydraulic pile-grabs operated with cable excavators or drilling buckets (or augers) operated with rotary drilling rigs are commonly used. Using the rotary method, the tool is rotated, lowered and pulled up with a telescopic Kelly bar. Casings are installed either with separately running casing oscillators or directly driven by the rotary drilling rig, see Figure 2.

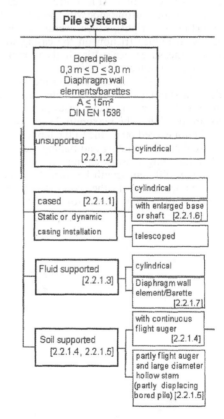

Figure 1. Bored piles systems, excerpt from German Geotechnical Society (2014).

In the following special attention is given to the manifold shapes and potentialities of the rotary drilling tools related to the different bored piles systems. The boring process with grabs is not further addressed in this paper.

3 SYSTEMS OF BORED PILES

3.1 Unsupported excavations

In sufficiently stable ground bored piles can be constructed without a casing or other bore support. In such cases, a protective casing is only necessary for stabilization of the top of the bore and for guidance of the tools.

If unstable strata are penetrated, this length of the bore should be supported.

The unsupported excavation method should only be employed for vertical and raked piles up to an inclination of 15:1.

3.2 Cased bored piles

In unstable ground casing shall be installed in advance of excavation. At its bottom, the casing is equipped with a cutting shoe with special, hardened teeth (Figure 3, right), which overbreaks the bore slightly to reduce friction on the casing wall, an essential measure when cutting in the primary piles of secant pile walls.

EN 1536 applies to vertical and inclined bored piles. The standard range of diameters lies between 0,6 m and 2,5 m, in special cases greater diameters are possible. The maximum length of casing should not exceed 60 m due to the limited force of the oscillator for rotating and pulling the casing.

For handling reasons casings are used in sections with different length between 1 m and 6 m each, composed up to the required total lengths by means of casing joints with conical bolts, see Figure 4.

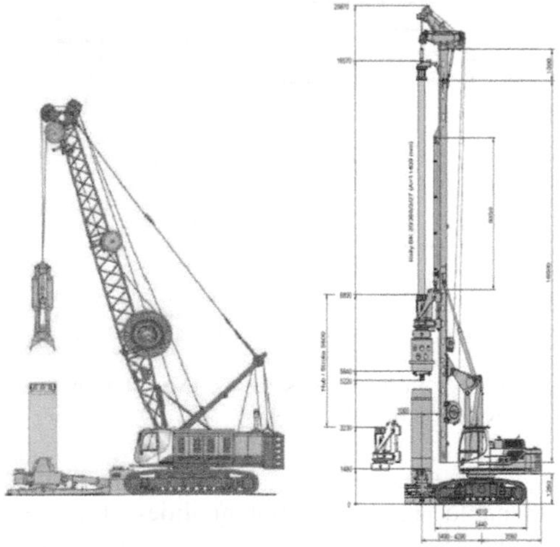

For excavation with rotary drilling rigs, a wide variety of tools is available to make allowance for the different soil, rock and groundwater conditions.

Augers are used for dry conditions in all types of soil and rock. Various teeth configurations and teeth arrangements give the opportunity to find the best fitting tool under the existing conditions. The augers may have one or two cuts (for bigger diameters) and flat teeth or round shank chisels or a combination of both. Figure 5 shows such a double cut auger with a combination of flat teeth and round shank chisels.

In hard rock tapered rock augers are used. Equipped with round shank chisels the tool gets caught in the weaker zones and fissures of the rock to crack the material and break it in small pieces.

Under water and in soft soils drilling buckets are the best choice for excellent performance. Analogue to the augers those tools are equipped with teeth and round shank chisels. The soil is scratched into the bucket by rotating the tool

Figure 3.   Drilling rig with casing and auger tool (left), cutting shoe with special, hardened teeth (right).

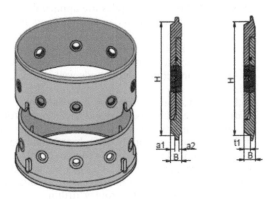

Figure 4.   Casing joint with conical bolts.

Figure 2.   Cable excavator with pile grab and oscillator (left). Rotary drilling rig with oscillator (right).

clockwise. After filling, the bottom of the bucket is closed by turning the tool counter-clockwise. The bottom tightens the bucket sufficiently, so weak soil and mud is enclosed in the bucket as well as sand and gravel under water. After pulling out the tool from the bore the bottom of the bucket can be opened by a special mechanism integrated in the tool to discharge the bucket at surface, see Figure 6.

Before concreting, the reinforcement cage equipped with a sufficient number of spacers is installed. The piling procedure is finalized by concreting under dry conditions using a free fall manner, while under the water table using the tremie method with tremie pipes.

For strong fissured rock (up to 100 MPa), drilling in boulders or in concrete, core barrels with tungsten carbide armoured blocks or with round shank chisels are available, see Figure 7. If the rock strength exceeds 100 MPa a very special tool, the roller bit core barrel is used, see Figure 8. All types of core barrels have one thing in common, that in a first step an annular groove is cut and in a second step the remaining core has to be collected in one piece with another special tool or destroyed with a rock auger or cross cutter, see Figure 8.

The whole working sequence of fully cased piles with rotary drill is shown in Figure 9.

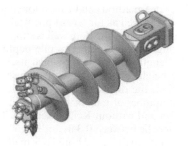

Figure 5. Double cut auger with combination of flat teeth and round shank chisels.

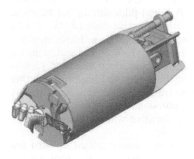

Figure 6. Double cut drilling bucket with flat teeth.

Figure 7. Core barrel with tungsten carbide armoured blocks (left) and round shank chisels (right).

Figure 8. Roller bit core barrel (left), cross cutter (right).

3.3 Fluid supported bored piles

For bigger diameter piles (2,0 m up to >3,0 m) and/or for greater depth fluid supported excavation is the method of choice. According to EN 1536 those piles can have either circular cross-sections or cross-sections composed of rectangles. The latter are also known as barrettes. Bentonite suspensions, polymer-solutions or a combination of both are generally employed as support fluids to stabilise the bore whole walls.

When work commences, a lead in tube (circular) is placed or a guide wall (barrette) installed. Then the soil is excavated under the support fluid. A sufficient high fluid overpressure against the outer ground water level has to be held. The stability and the characteristics of the supporting fluid shall be analysed compliant with EN 1538.

For circular shaped excavation drilling buckets as shown in Figure 6 are used. Diaphragm wall grabs or hydro fraises are generally used for non-circular cross-sections (Figure 10).

For fluid supported excavations, in particular for the hydro fraise method, a bentonite mixing and supporting fluid cleaning plant (desander) with pumps and pipes between plant and cutter has to be installed (Figure 11) A reinforcement cage is generally inserted after reaching the final depth. The concrete is subsequently placed by tremie method after cleaning the support fluid

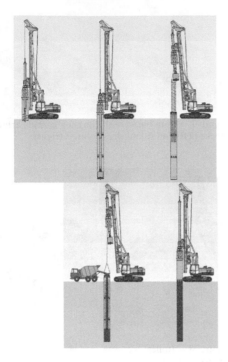

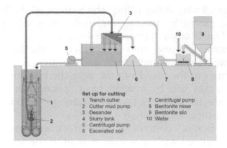

Figure 11. Set up for hydro fraise method.

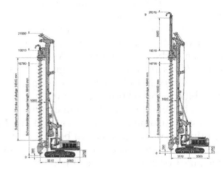

Figure 9. Working sequence fully cased pile with rotary drill.

Figure 12. CFA-drilling system without (left) and with kelly extension.

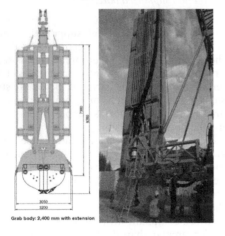

Figure 10. Diaphragm wall grab (left), hydro fraise (right).

from debris with desanders consisting of sieves and hydro-cyclones.

3.4 Soil supported bored piles

Soil supported, Continuous Flight Auger (CFA) bored piles are constructed by rotating a continuous flight auger into the ground and pumping in concrete through the hollow stem as the auger is retracted the piles can be installed with or without reinforcement.

The CFA method can be used for single foundation piles as well as for secant pile walls.

During boring, the bore wall is supported by the soil lying on the flight. The pile depth is limited to the length of the auger which itself is limited by the height of the rig. With the use of a Kelly extension, the drilling depth can be lengthened a certain number of meters. Figure 12 shows drilling rigs with and without Kelly extension. Typical pile diameters lie between 0,3 m and 1,2 m.

Contrary to the cased bored pile method (see 3.2) which can be applied in all types of soil and rock, EN 1536 gives numerous restrictions according to the soil properties in which the CFA method is applicable. These include the following:

CFA bored piles should not be installed in uniform, non-cohesive soils ($d_{60}/d_{10} < 1,5$) below the groundwater table, or in loose, non-cohesive soils with densities $D < 0,3$ or in soft, cohesive soils with undrained shear strengths $c_u < 15$ kN/m^2, unless the feasibility of the installation method has been demonstrated before commencing works. Uniform, non-cohesive soil where $1,5 < d_{60}/d_{10} < 3,0$ can be sensitive when situated below the ground water table.

To provide uncontrolled upward soil transport the advance speed of the tool must correspond with the rotation speed of the auger, if not, the ground support may be lost. It is obvious that this

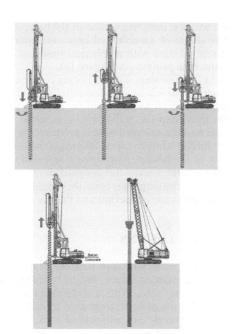

Figure 13. Working sequence CFA method.

situation will occur, when a rock stratum underlying a loose soil stratum has to be penetrated.

Retracting the auger should be executed without rotation. Due to the limited extraction force of the winches difficulties may occur in retracting the auger. In this case a slight rotation in the same direction as applied downwards is allowed, to reduce the friction and prevent soil transport into the fresh concrete.

Once the auger has been retracted, the pile top is cleaned from remaining soil and the reinforcement is pushed into the fresh concrete, in some cases by using a small vibrator. Special attention should be taken to the stability of the reinforcement cage to allow the handling of the cage in one piece and to insert the cage in the fresh concrete without demolition of the cage. Also the required spacers and centralizers should be rigid and properly fixed at the cage to survive the installation process.

To provide bleeding, dewatering and segregation, which would result in problems pushing in the reinforcement cage, special care with respect to the stability and workability of the fresh concrete should be taken.

The working sequence of the CFA method is shown in Figure 13.

3.5 Double head system

A combination of the CFA piles method and the cased bored piles method is the so called VDW technique with the double head system. It combines the advantages of both without having the restriction of the CFA method concerning the soil conditions. The main area of application is the construction of all types of pile walls, primarily in restricted site conditions and adjacent to existing buildings.

During the drilling process, a longer outer casing tube and an inner continuous flight auger are rotated concurrently in opposite directions by two independent rotary drives. The continuous flight auger, which is rotated clockwise by the upper rotary drive, and the drill casing, which is rotated anti-clockwise by the lower rotary drive, are drilled down to the required depth in a single operation.

The drill spoil is conveyed upwards by the auger flights and either ejected through apertures below the rotary drive or ejected into a spoil chute system.

A crowd unit attached to the rotary heads facilitates a vertical movement to the continuous flight auger of around 300 mm relative to the casing tube. This relative movement enables the drilling process to be adjusted to cope with varying ground conditions. To reduce the risk of base failure or to cut into primary piles of a pile wall, the drill casing is generally advanced ahead of the auger, whereas in dense soil formations drilling can also be carried out with the continuous flight auger being advanced ahead of the casing tube.

After having attained the terminal depth, concrete is pumped through the hollow stem of the

Figure 14. VDW rig adjacent to a building.

auger as the drill string (auger and casing) is simultaneously withdrawn. The reinforcing cage can subsequently be inserted into the fresh concrete.

3.6 Bored piles with enlarged base

When constructing base resistance piles it may be very useful to enlarge the base of the pile. An enlarged base allows improved utilisation of concrete strengths, and thus allows material savings in the pile shaft, because the adopted base resistances are generally considerably lower than the concrete strengths.

Enlarged bases can only be installed in stable soils or soils, that can be suitably stabilised for the purpose, e.g. using bentonite suspension or cement slurry. Construction of enlarged bases is generally only possible for vertical piles and in cased pile bores. It requires special belling-tools; these include special, extended belling buckets attached to the kelly bar and fitted with retractable reaming wings (Figure 16).

The reaming wings can generally be extended by the advance of the Kelly bar in conjunction with a scissor mechanism inside the belling bucket. The extension of the wings correlates to the controlled advance of the Kelly bar. The bucket section below

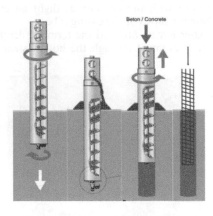

Figure 15. Working sequence double head system (VDW).

Figure 16. Belling bucket (left), exposed belly (right).

the wings is designed to accept the loosened spoil. The shape of an enlarged base corresponds to a truncated cone with a spherical base and a lower, cylindrical protrusion (Figure 16).

In the first step, a cased boring is carried out as described in 3.2. In a second step the casing is then retracted as far as the top of the reaming wings. In the third step the belling tool is inserted into the bore and the enlarged base is produced by rotating the bucket while the spoil falls into the bottom section of the bucket. Step three is repeated until the enlargement reaches the designed size.

After cleaning the bore thoroughly the reinforcement cage is inserted and the pile concreted.

4 CONCLUSIONS

For all types of soils and rock under nearly every imaginable circumstances several bored piles methods are available. With a wide range of drilling tools all sorts of soils, from mud to stiff soil, from sand to gravel- and all sorts of rock, from weak to hard, from weathered to solid- and under all groundwater conditions, from dry to under water table, is available. In the same sense a wide range of drilling rigs, from small to big, is available for all site conditions. A selection of these is shown in the paper. The features of the bored piles methods are named.

Both, the European Standard EN 1536 and the German Recommendations on Piling (EA-Pfähle) describe the bored piles methods and give principals and recommendation for the execution, which are mentioned in this paper.

ACKNOWLEDGEMENT

The pictures in this paper are taken from BAUER Maschinen brochures. Since the author is also co-author of the Recommendations on Piling many passages are taken from the Recommendations without marking them in particular.

REFERENCES

EN 1536:2015, *Execution of special geotechnical work—Bored piles.* Brussels: Comité Européen de Normalisation (CEN).

EN 1538:2015, *Execution of special geotechnical work—Diaphragm walls,* Brussels: Comité Européen de Normalisation (CEN).

EN 12699:2015, *Execution of special geotechnical work—Displacement piles.* Brussels: Comité Européen de Normalisation (CEN).

German Geotechnical Society 2014, *Recommendations on Piling (EA-Pfähle).* Berlin: Ernst & Sohn.

Proceedings of the first Southern African Geotechnical Conference – Jacobsz (Ed.)
© 2016 Taylor & Francis Group, London, ISBN 978-1-138-02971-2

Foundation engineering on highly variable ground conditions

J. Potgieter
University of Pretoria, Pretoria, South Africa

J. Breyl
Jones & Wagener, Johannesburg, South Africa

ABSTRACT: A multi-storey office building with four to six basement levels is under construction in Sandton, Johannesburg. As a result of the limited geotechnical investigation carried out, unexpected ground conditions were encountered at founding level. Whereas it was anticipated that most of the column foundations would be founded on rock, rock was only encountered in places and most of the base of the excavation was underlain a deeply weathered granite and diabase profile. In places, residual soils and highly weathered rock, incapable of supporting the design bearing pressures, extended several metres below the intended founding level. As a result the spread footings had to be modified by deepening or widening using mass concrete or by combining adjacent footings in order to achieve acceptable levels of differential settlement. This paper describes the range of *in situ* tests employed including CSW tests, plate load tests, DPSH testing, rotary core drilling with SPT tests and visual inspections. A comparison of various field testing techniques in evaluating soil stiffness is presented.

1 INTRODUCTION

Sandton Central, the business hub of Johannesburg, is host to many high-rise buildings and associated deep basements. On the north-eastern quadrant of the Rivonia Road and West Street intersection, a new multi-billion-rand development comprising six basement floors and at least 12 stories above ground level is being constructed.

An initial auger trialhole geotechnical investigation was carried out by geotechnical consultants in 2007. This was supplemented by a rotary core drilling investigation in 2014 done by the specialist lateral support contractor. Both these investigations were done from surface. The auger trial holes were drilled to refusal in accessible areas between existing structures and the rotary core boreholes were terminated at depths near the bulk excavation level of the proposed excavation.

Based on the assumption that the rockhead would be above or close to the base of the excavation, the structural engineers designed the foundations for the multi-storey structure as spread footings with an allowable bearing pressure of 2 MPa.

During the bulk excavation process, hard rock granite with adequate bearing capacity and requiring blasting for excavation was encountered, mainly in the north-western corner of the site. However, along the southern boundary and central portions

of the site, the material at bulk excavation level was a soil or decomposed rock that could be excavated with ease, indicating that the expected allowable bearing pressures of 2 MPa would not be attained. Soon after the foundation excavations at the column positions were commenced, the incompetent and highly variable nature of the material in these areas of the site became apparent. A reassessment of the founding conditions and modifications to the foundation geometry was required at a time when the construction of foundations was on the critical path and the reinforcement cages for many foundations had already been detailed and delivered to site.

This paper describes the additional geotechnical investigation and *in situ* tests carried out to identify the soil properties and the process followed to ensure the foundation design of each footing is satisfactory. This process is summarised in Figure 1.

2 INVESTIGATION

Closer inspection of the already excavated footings revealed that the southern and central portions of the site were underlain by residual granite and a deeply weathered diabase intrusion. In both the granite and diabase a high degree of variability was found. In some areas, the residual granite was a homogenous, friable, very dense sand whilst in

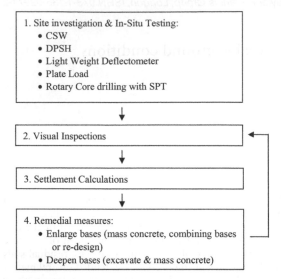

1. Site investigation & In-Situ Testing:
   - CSW
   - DPSH
   - Light Weight Deflectometer
   - Plate Load
   - Rotary Core drilling with SPT

↓

2. Visual Inspections

↓

3. Settlement Calculations

↓

4. Remedial measures:
   - Enlarge bases (mass concrete, combining bases or re-design)
   - Deepen bases (excavate & mass concrete)

Figure 1. Engineering procedure for investigating spread footings.

other areas highly weathered pegmatitic granite with rich quartz veins was present.

The diabase intrusion was a firm residual silt in some areas whilst very closely jointed, very soft rock excavating as a gravel was found in other instances.

The purpose of the additional geotechnical investigations were to identify the extent of the diabase intrusion and to obtain more information on the improvement in soil consistency with depth for settlement calculations. Several tests were employed in order to determine the soil characteristics applicable for bearing pressure and settlement calculations.

Each of the *in situ* tests and investigation methods are briefly described.

2.1 CSW tests

Continuous Surface Wave (CSW) tests were carried out to obtain the small-strain shear stiffness, G_o, of the material with depth.

A G_o value of 250 MPa represents the transition from soil to soft rock. G_o values in excess of 1000 MPa are indicative of hard rock. Seven CSW tests were carried out as shown in Figure 2, distributed on three different levels of the bulk excavation.

CSW tests in close proximity to each other showed varying results for G_o, confirming the highly variable rock levels and suggesting that differential settlements could approach total settlement.

2.2 DPSH tests

On completion of the CSW tests, DPSH tests were recommended in order to obtain a wider understanding of the material at depth. Twenty DPSH tests were carried out in the southern area of site. The test determined the number of blows required to drive a 50 mm diameter cone through successive 300 mm intervals using a standard SPT hammer.

DPSH tests refused at 0.3 to 5.7 m below the bulk excavation levels. At the time, the extent of the diabase dyke was not yet known. In hindsight, the DPSH refused on diabase between 3.6 m and 5.7 m. Generally, shallow refusal was seen on granite materials. The DPSH tests were used as a crude approximation of material stiffness with depth outside of the regions of CSW testing.

2.3 Light weight deflectometer

Two bases had already been enlarged and deepened in order to reduce bearing pressures and settlements. However, even at the reduced foundation level, rock was not encountered. Light Weight Deflectometer (LWD) tests were conducted on residual diabase and granite at two critical positions to determine the stiffness of the underlying soils. The Zorn ZFG 2000 LWD was used to conduct the tests. The ease of operation makes LWD tests a convenient and quick method to estimate the dynamic stiffness of the subgrade by measuring the deflection a plate undergoes with a standard drop energy. Tompai (2008) published several correlations between the dynamic subgrade modulus, E_{vd}, and elastic modulus, E_{v2}. The correlation used in this study was:

$$E_{v2} = E_{vd}/0.83 \qquad (1)$$

The LWD showed the stiff, residual diabase and the dense granite having a modulus, E, between 40 and 80 MPa.

2.4 Plate load test

A single plate load test was conducted on homogeneous residual granite, similar to the material on which one of the LWD tests was conducted. The test was conducted against the side walls close to the base of a test-pit excavated at the position of a shear wall.

Due to the influence zone of the 0.2 m plate being no more than 0.4 m, the test was used to determine the stiffness of the residual granite on the surface only. The plate load test resulted in a stiffness, E, of between 40 and 100 MPa which was calculated by elastic theory as shown in Eq 2.

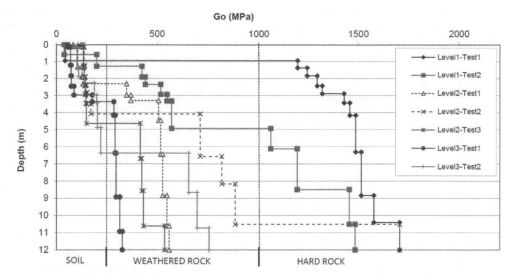

Figure 2. CSW test results showing small strain shear modulus against depth.

$$E = \pi \frac{D}{4} \frac{\sigma}{\rho} \left(1 - v^2\right) \qquad (2)$$

where, D = plate diameter; v = Poisson's ratio; σ = measured stress; and ρ = measured displacement.

2.5 Rotary core drilling with SPT tests

Finally, rotary core drilling was carried at 18 positions until hard rock was reached at an average depth of 10 m below the bulk excavation level. The borehole logs provided a better understanding of the soil profile with depth, in particular the depth to bedrock and the areas where granitic zones were underlain by residual silts from the diabase intrusion at a shallow depth below founding level as shown in Figure 3.

The SPT values in residual material above rock provided guidance as to the soil consistency to estimate the stiffness of the materials for settlement calculations.

Stroud (1989) recognised that soil stiffness is strain dependent and soil stiffness was estimated using the SPT-N value for an over-consolidated sand as shown in Figure 4.

The rotary core drilling revealed a strengthening profile down to bedrock at most positions. A localised water table was seen approximately 4 m deep on the western portion of the site.

2.6 Visual inspections

Almost all bases were inspected by the authors to evaluate the soil conditions prior to approval being

Figure 3. Photograph showing diabase underlying granite.

75

given to cast the blinding. The stiffness was estimated from the visual inspection (in combination with other in-situ tests carried out) and the settlement was calculated from the estimated stiffness.

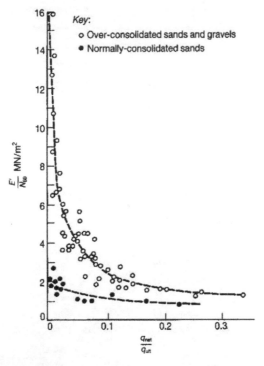

Figure 4. Relationship between stiffness, penetration resistance and degree of loading for sand (Stroud, 1989).

Good record keeping was essential in the process. Layout drawings of the foundations inspected was continuously marked up to develop an understanding of the position of the diabase dyke and sill. This assisted greatly in identifying bases where the apparently competent soft rock granite was underlain by residual diabase, necessitating deepening and/or widening of the footing excavations. The site geology at the time of writing is shown in Figure 5.

Many of the bases were inspected on more than one occasion due to changes in the depth or size of the base that had to be undertaken. Records were kept of each inspection noting the dimensions, founding material, additional measures required and ultimately approval for the blinding to be cast.

Settlement was the main criteria to approving bases. Bases that were predicted to settle in excess in 8 mm, were either resized to reduce the bearing pressure or deepened in order to avoid compressible layers at the surface.

After remedial measures had been carried out, bases were re-inspected and approved or further remedial work carried out.

Table 1 shows estimated stiffness values for granite and diabase derived from tests discussed above and experience from visual inspections.

3 SETTLEMENT

The ultimate soil bearing capacity was evaluated by Terzaghi's (1943) solution with appropriate

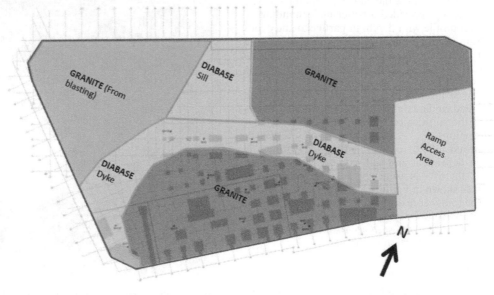

Figure 5. Site geology after in situ testing and visual inspections.

Table 1. Stiffness values, E, for granite and diabase.

MCCSSO consistency descriptor	Granite	Diabase
	MPa	MPa
Very dense/stiff	40–80	40–80
VD – VSR		100
VSR	200	150
VSR – SR	300	200
SR	400	300
MHR	600	500
HR/VHR	2000	1000

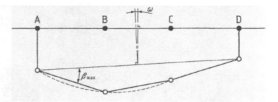

Figure 6. Definition of angular distortion, β_{max}, from EN1997-1:2004.

bearing capacity factors. Due to the size and depth of the foundations, bearing capacity in this study is not the critical factor but rather settlement. The settlement criteria is governed the allowable angular distortion.

3.1 Settlement tolerances

Although SANS 10160–1 provides guidance on the limiting differential settlement of structures, Annexure H of Eurocode 7 was used to provide guidance on the limiting deformation a structure is allowed to undergo.

Limiting differential settlement aims to limit angular distortion (or rotation). Structures can typically undergo angular distortions of between 1/2000 and 1/300 before serviceability issues such as cracking. An angular distortion of 1/150 is likely to cause structural instability. For the proposed development, an angular distortion of 1/500 was regarded as acceptable for the structure (Bjerrum, 1963).

Differential settlement is usually estimated as half of the total settlement. By referring to Figure 6, if point B settled by 8 mm and point A settled 4 mm, this would result in an angular distortion of 1/2000 for a grid spacing of 8 m applicable to the structure. If point B were to settle 8 mm and both points A and C were to settle 4 mm, this would result in angular distortion of 1/1000 about point B. Due to the highly variable ground conditions and the possibility of residual material adjacent to hard rock material as shown in Figure 7, the worst case scenario has to be applied if point B had to settle 8 mm and both points A and C did not settle. This would result in an angular distortion of 1/500. For the purposes of the investigation, 8 mm was considered as an acceptable settlement criteria.

3.2 Methods of calculation

3.2.1 CSW tests
Due to the urgency of the task, initially settlements were calculated based on CSW results. Settlements were predicted using the *Versak 2.0* (as cited in Archer & Heymann, 2015), specialised software developed for the settlement analysis of CSW test results. The software uses a stiffness degradation curve to estimate the settlement of bases from the *in situ* small-strain shear stiffness. A maximum bearing pressure for limiting settlement to 8 mm at founding depth of 3 m was calculated. This was used as initial criteria before more rigorous tests such as rotary core drilling was carried out.

3.2.2 Elastic settlement
The CSW tests gave a good indication of the material stiffness and hence settlement prediction. However, the seven CSW tests were confined to limited area and even within the area of testing it is noted that the soil conditions are highly variable. After carrying out a variety of tests across the site in attempt to evaluate the soil stiffness, visual inspections remained the only reliable method to evaluate the founding conditions (only at surface) of individual bases.

Settlements were calculated using Boussinesq (1885) generalised elastic theory which was later modified by Bowles (1988) with contributions to the influence factor. A depth correction factor (Fox, 1948) of 0.8 and a rigidity factor of 1.0 were applied to the settlement calculations. The applicability of this method largely depends on the accuracy of determining the soil stiffness, E. The emphasis should remain in the determination of soil stiffness which is more onerous that the analysis method used for calculating stiffness.

4 SOLUTIONS

Bases that were excavated on inadequate founding material had to either be enlarged or deepened. Enlarging the bases reduces the bearing pressure and hence the settlement. Deepening the base has a greater effect on reducing settlement than enlarging the base and was regarded as the primary alternative. Deepening bases has the effect of reducing the amount of poor material between the underside of the footing and the rockhead. However,

Figure 7.   Photograph showing highly variable conditions. Hard rock granite on the left, and very soft rock diabase on the right.

bases could only be excavated to the maximum reach of an excavator with additional safety concerns of the workers having to clean the floor of the base. If necessary, due to safety or accessibility, bases were enlarged by either filling mass concrete or redesigning the base. The minimum depth of mass concrete between the floor of the excavation and required underside of the base was governed by a 45° load spread. Several bases in close proximity were redesigned as combined bases due to enlargement.

Due to the highly variable soil conditions and residual soils many of the bases required enlargement or deepening. Remedial measures to bases required careful liaison between the Engineer, geotechnical engineers and Contractors to provide a cost effective solution that is pragmatic and satisfies the settlement criteria.

5   CONCLUSIONS

The limited geotechnical information available at the time of construction necessitated additional investigations with the purpose of determining soil founding conditions. Due to on-going construction, the additional investigation could not be comprehensively conducted due to restricted access and limited time.

Several methods were employed in order to evaluate soil stiffness and hence to estimate the soil stiffness and settlement.

Delays to construction were inevitably experienced due to constant back and forth interactions between the Structural Engineers, the Geotechnical Engineers and the Contractor having to conduct remedial measures such as deepening and widening bases.

The necessity of conducting thorough geotechnical investigations prior to construction cannot be overstressed.

ACKNOWLEDGEMENTS

The authors would like to thank Messrs A. Elmer and W. Strydom of WSP Parsons Brinckerhoff for participation in the project.

REFERENCES

Archer, A. & Heymann, G. 2015. Using small-strain stiffness to predict the load-settlement behaviour of shallow foundations on sand. *Journal of the South African Institution of Civil Engineering.* (57)2: 28–35.
Bjerrum, L. 1963. Discussion in *Proc. European Conference SMFE*, Wiesbaden, 3: 135–137.
Bowles, J.E. 1987. Elastic foundation settlement on sand deposits. *Journal of Geotechnical Engineering, ASCE,* 113(8): 846–860.
British Standards Institution, 2004. *Eurocode 7: Part 1, General rules (EN 1997-1:2004).* London: BSI.
Fox, E.N. 1948. The mean elastic settlement of a uniformly loaded area at a depth below the ground surface. *Proceedings, 2nd International Conference on Soil Mechanics and Foundation Engineering*, Rotterdam, 1: 1, 29–32.
Jennings, J.E., Brink, A.B.A. & Williams, A.A.B. (1973). Revised Guide to Soil Profiling for Civil Engineering Purposes in Southern Africa. *Trans. South Africa Inst. Civil Engineering*: 3–15.
SANS 10160–1:2011. *Basis of structural design and actions for buildings and industrial structures. Part 1— Basis of structural design.* S.A. Bureau of Standards, Pretoria.
Steinbrenner, W. 1934. *Tafeln zur setzungsberschnung.* Die Strasse, 1: 121–124.
Stroud, M.A. 1989. The Standard Penetration Test—Its Application and Interpretation. *Proc. I.C.E. Conf. on Penetration Testing in the U.K.:* 29–46.
Terzaghi, K. 1943. *Theoretical Soil Mechanics.* New York: John Wiley & Sons.
Tompai, Z. 2008. Conversion between static and dynamic moduli and introduction of dynamic target values. *Periodica Polytechnica* 52(2): 97–102.

Mining and tailings

Proceedings of the first Southern African Geotechnical Conference – Jacobsz (Ed.)
© 2016 Taylor & Francis Group, London, ISBN 978-1-138-02971-2

Rehabilitation design options analysis and geotechnical investigation of an abandoned coal mine

E. du Toit
SRK Consulting (Pty) Limited, Johannesburg, South Africa

ABSTRACT: The abandoned coal mine is situated in a rural part of Kwa-Zulu Natal in a residential area. The primary objective of the design was to rehabilitate the site to enable the construction of low cost single storey housing as intended for the post closure land use. The main consideration to the long term success of the rehabilitation is to prevent future mining of the coal seam. To determine the best overall rehabilitation option, a detailed options analysis was used to assess the continuation of acid mine drainage, spontaneous combustion and subsidence. This was then coupled to the cost of implementing the option and future costs that may be incurred if the option is found unsuccessful. A detailed geotechnical site characterisation was undertaken, which comprised the excavation of test pits, advancement of rotary core boreholes, installation of groundwater piezometers along with in-situ and ex-situ (laboratory) geotechnical testing. The final geotechnical design ultimately included the removal of the coal seam, extensive earthworks to re-contour the entire site, high-wall stabilisation, ground and surface water management.

1 INTRODUCTION

In South Africa there is an extensive amount of mines that were abandoned once the orebody was deemed depleted or the profitability of the mine ceased. The result is that these mines pose a hazard to all forms of life due to the remaining features and potential leachate of the exposed orebody. One such example is the abandoned coal mine situated in a rural part of Kwa-Zulu Natal in a residential area. The past (assumed formalised) and current (artisanal) mining processes resulted in a central high-lying area (saddle) with two adjacent pits with the coal seam exposed to the elements in the pit floors and high-walls. This exposure results in acid mine drainage, spontaneous combustion and subsequent subsidence. The artisanal miners are extracting the materials from the two pit floors and the face of the high lying area. The consequences of the informal mining methods employed by these miners comprise undermining of the high-walls, which results in slope failures, encroachment of the high-wall towards the school (located on saddle), structural damage of the school and alleged fatalities; limited access into and around the pit area; undulating pit floors and an accumulation of Acid Mine drainage (AMD) ponds on the uneven pit floors. Figure 1 shows a general cross-section of the terrain.

2 REHABILITATION ELEMENTS

2.1 Coal seam

The geotechnical investigation determined that the remaining coal seam ranges in thickness up to 6 m and has been mined out of the adjoining pits to the extent where it is now approximately only 0.1 m thick. Efforts to access the remnants of the coal seam within the saddle has resulted in exposure of the coal seam to the elements. This was caused by undermining the high-walls, digging up the pit floors and creating cavities in the high-walls which have led to most of the challenges associated i.e. AMD, spontaneous combustion, collapse of the high-walls, structural damage to the school, subsidence, uneven topography, access and water pollution.

2.2 Collapsing high-walls and structural distress at school infrastructure

Deep tension cracks were observed along the northern crest of the saddle's high-walls. A representative slope stability analysis indicated imminent failure—which indeed occurred in 2014. After this failure, the high-wall remained in an unstable state with additional tension cracks and slope angles in sections over-steepened to greater than 90° (measured from horizontal).

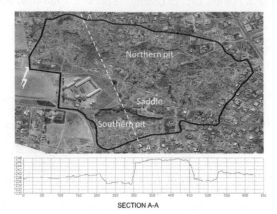

SECTION A-A

Figure 1. General site layout and cross section.

2.3 Subsidence

The areas of subsidence observed on the saddle surface were initially attributed to a bord collapse due to spontaneous combustion. However, underground mining may equally likely be the cause. A mining method previously generally deployed in KwaZulu-Natal comprised the excavation of hill side slopes to expose the coal seam. Bord and pillar mining often then proceeded in order to extract the coal seam remaining in the resulting saddle. Coal pillar collapse/stripping (due to either poor design or pillar stripping) can lead to areas of regular subsidence—as observed to an extent at the mine.

2.4 Exposed cavities

The numerous exposed cavities (possible adits) observed on site were mostly in a state of collapse and filled with rock debris. Three cavities were identified in northern pit, the largest of which was located in the central southern section of northern high-wall and resembled a declined shaft that may have collapsed.

2.5 Acid mine drainage

The evidence observed on site was in the form of red coloured water (ferric iron) in the north western corner of the southern pit and a white precipitate in large areas of both pits observed during the dry season. The acid generation from coal generally occurs when there is a sufficient amount of sulphides present, possibly as pyrites. The sulphides generate acid as the coal weathers due to exposure to moisture and oxygen i.e. oxidises. SRK's coal experience in the region indicates that the pit could remain an enduring source of AMD, especially if the geometry is unaltered and the coal seam remains in-situ.

2.6 Spontaneous combustion

This phenomenon comprise heat emanating from the high-walls along with a sulphurous smell and results from the oxidation of the coal seam where the three required elements are present (fuel source (coal), oxidising agent oxygen and a source of heat). The oxidation of the pyrite in the coal occurs spontaneously when it comes into contact with the atmosphere. This reaction is exothermic and produces the required heat source needed to start spontaneous combustion. The current cavities in the high-walls formed due to collapse or mining activities have created oxygen supply paths to the coal—thus feeding the self-initialised process.

3 IDENTIFICATION OF PREFERRED REHABILITATION OPTION

The primary objective of the design was to rehabilitate the site to enable construction of low cost single storey housing as intended for the post closure land use. The main consideration to the long term success of the rehabilitation is to prevent future artisanal mining of the coal seam. To meet the primary objective the following secondary objectives were addressed in the design:

- Preventing future AMD production in the site area;
- Preventing the production of spontaneous combustion in the site area;
- Preventing the occurrence of subsidence;
- Preventing future mining of the coal seam;
- Providing long term stability of the founding layer;
- Re-contouring the area to enable township development restricted to single storey housing; and,
- Provide ground and stormwater drainage management.

Four possible rehabilitation options were identified as follows:

Option 1: This alternative only allows for geometric landscaping without removing the coal seam and forms the shallow foundation rehabilitation cost;

Option 2: Dynamically compact the entire saddle prior to shallow foundation treatment;

Option 3: Fill all cavities by pressure grouting the entire seam remnant prior to shallow foundation treatment; and,

Option 4: Excavate and stockpile the remaining coal seam in the saddle prior to shallow foundation treatment.

Options 2 to 4 addresses the deep foundation rehabilitation.

3.1 Options analysis

A decision tree was assembled to assist in identifying the preferred rehabilitation option. This decision tree considered the possible effect of spontaneous combustion and subsidence occurring in the saddle, as reflected in an estimate of total risk cost. This risk cost comprise the damage risk cost (structural rehabilitation cost of area affected), baseline shallow foundation rehabilitation cost, deep foundation rehabilitation cost (specific costs for the options additional to the shallow foundation cost) and treatment cost of the AMD. The calculation of the damage risk cost was completed as follows:

1. A randomly selected percentage from a uniform distribution is assigned to whether spontaneous combustion will occur in the saddle;
2. A randomly selected percentage from a uniform distribution is assigned to the possibility that subsidence will occur if spontaneous combustion in the saddle continues or not;
3. The branch probability of failure is calculated by multiplying the results of steps 1 and 2;
4. The possible affected saddle areas are assigned 0 m² for no area affected or 30,000 m² (surface area of coal seam gleaned from geotechnical investigation) for a widespread affected area;
5. The structural rehabilitation rate is a random value from a normal distribution of rates currently associated with the construction cost of a low cost housing unit;
6. The damage cost is then calculated by multiplying the branch probability of failure by the area affected (4.) and the structural rehabilitation rate (5.); and,
7. The damage risk cost is finally the sum of all the damage costs calculated for the option in question.

The options are firstly ranked, 1 to 4, in terms of their total risk cost and whether constructing houses on the saddle may be authorised.

The design of treatment and holding facilities for the AMD are excluded from the 'Total cost excl. AMD treatment & unnecessary risk cost'. In addition to this the damage risk cost is also excluded from this cost ('unnecessary cost') for Option 1 as houses are not authorised for construction on the saddle and therefore no structural rehabilitation cost can be realised. The options are then ranked in terms of their resultant actual costs (stripped of inapplicable risk elements) and whether the housing may be authorised in each specific instance or not.

3.2 Randomisation of decision tree branch probabilities

Random selection of the input variables was undertaken since probability distributions descriptive of their respective behaviour were not available. However, upper and lower range bounds to each of the randomly selected input variable—with each step of a 1000 run Monte Carlo simulation. With run the ranking of the options was re-determined with the results of all 1000 runs captured to summary, from which the overall average rank of each option was calculated.

The following variables were assigned ranges for the sensitivity analysis:

- The probability that spontaneous combustion will continue in the saddle or not:

The range for each option varies as the rehabilitation plans can either be deemed more or less effective to curb spontaneous combustion. High probabilities, 95 to 100%, are assigned to Option 4 where spontaneous combustion will most likely not continue as the production will cease on removal of the coal seam. In the case of Option 2 it is estimated that spontaneous combustion is likely not to continue therefore a range bound of 40 to 90% (from which a random selection is made on each run) is assigned. For Option 3 it is estimated that spontaneous combustion assumes a greater than even probability not to occur (with voids pressure grouted) and therefore a range of 50 to 100% was assigned to the 'No' occurrence. Finally Option 1 is assigned a selection between 50 and 100% probability, that spontaneous combustion will continue as none of the voids in the coal seam are removed, filled or closed up.

- The potential that subsidence may continue preceding the combustive condition in the saddle:

Option 4 is again assigned the highest probability of non-occurrence, 97.5 to 100% range, as all the potential cavities and mechanisms for both spontaneous combustion and subsidence disappear on coal removal. Alternatively Option 1 was assigned 0% that subsidence will cease to occur as no cavities are filled, compacted or removed that can inhibit subsidence or spontaneous combustion from occurring. However, without data to support a positive assertion, the potential for subsidence to occur may fall (randomly) between 1 and 100%. For Options 2 and 3 there is relative certainty that subsidence has a less than 50% probability for occurrence therefore upon continuation of spontaneous combustion the probability ranges in both instances are set to between 5 and 50%. The possibility that subsidence will occur should spontaneous combustion not continue is deemed low and is thus set at 0 to 20% and 0 to 35% respectively for Options 2 and 3. A 10% higher range limit is set for Option 3 as pressure grouting is considered to present a marginally greater probability of success in sealing up coal voids;

A component of the primary rehabilitation objective is to prevent future mining activities. For this reason the potential for mining to continue was assessed and only Option 4 will completely prevent it as there will be no coal to mine. Option 3 has the potential to curb future mining as the coal seam will be covered in hardened cement that may be difficult for the community to mine through. Alternatively Options 1 and 2 will present the opportunity for artisanal mining to continue as the coal seam will remain in-situ. Option 4 was the preferred rehabilitation option and ranked in the number 1 position of the Monte Carlo simulation options analysis.

4 GEOTECHNICAL INVESTIGATION

Option 4's rehabilitation elements comprises mostly geotechnical design. For this reason a detailed investigation was completed to determine the geological profile, material strength parameters and groundwater location and recharge rate.

4.1 Test pitting and boreholes

In total 14 test pits were excavated up to refusal, profiled and sampled in and around the pits and saddle. Both of the pits are overlain by a layer of made ground composed of colliery spoil. This material is typically between 0.5 to 3.8 m thick and is generally composed of dark brown, light brown and black, loose, clayey gravel with gravel and cobble sized (occasionally boulder size) fragments of shale, carbonaceous siltstone, sandstone and coal with local zones of soft, gravelly clay. Locally, the colliery spoil is comprise dark brown and black, loose, very sandy gravel with gravel and cobble sized fragments of the above. The colliery spoil is either underlain by sandstone, shale or carbonaceous shale/coal depending on the location of the testpit. The entire saddle area is covered with a layer of colluvial soil that varies in thickness between 0.4 and 1.0 m. The colluvium is generally composed of light and dark brown, loose to medium dense sand with occasional sub angular to sub rounded fine and medium grained quartz and ferricrete gravel.

Five rotary core boreholes were advanced in the study area. The main purpose of the drilling was to establish the geological profile of the study area. NXC core size was generally used in the upper 0 to 2 m of each borehole while the remainder of each hole was completed using HQ sized drilling. The core was logged and sampled for laboratory testing. In the saddle area (BH 1), up to 6.8 m of coal was encountered at an elevation of 1420.6 to 1233.8 m above average mean sea level (amsl). The

top of the remaining coal in the pits are between 1231 to 1234 m above amsl. Coal occurs as black, slightly weathered, massive, soft rock, pyritic coal. Based on the geological profile of the borehole on the saddle, it seems likely that the saddle area (in the vicinity of the school) is underlain by some 7 m thick coal that likely belong to the Alfred coal seam.

4.2 Laboratory testing and results

The laboratory tests that were done on the soil samples include foundation indicators (sieve, specific gravity, hydrometer and Atterberg Limits), MOD compaction density, CBR and remoulded permeability. The rock samples were subjected to UCS tests.

The colliery spoil is typically composed of variable material types in that the soil classifies as either sandy lean clay, silty sand, silty/clayey gravel with sand or plastic clay with sand under Unified Soil Classification System (USCS). The grading envelope of the material confirms the variability in the composition of the colliery spoil in that the maximum and minimum grain sizes expressed as a function of the percentage of material passing a particular sieve size shows high variability between coarse and fine fractions for the sample population ($n = 8$). The clay content of the material varies between 3 and 27%, while the Plasticity Index (PI) shows a similar high variability of between 8 and 27. The colliery spoil is expected to vary in potential expansiveness between low to high.

Laboratory analysis of the three residual shale samples show the soil to have a fairly homogenous grading and the three samples all classify as sandy lean clay with a clay content range of 15 to 25% and a PI range of 14 to 26. The residual shale is potentially medium to highly expansive.

The residual sandstone classifies as either clayey sand, sandy lean clay, silty clayey sand, silty sand or clayey sand with gravel due to the range of properties obtained from the different samples. The material has a lower clay content and PI range of between 4 and 14% and 3 and 18 respectively. The bulk of the residual sandstone samples classify as having a low heave potential, with only sample TP 12 (1.7 to 2.0 m) returning a borderline medium potential heave classification.

The average Maximum Dry Density (MDD) of the colliery spoil is 1720 kg/m³ (range between 1640 and 1850 kg/m³), while the average Optimum Moisture Content (OMC) is 14.1% (ranging between 11.4 and 16.0%). The average MDD of the shale is 1820 kg/m³ (ranging between 1780 and 1880 kg.m³) with the average OMC being 13.7% (ranging between 10.2 and 16.4%). For the sandstone, the average MDD is slightly higher at 1947 kg/m³ (ranging between 1820 and 2060 kg/m^3) and the average OMC is 10.1% (ranging between 8.0 and 13.2%). SG's for the colliery spoil were quite low, ranging between 2.21 and 2.55.

The CBR values of the colliery spoil vary considerably but are generally high (excluding TP 14, 1.0 to 1.5 m). Samples TP 4 (0.5 to–1.0 m) and TP 16 (0.5 to 1.0 m) comply with the criteria for G6 material, while samples TP 6 (0.3 to 1.0 m) and TP 13 (0.1 to 0.6 m) classify as G8 and G9 materials. Elevated PI's of the samples from TP 8 do not meet the minimum PI screening criteria for G9 material.

Regarding the residual sandstone materials, although TP 11 (0.5 to 1.0 m) returned high CBR values, the elevated percentage swell does not meet the minimum G9 criteria of <1.50% swell. The low CBR and elevated PI of sample TP 12 (1.7 to 2.0 m) differs from that required for G9 material. Similarly, the PI of TP 20 (2.0–2.5 m) (PI = 18) is higher than the required minimum criteria for G9. The remainder of the residual sandstone samples classify as G6 to G8 material.

Samples were reconstituted to a target density of 98% of MOD density. The remoulded soil permeability is low for the colliery spoil, residual sandstone and shale. Average permeability of the colliery spoil is 7.3×10^{-7} m/s while the average permeability of the residual sandstone is 3.4×10^{-7} m/s. Remoulded permeability tests were only performed on one sample of residual shale (TP10 2.0–2.5 m) which returned a low permeability of 1.5×10^{-9} m/s.

Uniaxial Compressive Strength (UCS) tests results reporting modulus and Poisson's ratio (UCM) were performed on three samples of sandstone. The density of the samples ranged between 2240 to 2380 kg/m^3. The UCS results of sandstone samples retrieved from SRK BH 1 varied between 16.0 and 17.2 MPa and the sandstone therefore classifies as a medium hard rock.

4.3 In-situ testing

Eight dynamic plate load tests were performed at accessible locations within the pits to gauge the deformation modulus of the upper 300 mm of the colliery spoil. The test results confirm the results of the DCP tests in that loose to very loose (corresponding to E modulus of <10 MPa) consistency material is present in the upper 300 mm of the colliery spoil soil profile.

Ten Dynamic Cone Penetration (DCP) tests were performed adjacent to the test pit locations excavated within the pits. DCP tests consist of a cone (60° cone of 20 mm diameter) fixed to the bottom of a vertical rod. An 8 kg hammer weight is repeatedly lifted and dropped through a height of 575 mm onto a coupling at the mid-height of the rod to deliver a standard impact on to the cone that is driven into the underlying material. A vertical scale alongside the rod is used to measure the depth of penetration of the cone in mm per blow. Penetration depth less than 1 mm and exceeding 20 blows is considered as refusal. The DCP tests were performed to assess the in situ soil consistency after the classification system of Brink, Partridge & Williams (1982) that relate the penetration rate (mm/blow) to the soil consistency. It is evident that loose to very loose soil consistency prevails across the bulk of the pit floor covered with colliery spoil. In its present state, this material will likely be subject to settlement should housing development be constructed without in situ re-engineering of the colliery spoil.

From field observations and permeability characterisation inferred from the geotechnical investigation, it is estimated that surface water recharge of groundwater occurs relatively slowly with the permeability of the colliery pit spoil not exceeding 7.3×10^{-7} m/s.

The AMD contaminated groundwater coincides with the pit floor during the rainy season which leads to contamination of ponding surface water. The potential of contaminated AMD decant to the surrounding environment needs to be eliminated in order to adhere to section 7 of the National Water Act 36 of 1998 which requires clean water to be kept separate from contaminated water.

5 REHABILITATION DESIGN

Following the options analysis and geotechnical investigation the following comprised the rehabilitation design:

- Demolition of structures currently situated in the rehabilitation area;
- Excavation of coal seam overburden;
- Temporary high-wall stabilization using soil nails and rock bolts as well as shotcrete. Geotechnical parameters of the various layers used in the numerical analysis of the slope stabilisation;
- Coal seam excavated and relocation off-site. Due to the strength of the coal it is estimated that blasting will be required;
- Installation of sub-surface drains in the prepared surface (pits and coal excavation floor) to prevent the fluctuating groundwater table from saturating the final founding layer and collect residual AMD;
- Back-filling and terracing the terrain. The material removed from the coal seam overburden excavation and slope profiling of the pit boundaries will be used to terrace the final geometry to allow

sufficient stormwater run-off and construction of the low cost housing development.

REFERENCES

Brink, A.B.A., Partridge, T.C. and Williams, A.A.B., (1982), *Soil survey for Engineering. Monographs on soil survey*. Oxford University Press, New York.

SAICE (1995). *Code of practice: Foundation and super structures for single storey residential buildings of masonry construction*. Joint Structural Division: South African Institution of Civil Engineers and Institution of Structural Engineers.

TRH 14 (1985). *Guidelines for Road Construction Materials. Technical Recommendations for Highways*, Department of Transport, Pretoria.

Visser, D.J.L., (1989). *Explanation of the 1:1 000 000 geological map, 4th edition 1984*. Council for Geoscience, Pretoria.

Proceedings of the first Southern African Geotechnical Conference – Jacobsz (Ed.)
© *2016 Taylor & Francis Group, London, ISBN 978-1-138-02971-2*

Rock fall risk with analytical hierarchy process

D.J. Avutia Pr Eng
SRK Consulting, Johannesburg, South Africa

ABSTRACT: The geological variability and geotechnical uncertainty of fractured bedrock outcrops necessitates the use of deterministic software and quantitative tools to calculate the rock fall risks on mountainous regions. The deterministic analysis carried out with Trajec3D assesses the anticipated rock fall trajectories affecting infrastructure at the foot of the mountain. Quantitative analysis of the hazard factors such as boulder source areas, boulder potential energy, slope angle and vegetation are evaluated using Geographic Information System (GIS) techniques and Analytical Hierarchy Process (AHP). The GIS hazard maps and AHP are integrated to categorise the rock fall risks and provide sustainable solutions. The low to moderate risk categories are assigned catch berms and gabion mitigation solutions and the moderate to high risk categories are remediated with preventative meshing and cable-anchoring solutions for boulder displacement. The quantitative risk methodology discussed in this paper is well suited to preliminary evaluation of rock fall remediation and mitigation strategies.

1 INTRODUCTION

A mountainous region situated adjacent to a town has experienced underground mining blasting and continuous geological degradation resulting in major rock fall events. These rock fall events are triggered by the reduction in shear strength of the fractured bedrock outcrops which lead to large boulders descending the slope towards a town at the foot of the mountain. The rock fall risk and boulder trajectories are subsequently evaluated through the consideration of the geological variability and geotechnical uncertainty of fractured bedrock outcrops. This paper presents an example of the anticipated rock fall trajectory and calculates the inherent risk posed to infrastructure at the foot of the mountain.

2 ROCK FALL ANALYSIS SOFTWARES

The simulation of realistic rock fall trajectories is extremely difficult due to the uncertainty associated with the definition of input parameters such as the boulder geometry, slope topography, initial velocity and material friction angles. Rock fall analysis programs such as Trajec3D (Basson, 2012), RocFall (Rocscience, 2015) and RFall_3D (Gibson, unpubl.) are utilised to simulate boulder trajectories along varying slope topographies. Three main input parameters, namely the dynamic/static friction angle, normal coefficients of restitution and tangential coefficients of restitution are required to carry out rock fall analysis. The ability

of surface topography to impede the movement of a boulder is characterised by the Coefficients of Restitution (CoR). A CoR of zero suggests perfectly plastic collision while a CoR of one suggest a perfectly elastic collision.

2.1 Trajectory 3D (Trajec3D)

The Trajec3D rock fall analysis software is a three-dimensional and rigid body dynamics analysis program that simulates the trajectory of volumetric bodies during free fall, bouncing, sliding and rolling. The physics interaction between materials requires only three measurable and intuitive input parameters between the materials: the coefficient of restitution, the static friction angle, and the dynamic friction angle (Basson, 2012). A Trajec3D rock fall analysis model is presented and discussed in this paper to illustrate the capabilities and limitations of deterministic methods in rock fall studies.

2.2 Rocscience RocFall

The Rocscience 2D RocFall analysis software is well suited for the statistical analysis of rock fall risk. RocFall evaluates detached boulder characteristics, by calculating the energy, velocity and bounce height of various rocks (Chai *et al.*, 2013). The interaction response of dynamic boulders and static topography is defined in the slope material library, where the normal CoR, tangential CoR and friction angles are defined. These parameters may be assigned to a specific segment of the slope, as different materials are encountered throughout the slope. RocFall defines the boulder geometry, density and energy characteristics in the seeder rock type library, where 22 boulder shapes are available to the operators.

2.3 *RFall_3D*

The RFall_3D (Gibson, unpubl.) software evaluates boulder displacements in three dimensions, through deterministic and probabilistic analyses of the boulder velocity, moment of inertia and energy. The RFall_3D methodology resembles the Lump Mass Analysis (LMA) method using the moment of inertia instead of the actual 3D geometry to simulate rock fall trajectories. Vijayakumar *et al.* (2011) emphasise the importance of the boulder geometry towards rock fall trajectories. The host rock outcrops and structural orientations are the main factors controlling the rock boulder geometry and magnitude.

A comparison of the three programs is presented in Table 1 and all users should select a specific program based on the requirements of their study. RocFall has the most input parameter options coupled with the probabilistic functions; however the 2D analysis limitations where unsuitable for the boulder simulations in this study. The RFall_3D software is characterised by robust command driven inputs but the moment of inertia analysis method limits the visualisation of the rock fall trajectories, similar to the dimensionless point incorporated into the LMA analysis instead of the actual geometrical rock shape. The Trajec3D program was the most practical and visually astute software for the simulation of the dislodged boulders trajectories in this study.

3 CASE STUDY TRAJEC3D ANALYSIS

The potential hazardous rock fall bodies are categorised into two distinct shapes namely angular

Table 1. Rock fall software comparison.

Features	Trajec3D	RocFall	Rfall_3D
Rigid Body	√	√	√
Boulder geometry	√	√	x
Probabilistic	√	√	√
3 dimension	√	x	√
Multiple Segment Characterisation	x	√	√
Normal Restitution	√	√	√
Tangential Restitution	√	√	√
Static Friction	√	√	x
Dynamic Friction	√	√	√
Results Plotting	x	√	√
Importing of DXF's	√	√	x

spheres and angular ellipses. The smaller boulders tend to be angular spheres (<5 metres) while the bigger boulders (>5 metres) detach as angular ellipses, however the boulder geometry is highly dependent on the structural orientation of the rock fall source. The angular spheres have less geometry rolling resistance while the angular ellipses lack continuity during motion due to their irregular boulder geometry. The size or mass of the rock fall bodies controls the potential and kinetic energies generated by the detached rocks, which result in the larger diameter rocks descending at faster rock fall velocities than the smaller rocks (Basson *et al.*, 2013). The variability of the slope topography namely the vegetation, undulating terrain and flat roadway surfaces retard the trajectory of the rocks and are accounted for through the tangential CoR (Chai *et al.*, 2013).

The model input parameters were established during an in-situ assessment of the terrain and vegetation of the slope. The CoR was estimated to be 0.05 due to the trees and bushy vegetation predominantly covering the slope. This value was verified with the RocFall tangential CoR (block-field with bushes and small trees). The height CoR for rock rebounds of various shapes and sizes were explored by Basson *et al.*, (2013) where it was identified that more than 80% of catch berm CoR were below 0.1. The static and dynamic friction angles parameters are used to account for the boulder interactions with the terrain. The small trees and bushes on the terrain necessitate the high friction angle values of 65° and 60° for the static and dynamic analysis respectively. Table 2 summarizes the material behaviour input parameters used in the analysis.

The rock fall trajectory analyses were carried out by dividing case study area into 100 meter horizontal intervals with the assumption that boulders would not move laterally by more than 100 meters as observed during historic rock fall events. The case study slope area with infrastructure at the foot of the mountain was divided into eleven representative zones and the distance to each infrastructure point was measured in Figure 1.

Zone 2 of the case study area (Figure 1):

A potential rock fall source area was identified towards the top of zone 2 with a small boulder

Table 2. Trajec3D material parameter.

Properties	Value
Coefficient of restitution Velocity	0.1 m/s
Coefficient of restitution Height	0.001 m
Coefficient of restitution Velocity	0.05
Static Friction angle	65°
Dynamic Friction angle	60°

(<5 m) being anticipated from the walkover survey. Boulders (<5 m) are likely to be stopped by the vegetation prior to building up enough energy as shown in Figure 2. The boulder was assumed to be having a similar host rock density of 3100 kg/m3 resulting in a boulder weight of approximately 100 tons.

The Trajec3D analysis of 5 meter angular sphere boulders in Figure 3 illustrates that boulders were highly likely to reach the foot of the hill. The boulders descend the slope for 400 meters at an average velocity of 11 m/s. These results may be attributed to the lack of ability to account for the small trees and bushes vegetation in the Trajec3D software. The rock fall rolling distances achieved on Trajec3D were corrected with the 2D Rocfall program which resulted in the boulders being arrested by the small trees and bushes halfway down the slope.

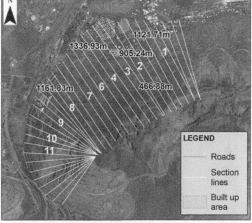

Figure 1. Trajec 3D analysis sections.

4 GIS TECHNIQUE

The dislodgement of rock boulders from the terrain involves multiple geological processes coupled with mine-induced damage, thus complicating the application of deterministic methods to determine the boulder dislodgement and trajectory based on the current relationship of the case study terrain parameters. The friction angle limitations in the Trajec3D, tree dimension limitations in the Rocfall and visual limitations in the RFall_3D software together with unavailable geotechnical and geological data resulted in inconclusive deterministic results, which necessitated the application of a Geographic Information System (GIS) indexing approach to quantify the rock fall risk as shown in Figure 4. The interpretation of aerial photograph, field surveys and the digitising of base maps in the GIS software is a better approach for identifying the top factors and locations contributing towards rock fall failure. Similar landslide hazard assessments utilising the GIS indexing approach have been published by Carrara *et al.* (1991) and Leroi (1996).

The rock fall hazard factors such as boulder source areas, boulder potential energy, slope angle and terrain vegetation are plotted as diagrams using GIS techniques. Each map classified the rock fall risk into four significant classes, namely very

Figure 2. Zone 2 terrain and potential rock fall source.

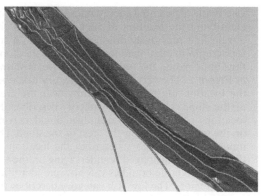

Figure 3. Trajec3D zone 2 rock fall analysis.

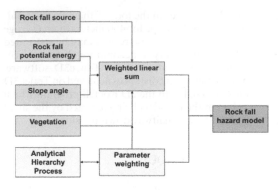

Figure 4. GIS multi-criteria evaluation for rock fall hazard assessment.

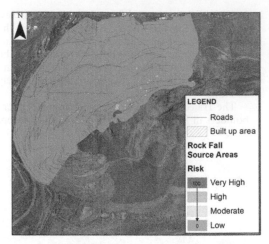

Figure 5. Rock fall source hazard map.

high, high, moderate and low. Numerical values were assigned to each risk class incrementally from zero to a hundred to facilitate the quantification of the each rock fall hazard map.

4.1 Rock fall source area

The rock fall source area map is created by identifying the rock outcrops in the terrain and previously dislodged boulders on the topography as shown in Figure 5. The very high risk category describes the rock outcrops that potentially dislodge boulders greater than 15 metres. The high risk category describes rock outcrops that potentially dislodge boulders in between 10 and 15 metres. The moderate risk category describes previously dislodged boulders lying on the terrain and rock outcrops that potentially dislodge boulders in between 5 and 10 metres. The low risk category describes areas with no potential rock fall source areas.

4.2 Rock fall potential energy

The boulder potential energy map is generated by identifying the location of rock outcrops in the terrain and previously dislodged boulders on the topography in relation to the elevation of the slope, thus anticipating the trajectory distance as shown in Figure 6. The very high risk category describes the location of the rock outcrops near the crest of the slope. The high risk category describes the location of the previously dislodged boulders lying on the terrain just below the crest of the slope. The moderate risk category describes the location of the previously dislodged boulders lying on the terrain, but no large boulders (>5 m) were identified in this region. The low risk category describes the areas with flat rock fall gradient at the foot of the mountain.

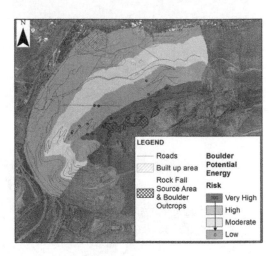

Figure 6. Rock fall potential energy.

4.3 Slope angle hazard

The slope angle hazard map is produced by measuring the change in slope angle of the topography as shown in Figure 7. The very high risk category describes the areas with steep gradients with angles ranging for 45 to 60 degrees. The high risk category describes the locations with angles ranging for 30 to 45 degrees. The moderate risk category describes the locations with angles ranging for 15 to 30 degrees. The low risk category describes the areas with flat gradients with angles ranging for 0 to 15 degrees. The relative importance of each hazard factor will be discussed in the Analytical Hierarchy Process (AHP) section of the paper.

4.4 Vegetation hazard

The case study slope is predominantly situated in the banded-iron formation geology, but the rock fall events are significantly affected by the vegetation cover characterised by the CoR and friction angle parameters in the Trajec3D analyses. The boulder trajectory distances are controlled by the ability of the vegetation to retard boulder movements. The map shown in Figure 8 is created through the identification of variable vegetation cover. The very high risk category describes the location of the rock outcrops and boulders near the crest of the slope with no vegetation. The high risk category describes the location of the previously dislodged boulders

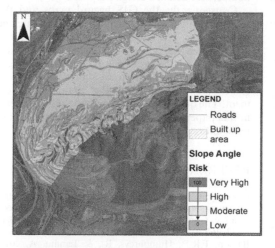

Figure 7.   Slope angle hazard map.

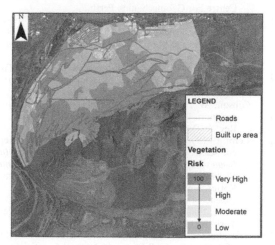

Figure 8.   Vegetation hazard map.

lying on the bare terrain just below the crest of the slope. The moderate risk category describes the location where previously dislodged boulders are situated in the sparse vegetation consisting of small trees. The low risk category describes the areas with dense vegetation in the form of small trees and bushes.

5   ANALYTICAL HIERARCHY PROCESS (AHP)

The valuation of the rock fall hazards primarily involves the identification of rock fall source areas, varying slope angles and the varying slope vegetation in the case study area. Subsequent to the identification of the pertinent hazards, the grading of each hazard based on its contribution to the rock fall risk is carried out. Preferential values ranging from 0.111 (extremely less important than) to 1 (equally important as) are assigned to each rock fall hazard factor and the inverse 9 (extremely more important than) factors are calculated in the square reciprocal matrix through a pair-wise comparison relating rows relative to columns as shown in Table 3.

The AHP by Saaty (1980) is used to ensure the consistency of the assigned weights for all the rock fall factors in Table 4. The size of a detached boulder is controlled by the rock fall source area which is the most influential factor with a weighting of 39%. The vegetation has minimal influence on very larger boulders thus this factor was assigned the least weighting of 15%. The AHP weighting method calculates a Consistency Ratio (CR) to check the reliability of the rock fall hazard weights. A CR less than 0.10 is recommended to confirm the degree of consistency implemented during the assignment of the hazard factor weights.

The four GIS maps and AHP weights are integrated to produce the rock fall hazard map shown in Figure 9. The very high risk zone is situated at the crest (south) of the slope where no vegetation, steep slope angles, high potential energy and prominent rock outcrops are present. The low risk zone situated at the foot (north) of the mountain is characterized by small trees/bushes, flat slope gradients, low potential energy and no rock outcrops.

The GIS hazard maps illustrate and quantify the rock fall risk, while the AHP methodology ensures consistent assignment of the hazard weightings for the rock fall risk hazards. The construction of the hazards maps and simulation of boulder trajectories results in the development of risk maps to categorise the boulder risk as low, moderate or high risk.

Table 3. Assigned values in Analytical Hierarchy Process (AHP).

No.	Factor	Rock fall source areas	Boulder potential energy	Slope angle	Vegetation
1	Rockfall source areas	1	1.4	2.0	2.5
2	Boulder Potential Energy	0.70	1	1.5	2.0
3	Slope angle	0.50	0.65	1	1.3
4	Vegetation	0.40	0.50	0.75	1
	Total	2.60	3.58	5.29	6.83

Notes: 9 Extremely more important than 1 Equally important as 1/9 Extremely less important than.

Table 4. Normalized scores in analytical hierarchy process.

No.	Parameters	Cumulative normalized score	Weights	CM
1	Rockfall source areas	1.53	0.38	4.00
2	Boulder potential energy	1.13	0.28	4.00
3	Slope angle	0.76	0.19	4.00
4	Vegetation	0.58	0.15	4.00
	Total	4.00	1.00	

Notes: CM Consistency measure   CI   0.00
CI Consistency index   RI   0.90
RI Random index   CR   0.00
CR Consistency ratio, recommended to be <0.10.

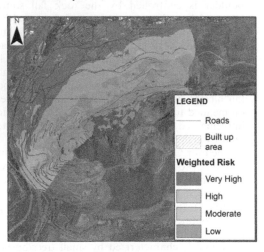

Figure 9. Comprehensive rock fall hazard map for sovereign hill.

6 DISCUSSION AND RECOMMENDATIONS

Fractured rock fall outcrops releasing boulders are major hazards in slope engineering. The host rock outcrops and structural orientations are the main factors controlling the rock boulder geometry and magnitude. Additionally, the trajectories of the boulders are controlled by the slope geometry and vegetation cover of the slope characterized by the CoR. The Trajec3D, RocFall and RFall 3D rock fall programs are able to represent the boulder geometries but unable to accurately simulate the energy dissipation (CoR) and lateral displacement (2D limitations) of boulders for detailed design purposes.

Consequently, the GIS hazard maps and the AHP are integrated to identify and calculate the rock fall risk. The low to moderate risk categories are assigned catch berms and gabion mitigation solutions and the moderate to high risk categories are remediated with preventative meshing and cable-anchoring solutions for boulder displacement. The quantitative risk methodology discussed in this paper is well suited to preliminary rock fall remediation and mitigation strategies.

REFERENCES

Basson, F.R.P. 2012. "Rigid body dynamics for rock fall trajectory simulation." American Rock Mechanics Association, *Geomechanics Symposium* held in Chicago-USA,. 12–267.

Basson, F.R.P., Humphreys, R., & Temmu, A. 2013. "Coefficient of restitution for rigid body dynamics modelling from onsite experimental data". Australian Centre For Geomechanics, Perth,.

Carrara, A., Cardinali, M., Detti, R., Guzzetti, F., & Reichenbach, P. 1991. "GIS techniques and statistical models in evaluating landslide hazard". *Earth Surface Processes and Landforms*, Vol. 16, pp.427–445.

Chai, S., Yacoub, T., & Charbonneau. K. 2013. "The Effect of Rigid Body Impact Mechanics on Tangential Coefficient of Restitution. *Geo-Montreal*, Canada.

Gibson, W. 2015. Three dimensional rock fall analysis program. Unpublished RFall_3d Manual Version 3.0 Rev_A. Perth, Australia.

Leroi, E. 1996. "Landslide hazard—Risk maps at different scales: Objectives, tools and developments". *In: Senneset (ed) Landslides, Proceedings of 7th International Symposium on Landslides*, Trondheim, Norway, Rotterdam: Balkema pp. 35–51.

Rocscience. Inc, 2015. RocFall Toronto, Canada.

Saaty, T. 1980. "*The Analytical Hierarchy Process*" New York: McGrawHill.

Vijayakumar, S., Yacoub., & T., Curran, J.H. 2011. "Effect of Rockfall Shape on Normal Coefficient of Restitution" *Proceedings of the US. Rock Mechanics Symposium* (ARMA) San Francisco, CA, USA.

Proceedings of the first Southern African Geotechnical Conference – Jacobsz (Ed.)
© 2016 Taylor & Francis Group, London, ISBN 978-1-138-02971-2

Defining geotechnical design sectors: Apparent dip analysis at Jwaneng Mine, Botswana

I.J. Basson & P.K. Creus
Tect Geological Consulting, Steenberg, South Africa

K. Gabanakgosi & O. Mogorosi
Debswana, Jwaneng Mine, Jwaneng, Botswana

M. Bester
Anglo American Kumba Iron Ore, Centurion, South Africa

ABSTRACT: This contribution describes a methodology for assessing the interaction of dominant bedding planes within anisotropic materials (laminated, quartzitic and carbonaceous shales of the Griqualand West Supergroup at Jwaneng Mine) with a design pit. Lithological contacts from implicit 3D lithological and structural models are used as proxies for bedding. Apparent dips are derived by combining the orientations of individual triangles on these contacts with orientations of individual triangles on a design surface, using a nearest-neighbour algorithm. Apparent dips may be contoured, binned into categories that show inward or outward dipping attitudes, or areas where critical friction angles are exceeded. *A priori* definition of specific design sections is minimized, design sectors are more intelligently defined and reliance on average or median dips for a simplistic, wedge-shaped design sectors, is obviated. This is particularly important where simplistic sectors contain significant changes in dip direction of the main plane of anisotropy/bedding.

1 INTRODUCTION

Data from geological, structural, rock mass and hydrogeological models provide inputs into design sectors for geotechnical models and failure modes. It is often difficult to determine the size and exent of each design sector or domain, which contains structurally-controlled and/or anisotropic materials, prior to analysis or early in the analysis process. Furthermore, geotechnical drilling, structural mapping data, scanline data and geotechnical mapping data utilizing Sirovision™ software are either sparse or are focused in narrow windows around fixed vertical 2D design sections.

The orientation and location of these sections are predicated on the geometry of a conceptual shell or design, in turn based on an economically-driven optimization process by utilizing Whittle optimization. Sections are typically radially-oriented around the center of a pit or sub-pit. These sections seek to show the interaction of the true dip of planes of anisotropy (e.g. bedding, foliation) with the slope. At Jwaneng, a structurally-controlled stepped path failure mechanism (interaction between bedding, joints, and rock bridges) is anticipated. Pre-defined design

sectors and 2-dimensional design sections, however, limit the Geotechnical Engineer to "snapshots" of the *spatially continuous and variable interaction of geology with the existing or planned pit surface.* The analysis and slope design of volumes between these vertical sections becomes a matter of interpolation and could result in increased geotechnical risk should possibly unfavourable interaction of geology with slope geometry be overlooked.

In order to deliver practical, sector-specific design parameters as input for mine design, a methodology was derived to improve the definition of these geotechnical design sectors early in the process, thereby providing a robust *a priori* definition of potential design sectors for subsequent analysis. This contribution augments the use of vertical 2D design sections by extracting the spatial variability of the abovementioned interaction, by means of apparent dip maps. The methodology has been successfully applied at Kumba's Sishen and Kolomela Mines (in the Transvaal Supergroup). The recent construction of a comprehensive, implicit 3D model of the country rock around the Jwaneng kimberlites has made such an analysis possible, while ongoing issues in a particular part of the pit (the NE Corner) have necessitated the extraction of as much relevant information as possible from available 3D models and existing data. The use of macros allows for any changes in design to be analysed according to the methodology presented herein, thereby resulting in both a dynamic process and a direct input into mine planning. The methodology may be applied to most planar features throughout any volume of interest.

2 GEOLOGY

2.1 Stratigraphy

Jwaneng Mine is in the Okavango and Shakwe Zone(s) (Carney *et al.* 1994), covered by Palaeo-proterozoic (*circa* 2300–2100 Ma) sediments of the Griqualand West Supergroup of southern Africa. The base of the sequence consists of the *c.* 2.552 Ga Malmani dolomite. The contact between the Malmani Subgroup and the Rooihoogte/Duitschland Formation has been delineated in several very deep drillholes. The overlying (Middle) Rooihoogte/Duitschland Formation consists of laminated sandstone, mudstone, greywacke and black shale. The upper and lower portions of the (Middle) Rooihoogte/Duitschland Formation are subdivided into Quartzitic Shale (QS2) and Carbonaceous Shale (CS2). A chert pebble conglomerate, consisting of angular to rounded chert clasts in a gritty matrix, separates the upper part of the Rooihoogte/Duitschland Formation from its lower part, representing an 80 Ma hiatus in the depositional history. This unit is termed the Bevets Conglomerate (BVT) on Jwaneng Mine, although Coetzee (2001) found that the *base* of the Duitschland Formation near Thabazimbi (South Africa) is occupied by a "flat" chert-pebble conglomerate, which is termed the Bevets or Bevetts Conglomerate at this location.

The upper part of the Rooihoogte or Duitschland Formation consists of fine-grained to medium-grained, poorly-sorted argillaceous quartzite, greywacke, subgreywacke and silty mudstone. Quartzitic Shale (QS1) is succeeded by a dia-chronous, transgressive (erosional) surface. Above this erosional surface, the 2.264 Ga Timeball Hill (or Lower Ditlhojana) Formation consists of deep marine laminated sandstone, mudstone, greywacke and black shale interbedded with thin, mafic tuff bands. The upper part of the Timeball Hill (or Upper Ditlhojana) Formation consists of laminated sandstone and black, Laminated Shale (LS). The 2.184 Ga Boshoek Formation, which overlies the upper Timeball Hill Formation, consists of felsic volcanics and graphitic or carbonaceous shale with sandstone intercalations. Together with the Upper Timeball Hill Formation, these comprise the Laminated Shale unit (LS).

The Kalahari Sequence or Group consists of calcrete, sand and scree, which includes alluvial material of unspecified or undocumented origin. The lower part of the Kalahari Group in the area of interest consists of calcrete or caliche, which comprises calcium carbonate-cemented surficial gravel, sand, clay and silt.

2.2 Deformation events

The deformation sequence and geometry in 3D reveal an early compressional deformation (D) event directed towards a bearing of 350° or almost due north (Figure 1). This event is expressed by a series of mesoscale folds and associated low-angle thrusts that show strike extents on the order of 100 m. Thrusts merge laterally and down-dip into bedding, expressing "distributed" strain throughout the sequence. This early compressional event was defined by Dietvorst (1988) in the Ramotswa-Lobatse area, 120 km ESE of Jwaneng (Figure 1). In that study, regional folding around NE- to ENE-trending, gently plunging (0°–20°) open folds preceded extension and the development of normal faults. A similarly-oriented compressional event is evident in the Thabazimbi area in South Africa, approximately 250 km to the east of Jwaneng Mine. At Thabazimbi, a maximum age of thrusting is constrained by a quartz porphyry age of 2054 ± 4 Ma (essentially Bushveld Complex age), from the basal unconformity of the Waterberg Group (Dorland *et al.* 2006). The Ramotswa-Lobatse area shows

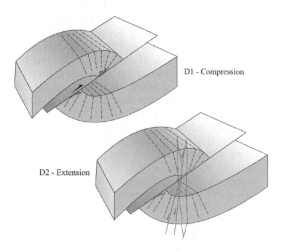

Figure 1. Schematic of the deformation events at Jwaneng Mine: D_1 compression forming breached folds and low-angle thrusts; D_2 extension exploiting fanning axial planar cleavage.

large-scale, normal faulting that sub-divides the Transvaal Supergroup into blocks showing characteristic dips and F_1 fold axis orientations (Dietvorst, 1988). Based on Dietvorst (1988), an overview of deformation in the Thabazimbi area, and mapping performed at Jwaneng Mine, regional NE- and ENE-trending faults exploited axial planar cleavage around D_1 anticlines and synclines. In effect, the orientations of D_2/D_3 are controlled by the macrostructural "fabric" imposed on the sequence due to early development of this fold-and-thrust belt. The geometry resulting from this sequence of deformation events is a series of NW-dipping, NE-SW elongated, fault-bounded, wedge-shaped blocks. These define structural domains, each with a characteristic average bedding dip, determined from robust datasets. Relatively minor E-W trending faults within these blocks may subdivide the geology into rhomboid-shaped blocks.

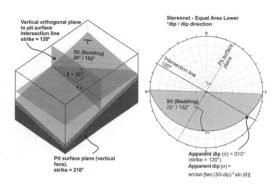

Figure 2. Schematic showing the calculation of apparent dip.

3 METHODOLOGY

Standard mapping data, downhole oriented bedding from acoustic or optical televiewer (A/OTV) logs and well-constrained, high-confidence 3D models, which are built using implicit modelling techniques, form the basis for the analysis. Based on well-constrained mapping and oriented geotechnical drilling, lithological contacts are predominantly parallel to bedding in their related volume(s), with the exception of volumes that have tectonic or erosional unconformities against their upper limits. Apparent dips of a given lithology, at either an existing or planned pit surface, may therefore be derived from a combination of the dip and dip direction of individual triangles on finely-triangulated lithological surfaces (*viz.* sedimentary contact orientations represent bedding orientations), with the dip direction of individual triangles on pit surfaces, on a pre-defined grid basis. The relationship between the orientation of contacts and the orientation of bedding may be tested via stereonet analysis. This produces an *apparent dip* value that represents the continuous variation in the interaction of the main plane of anisotropy with the pit surface or design surface. Several steps are adopted in Micromine™ to create apparent dip maps (Figures 2, 3):

- Extraction of triangle orientations (dip and dip direction) from wireframed or triangulated surfaces, assigned to structural domains (viz. fault-bounded blocks). Surfaces of specific lithological contacts are proven and/or reasonably assumed to be parallel or sub-parallel to bedding within their respective structural domains, thereby providing a "proxy" to bedding orientation;

- Erosional or tectonic unconformities, which violate one of the basic assumptions of the analysis, are filtered out;
- Data from triangles on the sides of fault-bounded blocks, which would provide anomalously steep dips where they run up against steep or vertical bounding faults, are filtered out;
- Using a nearest-neighbour algorithm, all surface orientation data, with respect to each structural domain, are assigned to a pit or design. The proximity over which this projection is done from off-(pit)-surface varies, but it is usually advisable to refer to the thickness, lateral extent, and internal coherence of the unit from which data are taken and, by default, through which data will be extrapolated. It was found that using true dip data within a vertical distance of 20 m–50 m of the pit surface was acceptable;
- Lastly, an apparent dip is calculated by spatially coinciding, on a grid basis, the proxy to bedding surface orientation (dip and dip direction) with face orientation of the design pit;
- Apparent dips are displayed only where units intersect the design pit. This was achieved by clipping modelled lithological volumes against the design pit, to show where these intersect the design pit;
- Areas on ramps and berms are filtered out;
- Apparent dips are re-gridded using an inverse distance interpolator to give an accurate representation of the spatial variation in apparent dip for each anisotropic lithological unit, within each fault-bounded block. Ranges may be arbitrary or plots could show individual units with ranges or bins based on various critical friction angles, or fail/no-fail maps (e.g. Figures 3c and d).

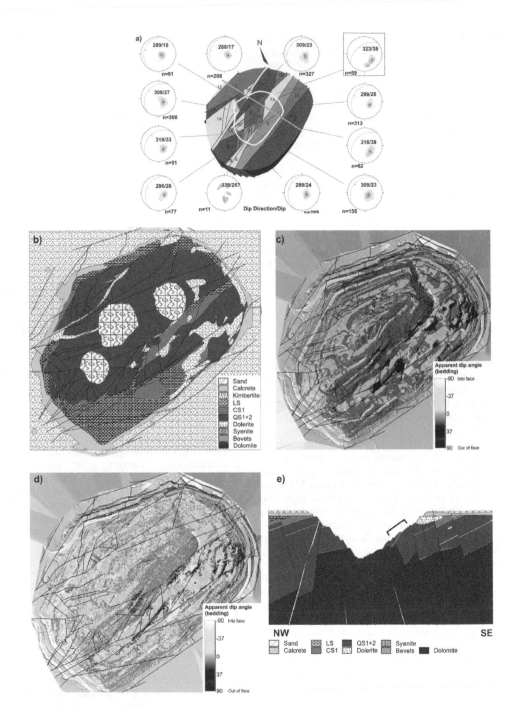

Figure 3. a): Sub-domaining of bedding readings from pit mapping into fault- bounded blocks. Anomalously steeper dips from Domain 7A are highlighted. Approximate extent of the pit surface is shown; b): 3D geological model clipped to pit surface. Shows daylighting or intersecting country rock lithologies; c): Contoured plot of apparent dip for QS, CS, LS (dol. sill is excluded), binned according to average friction angle (37 deg) and sense of dip (into vs out of face). The plot highlights a zone of relatively steeper bedding, compared to the underlying design surface, within Domain 7A and trending NE-SW through Domains 8 and 9; d): Dip direction and dip symbols, binned according to the chosen friction angle and coloured according to apparent dip angle and sense of dip; e): NW-SE trending cross-section showing the 3D model clipped to the design surface and the relatively steeper bedding for Domain 7A (indicated).

4 DEFINITION OF GEOTECHNICAL DESIGN SECTORS

By incorporating data obtained from apparent dip maps and utilizing various strength models according to conceptual failure models (with respect to lithological layers), geotechnical design sectors may be defined. Design sectors are first identified and grouped according to lithologies with similar strength properties, *i.e.* Hoek-Brown strength parameters to represent the rock mass strength. For instance, quartzitic shale is very different from laminated shale or carbonaceous shale at Jwaneng. Zones where the apparent dip of bedding is into the design pit, but importantly, with apparent dips regularly exceeding 37°, are considered for the delineation of separate design subsectors (Figure 3). Conversely, large areas where the apparent dip is more favourable (*viz.* away from the pit centre or into the pit face), may be delineated, which provides opportunity for optimizing the design.

This methodology allows for a significantly improved definition of design sectors without using averages or median dips of planes of bedding or other planes of anisotropy for a whole domain or sector. When combined with strength and failure criteria, this allows for the early subdivision of domains into practical design sectors as individual 3D volumes. When utilized early in the mine design process, this methodology optimizes (*viz.* usually reduces) the required number of 2D design sections that need to be analysed. Time and expense required to perform the iterative process of geotechnical stability analysis and risk assessment is also optimized and reduced.

A case in point is the NE Corner of the pit at Jwaneng Mine. An apparent dip analysis highlighted potentially problematic blocks (Block 7A and 8A) (Figures 3a, c), which appeared to show steeper bedding dips compared to adjacent blocks (an additional 8°–10°), based on an average within a DIPs stereonet window (Figures 3a, c, d and e). The identified anomalously steep bedding, highlighted by the apparent dip analysis, has over time led to planar sliding failures affecting single to double benches. The failure of such benches in planar sliding leads to pit-ward dipping bedding surfaces that extended over 3–4 benches. Particularly to the NE part of the pit, access ramps were at risk of failure due to the locally steep bedding. Using the methodology described herein, the risk to the ramp was identified early and detailed stability analysis was thereafter carried out to determine the Factor of Safety (FoS) and probability of failure for the area. The results of the stability analysis indicated lower than acceptable FoS and, as mitigation, slope support designs were determined. The support which

was designed and successfully installed in the area comprised concrete piles primarily aimed at preventing planar sliding of the blocks (Figure 4).

The methodology has greatly assisted the early identification of the stability condition. It has allowed for forward prediction of the stability implications, based on the still-to-be-mined benches with respect to the mine design. Business risks associated with personnel and equipment safety as well as loss of the access ramps could have had devastating impacts on the mine's sustainability.

5 CONCLUSION

The methodology and example presented herein augments the use of 2D design sections by improving the early definition of design sectors, as inputs to the mine planning process or into focussing on various areas. This is particularly important where the main plane(s) of anisotropy show significant changes in dip or dip direction, across various domains.

This is the case at Jwaneng, where each fault-bounded block exhibits its own particular dip and dip direction due to early folding and variable rotation of fault-bounded blocks. In addition, the potentially problematic Blocks 7A and 8A occurred where there is a change in trend of pit and the future design.

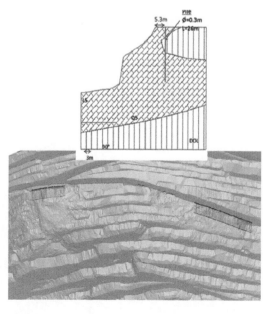

Figure 4. Pile reinforcement (slope support) installed on the ramp to ensure long term sustainability of the ramp.

This methodology allows for the incorporation of data from continuous mapping of interim or final faces. In this process, mapping, scanline and Sirovision data may be incorporated directly into the grid file, ideally followed by macro-driven re-analysis, thus allowing for continuous recalibration of the geotechnical model based on mapping data which are progressively more proximal to the final layout or design. Dedicated geotechnical drillhole information, which contains oriented downhole data, may be added to further refine and optimize the design at a very early stage. Indeed, any orientation data, whether desurveyed downhole bedding or fracture readings, surface mapping, Lidar or I-Site data, which falls within the distance defined by the nearest-neighbour algorithm, may be incorporated in the analysis.

This technique also overcomes the situation wherein certain parts of the pit inevitably show a dramatic change in orientation with respect to strike or dip direction *within* a domain. This has historically been problematic when using the more traditional means of average or median dips, which inevitably are not representative of the whole design sector. Improved design sector classification highlights possible areas—of any size or shape and not just vertical "wedges"—that show unfavourable interaction with slope geometries in future push-backs or on final faces. In turn, this results in a more robust and practical suite of sector-specific design parameters as inputs for mine design.

REFERENCES

Carney, J.N., Aldiss, D.T. & Lock, N.P. 1994. *The Geology of Botswana*. Technical Report, Geological Survey Department of Botswana.

Coetzee, L.L. 2001. *Genetic stratigraphy of the Palaeoproterozoic Pretoria Group in the western Transvaal*. M.Sc. Thesis (unpub), Rand Afrikaans University, Johannesburg, 206 pp.

Dietvorst, E.J.L. 1988. Early Precambrian Basement uplift and block faulting along the western margin of the Bushveld Complex, Southeastern Botswana. *Journal of African Earth Sciences*, 7(4): 641–651.

Dorland, H.C., Beukes, N.J., Gutzmer, J., Evans, D.A.D. & Armstrong, R.A. 2006. Precise SHRIMP U-Pb zircon age constraints on the lower Waterberg and Soutpansberg Groups, South Africa. *South African Journal of Geolog:*, 109, 139–156.

Proceedings of the first Southern African Geotechnical Conference – Jacobsz (Ed.)
© *2016 Taylor & Francis Group, London, ISBN 978-1-138-02971-2*

Waste dump and stockpile management

D.A. Olivier

SRK Consulting, Johannesburg, South Africa

ABSTRACT: Waste dumps represent significant structures on most open pit mining sites. The management of mine waste dumps has historically been assigned to the on-site mine planning and operations departments. In recent years, the focus has turned to a more risk based approach to waste dump management and the auditing of these man-made structures. A risk based classification study of mine waste dumps was carried out by the Mine Waste Rock Pile Research Committee of British Columbia (1992) and Blight & Amponsah-Da Costa (1999) carried out a study to determine the influence of erosion on slope stability. The British Columbia study identified that certain technical issues would remain unsolved, particularly where it is anticipated that one or more adverse conditions may occur. A risk based conceptual evaluation system to determine the likelihood of waste dump slopes being unstable was developed. This methodology or risk based audit procedure is continuously developing to encompass changing on site requirements on mining operations.

1 INTRODUCTION

Global resource pressures have resulted in larger and deeper open pit mines. This, along with environmental and social pressures has also created a premium on waste dumping sites and stockpiles on mine sites. The results are higher and more complex waste dumps and material stockpile structures than ever before.

Historically, the design and management of waste dumps and material stockpiles has been the responsibility of mine planning and mining operation engineers, with empirical design charts and rules of thumb forming the basis for the design and management of these structures. Substantial research was carried out to determine the significance of water and wind erosion off these structures, and the ideal overall slope angle to ensure long term stability and rehabilitation by Blight & Amponsah-Da costa (1999).

Recently the focus has shifted to a risk based approach to waste dump management and auditing, with caveats often resulting from inadequate design and implementation practices.

2 HISTORICAL BACKGROUND

The mine presented as a case study is located in the Democratic Republic of Congo (DRC). The large mining lease is located on the Copperbelt in the Katangan province.

Initial site inspections of waste dumps and stockpiles show two sites observed in 2009 with one waste dump and a low grade stockpile. Low grade ore is traditionally stockpiled over waste on this site. Figure 1 shows the low grade ore stockpiled on a waste dump, and Figure 2 shows the cracks that resulted in waste material failing at the toe and resulting in the contamination of the low grade ore stockpiled above.

The photograph shown in Figure 3 shows the failed waste section of the dump and the slumped, contaminated low grade ore.

No foundation tests were carried out, and a superficial site selection was accepted based on volumes required to be stockpiled and dumped. Only when a substantial volume of low grade material was contaminated due to the waste material failing, was geotechnical advice sourced. No regular visual examinations were carried out at this stage.

This failure marked the beginning of the implementation and continuous improvement in long term planning, short term onsite planning and

Figure 1. Low grade ore over waste.

Figure 2. Cracks in low grade stockpile material.

Figure 3. Failed waste area and slumped low grade ore.

monitoring of waste dumps and stockpiles on this lease area.

3 DUMP GEOMETRY

A typical cross section through an existing waste dump is included as Figure 4. Waste dump geometry varies considerably from site to site. The waste material on this site is end tipped and dozed, and initially rests at a repose angle of between 35° and 37°. The slopes inevitably slough down to angles of between 32° and 34° over time.

4 MANAGEMENT AND IMPLEMENTATION STRATEGY

During discussions with mine personnel, it became evident that research was required to design and implement an auditing system to ensure adequate management and safe construction of the large waste dumps and stockpiles structures planned for the operational site.

4.1 Methodology

The agreed methodology required research into industry accepted guidelines for assessing and quantifying dump stability hazards and the associated risks. The research material included:

– The American Society of Testing and Materials Standards;
– United State Army Corps of Engineers— Standard for Geotechnical Investigations 1991;
– The Stability of Waste Rock Embankments— British Columbia 1992; and
– Deformation and Monitoring of waste dump slopes—SRK.

During the research process it became apparent that the following requirements needed to be met during any audit procedure on dumps:

– Hazard identification system;
– Associated risk rating;
– Quantify risks and calculate a dump stability class characterization; and
– Apply the dump stability rating to all waste dumps and stockpiles on site.

A list of general hazards applicable to any waste dump or stockpile was sourced and is shown in Table 1.

The hazards are then rated according to certain criteria and a point value is assigned to the hazard. Each hazard is examined on site and the point rating determined for that site. The Dump Stability Rating (DSR) is then summed and a stability class is determined. The appropriate guidelines are set out per dump stability class and are included in Table 2.

4.2 Implementation

Once selected, the auditing system was applied to the waste dumps and stockpiles in the mining lease, and a base line assessment was compiled for

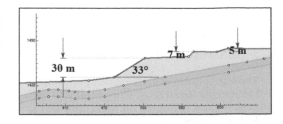

Figure 4. Typical cross section through an existing waste dump.

all active and inactive dumps and stockpiles. The results proved to be that the existing waste dump had historically presented periodic signs of failure when slope heights had reached approximately 15 m. There was clear correlation between the slope height, the onset of the wet season and dump instability.

Evaluating the parameters presented in Table 1, it became clear that the fundamental requirements during the planning stage of waste dump management had not been met. The following shortcomings during the base line assessment were identified:

– There was no foundation data;
– The Environmental and Social Impact Assessment (ESIA) topsoil removal requirement was not completed successfully;

Table 1. WRD and stockpile general hazard list.

Hazard	Action	Output
Foundation Slope	Contour Plans	Gradient Analysis
Foundation Material	Geological and geotechnical mapping Test pits Trenching	Lithology Geotechnical soil properties Bearing capacity assessment
Depth of Weak Material	Test pits	Lithological column
Foundation Sensitivity to Groundwater	Topography Mapping groundwater sources Flow Data	Hydrology map and hydrogeology
Foundation Sensitivity to Pore pressure	Boreholes Piezometers	Hydrology map and hydrogeology
Embankment Height	Survey	Embankment contour map
Embankment Slope	Survey Slope stability assessment	Embankment footprint Factor of safety determination
Embankment Volume at Risk Unconfined toe area Confined toe area	Survey and Mapping	Embankment contour map
Embankment Material	Drill core geotechnical mapping / logging	Waste material geotechnical properties
Embankment Construction	Observation and monitoring	Waste dump material map
Compliance to embankment design parameters	Observation and monitoring	Compliance report
Embankment Advance Rate	Survey Observation and monitoring	Waste dump advance rate plan
Embankment Sensitivity to Precipitation	Rainfall statistics	Probable Maximum Flood and Probable Maximum Precipitation graphs
Embankment Sensitivity to Pore Pressure	Piezometers	Hydrology map and hydrogeology
Embankment Sensitivity to External Source of Water	Survey Observation and monitoring	Mine plan/including waste dump and tailings storage facility (TSF)
Seismicity	Regional probabilistic seismic analysis	Peak Particle Accelerations

Note: this table was derived from the British Columbia Mine Waste Rock Pile Committee (1992) interim guidelines listed in the references

– There was no waste dump plan other than a volume based requirement to dump waste somewhere, and no benches were planned;
– The waste material was not segregated and large volumes of weak talc were being deposited on the dump, with no idea where in the dump the talc was located;
– The dump site was down valley over numerous deep water channels that became flowing streams during the wet season; and
– The advance rate was not monitored.

The above "non-conformance" to industry accepted standards had detrimentally affected the stability of the existing waste dump, and would continue to do so for the foreseeable future. The audit provided a window to observe what influence the criteria chosen would have on the stability of the dump.

The base line audit exposed flaws in the planning stage of waste dump and stockpile management, but it also provided a platform on which a strategy could be developed to improve the shortcomings and address them at the appropriate stage of expansion of the mining process, with the associated dumping volumes. Audit frequency was considered to be biannual, in line with industry standards.

A foundation database was developed to determine the foundation material strength characteristics. Ideally, all new dump sites would be identified according to a set of requirements, as listed in Table 1, well in advance of dumping in order to obtain foundation samples for testing to improve the database.

This proved to be a challenging task, as at least six dumps would be developed before mine planning would be able to select a site well enough in advance to facilitate foundation testing, However, it was eventually achieved, and planning protocols and monitoring have improved vastly on the mine.

4.3 Planning and construction

A strategy to implement a traceable audit procedure was implemented on the mine site after discussion with relevant parties, utilizing the data previously discussed.

The results of sustained focus on the audit criteria can be seen, as focus is sited on the waste dumps and stockpiles at least twice a year during the audit process. Mining personnel and dump truck operators are included in the biannual feedback meetings, and contribute positively towards waste dump and stockpile management and strategies for improvement.

Table 2. Dump stability class.

Dump Stability Class	Relative Failure Hazard	DSR Range	Guidelines for Preliminary Site Characterization
I	Negligible	<300	Basic site reconnaissance and baseline documentation Minimal laboratory testing of foundation soils and waste materials
II	Low	300-600	Basic site reconnaissance and baseline documentation, followed by: Thorough site investigation, including terrain mapping; Test pitting and sampling may be required to supplement mapping; and Limited laboratory index testing.
III	Moderate	600-1200	Basic site reconnaissance and baseline documentation, followed by: detailed phased site investigation; initial investigation phase including test pitting; advanced investigation phases including drilling, undisturbed sampling; some instrumentation (e.g. piezometers) likely required; and comprehensive laboratory testing program.
IV	High	>1200	Basic site reconnaissance and baseline documentation, followed by: detailed phased site investigation; initial investigation phase including intensive test pitting/trenching; advanced investigation phases likely requiring drilling, undisturbed sampling; instrumentation will be mandatory (piezometers, radar, extensometers); and comprehensive, phased laboratory testing program.

Note: this table was derived from the British Columbia Mine Waste Rock Pile Committee (1992) interim guidelines listed in the references

5 CONCLUSIONS AND RECOMMENDATIONS

5.1 *General*

Although the criteria selected for this particular mine site tends to be more qualitative than quantitative, the methodology in selecting the audit criteria is sound. It is based on many years of experience from the Mines Department of British Columbia involved with the development of a set of criteria that could be assigned a points system to assist with rating dumps and stockpiles. While the rating system currently only gives a snap shot of what the rating is at any one time, it does focus attention on the areas that can be improved.

The rating system has shortfalls, such as, future plans are not considered during the audit process and the ratings assigned only consider the status of the dump or stockpile on the day of audit. The audit process in itself provides the tools to develop a system that works for the personnel on mine sites that require assistance with the management of these large structures.

On this particular site, dump operators are carrying out daily visual inspections prior to going on shift, and these observations are being added to a database and shared on their SharePoint system. While these observations only have minor geotechnical input, it has spurred the geotechnical staff on to add their geotechnical observations to the existing database. Critical observations such as cracks, displacement or excessive slope height are communicated immediately, and the follow up report is added to the SharePoint system.

The audit process on this mine will continue to evolve as the mine develops, and more relevant information is sourced to improve the day to day management of these structures, improving their stability over the long term.

Often waste dump failures have resulted from inadequate design and implementation practices. Based on the experience gained from this case study and other similar case studies, and the research undertaken to provide the site with a comprehensive design and construction guide, the following design guidelines are proposed:

5.2 *Recommended guidelines*

A guideline document has been prepared for the mine site and comprehensively covers the suggested requirements. The guide is outlined briefly and includes the following:

- Planning
  - Basic design considerations
  - Mining factors
  - Physical constraints
  - Environmental impact
  - Short and long term stability
  - Social considerations
- Site characterization
  - Geotechnical
  - Hydrology and climate
  - Bedrock geology and tectonics
  - Surficial geology and soils
  - Hydrogeology
- Material properties and testing
  - Foundation soils
  - Foundation bedrock
  - Mine rock
  - Overburden and waste materials
- Mine dump classification
- Mine dump construction descriptions are presented as Figure 5
- Factors affecting stability
- Dump configuration
- Foundation slope and degree of confinement
- Foundation conditions

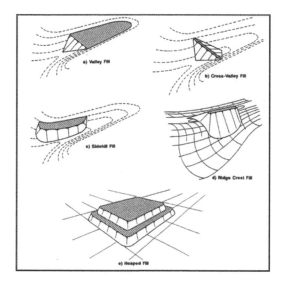

Figure 5. Mine dump descriptions.

– Bearing capacity estimations are vital on soil foundations—no matter what thickness, as this will be the weakest link
– Method of construction
– Piezometric and climatic conditions
– Dumping rate
– Seismicity and dynamic stability
– Classification scheme

– Dump Stability Rating (DSR)
– Dump stability class and the geotechnical investigation guidelines per class
– The application of the ratings during the design process—level of effort required
– Risk assessment
– Safety
– Risk to infrastructure
– Environmental risk
– Construction
– Foundation preparation
– Surface water control
– Construction methods
  – Ascending
  – Descending
– Stability analysis
– Embankment failures
– Foundation failures
– Analysis methods
– Interpretation of stability analysis results

REFERENCES

Blight, G. E. & Amponsah-Da Costa, F. 1999. *Improving the erosional stability of tailings dam slopes*.
British Columbia Mine Waste Rock Pile Research Committee. 1992. *Mined rock and Overburden Piles—Interim guidelines*.
US Army Corps of Engineers. 1991. *Geotechnical investigations*.

Proceedings of the first Southern African Geotechnical Conference – Jacobsz (Ed.)
© 2016 Taylor & Francis Group, London, ISBN 978-1-138-02971-2

Geotechnical characterization and modelling of the Mandena heavy mineral sand deposit

L.I. Boshoff & A.B. Bracken
SRK Consulting, Johannesburg, South Africa

F. Hees
Rio Tinto Iron and Titanium-RTIT, Johannesburg, South Africa

ABSTRACT: Biological cementation of the Mandena heavy mineral deposit in Madagascar was causing difficulty in dredge mining and hence in planning and control of production rates. An investigation was initiated to better evaluate the geotechnical properties of the orebody sands. The investigation was aimed at defining the distribution and geotechnical properties of the cemented strata, in order to generate a site specific geotechnical model for the Mandena deposit. The geotechnical characterisation was developed using an integrated approach of several ground investigation techniques such as piezocone (also known as CPTu) testing, geotechnical drilling, SPT testing as well as laboratory testing. The final outcomes of the investigation included a 3D geotechnical model showing the distribution and extent of the cemented areas as well as the typical geotechnical parameter ranges associated with each of the defined geotechnical units. The interpretations developed were incorporated into the mine and production planning to ultimately optimise operational planning and control.

1 INTRODUCTION

QIT Madagascar Minerals (QMM) is a heavy mineral sand mining operation at Mandena, near Fort Dauphin in Madagascar. During the final planning and commissioning stages of the project, the need to fully characterise the unusual geotechnical conditions was identified and a geotechnical investigation was initiated.

2 NATURE OF THE PROBLEM

During the commissioning phase it was found that the mining production was being impacted by biologically cemented/indurated sands layers encountered within the sands. Upon closer inspection of the indurated horizons encountered within the dredging face it was discovered that these layers possess a high loading of both bacteria and fungi. It appears that the bacteria produces an extracellular (outside the cell) polysaccharide i.e. sugar and carbon salts which form a jelling biofilm to trap nutrients. There are a number of different bacteria. The fungi on the other hand produce a network of fibrous strands that bind the sand grains together. Thus the "apparent cohesive" strength or binding of the indurated sand is probably contributed by both bacteria and fungi (Lynn, 2008).

The main challenge of this project was posed by the many unknowns surrounding the nature and extent of this biological cementation that was presenting a geotechnical problem for the mining process. Typically the following questions arose:

- Is this a biological, geological or geotechnical problem, or all of the above?
- How does one go about modelling when not necessarily governed by apparent geological processes or markers?
- What tools and techniques would be best utilised to cope with the variable conditions being investigated? How effective will they be?
- How does one ensure the tools and techniques used can provide useful data that can be verified?
- How does one combine all the available data collected to generate a practical geotechnical model that will assist in a mining production environment?
- How does one approach the modelling of geotechnics and relate this to the geological resource model and mine plan?

3 DATA GATHERING AND EVALUATION

3.1 General

The scoping of the investigation was completed during various brainstorming sessions considering existing data, the most crucial engineering parameters required for mine and dredge design and also which geotechnical investigation techniques available could potentially obtain the needed data effectively. The scoping study also included trials using different rotary and vibrocore drilling techniques to confirm which would be most appropriate to use.

This process culminated in an investigation methodology involving two phases. Phase 1 focused on high density geotechnical rotary core drilling, logging, sampling and Standard Penetration Testing (SPT) as well as piezocone with pore pressure measurement (otherwise known as CPTu) testing within a limited area. Soil laboratory testing and rock testing was also completed on samples retrieved from the boreholes. It should be noted that following the trials, it was decided not to use a vibrocore drill as it was found that the vibration of the drill broke the intergranular bonds in the indurated layers, resulting in the indurated materials being recovered as soils.

Phase 2 focused on a lower density of testing on the remainder of the ore body. During Phase 2 all available historic information was used, including existing exploration borehole log data, where appropriate, to generate a 3D geotechnical model and typical geotechnical parameter ranges for the most important geotechnical properties.

In-situ measurements using a piezocone (CPTu) rig provided direct measurement of cone resistance (q_c), sleeve friction (fs) and pore water pressure (u), which can be used to derive various geotechnical parameters. Due to the expected variable conditions refusal of the CPTu equipment was expected on the hardest indurated layers. A standard drilling rig or hydro vibracore rig (sonic method) was therefore provided to advance through the hardest layers when encountered.

Selected CPTu locations were paired with high quality rotary core drilling, using a triple tube barrel and a mud flush intended to obtain cores of indurated sands and more cohesive/dense soil layers. Limited SPT testing was also completed in the holes and used to provide a comparison with the CPTu data. Detailed geotechnical logging of the core was completed. Sampling from the core for laboratory testing was conducted to confirm parameters derived from the CPTu interpretations and logging and to determine parameters that could not be derived from the CPTu testing and logging.

The combined Phase 1 and Phase 2 geotechnical database consisted of 147 CPTu locations, 48 Borehole and SPT test locations. More than 200 disturbed samples and approximately 80 core samples were tested in the laboratory.

3.2 Field logging of rotary drilling samples

Poor recoveries obtained during drilling provided little material within the un-indurated zones and this posed various challenges during field logging.

To ensure consistent capturing of information that could later be calibrated and refined when combined with the other data gathered (such as the CPTu and laboratory information) a site specific approach to logging was adopted to ensure a consistent account of the indurated materials were captured. Generally zones registering SPT N_{field} values of over 50 blow counts (noted as "Refusal") were deemed to be of very dense consistency or possibly of extremely soft to soft rock strength with a Uniaxial Compressive Strength (UCS) of 0.5–3 MPa. The presence of rock like indurated materials was logged only if core samples were available to confirm its presence, irrespective of the SPT N_{field} values recorded. Where it could be confirmed from core or SPT samples extracted, the thickness, spacing and depth of indurated beds were recorded.

Based on this approach, zones with similar geotechnical conditions were identified and provided with an overall description. This process was the first step in the generation of a site specific geotechnical framework/unit model.

3.3 Correlation of in-situ testing methods

A numerical correlation between the CPTu derived SPT N_{60} values, q_c and the SPT N_{field} values obtained in the paired locations was conducted.

To obtain a point value to correlate with the SPT N_{field} values recorded a linear average output value from the CPTu at the start and end of the SPT testing depths was assumed.

A graphical presentation of a typical CPT derived SPT N_{60} trend and SPT N_{field} values plotted with depth for test and drilling location No. 42 are shown in Figure 1 below.

Acceptable correlations were found to exist for 58% of the test positions providing correlation coefficients from 0.5 to 0.94. Visual assessment of the trends indicated that in several of the poorer correlating cases, trends were off set, but that overall visually the trends are similar.

Correction factors were applied to the SPT N_{field} values, but none of the standard correction factors (Clayton, 1995) (such as rod energy, overburden and anvil size) could provide a consistent improvement in the correlations obtained using the raw data.

3.4 CPTu interpretations and borehole logs vs laboratory results

Good correlation was found to exist between the rotary drill sample logging, laboratory results

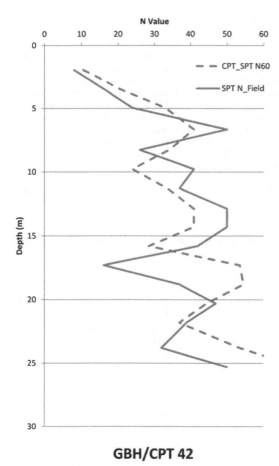

N Value

GBH/CPT 42

Figure 1.   Graphical comparison of CPTu derived SPT N_{60} with SPT N_{field} values-Test location No. 42.

and the soil behaviour classes as defined by soil behaviour charts developed by Robertson *et al.* (1988) for the CPTu.

The CPTu identified sands and silty sands dominating the profile with some materials, typically those found at the base of the probe holes, classifying as silts or sandy clay type materials.

The laboratory UCS test results agreed well with the rock strengths identified during borehole logging with most falling within the strength range (UCS of 0.5 to 3.0 MPa) predicted during logging (Extremely to Very Soft rock).

The permeabilities measured in the laboratory for the core samples were in line with permeability predictions based on the CPTu derived soil behaviour types (sands and silty sand).

However, some exceptions occurred when correlating CPTu information with samples extracted from the paired rotary drill holes indicating that significant variation in induration strength can exist over small distances.

3.5   *Generation of a geotechnical framework/unit model*

The processes described in Section 3.1 to 3.4 led to the definition of a geotechnical unit framework within which further data evaluation could take place. The framework focused mainly on defining zones that would have specific engineering implications for the mining operation.

General internationally accepted guidelines as provided in literature and summarized in the TRH10 (1987) document (Technical Recommendation for Highways-South Africa) relate CPT q_c to SPT N_{field} values and corresponding density descriptions as shown in Table 1 below.

Spikes in q_c values could be correlated with very dense to indurated layers located within generally medium dense to very dense silty sand layers. This finding agreed with what was observed during logging of the boreholes drilled. It also became clear that the CPTu was also more effective in picking up the presence of very thin indurated layers that would not be identified during drilling and SPT testing. The conceptual geotechnical unit framework determined considering all the gathered data available is shown in the Table 2 below.

Simple principles were developed to ensure a consistent evaluation, and optimisation of the geotechnical zone boundaries. These included:

– Adopting CPTu data as the more accurate and continuous record of in-situ conditions due to generally low sample recoveries in the boreholes.
– When delineating and optimising geotechnical zone boundaries density would have priority over descriptive properties (such as colour).
– Zone 3 conditions were inferred where significant spikes in the CPTu q_c could be observed of 30–40 MPa, where vibracore drilling was required or where recovery of core samples confirmed the presence of indurated layers.

Table 1.   Internationally accepted consistency descriptors as a function of SPT N and CPT q_c value ranges obtained (adopted from TRH10).

Density description	SPT N	CPT q_c (MPa)
Very loose	4	2
Loose	4–10	2–5
Medium dense	10–30	5–15
Dense	30–50	15–20
Very dense	>50	>20

Table 2. Geotechnical unit model derived from Phase 1 fieldwork data and observations.

Zone	Name	Geotechnical description	Distribution
1	Aeolian sands	Free flowing to non-free flowing loose to medium dense sands.	Usually the surface deposit. Occasionally absent.
2	Strandline sands	Non free flowing clean to organic, occasionally ferruginised, medium dense to very dense sands.	Usually underlying Aeolian Sands. Occasionally exposed at surface. Seldom absent, but possible.
3	Strandline sands with indurations	Non free flowing medium dense to very dense sands of interbedded with extremely soft (UCS 0.5–1 MPa) to very soft (UCS 1–3 MPa) Indurated sands /Sandstone.	Lenses generally within Zone 2. Occasionally at top or base. Apparently random distribution. Sometimes absent.
*4	Deposit base strandline sands, silts and clays	Interbedded sands, silts and clays of variable consistency.	At base of and occasionally interleaved with Zone2/3. Sometimes absent.
*5	Residual sandy clay	Firm clay, silt or clayey sand.	At base of Zone 2/3 or 4.

*Zone 5 represents the clay basement and is not part of the resource, whereas Zone 4 is a transitional zone between the resource and basement.

– If a section was logged in boreholes as a Zone 2 and the CPTu data indicated hard layers these Zones would be upgraded or adjusted to Zone 3 in accordance with CPTu data.

4 3D MODELLING APPROACH

During Phase 2, Phase 1 interpretations were verified. Phase 2 also included scrutinizing the geological exploration data gathered to aid in 3D modelling. This in itself provided a challenge as geological logging had been completed from a "geological and resource modelling perspective" whereas the "geotechnical zones" were based mostly on engineering properties of the materials as logged and measured using specific techniques.

The geotechnical zone modelling and numerical modelling were approached separately. The measured data was considered hard data, whereas the zone boundary interpretations made were based on a combination of the descriptive and in-situ measured data. Geotechnical zones were therefore modelled as separate "lithologies".

4.1 Geotechnical zone lithology modelling

The geotechnical zone volumes were interpolated using spheroidal variogram settings using constant Kriging and a range of approximately 300 m. The ellipsoid ratios were optimised to reflect the general dimensions of the volumes being modelled (i.e. disk shaped with small Z relative to ellipse radius).

Zones 1, 2 and 3 were modelled separately; however Zones 4 and 5 were combined and modelled as one composite region delineating the clayey deposit base consisting of either marine clays

and/or residual clayey sands. Areas not defined by the interpolants developed from existing geotechnical data were inferred as comprising of Geotechnical Zone 2 type sands.

The interpolations where modelled assuming the following age succession:

– Zone 3 is considered the youngest, generally cutting indiscriminately through all deposited sands, as cementation took place after deposition of the strandline deposits.
– Zone 1 to Zone 5 (excluding Zone 3) was assumed to be increasing in age, however Zone 2 could be interleaved with Zone 4 type materials (clays and silty sand lenses) due to the nature of deposition of the strandline sands.

Correlation between the geological database generated during exploration drilling and the Geotechnical database were considered. After some review it was confirmed that the merging of the two datasets will not be ideal and would possibly skew the presentation of the geotechnical information gathered. The two data sets were therefore kept separate from each other within the modelling software. However, the geological data had to be utilised to some extent during modelling to supplement the limited geotechnical data points available. To correlate indurated conditions captured within the geological database an approach introduced by project geologist Hees (2010) was used to include an "induration" interval to identify zones where described/indicated and inferred indurated conditions can be expected. This interval could be used to group key lithologies within the Leapfrog modelling software to reflect the distribution of the indurated conditions as per the geological database, which could then be compared with the geotechnical zone boundaries. In areas where little data was available the Zone 3 model was extrapolated by utilising the reviewed and grouped geological data as well as imported sections. However, areas existed where indurated conditions were identified that were not captured in the geological database. Geotechnical data therefore took precedence in most cases when interpreting the Zone 3 materials boundaries.

4.2 *Numerical data analyses in 3D*

A design chart was developed by the dredge designers matching SPT N number to output in tonnes per hour (tph) for the existing dredge used when production issues ensued. For this reason the first attempts in modelling of geotechnical properties was focused on the SPT N numerical data as known relationships existed. Numerical modelling of the SPT N_{60} and SPT N_{field} values were undertaken and run using similar settings to the lithological model.

A separate 3D interpretation was made using only the numerical data gathered to interpolate areas associated with specific numerical parameter thresholds within specific orebody areas. Numerical data within each modelled zone could also be extracted and analysed statistically. The numerical model can currently provide 3D contour maps of a specific geotechnical parameter modelled at any level or on any section through the orebody volume. The numerical interpretations do not agree in all areas with the zone interpretations as selected locations were modelled using only descriptive logging information with no numerical geotechnical background data (e.g. hydrogeological borehole positions or exploration data). Gaps in the numerical data coverage still exist however the process demonstrated that numerical modelling of a geotechnical parameter can be very useful in this case.

5 MODELLING RESULTS

Of most significance to the mining operation is the spatial distribution of Zone 3, which is the most indurated zone. Modelling of Zone 3 showed it to consist of relatively continuous, but gently undulating indurated beds and/or lenses across the majority of Mandena area. It can be absent/under-developed in certain locations. The macro appearance of the indurated zones is that of large to small scale lensing rather than the existence of distinct indurated horizons but these macro lenses can be made up of various indurated horizons of variable thickness.

Zone 3 conditions are expected to be concentrated around elevations ranging from +4 m msl to –1 m msl.

Table 3 provides the numerical information extracted for each geotechnical zone modelled providing parameter ranges and typical values of the SPT N_{field} and SPT N_{60} values measured within these areas.

Numerical data available for the indurated conditions (Zone 3) is likely to be less representative due to early refusal of the CPTu, and the relatively small number of boreholes drilled also limiting conventional SPT measurement.

Laboratory testing results have confirmed the expected UCS values associated with the most indurated Zone 3 materials. At present the maximum strength/hardness of the indurated materials recovered and tested can be assumed to be around 2 MPa, as measured during the Phase 1 testing campaign.

Although the properties of Zones 2 and 3 are highly variable, the median SPT N value exceeds 30 for Zone 2 for the majority of the orebody.

The Zone 3 median SPT N value is estimated at 53 (possibly higher, if adjustments are made for rock like materials on which the CPTu and SPT refused-Table 3 refers). Current measured SPT N values for Zone 3 ranged from as low as 3 to 100.

Table 3. Numerical SPT N_{field} and SPT N_{60} data extracted within each geotechnical zone.

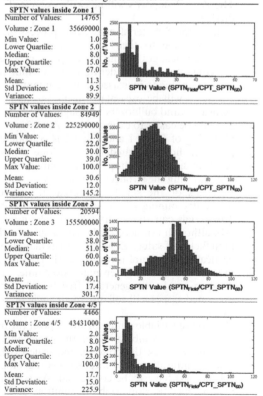

SPTN values inside Zone 1	
Number of Values:	14765
Volume : Zone 1	35669000
Min Value:	1.0
Lower Quartile:	5.0
Median:	8.0
Upper Quartile:	15.0
Max Value:	67.0
Mean:	11.3
Std Deviation:	9.5
Variance:	89.9

SPTN values inside Zone 2	
Number of Values:	84949
Volume : Zone 2	225290000
Min Value:	1.0
Lower Quartile:	22.0
Median:	30.0
Upper Quartile:	39.0
Max Value:	100.0
Mean:	30.6
Std Deviation:	12.0
Variance:	145.2

SPTN values inside Zone 3	
Number of Values:	20594
Volume : Zone 3	155500000
Min Value:	3.0
Lower Quartile:	38.0
Median:	51.0
Upper Quartile:	60.0
Max Value:	100.0
Mean:	49.1
Std Deviation:	17.4
Variance:	301.7

SPTN values inside Zone 4/5	
Number of Values:	4466
Volume : Zone 4/5	43431000
Min Value:	2.0
Lower Quartile:	8.0
Median:	12.0
Upper Quartile:	23.0
Max Value:	100.0
Mean:	17.7
Std Deviation:	15.0
Variance:	225.9

It is therefore clear from the data evaluation shown in Table 3 that the numerical data and the zone model has been successful in delineating distinct geotechnical zones that can be applied to the mine planning and design process.

6 APPLICATION TO MINING OPERATION

Based on existing design charts the dredge should be able to achieve on average around 1900 tph Zone 2 materials and no more than 1300 tph in the Zone 3 materials. By applying typical expected weighted SPT N values to certain mining blocks (e.g. by determining the weighted average volume each geotechnical zone represents in a specific mining block), one can more accurately predict advance rates and set realistic production targets and/or plan which areas should rather be earmarked for mechanical pre-conditioning to obtain the desired throughputs. This information can be/was also used to specify a larger dredge that would be able to deal with most of the geotechnical conditions predicted by the model.

Geotechnical parameters associated with the various zones modelled and the nature of the zones as reviewed in 3D have the following implications for the mining operations:

- The Zone 1 deposits will be easily excavated or dredged.
- Zone 4 and 5 materials may cause process inefficiencies due to their more cohesive nature causing blockages of, and build up on, processing equipment and hindering of heavy mineral separation due to increased slimes contents.
- All areas will be excavatable using mechanical excavators and bulldozers, although Zones 1 and 2 would be easier to excavate than Zone 3.
- Dredging of Zone 1 and Zone 2 is considered possible, however dredging of Zone 3 can result in significant drops in productivity when encountering the stronger indurations.
- In terms of the stability of excavated slopes in the dry mining pits it is expected that Zone 3 materials can stand vertically over long periods, although not necessarily with a sufficiently high factor of safety. Zone 2 materials may stand vertically for several metres over short periods but will collapse to lower angles over time. The Aeolian sands are generally free flowing and excavated slopes will collapse to an angle of 30° to 35°.
- Due to the variable nature of the deposit (introduced by the combination of the geotechnical zone thicknesses, the presence and scale of the Zone 3 indurations and localised hydrogeological factors) have to be considered. As a result no one specific slope stability solution or FoS can be assigned and slope stability management must be approached as a daily planning and production requirement requiring on-going assessment by a competent person.

7 CONCLUSIONS

The Mandena case study demonstrates the importance and necessity of geotechnical investigations in the early phases of a project to ensure understanding of the geotechnical parameters of an orebody and the potential limitations this could pose to the chosen mining process. Unfortunately the importance of these investigations in the early phases of mining projects (specifically referring to heavy mineral sand mining projects) are commonly forgotten or focus is only placed on civil or resource estimation applications rather than on the understanding of how geotechnical conditions could potentially impact on the mining process.

The case study further demonstrates the importance of using an integrated approach including various geotechnical and geological techniques to successfully characterize unusual and variable geotechnical conditions.

By using an integrated approach and combining data from geological and geotechnical sources it was possible to generate a geotechnical zone model in 3D into which any of the in-situ measured and/ or derived parameters obtained from CPTu testing can be imported, modelled and/or data extracted and analysed statistically for application to the mining requirements.

This integrated method of investigating, defining and presenting large scale, and unusual geotechnical problems allowed the Mandena operation to enhance its production planning capabilities.

REFERENCES

Boshoff, L. & Bracken, A. 2011. *QMM Mandena Phase 1 Geotechnical Investigation-Final Draft Report*, SRK Report Number 414630/2, Prepared for Rio Tinto.

Boshoff, L. & Bracken, A. 2012. *Project TiO₄- Mandena North West and North East Satellite Areas Geotechnical Investigation Draft Report*. SRK Report Number 437868/3.1. Prepared for Rio Tinto Management Services.

Boshoff, L. & Howell, G. 2011. *Pond level raising review—Mandena, QMM*. SRK Report Number 437868/1. Prepared for Rio Tinto Management Services.

Boshoff, L., Bracken, A. & Singh, R. 2012. *Desk top Review of available data for Project TiO₄- Goetechnical Aspects*. SRK Report Number 437868/2-Final Rev1. Prepared for Rio Tinto Management Services.

Campenella, R.G. & Robertson, P.K. 1988. *Current status of the piezocone test*. Penetration Testing, ISOPT-1, De Ruiter (ed.).

Clayton, C.R.I. 1995. *The Standard Penetration Test (SPT): Methods and use.* Construction Industry Research and Information Association Report 143, 143pp. London: CIRIA.

Committee for State Road Authorities. 1987. *Technical Recommendations for Highways, TRH10-Design of road Embankments.* Pretoria: Department of Transport.

Hees, F. 2010. *Orebody knowledge workshop "Biocrete Model Stream 2 Area-Mandena Project".* Presentation as prepared for Rio Tinto.

Lunne, T., Robertson, P.K, & Powel, JJM. 1997. *Cone Penetration Testing in Geotechnical Practice.*

Lynn, B. 2008. *Summary Report on the Results of Laboratory Testing Carried out on Indurated Sand Samples.* Ref 30141. Prepared for Rio Tinto.

Meigh, A.C. 1987. *Cone Penetration Testing Methods and Interpretations,* CIRIA.

Rio Tinto. 2005. *QMM feasibility report—Geological Investigation & Resource Estimate Report.* Rio Tinto.

Robertson, P.K. 1992. *Estimating in-situ soil permeability from CPT & CPTu.* Gregg Drilling and Testing Inc. California, USA: Signal Hill.

Watts, Griffis & McOuat.1988. *Evaluation of Heavy Minerals Deposits of the Fort-Dauphin Exploration Area.*

Proceedings of the first Southern African Geotechnical Conference – Jacobsz (Ed.)
© *2016 Taylor & Francis Group, London, ISBN 978-1-138-02971-2*

Tailings as hazardous waste: Implications on design and construction management

J. Thuysbaert & W. Kruger
SRK Consulting, Johannesburg, South Africa

ABSTRACT: Updates of regulations, guidelines and standards for the design, operation and general management of tailings storage facilities are under development in South Africa. The legislation of South Africa emphasises the need for all reasonable measures to be undertaken to prevent the pollution of any water resources from occurring, continuing to occur or re-occurring. This diverges from the (prevalent) approach of remedying the effects of a polluted system. Historically, legislation excluded or, rather, did not explicitly include all tailings as a potentially hazardous waste that needed to comply with the minimum requirements for the disposal of waste. Recently, however, amendments to legislation specifically classify tailings as a hazardous waste that has the potential to pollute resources. As a result, the disposal thereof must comply with the stipulated minimum requirements outlined by legislation. This paper presents a case study of a recently designed and constructed tailings facility that entails novel solutions to complexities similar to those that could arise in the future from the application of some of the most recent legislation, and the implications on the holistic approach to the design and management of tailings facilities.

1 INTRODUCTION

1.1 *Legislation, industry guidelines and standards*

The historic approach to waste facilities in South Africa was based on pollution mitigation methods published by the South African Department of Water and Forestry (DWAF) in the *Minimum Requirements for Handling, Classification and Disposal of Hazardous Waste 1998* which was the industry guideline until recently.

A significant increase in awareness of the limited renewable water resources in South Africa; continued experienced gained in the application of the guidelines and regulatory standards; and the gazetting of the *National Environmental Management Waste Act 2008*, which allows for the establishment of regulations, has led to the South African Department of Environmental Affairs (DEA) publishing the following regulations for the classification and containment of waste:

– *Waste classification and management regulations 2013 (reg 634)*
– *National norms and standards for the assessment of waste for landfill disposal 2013 (reg 635)*
– *National norms and standards for disposal of waste to landfill 2013 (reg 636)*

Over and above the changes to the legislation in South Africa, revisions are also being made to the *Code of Practice for Mine Residue Deposits 1998* which is the South African standard for management of residue deposits. Individual corporations are also updating their own safety and safety management standards. The emphasis of these revisions is a holistic management approach whereby an "Engineer of Record" is appointed for a residue deposits of which the responsibility for all aspects including the monitoring and continual design of the entire system as well as all its components falls within the ambit of this "Engineer of Record" appointment.

1.2 *Relevance to geotechnical engineering*

The repercussions of the publication of the regulations listed above on the mining industry and more specifically the civil engineers that design and manage mine Tailings Storage Facilities (TSFs) are that tailings material which had previously typically been classified as presenting a low risk to polluting water resources, and therefore required minimal to no containment of the material and its leachates, is now more commonly being classified as a hazardous waste which requires that the storage facilities of these materials now need to comply with stipulated minimum engineering design requirements that entail lining the facility with a containment barrier. As a result many civil engineers that work within mining and similar industries will now be faced with the complexities that these containment barriers will bring and will need to do so as not to exasperate the already potentially major economic impacts that arise with having to line these TSFs with containment barriers.

The design of the facility presented in this paper was done before the legislation discussed above had come into effect. The outcome of the geochemical investigation of the material that would be stored by the facility required that the facility be lined with a containment barrier and many of the design and construction challenges faced in the project are likely to be faced in larger TSF projects in the future. Also, the solutions found in both the design and management of the construction were novel and interesting in their own right.

2 GROOTEGELUK CYCLIC PONDS FACILITY

2.1 Project location

The Mine is located 25 km west of Lephalale in the Limpopo Province of South Africa. Eskom's (operational) Matimba and (under construction) Medupi coal power stations are located to the east and south of the Mine respectively and receives coal from Grootegeluk for power generation.

2.2 Project description

Raw coal is processed through a number of plants operated by the Mine, coal slurry and fine discard are by-products from this process. The existing process coal slurry disposal facility is a conventional TSF which is a combination of two TSFs known as Dams 1 and 2. In anticipation of the design life of the current facility nearing and the endeavour of the Mine to have zero surplus at the end of their coal production process, the Cyclic Ponds Facility (CPF) was conceived (refer Figure 1), which receives a slurry of coal fines from the Mine's plant via a slurry delivery pipeline. It is allowed to dry and once a suitable moisture content has been attained—the coal fines are mechanically reclaimed and hauled to the power plants for power generation.

The ponds are operated in a cyclic manner so that sufficient deposition space is always available. A sophisticated drainage system conveys seepage water to a Return Water Dam (RWD) complex for reuse by the Mine. The facility prevents seepage of contaminated effluent into the ground due to an advanced liner barrier system. Construction of the facility is performed by Aveng Grinaker LTA (AGLTA), commenced in October 2012 and concluded in November 2015. The design consultant is SRK Consulting.

Figure 1. Aerial photograph of the Cyclic Ponds Facility (CPF).

2.3 Geochemical investigation on coal slurry

Geochemical testing was done on the coal slurry that was to be stored in the CPF prior to commencement of the design (SRK Consulting 2011). The geochemical testing indicated that the coal slurry is likely to generate acidity and release inorganic salts, metals and trace organics into the environment, in the absence of an engineered containment barrier.

It was therefore recommended that the proposed plans incorporate an engineered lining system with leachate collection.

2.4 Liner barrier system

The CPF consists of four 10 m high ponds separated by three divisional walls each provided with a decant system and sophisticated above and under liner drainage systems which conveys decant and drainage water to the Return Water Dam (RWD) complex. The liner configuration installed on the Cyclic Ponds is shown in Figure 2.

Geosynthetic Clay Liner (GCL) is a high performance liner comprising a geotextile or geomembrane carrier component integrated (needled) with a layer of low permeability Bentonite.

The GCL is installed directly on top of subgrade in the Cyclic Ponds to retain seepage inside the facility. Low Linear Density PolyEthylene (LLDPE) geomembrane is a smooth, black, high quality geomembrane produced from specially formulated, virgin polyethylene resin that will waterproof the ponds. Bidim is a continuous filament, non-woven, needled punched polyester geotextile used as a puncture protection layer directly over the primary geomembrane liner. A graded sand bedding is provided to protect the geomembrane from direct pressure from the stone layer. These layers both serve as a filter which drains seepage water to above liner drains and also dissipates excess pore pressures during the deposition phase. The coal discard layer does not function as a filter but prevents erosion of the stone layer. Snow netting placed serves as a visual barrier during the reclamation phase.

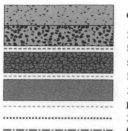

Coal Fines

200 mm Coal discard
Snow netting
500 mm stone layer
Bidim A6
300 mm graded sand
Bidim A6
2.0 mm LLDPE geomembrane
GCL liner

Figure 2.   Cyclic ponds liner configuration.

The RWD complex comprises two identical compartments of equal volume and geometrical configuration that can be isolated at the drainage collector splitter box during operation, de-silting and maintenance. The liner configuration installed on the RWD complex is shown in Figure 3.

HI-DRAIN which serves as a leakage detection layer, is a durable, heat formed multiple-cuspated synthetic sheet which provides a high capacity non-restricted drainage path and placed between the 1.5 mm (secondary) and 2.0 mm (primary) High Density Polyethylene (HDPE) geomembrane liners. The cusps (hollow parts) are filled with graded sand for additional structural support.

Due to the potential for silt to build up in the RWD compartments a robust 150 mm thick Oxy-fibre reinforced concrete protective layer is placed on top of the liner system.

2.5   Long term stability of the divisional walls

To reduce the risk of mechanical equipment damaging the liner during the reclamation phase the design of the Cyclic Ponds is such that the liner runs continuously underneath the base of the divisional walls.

Slope and slip stability assessments performed for conditions where one compartment is full and the adjacent compartment is empty requires the divisional walls to comprise of a profiled subgrade base as shown in Figure 4. This to resist the full lateral loading force applied by the coal slurry and also considers the presence of the weak layer comprising the liners below the embankment.

The matter of soil-geomembrane and geomembrane-geotextile interface friction is well documented all over the world. The slope stability is governed by the interface strength of the liner interface with the lowest friction angle. The interface friction angle between the LLDPE and geotextile (Bidim) would be similar or greater than that of the interface friction angle between the GCL and LLDPE because of the presence of bentonite in the GCL which lowers interface friction when hydrated.

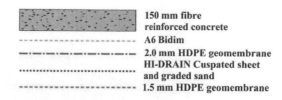

150 mm fibre
reinforced concrete
A6 Bidim
2.0 mm HDPE geomembrane
HI-DRAIN Cuspated sheet
and graded sand
1.5 mm HDPE geomembrane

Figure 3.   Return water dam complex liner configuration.

Shear tests performed on the GCL/LLDPE interface resulted in a lower bound effective stress peak friction angle of 10 degrees and a lower bound effective stress residual shear strength of 7.2 degrees. For the design, an effective friction angle of 10 degrees is used for the GCL/LLDPE interface. A check is performed using a friction angle of 6 degrees which is significantly lower than 7.2 degrees.

Due to the low permeability of the liner system, the phreatic surface can build up within the divisional wall resulting in compromised slope stability. Therefore a 10 m wide blanket drain and a 300 mm wide vertical drain are provided within each divisional wall. This internal drainage system will adequately draw down the phreatic surface and prevent seepage into adjacent ponds.

2.6   Liner Construction Quality Assurance (CQA)

To ensure the contractor maintains installation quality of liner works as per the design specification during construction, inspections are regularly performed by the Resident Engineer (RE) of which the following aspects receive specific attention (SRK Consulting 2015);

– Ensure the subgrade surface finish prior to the deployment of the liner system is even, compacted to the specified density and to the grade shown on the drawings. The RE inspects the earthworks prior to installation of any liner sheeting which includes inter alia the removal of any protrusions, stones, roots, vegetation and other objects that may be potentially detrimental to the integrity of the geomembrane.
– Approval of construction method statements which include panel layouts.
– Verify the integrity of the contractor's material inventory. This is performed when the material is rolled out to protect it from premature exposure to the elements prior to placement.
– Ensure geomembrane installation is in accordance with SANS/10409.
– Geosynthetic Clay Liner (GCL) overlap width—Bentonite paste is used at overlaps.
– Ensure the GCL is not exposed to sunlight for extended periods and not affected by wet conditions during deployment.

Figure 4.   Photograph of cyclic pond divisional wall no.1 during construction showing the profiled base.

– Ensure geomembrane rolls are strategically positioned such to minimize the amount of handling required and that seaming follows placing as soon as practicable.
– Ensure no bridging or 'trampolining' (the effect that occurs when temperature changes cause the geomembrane to contract and raise out of the depressions formed as part of the earth profiling and which can lead to stressed conditions under load) occurs in the geomembrane by incorporating suitable slack allowances.
– Ensure no transverse welding occurs on wall embankments.
– Ensure panel placement is not performed in the presence of excessive moisture, in an area of standing water or during high winds.
– Ensure the method and equipment used to deploy the panels used is such as to not damage the liner or the supporting surface.
– Seaming equipment in compliance with the requirements of Geosynthetic Research Institute (GRI) Test Method GM 19.
– Verify air pressure testing on hot wedge geomembrane seaming in accordance with GRI Test Method GM6.
– Verify spark testing on extrusion fillet geomembrane seaming in accordance with ASTM D 6365.
– Ensure defective or damaged geomembrane portions are replaced or repaired in accordance with SANS 10409.
– Requesting additional (laboratory) testing in accordance with GRI Test Method GM 19 when the RE is not satisfied with the repair works.
– Geomembrane seaming cannot proceed at material temperatures below 0°C or above 75°C. Suitable temperature measuring equipment is provided by the contractor.

2.6.1   *Cyclic ponds divisional wall base CQA*
Construction of the earth profiles at the base of the divisional walls is performed in accordance

with the contractor's method statement approved by the RE. The profiles are set out and excavated according to specified depth, grade and cross section. The edges of the earth profiles are trimmed to have a radius of 200 mm to allow the liner system to sit neatly without 'trampolining' and also allow seaming to occur smoothly. The RE approves the subgrade surface to receive the liner and ensures the depth of the earth profiles is to specification.

The liner is deployed from the eastern and western sides of the divisional walls towards the centre. The GCL is deployed after which the LLDPE is deployed following the same deployment direction as the GCL to cover the GCL. Placement of A6 Bidim follows subsequent to pressure and spark testing of the geomembrane witnessed by the RE. Sand bags are strategically placed in the depressions as shown in Figure 4 to reduce the effect of 'trampolining'. Welds are left open at the centre and outer edges of the divisional wall during the placement of fill material on top of the liner system to allow for free movement of the liner during filling. Articulated Dump Trucks (ADT) strategically end tip fill material on top of the of the liner system. Fill material is pushed by bulldozer in a northerly or southern direction at approximately 5 m intervals. The RE ensures no 'trampolining' or the generation of folds occurs during this process. Selected sand material is placed initially for the first 750 mm to protect the liner system from being punctured by loose stones during filling and compaction. Earth material is subsequently end tipped and compacted to specification in layers of 300 mm to the embankment crest level.

Due to the expansive nature of the LLDPE under heat, no filling on top of the liner occurs when the liner temperature exceeds 30 °C. Filling is performed upon a maximum of four longitudinal Bidim panels at a time to prevent the generation of folds and trampolines and to protect the liner system from being exposed to weather elements.

2.6.2 Return Water Dam (RWD) CQA

Placement of 30 MPa Oxyfibre reinforced concrete in the RWD compartments on top of the liner system is performed in accordance with the contractor's method statement approved by the RE.

The subgrade surface level is controlled by survey prior to deployment of the liner system. The RE ensures the orientation of the HI-DRAIN between the 1.5 mm (secondary) and 2.0 mm (primary) HDPE geomembrane liners is installed with the cusps (hollow parts) facing upwards and filled with graded sand.

A number of storm events caused rainfall water to pond in between placed liner layers during construction. After every rainfall event the RE performs a site assessment and reports all damages to the client. The contractor rectifies the damaged works which is inspected by the RE prior to new works commencing. The RE ensures no water is present between liner layers prior to placement of subsequent layers.

Due to the expansive and contractive nature of HDPE geomembrane under cyclic temperatures no concrete placement is permitted when geomembrane temperatures exceed 30°C. Prior to concrete placement the RE inspects every panel and strictly ensures no folds or 'trampolines', particularly at the toe of embankments, are visible.

Final floating is performed as soon as the concrete starts to set and a curing compound is applied subsequently. Shrinkage cracks are controlled by the provision of approximately 3 mm wide by 40 mm deep saw cut joints cut 4–12 hours after concrete placement, to create panels with a length to width ratio of less than or equal to 1.25. The contractor submits an as-built survey to the RE after concrete placement to ensure concrete tolerances in accordance with SABS 1200 are being adhered to. Visual inspections are also performed by the RE to ensure surface finishing is satisfactory and to verify as-built dimensions are to specification. Concrete panels with dimensions and/or alignments that do not conform to allowed tolerances are demolished and rebuilt to conform to design.

2.7 Environmental benefits of the CPF

There is no permanent storage of coal residue due to the ponds being operated in a cyclic manner therefore the footprint of the facility will not increase and is set for the life of the Mine. The facility will prevent seepage of contaminated effluent into the ground due to an advanced liner system below the footprint of both the Cyclic Ponds and the RWD complexes. A sophisticated drainage system conveys seepage water to a Return Water Dam (RWD) complex for reuse by the Mine.

Large operational pools on top of conventional TSFs cause significant loss of water through evaporation. The aim for the Cyclic Ponds is to dry the coal fines as soon as possible thus strictly limiting the size of the pool during the deposition phase.

2.8 Technological advances of the CPF

The reclamation of coal fines in a continuous operation is in its own right a first. The complex base construction of divisional walls in conjunction with the liner system draped over is a construction engineering feat (sequencing of difficult earthworks with the installation of the liner system during extreme temperatures as well as construction above it eventually).

The design of the evaporative drying of coal fines is based on modern and ongoing research on the subject and science of evaporative drying.

3 CONCLUSIONS

Recent changes in the South African legislation that deal with classification and containment of waste will require lining of the facilities that store tailings with containment barriers. This paper presents design solutions for foreseeable problems that arise from lining facilities. These complexities are primarily due to compromised slope stability and sliding resistance of the retaining walls as a result of the low friction in the weak layer which is the interface between the GCL and geomembrane for this particular case. The paper also presents Construction Quality Assurance (CQA) procedures that deal with the practical problems that arise from the sequencing of difficult earthworks and installation of the liner system during extreme temperatures as well as construction above it.

REFERENCES

Department of Environmental Affairs. (2008). National Environmental Management Waste Act. *Government Gazette*. Cape Town, South Africa.

Department of Environmental Affairs. (2014). National Environmental Management Waste Amendment Act. *Government Gazette*. Cape Town, South Africa.

Legge, K.R. et al. (2014). Progressive interaction in engineering review of license applications as at January 2014, s.l.: s.n.

SRK Consulting. (2011). Geochemical investigation for a proposed coal slurry cyclic dam at Grootegeluk. *Report number 43338/1*. Johannesburg, South Africa.

SRK Consulting. (2015). Construction closure report of the Cyclic Ponds Facility at Grootegeluk Mine in Limpopo. *Report number 43338/11*. Johannesburg, South Africa.

Modelling and design

Proceedings of the first Southern African Geotechnical Conference – Jacobsz (Ed.)
© 2016 Taylor & Francis Group, London, ISBN 978-1-138-02971-2

Geotechnical design using SANS 10160: A comparison with current practice

B.J. Barratt
HHO Consulting Engineers, Cape Town, South Africa

P.W. Day
University of Stellenbosch, Stellenbosch, South Africa

ABSTRACT: SANS 10160:2011 is a South African basis of design standard that uses the partial factor limit states design approach. It differs from more traditional working stress design methods that are based on a global factor of safety. This paper applies the approach advocated in SANS 10160-5 to several design examples, which include various shallow foundations, a retaining wall and an earth embankment. It then compares the results with those from working stress design. The probability of failure and reliability index are also determined for each example. Designs in accordance with SANS 10160 were reasonably consistent in terms of probabilities of failure, which were below 1% in all cases. Reliability indices varied between 2.5 and 2.9, falling short of the target reliability index of 3.0 in SANS 10160. The wide range of central factors of safety obtained for examples that yielded very similar reliability indices and probabilities of failure confirms that the factor of safety is a poor measure of structural safety.

1 INTRODUCTION

1.1 *Background*

Traditionally, South African geotechnical designers made use of local technical publications, internationally popular textbooks or references and certain international standards rather than relying heavily on national codes or standards (Day 2013). Although partial factor limit states design was incorporated into the 1989 version of the South African "loading code", no provision was made for geotechnical loading. Replacement of this code by SANS 10160 in May 2010 addressed this omission by adding an additional part to this standard, SANS 10160-5: *Basis for Geotechnical Design and Actions* (SANS 2011b), and in so doing opened the way to harmonisation of structural and geotechnical design standards in South Africa (Day & Retief, 2009).

Following minor revisions, SANS 10160 was re-issued in 2011 and is the current South African basis of structural design and actions for buildings and industrial structures. Part 5 of this standard (SANS 10160-5) provides the basis for geotechnical design and determination of geotechnical actions. SANS 10160 is not a design code, but is a basis of design that *"serves as a general standard to specify procedures for determining actions on structures and structural resistance*

in accordance with the partial factor limit states design approach" (SANS 10160-1). The design of slopes, embankments and free-standing retaining structures is not covered in SANS 10160-5, but the design principles are still applicable. In the absence of a South African geotechnical design code, Eurocode 7 (EN 1997-1) may be used as a design code in conjunction with SANS 10160-5 (Day, 2013).

In 2008, the Geotechnical Division of SAICE agreed that South African geotechnical engineers would use EN 1997-1 in conjunction with SANS 10160 on a trial basis for five years (Day, 2013). On expiry of this trial period, the Geotechnical Division elected to draft a South African geotechnical design code based on SANS 10160 and EN 1997-1. This activity is presently underway.

1.2 *Comparison of design methods*

Working Stress Design (WSD), which has been the geotechnical industry norm for decades in South Africa, calculates the loads applied to a structure (P_i) and the ultimate capacity of the structure to resist loading (R_u) using best estimate values for input parameters. Provision for safety is made by applying a factor of safety (F) such that:

$$R_u / F \geq \Sigma P_i \text{ or } R_u / \Sigma P_i \geq F \qquad (1)$$

The value of F varies depending on the nature of the problem, being around 1.3–1.5 for slope stability and 2.5–3.0 for bearing capacity.

SANS 10160-5 uses a partial factor Limit States Design (LSD) approach in which the requirement is that the design resistance of the structure (R_d) should be greater than the design effect of actions (E_d), or:

$$R_d \geq E_d \qquad (2)$$

Provision for safety is made in two ways. Firstly, input parameters used in the calculation of both actions and resistances are selected as cautious estimates of the value affecting the occurrence of the limit state under consideration (SANS 10160-5). These are known as characteristic values. Secondly, partial factors are applied to the characteristic values of actions and material properties (and/or resistances) to obtain design values. These partial factors typically range between 1.0 and 1.6.

Day (2013) briefly compared the application of SANS 10160:2011 to WSD for typical spread footing and earth pressure problems while Chang (2015) compared WSD to LSD for pile foundations. Parrock (2015a; 2015b) also compared results of WSD, LSD and reliability-based design in order to stimulate debate amongst a working group set up by SABS TC98 SC06 to develop the South African geotechnical design code.

Whatever the nature of a geotechnical structure, a geotechnical practitioner needs to have an understanding of the design methods and confidence in the results. Comparisons between solutions determined using SANS 10160 and those from WSD are seen as a vital step towards realising that confidence (Barratt, 2015). This paper provides such comparisons by examining a few relatively simple design examples and is based on the work of Barratt (2015).

2 METHODOLOGY

2.1 Overview

An International Workshop on the Evaluation of Eurocode 7 was held in Dublin in 2005. Prior to this workshop, ten geotechnical design examples were distributed to selected organizing committee members in various European countries to be analysed in accordance with EN 1997-1 (Orr, 2005a). An analysis of the solutions received from different European geotechnical engineers as well as a set of model solutions to these design examples was prepared by Orr (2005b).

Three of the five design examples in this study are based on the examples used by Orr (2005a;

2005b). Although some soil parameters were changed, the loads and foundation details were typically retained. The main reason for this was to have model solutions available, albeit for Eurocode designs, to check correctness of the calculations. Design examples (b), (c) and (d) of this study follow Orr's Examples 1, 2 and 5 respectively. Design example (a) was added to include a strip foundation example, while example (e) was added to provide an earth embankment example. Orr's Example 10, a road embankment on soft ground, was useful in checking the solution to design example (e).

2.2 Design examples

In design examples (a), (b) and (c), both granular and cohesive soil conditions were assumed. For granular soils, the mean values of cohesion, friction angle and unit weight of the overburden and foundation soil were taken as 0 kPa, 35° and 18 kN/m³ respectively. For cohesive soil conditions, the mean undrained shear strength of the soil is 100 kPa and its mean unit weight is 18 kN/m³.

Design example (a), illustrated in Figure 1, comprised a concentrically loaded strip foundation with vertical dead load (permanent action) and live load (variable action). Dead and live loads (taken as characteristic values) are 900 kN and 600 kN respectively.

Design example (b), illustrated in Figure 2, is a square footing carrying concentric vertical loading comprising dead and live loads of 3000 kN and 2000 kN respectively.

Design example (c), illustrated in Figure 3, is a square foundation carrying both vertical and horizontal loads. A horizontal wind load of 400 kN acts at a height of 4.0 m above ground level. The dead and live loads are 3000 kN and 2000 kN respectively.

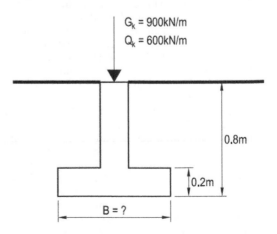

Figure 1. Design example (a)—Strip footing.

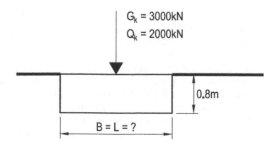

Figure 2.   Design example (b)—Square footing.

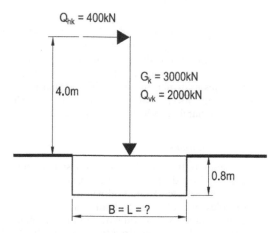

Figure 3.   Design example (c)—Square footing with variable horizontal load.

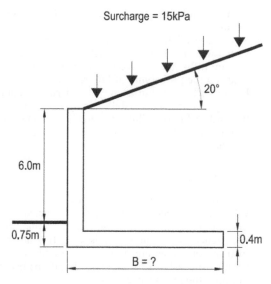

Figure 4.   Design example (d)—Cantilever retaining wall.

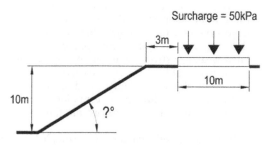

Figure 5.   Design example (e)—Earth embankment.

Design example (d), illustrated in Figure 4, is a 6 m high cantilever retaining wall. The wall is founded at 0.75 m depth. The material behind the wall slopes at 20° and carries a 15 kPa surcharge load.

In design example (d), only granular soil conditions were considered. Mean values of cohesion, friction angle and unit weight of the backfill and foundation soil were 0 kPa, 35° and 18 kN/m³ respectively.

Design example (e), illustrated in Figure 5, is a 10 m high earth embankment with a 10 m wide surcharge load of 50 kPa acting 3 m from the crest of the embankment. This is based on the Case 2 example of Länsivaara & Poutanen (2013), which was further examined by Parrock (2015a; 2015b). Mean values of cohesion, friction angle and unit weight of the embankment soil are 8 kPa, 26° and 18 kN/m³ respectively.

2.3   Methodology

The first step was to determine the geometry of the structure (size of footing or slope angle) required to satisfy the various limit states (STR, STR-P, GEO and EQU) given in SANS 10160, supplemented where necessary by reference to EN 1997-1. This step required selection of appropriate characteristic values for soil parameters, as discussed below. For the foundation examples, bearing resistance and sliding resistance were considered. Although SANS 10160-5 does not cover free-standing retaining walls or embankment slopes (SANS 10160-5), the principles still apply. In design example (d), the retaining wall foundation width that satisfies the various limit states given in SANS 10160 for both bearing and sliding resistance was determined. In design example (e), the maximum slope angle that satisfied the GEO limit states for global stability was determined using the GeoStudio SLOPE/W computer program.

The second step was to calculate the central factor of safety (F) in terms of WSD using the foundation dimensions and embankment slope determined in Step 1. Mean values of parameters

123

with unfactored loads were used in these calculations.

The central factor of safety is not necessarily a measure of safety of a structure due to the significant influence of uncertainties in materials and loads on the probability of failure (Russelli, 2008). For this reason a third step was undertaken, comprising a simplified reliability analysis to evaluate the probability of failure. The method of Duncan (2000) was selected for this step because of its simplicity.

2.4 *Selection of characteristic values*

According to SANS 10160-5, the characteristic value of a geotechnical parameter is selected such that the possibility of a worse value governing the occurrence of the limit state is <5%. If a normal distribution is assumed, the 5% fractile corresponds to a value 1.64 standard deviations below or above the mean value.

The standard differentiates between situations where behaviour of a structure at the limit state is governed by the lowest (or highest) value of the ground property and those where behaviour is governed by the average value. An example of the former is an end bearing pile where an isolated weak (or strong) zone of soil below the toe of the pile could govern its load carrying capacity. A friction pile, on the other hand, is an example of the latter as failure must occur along the entire length of the pile shaft before the limit state is reached.

Schneider (1997), cited by Simpson (2012), suggests that where (a) the limit state depends on averaging of properties of a large amount of ground, (b) ground properties vary in a homogeneous, random manner, and (c) at least ten test results are available, the characteristic value should be selected as 0.5 standard deviations below (or above) mean value. Thus, depending on the geometry of the problem and extent to which local or average values govern the performance of the structure at the limit state, the chosen characteristic value would generally lie somewhere between 0.5 and 1.64 standard deviations from mean value.

For purposes of this study, the characteristic values of soil parameters used in the foundation design examples (a) through (c) and the retaining wall example (d) were selected to be 1.0 standard deviations below the mean value. This was considered appropriate as some averaging of soil properties will occur, albeit over a limited volume of soil. For the embankment slope in design example (e), soil properties will be averaged over the full area of the slip surface. In this case, characteristic values of soil parameters were selected to be 0.5 standard deviations below the mean value.

It is generally acceptable to use estimates of the standard deviation of a parameter based on published values, which are most commonly expressed in the form of the Coefficient Of Variation (COV) (Duncan, 2000). Duncan further stresses the importance of using engineering judgement when selecting values for COV from published values, as the range in values is often large. The coefficients of variation selected for determination of characteristic values in this study are presented in Table 1.

2.5 *Reliability analysis*

Duncan's method was selected to determine the reliability of the structures. Using the COVs in Table 1 and a Taylor series technique as described in Duncan (2000), the most likely value and the coefficient of variation of F were determined. From this information, the likelihood that F could be less than 1.0 (probability of failure) and the lognormal reliability indices were calculated.

2.6 *Limitations*

Although both granular and cohesive materials were considered, the parameters assigned to these soils were selected to be uniform across all design examples except the embankment example(e), which assumed a c'/φ' soil with unique parameters. Further studies are required to assess the effects of varying soil parameters over the range of expected values.

In the foundation design examples, only bearing capacity was assessed. The effect of settlement, which is often a limiting criterion in foundation designs, was not considered in this study.

El-Ramly *et al* (2001) noted that simplified reliability analyses which do not take account of reduction in uncertainty of soil variability due to spatial averaging, may be inaccurate. In their comparison, simplified analyses overestimated the probability of unsatisfactory performance by a factor of five when compared to probabilistic and First Order Second Moment methods. Further studies should consider more rigorous methods of reliability analysis than the simplified method used in this study.

Table 1. Selected coefficients of variation.

Parameter	Unit	COV (%)
Unit weight	kN/m^3	5
Effective stress friction angle	deg	10
Effective cohesion	kN/m^2	40
Undrained shear strength	kN/m^2	25

3 RESULTS

The solutions to the design examples and further results of this study are given in Tables 2 to 6.

Table 2. Design example (a) solutions and results.

	Granular soil	Cohesive soil
Foundation width (m)	3.54	5.84
Factor of safety using WSD	4.37	2.03
Lognormal reliability index	2.46	2.84
Probability of failure (%)	0.69	0.23

Table 3. Design example (b) solutions and results.

	Granular soil	Cohesive soil
Foundation width (m)	3.55	4.08
Factor of safety using WSD	4.25	2.08
Lognormal reliability index	2.47	2.92
Probability of failure (%)	0.67	0.18

Table 4. Design example (c) solutions and results.

	Granular soil	Cohesive soil
Foundation width (m)	4.22	4.03
Factor of safety using WSD	5.02	2.03
Lognormal reliability index	2.79	2.82
Probability of failure (%)	0.26	0.24

Table 5. Design example (d) solutions and results.

	Granular soil	
Foundation width (m)	7.67	
Ultimate limit state	Bearing	Sliding
Factor of safety using WSD	8.31	3.00
Lognormal reliability index	2.57	6.56
Probability of failure (%)	0.50	0.00

Table 6. Design example (e) solutions and results.

	c'/φ' soil
Embankment slope (°)	27.8
Factor of safety using WSD	1.44
Lognormal reliability index	2.93
Probability of failure (%)	0.17

4 DISCUSSION

Design examples (a) to (c), which represent bearing capacity of shallow foundations, yield very similar results. For granular conditions, factors of safety varied between 4.25 and 5.02, well above the accepted range of 2.5 to 3.0 with which most engineers would be comfortable (Duncan, 2000). In terms of reliability, the resulting lognormal reliability indices ranged between 2.46 and 2.79. Probabilities of failure were 0.3–0.7%. Despite the high factors of safety, the lognormal reliability index and the probability of failure place the foundations in the "below average" range of expected performance according to USACE (1995).

For cohesive (undrained) conditions, factors of safety varied between 2.03 and 2.08. These factors of safety are lower than the traditionally accepted range of 2.5 to 3.0. However, in terms of reliability, the resulting lognormal reliability indices of 2.82–2.92, are quite similar to those for granular conditions. Probabilities of failure were all around 0.2% with the expected level of performance falling between categories "below average" and "above average" (USACE, 1995).

The significant difference in F between granular and cohesive conditions yet reasonably similar reliability indices and probabilities of failure illustrates the point made by Russelli (2008) that traditional deterministic factors of safety do not adequately quantify safety of a design.

For the retaining wall in design example (d), the factor of safety for bearing capacity is 8.31 and for sliding is 3.0, both well above the commonly adopted value of 1.5 for retaining walls (Krey, 1926 cited in Meyerhof, 1994). Despite the very high factor of safety for bearing capacity, the lognormal reliability index was found to be 2.57 and probability of failure was determined as 0.50%, again in the USACE "below average" range of expected performance.

For design example (e), the slope angle satisfying the global stability limit state was determined as 28 degrees. This slope angle equates to a conventional safety factor of 1.44, which is marginally below the commonly accepted value of 1.5 for permanent slopes (Krey, 1926 cited in Meyerhof, 1994). The resulting lognormal reliability index of the factor of safety is 2.93 and probability of failure was determined to be 0.17%. This reliability index is very near to the target reliability index of 3.0 for SANS 10160 and falls just short of the USACE "above average" range.

In all examples, the calculated log-normal reliability index fell below the target value of 3.0 implied by SANS 10160. A possible reason for this is that no consideration was given to the statistical distribution of applied loads in this study. The same applied loads were assumed in both the limit states and working stress designs. This amounts to an assumption that characteristic load is equal to mean (average) load, which is not correct. If the statistical distribution of applied loads is included in the analysis, characteristic values for the applied loads used in the LSD analysis will be higher than the mean value resulting in a more conservative geometry from Step 1 of the methodology. This will result in a reduction in the probability of failure and an increase in the reliability index. The statistical distribution of applied loads should be taken into account in future studies of this nature.

As pointed out by Retief & Dunaiski (2009), both EN 1990 and ISO 2394 permit assignment of separate reliability targets to actions and resistances by making use of sensitivity factors. In the case of resistances, the sensitivity factor $(\alpha_R) = 0.8$, which gives a target reliability for resistance $(\beta_{T,r})$ of $0.8 \times 3.0 = 2.4$. The reliability indices given in Tables 2 to 6 compare well with this target value.

5 SUMMARY AND CONCLUSIONS

A comparison of geotechnical design using SANS 10160 with that of current practice was carried out. Five design examples were selected as the basis for this comparison, with shallow foundation examples being analysed for both granular (drained) and cohesive (undrained) soil conditions. The methodology comprised determination of the geometry of the structures required to satisfy requirements of SANS 10160 in conjunction with EN 1997-1, followed by determination of the probability of failure and the log-normal reliability index. In addition, F was calculated using traditional WSD methods.

The authors acknowledge that the accuracy of simplified reliability analyses may be limited (El-Ramly et al, 2001), and that the effect of applying more rigorous reliability analysis methods must still be determined.

In the foundation design examples, similar ranges of lognormal reliability indices and probabilities of failure were calculated. Despite this, the factors of safety obtained for granular soils (3.5–4.4) differed significantly from those obtained for cohesive soils (2.0–2.1). This lack of correlation between reliability index and factor of safety was further illustrated in the retaining wall design example where the factor of safety against bearing capacity was much higher than that for sliding, yet the lognormal reliability index for bearing capacity was significantly lower than that for sliding. This clearly demonstrates the inability of the central factor of safety to indicate the true safety (reliability) of a design.

For the earth embankment design example, the factor of safety (1.44) is slightly below that which would be considered adequate by most engineers but the reliability index (2.93) is only marginally below the target value of 3.0 implicit in SANS 10160.

The reasonably consistent reliability index of 2.46 to 2.93 and probabilities of failure of under 1% for all design examples in this study suggest consistency of designs in accordance with SANS 10160 for different types of structures. However, these reliability indices fall short of a target reliability index of 3.0 that is implied in SANS 10160 (Retief & Dunaiski, 2009; Retief et al, 2009). This is probably because the statistical distribution of applied loading was ignored in this analysis.

Characteristic values of soil parameters for this study were selected as 1.0 standard deviations below the mean value for design examples (a) to (d) and 0.5 standard deviations below the mean for design example (e). This was based on the extent of the failure surface and the degree to which soil properties will be averaged. Consistency of results and reasonable levels of reliability obtained, suggests that this selection of characteristic values is reasonable.

REFERENCES

Barratt, B.J. 2015. *A comparison of geotechnical design using SANS 10160 with that of current practice through design examples.* MEng Thesis, Stellenbosch University.

Chang, N. 2015. Comparison of working stress design and limit state design for pile foundations. *Civil Engineering: Magazine of the South African Institution of Civil Engineering.* 19(3):18–20.

Day, P.W. 2013. *A Contribution to the Advancement of Geotechnical Engineering in South Africa.* DEng Thesis, Stellenbosch University.

Day, P.W. & Retief, J.V. 2009. Provisions for geotechnical design in SANS 10160. In Retief, J.V. & Dunaiski, P.E. (eds.), *Background to SANS 10160.* Stellenbosch: AFRICAN SUN MeDIA:189–204.

Duncan, J.M. 2000. Factors of Safety and Reliability in Geotechnical Engineering. *Journal of Geotechnical and Geoenvironmental Engineering.* 126:307–316.

El-Ramly, H., Morgenstern, N.R. & Cruden, D.M. 2001. Probabilistic slope stability analysis for practice. *Canadian Geotechnical Journal.* 39:665–683.

EN 1990:2002. *Eurocode: Basis of Structural Design.* CEN, Brussels.

EN 1997-1:2004. *Eurocode 7: Geotechnical design—Part 1: General rules.* CEN, Brussels.

ISO 2394:1998. *General Principles on Reliability of Structures.* ISO, Geneva.

Länsivaara, T. & Poutanen, T. 2013. Safety concepts for slope stability. In L'Heureux, J., Locat, A., Leroueil, S., Demers, D. & Locat, J. (eds.), *Landslides in Sensitive Clays: From Geosciences to Risk Management.* Dortrecht: Springer.

Meyerhof, G.G. 1994. *Evolution of Safety Factors and Geotechnical Limit State Design.* The Second Spencer J. Buchanan Lecture, Texas A&M University College Station, Texas, 4 November 1994.

Orr, T.L.L. 2005a. Design Examples for the Eurocode 7 Workshop. In Orr, T.L.L. (ed.), *Proceedings of the International Workshop on the Evaluation of Eurocode 7.* Dublin, 31 March and 1 April 2005:67–74.

Orr, T.L.L. 2005b. Model Solutions for Eurocode 7 Workshop Examples. In Orr, T.L.L. (ed.), *Proceedings of the International Workshop on the Evaluation of Eurocode 7.* Dublin, 31 March and 1 April 2005:75–108.

Parrock, A. 2015a. Letter. 16 March. Limit State Design Code for Geotechnics.

Parrock, A. 2015b. Letter. 4 July. Limit State Design Code for Geotechnics.

Retief, J.V., Dunaiski, P.E. & Day, P.W. 2009. An overview of the revision of the South African Loading Code SANS 10160. In Retief, J.V. & Dunaiski, P.E. (eds.), *Background to SANS 10160.* Stellenbosch: AFRICAN SUN MeDIA:1–24.

Retief, J.V. & Dunaiski, P.E. 2009. The Limit States Basis of Structural Design for SANS 10160-1. In Retief, J.V. & Dunaiski, P.E. (eds.), *Background to SANS 10160.* Stellenbosch: AFRICAN SUN MeDIA:25–56.

Russelli, C. 2008. *Probabilistic Methods applied to the Bearing Capacity Problem.* PhD Dissertation, Institut für Geotechnik der Universität Stuttgart.

Simpson, B. 2012. *Eurocode 7—fundamental issues and some implications for users.* Keynote lecture, Nordic Geotechnical Conference, Copenhagen, July 2012.

SANS 10160-1:2011. *Basis of structural design and actions for buildings and industrial structures Part 1: Basis of structural design.* SABS Standards Division, Pretoria.

SANS 10160-5:2011. *Basis of structural design and actions for buildings and industrial structures Part 5: Basis for geotechnical design and actions.* SABS Standards Division, Pretoria.

U.S. Army Corps of Engineers. 1995. *Introduction to Probability and Reliability Methods for Use in Geotechnical Engineering.* Washington, DC: Dept. of the Army.

Proceedings of the first Southern African Geotechnical Conference – Jacobsz (Ed.)
© 2016 Taylor & Francis Group, London, ISBN 978-1-138-02971-2

Semi-probabilistic geotechnical limit state design: Salient features and implementation roadmap for southern Africa

M. Dithinde
University of Botswana, Gaborone, Botswana

J.V. Retief
University of Stellenbosch, Stellenbosch, South Africa

ABSTRACT: For quite some time now, geotechnical design worldwide has been undergoing a transition from the traditional working stress design to limit state design. Geotechnical limit state design has now matured to a semi-probabilistic design approach. Despite this international trend, geotechnical design in Southern Africa and arguably the entire African continent is still based on the working stress design philosophy. This is attributed to genuine lack of appreciation of the principles and benefits of this new and more rational design philosophy. To keep pace with current international developments, the region should convert to semi-probabilistic limit state design as a matter of urgency. Accordingly, this paper presents salient features of geotechnical limit state design, demonstrate through a design example its benefits, and discuss an implementation roadmap.

1 INTRODUCTION

The development of systematic and rational treatment of risk to implementation of reliability-based design through codes and standards is the subject matter of ISO2394:2015. However, for code development the semi-probabilistic limit state approach such as the partial factor method has been internationally accepted as the standard basis on which the new generations of geotechnical codes are being developed today (e.g. Ontario Highway Bridge Design Code 1983; Canadian Foundation Engineering Manual, 1992; EN 1997-11994). Furthermore, the new informative Annex D of ISO2394:2015 identifies and characterises critical elements of the geotechnical reliability-based design process.

Despite the worldwide initiatives to bring geotechnical design within the semi-probabilistic framework, the African continent appears not to be too concerned about these developments. The participation of the continent in international activities on the subject is minimal. Even the Regional Conferences for Africa on Soil Mechanics and Foundation Engineering do not feature this important subject. The rather passive attitude towards geotechnical limit state activities can be attributed to: (i) the culture of adopting foreign design codes and (ii) genuine lack of appreciation of the principles and benefits of this new design philosophy. Regarding adopting foreign codes, the tradition in Southern Africa has been to adopt the British standards (e.g. BS 8004 for foundation design). Now that British standards have converted to the Eurocodes, it appears that the region may simply adopt Eurocode 7 for geotechnical design. A caution should be sounded here that, a semi-probabilistic code cannot be simply adopted as it is generally calibrated to the local conditions (e.g. local design practice, experience and environment such as local geology, soil type and conditions, site investigation practices). Also semi-probabilistic design codes are calibrated to a target level of reliability as expressed by a target reliability index value which varies with the economic status of a region or country. For example the target reliability index for common structures in South Africa is 3.0 while in Europe it is 3.8. On this account, most of the Southern Africa countries are poor and cannot afford to design for a target reliability index of 3.8 embodied in Eurocode 7. Even if this technical aspect is ignored, there is a question of developing a national annex which addresses the nationally determined parameters (partial factors, design approach, model factors, target reliability index, etc.). Therefore the apparent easy route of just adopting a foreign limit state design code is not feasible this time around.

Since adoption of a foreign code is out of the equation, the only alternative left to the Southern African countries is to develop their own limit state design codes. Already some initiatives in this direction are underway in South Africa. Provision for geotechnical limit state design has been made in the recently published revised South African Loading Code (SANS 10160:2011 Basis of structural design and actions for buildings and industrial structures). Under the auspices SADCSTAN (a body mandated to coordinate standardisation activities in the SADC region), most Southern African countries will adopt this standard. Nonetheless, a smooth transition to semi-probabilistic limit state design requires an in depth understanding of this new design philosophy as well as an appreciation of its benefits. Accordingly this paper provides the salient features of this design philosophy, and recommends an implementation roadmap for Southern Africa.

2 DEVELOPMENT OF SEMI-PROBABILISTIC GEOTECHNICAL LIMIT STATE DESIGN

Limit state design requires that the designer should check the adequacy of the structure against collapse and serviceability. Many geotechnical engineers (e.g. Boden, 1981; Becker, 1996) are of the opinion that this approach has always been used in one form or the other in geotechnical design and therefore limit state design is not a radically new method. The first limit state code of practice was the 1956 Danish Standard for foundations. This resulted from work by Taylor and Hansen who in 1948 introduced separate factors of safety on the cohesive and frictional components of the shear strength parameters in the analysis of stability of slopes. The approach was generalised by Hansen (1965) when he proposed partial factors on different type of loads, shear strength parameters and pile capacities for ultimate limit state design of earth retaining structures and foundations. It appears that in earlier days, geotechnical engineering was ahead of structural engineering in the knowledge and application of limit state design philosophy. However it should be noted that the partial factors were developed purely on the basis of intuition and judgement. Furthermore the new approach was required to produce designs similar to the existing working stress design approach.

Limit state design in the partial factors format became the general design approach in structural practice in the 1970s. This reignited the interest in this design format to the geotechnical profession. Christian (2003) asserts that the interest was driven by the desire to apply the same mathematical insights to geotechnical practice that have proved successful for structural engineering. The move was also motivated by the desire to achieve compatibility between geotechnical and structural

engineering. It might be said that conceptually at that stage the approach was similar to Hansen's proposal hence there were a severe criticism about the use of partial factors in geotechnical design. Some of the reservations expressed by the geotechnical profession (e.g. Boden, 1981; Semple, 1981) include:

- Not too many foundations and substructures are failing at the moment or appear to have excessively conservative designs, so why the fuss?
- The method is cumbersome to use, with a multiplicity of coefficients and increases the chances for computational errors;
- Results of analysis using partial factors must fit experience and therefore do not produce substantial differences in overall safety factors;
- Prescribed factors applied to soil properties might define soils which could not possibly exist. Under such circumstances the design could hardly be considered to be realistic;
- Partial factors can lead to probability theory and encourage statistical assessment of measured data. In geotechnical engineering, these provide little insight and divert attention from reality. Statistical analysis may cause major errors in selection of soil parameters, there being no comparable problem with structural design in manufactured materials;

The controversy and severe criticism on the application of limit state design with partial factors to geotechnical practice appear to stem from the lack of a theoretical basis on which the partial factors were developed. Just like the global factors, the partial factors were developed purely on basis of intuition and judgement. Accordingly in recent years, probability and reliability theory have been integrated with the original deterministic limit state design leading to the semi-probabilistic limit state design. In accordance with ISO2394:2015, the semi-probabilistic approach, such as the partial factor method requires sufficient understanding of failure modes and uncertainties to be categorized and modelled in a standardized manner.

3 SALIENT FEATURES OF THE SEMI-PROBABILISTIC LIMIT STATE DESIGN

3.1 Reliability basis

Reliability theory serves as the theoretical basis for Semi-probabilistic limit state design approach. The general reliability problem consists of a performance function $G(x_1, x_2,..x_n)$ and a multivariate probability density function $f(x_1, x_2,..x_n)$. The former is defined to be zero at the limit state, less than zero when the limit state is exceeded (i.e. fail), and greater than zero when the design is on the safe side. The basic measure of reliability is the probability of failure which is the probability of the limit state violation and is mathematically expressed as:

$$P_f = P[g(X) < 0] \qquad (1)$$

For the time-invariant reliability problem described by a time independent joint probability density function, the probability of failure can be determined using the following integral:

$$P_f = \int ... \int_{g(x) \leq 0} f_x(X)dx \qquad (2)$$

in which $f_x(X)$ is the joint probability distribution function of the n-dimensional vector X.

For practical purposes, analytical approximation of the integral are employed to simplify the computations of the probability of failure. The most common methods of approximation are the First Order Second Moment reliability methods (FOSM) also known as level II reliability methods. In the FOSM approximation, the probability of failure is expressed in terms of the reliability index (β). For a simple performance function with only two variable R and Q representing the resistance and load effect respectively β is given by:

$$\beta = \frac{\mu_R - \mu_Q}{\sqrt{\sigma_R^2 + \sigma_Q^2}} \qquad (3)$$

where μ_R = mean resistance, μ_Q = mean load, σ_R = standard deviation of the resistance, σ_R = deviation of the load.

For complex performance functions, detailed procedure for computation of β can be found in structural reliability text books (e.g. Holicky 2009).

3.2 Limit states

Limit states are states beyond which the structure no longer satisfies the design performance requirements (EN 1990). For a given structure several failure modes or limits states can be defined. However the consequences of various failure modes are not of equal magnitude. Some failure modes lead to outright collapse while others only affect the functioning of the structure. In general three limit states are considered:

- Ultimate Limit State (ULS), describing outright collapse of the structure (e.g., strength, ultimate bearing capacity, overturning, sliding, etc.
- Serviceability Limit State (SLS), describing loss of function of the structure without collapse.

Serviceability limit states concern the functionality of the structure, the comfort of people and aesthetic appearance of the structure. Examples include deformations, settlement, vibrations, cracks, and local damage of the structure in normal use under working loads such that it ceases to function as intended.

- Accidental Limit State (ALS), describing failure induced by accident conditions (e.g. collisions, explosions).

The acceptable probability of failure varies with the consequence of failure. Generally the highest safety requirements are set for ULS. Accepted probability of failure for SLS can be considerably high, especially if the effects of failure are irreversible.

3.3 Design format

The design format is given by Eq (4):

$$\frac{R_k}{\gamma_R} = \sum \gamma_i Q_{ni} \qquad (4)$$

in which R_k = characteristic resistance, γ_R = resistance partial factor, Q_{ni} = nominal value for the ith load component, γ_i = load partial factor for the ith load component.

The fundamental requirement is that the factored strength is greater than the factored loads. However the distinguishing feature is the derivation of the partial factors on the basis of reliability theory and explicitly and systematically incorporating the major sources of uncertainties. Essentially, semi-probabilistic limit state is a level 1 reliability based design method in which appropriate levels of reliability are provided on the structural component by the use of prescribed partial factors. The use of such partial factors ensures consistent level of reliability over a range of structures. Although the approach is set within a probabilistic framework, it does not require explicit use of the probabilistic description of the variables. This is an advantage in that even engineers with no knowledge of probability and reliability theory can produce designs at a prescribed level of reliability.

Conceptually, the determination of partial factors in the reliability framework is the reverse of the process for computing the reliability index. Common reliability calibration methods include: (a) Advanced First-Order Second Moment approach (A-FOSM), (b) Design value method, (c) Mean Value First-Order Second Moment approach (MV-FOSM) and (d) Approximate Mean Value First-Order Second Moment approach (Approx-MVFOSM). Dithinde and

Retief (2015) investigated the four methods with respect to pile foundations and concluded that γ_R values from the principal calibration method (i.e. A-FOSM) are comparable to results obtained from the Design Value and the approximate MV-FOSM methods. This implies that the approximate methods yield reasonable results, further suggesting that reliability calibration can as well be based on the simple approximation procedures.

4 ILLUSTRATIVE EXAMPLE

The conservatism implied by the working stress approach as compared to the semi-probabilistic limit state design is demonstrated by a pile design example shown in figure 1. For the given design situation, it is required to determine the pile length using the two design philosophies. The solution to the problem is presented in Table 1. In the WSD approach, a customary factor of safety of 2.5 was applied to the pile capacity. For the semi-probabilistic limit state design, resistance partial factors from a reliability calibration study by Dithinde (2007) were used. It is evident from the solution that the semi-probabilistic limit state design approach produces a pile length that is 22% shorter than that of WSD. For a large piling project, the reduction of pile lengths by 22% constitutes a significant saving in materials and hence overall cost.

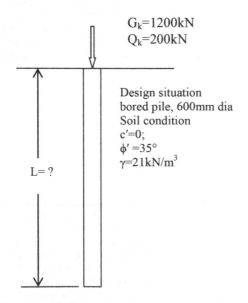

G$_k$=1200kN
Q$_k$=200kN

Design situation
bored pile, 600mm dia
Soil condition
c'=0;
ϕ' =35°
γ=21kN/m^3

L= ?

Figure 1.   Pile design example.

5 IMPLEMENTATION ROADMAP

5.1 *Phase 1: Adoption of BS EN 1997-1*

While still collecting statistical data for calibration of both the load resistances partial factors, consideration should be given to adopting an existing deterministic limit sate design code. In Southern Africa, many technical rules and standards on civil engineering works are linked to British standards and therefore it is only reasonable to adopt the BS EN version of Eurocode 7.

Adopting an existing limit state design code alleviates manpower and financial recourses constraints associated with the development of a code from the first principle. The key to implementation of the Eurocodes is the formulation of national annexes. The national annex present information on parameters left open in the Eurocode for national choice referred to as Nationally Determined Parameters. The clauses and sub-clauses permitting national choices are detailed in EN 1997-1.

It is at the national annex stage that local design experience base and conditions are taken into account. The long tradition of using British standards has resulted in local design practices having significant similarities with the UK design practices. Accordingly a UK national annex could be adapted. However, adopting the entire Annex is not feasible as the partial factors are based on a reliability index of 3.8 which is extremely high for the Southern African economies. Therefore some degree of calibration of EN 1997-1 to local practice is necessary.

The aim is to ensure that design to EN-1997-1 is essentially identical to that attained by the existing practice in terms of the overall factor of safety. For a given design situation, the procedure to achieve this entails:

- Carrying out the design in accordance with EN 1997-1 and partial factors given in the UK national annex,
- The design so produced is checked using conventional methods to see its global factor of safety satisfies the stipulated minimum value for the respective design case,
- If the design meets the set criteria, the partial factors are adopted. If not the values are adjusted till the desired factor of safety is achieved.

Adopting of the Eurocodes by third countries enjoys support from European countries. The European companies have asked the European Commission to promote the use of the Eurocodes outside the EU, since a standardisation environment as much as possible in line with the European system will foster their competitive advantage

Table 1. Design example solution.

	WSD Approach	Semi-probabilistic LSD
Actions	(not factored)	(only variable actions factored)
Permanent (column load)	1200 kN	$1200 \times 1.0 = 1200$ kN
Variable (from column)	200 kN	$200 \times 1.3 = 260$ kN
Permanent (Pile weight)	6.79 LkN	$6.79 \times 1.0 = 6.79$ LkN
Total (E_d)	1400 + 6.79 L	1460 + 6.79 L
Geotechnical parameters	(unfactored)	(unfactored)
Angle of internal friction	35	35
Unit weight	21	21
Design parameters		
N_q	75	75
δ	35	35
K_s	0.9	0.9
Characteristic resistance (R_k)	$A_b(N_q\sigma_v) + A_s(0.5\,K_s\sigma_v \tan\delta)$	$A_b(N_q\sigma_v) + A_s(0.5K_s\sigma_v \tan\delta)$
	$445.73\ L + 12.47\ L^2$	$445.73\ L + 12.47\ L^2$
Safety factors		
Global factor of safety	2.5	N/A
Resistance partial factor	N/A	1.8
Design resistance (R_d)	$R_k/2.5$	$R_k/1.8$
	$4.98\ L^2 + 178.29\ L$	$6.91\ L^2 + 247.63\ L$
Equilibrium equation ($R_d = E_d$)	$1400 + 6.79\ L = 4.98\ L^2 + 178.29\ L$	$1460 + 6.79\ L^2 =$
	$4.98\ L^2 + 178.29\ L$	$6.91\ L^2 + 247.63\ L$
	$4.98\ L^2 + 171.5\ L\text{-}1400 = 0$	$6.91\ L^2 + 240.84\ L\text{-}1460 = 0$
L (m)	6.81	5.27

in the high value segment of project design and management and in the areas of niche specialisation (Andersson et al. 2003).

The European Commission has always seen the promotion of the Eurocodes in third countries as an important objective (Andersson et al, 2003). In Guidance Paper L6 (European commission, 2003) on application and use of the Eurocodes, it is clearly stated that one of the major intended benefits and opportunities of Eurocodes is to increase the competitiveness of the European construction Industry in its world-wide activities.

5.2 Phase II: Semi-probabilistic limit state

The main distinguishing feature of this phase is the calibration of the partial factors in a probabilistic framework. The process of reliability calibration will entail:

- Gathering of resistance and load statistics: The calibration process requires the resistance and load statistics (mean, standard deviation and distributions) as input. The required statistics are generated from the measured value of the random variable (resistance or load) and the predicted value yielded by the theoretical design model. Therefore the statistics are generally represented in terms of the ratio of the measured to

predicted values. Since in geotechnical practice, there are many theoretical and semi-empirical models for evaluating a given limit state (e.g. bearing capacity), the statistics need to be determined for each model in use by a given country.

- Design and construction specification: To enable the formulation of performance functions for the purposes of computing reliability indexes, up-to-date local design equations for various geotechnical limit states are required.
- Characteristic values of soil parameters: In the level I reliability based-design approach, the design values of various parameters including soil properties are obtained by multiply or dividing the characteristic value with the appropriate partial factor. However, currently there is no universally acceptable approach for determining characteristic values of soil properties. Therefore on the basis of the engineering geology of the region, clear procedures for determining the characteristic soil parameters to be used in the design calculations should be developed.
- Target reliability index for geotechnical design in general: The application of the derived partial factors yields designs that meet the prescribed target reliability index. Accordingly the target reliability index (β_T) is an important input in the calibration process. Several approaches for setting

the target reliability index are available. However, currently the most widely used selection criteria is the range of beta values implied in the past or current practice. Reliability indexes inherent in the current design practice capture the long successful local design experience. Therefore studies to establish reliability indexes inherent in the current practice for various geotechnical designs should be carried out.

- Calibration of resistance factors: With all the required input data collected, the final stage is to carry out the actual reliability calibration of partial factors. It should be noted that calibration of partial factors is dependent on local design practice, experience and environment such as local geology, soil type and conditions, site investigation practices (extend, methods, standards, equipment advances). Since these factors generally differ from one country to another, values developed for a specific country cannot be simply adopted by another country. Therefore there is need to calibrate geotechnical resistance factors for different applications utilising local databases. A typical example of such calibration studies can be found in Dithinde 2007.

6 CONCLUSIONS

The paper discussed the salient features and roadmap for implementation of semi-probabilistic geotechnical limit state design. The following conclusions can be drawn from the discussion:

- Despite the initial controversy and severe criticism on the application of limit state design with partial factors to geotechnical practice, the profession has now embraced limit state design philosophy as demonstrated by existence of some geotechnical limit state design codes.
- The semi-probabilistic limit state design philosophy provides a framework for harmonising geotechnical design rules with that of structural design involving other materials (e.g. performance requirements, specification of the limit states, design situations to be checked, reliability requirements, and treatment of basic variables (actions, materials properties, and geometric data).
- Development of a full semi-probabilistic geotechnical limit state design requires gathering of resistance and load statistics, establishment of the target reliability index for geotechnical design, reliability calibration of partial factors, development of design and construction specifications, and definition of characteristic values of soil parameters.
- Gathering of input data for reliability calibration of partial factors required for a semi-probabilistic

limit state design is lengthy process. At this stage such data is not available hence the need for an interim implementation of limit state design in a deterministic framework. For this phase, Eurocode 7 can be adopted with the UK annex. However the UK annex need to be calibrated to the local design practice.

REFERENCES

Andersson, C., Dimova, S., Géradin, M., Pinto, A. & Tsionis, G. 2003. *Eurocodes promotions in third countries*. European Commission Research. Guidance paper L—application and use of Eurocodes. Brussels.

Becker, D.E. 1996. Eighteenth Canadian geotechnical colloquium: limit state design for Foundations. Part 1. An overview of the foundation design process. *Canadian Geotechnical Journal*, 33, 956–983.

Borden, B. 1981. Limit state principles in geotechnics. *Ground engineering*, Vol. 14, No. 6, pp. 21–26.

British Standards Institution 1996 BS 8004 *Code of Practice for Foundations for foundation design*.

Canadian geotechnical society. 1992. *Canadian foundation engineering manual*. Richmond, B.C.

Christian, T. 2003. Geotechnical acceptance of limit state design methods. *LSD2003: International workshop on limit state design in geotechnical engineering*, Massachusetts, USA.

Dithinde, M. 2007. *Characterisation of model uncertainty for reliability based design of pile foundations*. PhD dissertation submitted to the University of Stellenbosch, unpublished.

Dithinde, M. & Retief, J.V. 2015. Comparison of methods for reliability calibration of partial resistance factors for pile foundations: *Fifth international symposium on geotechnical safety and risk*, Rotterdam, Netherlands, 467–472.

EN 1990 2002. *Eurocode: Basis of structural design*. European committee for standardization (CEN): Brussels.

EN 1997 2004. *Eurocode 7: Geotechnical design—Part 1: General rules*. European committee for standardization (CEN): Brussels.

Hansen, J.B. 1965. The philosophy of foundation design: design, criteria, safety factors and settlement limits. *Proc. of a symposium on bearing capacity and settlement of foundations*, Duke University, pp. 9–13.

Holický, M. 2009. Reliability analysis for structural design. *SUNMeDIA Press*, Stellenbosch, ISBN 978-1-920338-11-4.

ISO 2394 2015. *General principles on reliability for structures*. International organisation for standardisation, Geneva.

Ontario Highway Bridge Design Code 1983, *Ontario Ministry of Transportation and Communications*, Highway Engineering Division.

SANS 10160-5:2011. *South African national standard Part 5: Basis for geotechnical design and actions*. SABS standards division, Pretoria.

Semple, R.M. 1981. Partial coefficient design in geotechnics. *Ground engineering*, 14(6), 47–48.

Proceedings of the first Southern African Geotechnical Conference – Jacobsz (Ed.)
© *2016 Taylor & Francis Group, London, ISBN 978-1-138-02971-2*

Reliability-based tools for quantifying risk on geotechnical projects

S.M. Gover

Golder Associates Africa (Pty) Ltd., Midrand, South Africa

ABSTRACT: Geotechnical risk is the risk to infrastructure or construction work created by the ground conditions. In times of economic downturn, funding is often limited early in the project life cycle which results in cost-cutting by project stakeholders. These cost-cuts inevitably have a knock-on effect to the design, construction and operation stages of the project, and often result in cost over runs rather than savings. This paper introduces a number of risk/reliability-based tools for quantifying the risk imposed on geotechnical projects, and by means of an example, illustrates how these tools can optimise the level of site investigation required to minimise the overall project risk.

1 INTRODUCTION

Geotechnical risk is the risk to infrastructure or construction work created by the ground conditions. It can be defined as a combination of the likelihood of a geotechnical hazard occurring, and the consequences of that hazard. Examples of geotechnical hazards can include soft ground, expansive, collapsible, dispersive or liquefiable soils and slope instability to name a few.

In times of economic downturn, funding is often limited early in the project life cycle which results in cost-cutting by project stakeholders. For example, client's will find it easier to focus funds on operational costs rather than capital, and therefore look to carry out only the minimum-required investigatory work; and consultants may look to sacrifice certain expensive scope of work items in order to win tenders and secure business. These cost-cuts inevitably have a knock-on effect to design, construction and operation stages of the project, and often result in cost over runs rather than savings.

Risk engineering and reliability-based analysis provide tools for quantifying the risk imposed on geotechnical projects, as well as the language to relate this important message to critical project stakeholders. This paper discusses these tools and illustrates how the intensity of site investigation has a direct influence on the level of uncertainty in geotechnical parameters, and subsequently the overall risk related to the project.

2 PROBLEM DEFINITION

2.1 *Effect of site investigation cost on projects*

According to Kulhawy (2000), there are four ways to pay for geotechnical information:

1. In proper site evaluations
2. If not in 1, then in over-design
3. If not in 1 or 2, then in over-construction
4. If not in 1, 2 or 3, then in claims and legal proceedings.

Ground-related problems adversely affect project costs, completion time, profitability, health and safety, quality and fitness-for-purpose, and can lead to environmental damage (Clayton, 2001). One of the most common reasons for project overruns is the insufficiency of site investigation data. Figure 1 shows results of Clayton's study which compared the cost of site investigation work (expressed as a as a percentage of the tender cost) with the magnitude of the project overrun caused by geotechnical problems.

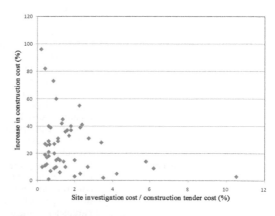

Figure 1. Geotechnical construction risk mitigated by investment in site investigation.

2.2 *Relation to risk*

While the aforementioned quote and chart convey an important message towards appreciating the value of site investigations, the level of risk inherent to the project or infrastructure should, however, be properly understood and related to information requirements of the project. For example, higher probabilities of failure may be found to be acceptable for a water retaining dam in an unpopulated area, than one immediately upstream of a large city.

3 RISK AND RELIABILITY-BASED TOOLS

3.1 *Levels of site investigation*

There is much literature and country-specific standards/codes of practice etc. which define the level of site investigation required for various stages of the project (i.e. prefeasibility, design *etc.*). These documents are important and ensure that the practice of geotechnical engineering is regulated and projects/sites do not go "under-investigated".

For the purpose of illustration in this paper, the following levels of site investigation are defined:

Table 1. Illustrative levels of site investigation.

I.	Desk study	No intrusive investigations or laboratory testing. Based on observations, engineering judgment and available literature.
II.	Indicative	Random/sparse intrusive investigations using rudimentary in-situ and/or laboratory testing techniques.
III.	Specific	Investigations based on suggested quantities/ grid spacing. In-situ and laboratory testing techniques designed to provide quantitative and qualitative information for the expected site conditions.
IV.	Confirmatory	Advanced testing specifically designed to confirm site conditions or geotechnical parameters critical to the design.

3.2 *Levels of uncertainty*

A useful means to represent variation in a parameter is by the Coefficient Of Variation (COV), a dimensionless ratio that is independent of the units of the parameter and can therefore be used to compare the variations of many different parameters:

$$COV = \frac{\sigma}{\mu} \qquad (1)$$

where σ = standard deviation of the parameter; and μ = mean of the parameter.

Relevant COV values are given in Table 2:

An observation made from this data is that the COV for geotechnical materials is generally higher than for, say, structural materials, i.e. anything from 5% to 80% where structural materials generally have variations in the order of 10% to 20%. This can be attributed to both natural variability present within the soil, and 'measurement error' which is a combination of factors such as sample disturbance, accuracy of testing methods and equipment, and human error in the testing process.

To illustrate the impact 'measurement error' has on uncertainty in geotechnical parameters, consider the degree of variation one might expect when evaluating the friction angle of a soil. Two

Table 2. Coefficients of variation for certain geotechnical parameters (Extracted from Baecher & Christian, 2003).

Geotechnical parameter	COV range (%)
Angle of friction (sands)	5–20
Angle of friction (clays)	12–56
CBR	17–58
Cohesion (undrained) (clays)	20–50
Cohesion (undrained) (sands)	25–30
Compaction (OMC)	11–43
Compaction (MDD)	1–7
Compressibility	18–73
Consolidation coefficient	25–100
Density (apparent or true)	1–10
Elastic modulus	2–42
Linear shrinkage	57–135
Liquid limit	2–48
Moisture content	6–63
Permeability	200–300
Plastic limit	9–29
Plasticity index	7–79
Standard Penetration Test	27–85
Tensile strength	15–29
Unconfined compressive strength	6–100
Void ratio	13–42

extremes in accuracy exist whereby the friction angle could be determined by either triaxial testing or Standard Penetrometer Testing (SPT). Triaxial testing is understood to be highly accurate due to the controlled nature of the test and ability to accurately measure strains and pore pressures; whereas, the variation in the SPT N number achieved (which would be used to infer a friction angle) can be expected to vary by up to 100% between two separate SPT tests at the same location (CIRIA, 1995).

COV for the friction angle could therefore be expected to be approximately 5% from triaxial tests and 15% to 20% from the SPT test. This would relate to friction angles of, for example, $30° \pm 1.5°$ (mean $\pm$ standard deviation) compared to $30° \pm 6.0°$.

A number of geotechnical parameters from Table 2, which might relate to the level of site investigation carried out, are plotted against the levels of investigation (see Table 1) in Figure 2 for illustrative purposes. Note that a linear relationship was arbitrarily assumed between COV and the related level of investigation, although it is expected that some definitive relationship would exist if the levels of site investigation were more accurately defined.

Based on the assumed linear relationship, it could be expected that for each of these parameters, the COV could be approximately halved by carrying out a Level III site investigation as opposed to a Level I. It must be noted again that little science was applied to obtain this relationship at this point, and no fieldwork quantities can yet be extrapolated for the Level III site investigation, but the preliminary message obtained is rather significant. This will be illustrated in Section 4, where it is demonstrated how a decrease in COV drastically effects design.

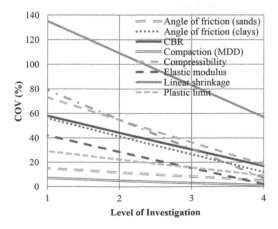

Figure 2. Approximate relationship between level of site investigation and COV for selected parameters.

3.2 Reliability-based analysis

3.2.1 Background

Reliability-based analysis uses the uncertainty present in loads and resistance, or demand and capacity, of a particular system to evaluate the uncertainty in the resulting safety margin or safety factor. This uncertainty can be represented as a reliability index which can subsequently be related to a probability of failure using the standardised Normal distribution.

The reliability index, β, may then be calculated by the following equation (Baecher & Christian, 2003):

$$\beta_F = \frac{E[F] - 1}{\sigma_F} \qquad (2)$$

where $E[F]$ = expected value/mean of the safety factor distribution; and σ_F = standard deviation of the safety factor distribution.

The evaluation of uncertainty in the calculated safety factor is a special case of error propagation, and requires the solution of a Taylor series expansion of the performance function. Hasofer & Lind (1974), however, proposed an alternative approach to the Taylor series expansion based on a geometric interpretation of the reliability index as the distance in dimensionless space between the peak of the multivariate distribution and a function defining the failure criterion in terms of matrix formulation, as given in (Ditlevsen, 1981):

$$\beta = \min_{x \in F} \sqrt{(x - \mu)^T C^{-1}(x - \mu)} \qquad (3)$$

where x = vector representing the set of random variables, x_i; μ = vector representing the set of mean values, μ_i; C = covariance matrix; and F = failure region.

This method is conceptually and computationally difficult and Low & Tang (1997) therefore introduced an alternative by using the perspective of an expanding ellipsoid in the original space of the random variables. By this perspective, the quadratic form in Ditlevsen's equation $[(x - \mu)^T C^{-1}(x - \mu) = 1]$ represents an ellipsoid with failure occurring at the first point where the expanding ellipsoid touches tangent to the failure surface, as shown graphically in Figure 3 for two random variables x_1 and x_2.

The expanding surfaces of the ellipsoid can be seen as decreasing probability values and the reliability index β is the axis ratio of the ellipsoid that touches the failure surface (R) and the one-standard-deviation dispersion ellipsoid (r).

Low & Tang (2007) extended the method to develop a more efficient and robust technique.

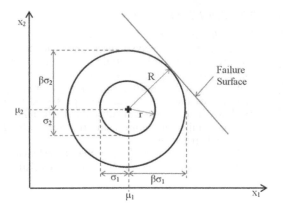

Figure 3. Illustration of the Low & Tang (1997) expanding ellipsoid perspective.

This method proposed varying the row and column vectors as dimensionless numbers and back-calculating values of x_i:

$$\beta = \min_{x \in F} \sqrt{(n)^T R^{-1}(n)} \qquad (4)$$

And,

$$x_i = F^{-1}[\phi(n_i)]$$

where n = vector of n_i dimensionless variables; R = correlation matrix; and $\phi(n_i)$ = standard normal cumulative distribution.

The first and third terms under the square root sign in Equation (4) represent the equivalent standard normal vector, which is automatically varied as numerical values (void of equations) during the spreadsheet-based constrained optimization search. In simple terms, this vector represents the size of the expanding ellipsoid, which is optimised to locate the smallest ellipsoid (or combination of variable parameters) that touches the tangent to the failure surface. The shape of the distribution is governed by the second term in the equation, i.e. the correlation matrix.

3.2.2 Spreadsheet formulation of the Low & Tang (2007) FORM method

The matrix notation given in equation (4) can be implemented in Microsoft Excel as illustrated in Figure 4 (bottom of page) for a slope stability problem comprising four uncertain parameters:

The matrix formulation was then solved by object-oriented constrained optimisation using the Solver tool in Excel as follows:

a. n_i values are set equal to zero initially (cells g1:g4 in Figure 4);
b. The transpose of the column vector is calculated;
c. The inverse of the correlation matrix is calculated (cells h1:k4);
d. The matrix product of the vectors is computed;
e. The square root of the matrix product is calculated to solve for β based on n_i values;
f. x_i values are back-calculated based on n_i values and the user-created function "x_i" provided in (Low & Tang, 2007);
g. The performance function is solved based on x_i values; and
h. Solver is invoked to minimise β by changing the n_i values (subject to $g(x) = 0$) and steps b) through g) are repeated.

3.3 Levels of acceptable risk

Some risks we take on voluntarily, like participating in sports or driving an automobile. Others we are exposed to involuntarily, like a dam failing upstream of our home or disease due to air pollution. Just being alive in the US or Europe carries a risk of dying of about 1.5×10^{-6} per hour, and some risk analysts consider this number (approximately 10^{-6}) a baseline to which other risks might be compared (Stern et al. 1996).

In order to gain an appreciation of levels of risk, or acceptable probabilities of failure, Table 3

	a	b	c	d	e	f	g	h	i	j	k
	Material	Property	Distribution	μ	σ	x_i	n_i	ϕ_1	c_1	ϕ_2	c_2
1	Embankment	Phi	Normal	37.00	5.55	31.19	-1.047	1	-0.4	0	0
2	Embankment	Cohesion	Normal	5.00	1.00	5.36	0.363	-0.4	1	0	0
3	Clay	Phi	Normal	28.00	4.20	20.27	-1.839	0	0	1	-0.4
4	Clay	Cohesion	Normal	10.00	2.00	10.76	0.381	0	0	-0.4	1

Optimised vector   Correlation matrix [R]

`=SQRT(MMULT(TRANSPOSE(g1:g4),MMULT(MINVERSE(h1:k4)),(g1:g4)))`

g(x):	2E-08
β:	2.152
Prob.Fail:	1.57E-02

Figure 4. Spreadsheet formulation of the Low & Tang (2007) slope stability analysis.

Table 3. Average risk of death to an individual from various human-caused and natural accidents (Extract from US Nuclear Regulatory Commission, 1975).

Accident type	Total number	Individual chance per year
Motor vehicle	55,791	1 in 4,000
Falls	17,827	1 in 10,000
Drowning	6,181	1 in 30,000
Firearms	2,309	1 in 100,000
Air travel	1,778	1 in 100,000
Electrocution	1,148	1 in 160,000
Lightning	160	1 in 2,500,000
Tornadoes	91	1 in 2,500,000

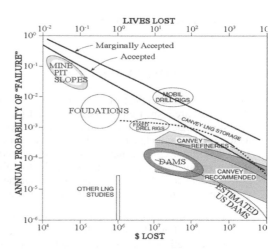

Figure 5. Chart showing average annual risks posed by a variety of traditional civil facilities and other large structures or projects (Baecher 1982).

presents statistics published by the US Nuclear Regulatory Commission (1975) of deaths caused by various human-caused and natural accidents.

Whitman (1984) presented the chart in Figure 5, which was modified from data published in (Baecher, 1982), plotting probabilities of failure of various engineering projects against costs that society, or at least some operating part of society, has implicitly found acceptable.

In certain countries, acceptable limits are already prescribed for various types of infrastructure, but these are not presently available in South Africa. The information presented above could therefore be used to inform project stakeholders of the approximate level of risk applicable for their project.

3.4 Decision tree analysis

A variety of methods are available for analysing engineering risks, but decision trees have become the common approach for complex geotechnical systems. They organise information needed in a logical manner, allow for a repeatable and consistent process, provide an explicit consideration of uncertainty and risk, incorporate input from stakeholders in the decision, and present and explain the basis for the decision.

There are two basic components in a decision tree analysis, namely alternatives and outcomes. A simple example is shown in Figure 6, where two alternatives for a possible design are illustrated, each with adequate or inadequate performance outcomes.

Once all outcomes are analysed for the various alternatives, expected consequence can be calculated by multiplying the probabilities along each branch of the tree. This cumulative probability can subsequently be multiplied by an expected cost of the respective consequence to give an

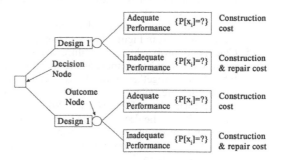

Figure 6. Decision tree illustration.

expected project value to be used in decision making.

When assessing levels of investigation required for a particular project, reliability-based analysis can be used to quickly and effectively estimate probabilities of failure for various COV values of the geotechnical parameters required. If the repair/claim cost for the particular failure is understood (see Figure 5), this can be multiplied by the probability to obtain an event cost.

If it is subsequently assumed that each outcome of the various alternatives represent minimum and maximum event costs, and that this range of values can be represented by a normal distribution, then using the (Dai & Wang, 1992) "Three-sigma rule" (which stems from the fact that 99.73% of all values in a normal distribution fall between 3 standard deviations either side of the mean), we can plot each result of the decision tree analysis as probable cost distributions for comparative purposes. This is better illustrated in the section following.

4 ILLUSTRATIVE EXAMPLE

In this example, the effect of two site investigation alternatives are compared to assist in the selection of a slope angle for a water retaining embankment upstream of a populated area. Tender costs for a 1:2.5 and 1:3.0 embankment slope were selected from a recent project as $25 million and $30 million respectively. These costs were used, together with the tools in Section 3, to infer the following:

a. The site investigation alternatives are either Level II or Level III, which, based on Figure 1 were selected as 2% and 6% of the tender costs respectively.
b. The range of COV for the friction angle of the embankment material (sand) is 5%–20%. For the Level II investigation the COV is assumed to be 15%, and for the Level III investigation 10% (see Figure 2).
c. The potential cost associated to failure of the dam is taken from Figure 4 as $500 million.
d. Calculated probabilities of failure determined using reliability-based analysis are listed in the table below, together with the assessment as to whether the failure is acceptable or not (for more information refer to (Gover, 2015)):
e. The results of the decision tree analysis (minimum and maximum probably event costs) are plotted as Normal probability distributions functions in Figure 7.

Based on the results presented in Figure 7, the following observations are made:

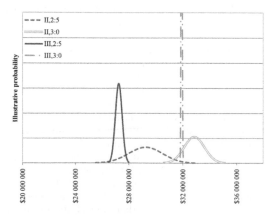

Figure 7. Results of the decision tree analysis plotted as normal probability distribution functions.

Table 4. Results of the reliability-based stability analysis.

Design case	Probability of failure (Pf)	Acceptable Pf?
Cov 15%, 1:2.5 slope angle	1.5 E-02 (1.50%)	Unacceptable
Cov 15%, 1:3.0 slope angle	6.0 E-03 (0.60%)	Acceptable
Cov 10%, 1:2.5 slope angle	3.0 E-03 (0.30%)	Marginally acceptable
Cov 10%, 1:3.0 slope angle	4.5 E-04 (0.045%)	Acceptable

a. A level II investigation and 1:2.5 slope angle would result in the cheapest construction cost, but the probability of failure is unacceptable.
b. A level II investigation and 1:3.0 slope angle has an acceptable probability of failure, but should a failure occur, the repair/ claim costs would be the highest.
c. A level III investigation and 1:2.5 slope angle is only marginally acceptable, but the range of possible costs is relatively narrow and the mean is approximately $5.0 million less than the final case.
d. A level III investigation and 1:3.0 slope angle results in the narrowest possible cost range due to the very low probability of failure, but is the most expensive.

Different Clients may prefer either of b), c) or d), depending on their preferred level of risk aversion, for example, a Client may be inclined to pay higher costs for the lowest probability of failure (option d)). The ideal case would likely be to carry out slightly more than $1.5 million (6% of the tender cost) worth of Level III site investigations in order to bring the COV of the embankment material to below 10%, and therefore return an acceptable probability of failure.

5 CONCLUSIONS

The reliability-based concepts/tools discussed in this paper are shown to provide valuable insight into the level of risk associated with geotechnical projects.

An example was presented whereby the various tools are utilized to illustrate the risk, particularly in terms of cost, associated to the level of investigation carried out for an embankment dam. Based on the results presented, an optimum level of site investigation may be inferred in order to minimize the project risk.

Some rudimentary assumptions and relationships were utilised, but the author is confident that with further study, the necessary relationships (such as the relationship between level of investigation and COV for different geotechnical parameters) can be ascertained so that detailed decision tree analyses may be carried out.

REFERENCES

Ang, A.H.-S. & Tang, W.H., 1975. Probability Concepts in Engineering Planning and Design. Vol.1 ed. New York: John Wiley & Sons.

Baecher, G.B., 1982. Simplified Geotechnical Data Analysis. Bornholm: s.n.

Baecher, G.B. & Christian, J.T., 2003. Reliability and Statistics in Geotechnical Engineering. West Sussex: Wiley.

CIRIA, 1995. Report 143—The Standard Penetration Test (SPT): methods and use, London: Construction Industry Research and Information Association.

Clayton, C.R.I., 2001. Managing geotechnical risk: improving productivity in UK building and construction, Thomas Telford.

Dai, S.-H. & Wang, M.-O., 1992. Reliability analysis in engineering. New York: Van Nostrand Reinhold.

Ditlevsen, O., 1981. Uncertainty Modelling: with Applications to Multidimensional Civil Engineering Systems. New York: McGraw-Hill.

Gover, S.M., 2015. Reliability-based slope stability analysis and associated severity of failure of tailings dams, University of Pretoria.

Hasofer, A.M. & Lind, N.C., 1974. Exact and invariant second-moment code format. Journal of Engineering Mechanics, 100(1), pp. 111–121.

Low, B.K. & Tang, W.H., 1997. Efficient reliability evaluation using spreadsheet. *Journal of Engineering Mechanics, ASCE, New York*, 123(7), pp. 749–752.

Low, B.K. & Tang, W.H., 2007. Efficient Spreadsheet Algorithm for First-order Reliability Method. *Journal of Engineering Mechanics*, ASCE, 133(12), pp. 1378–1387.

Stern, P.C., Fineberg, H.V. & National Research Council (U.S.). 1996. Understanding Risk: Informing Decisions in a Democratic Society. Committee on Risk Characterization, National Academy Press.

Whitman, R.V., 1984. Evaluating Risk in Geotechnical Engineering. Journal of Geotechnical Engineering, ASCE, 110(2), pp. 145–188.

Proceedings of the first Southern African Geotechnical Conference – Jacobsz (Ed.)
© *2016 Taylor & Francis Group, London, ISBN 978-1-138-02971-2*

Neuro-genetic approach for immersed CBR index prediction

M.A. Bourouis, A. Zadjaoui & A. Djedid
Faculty of Technology, University of Tlemcen, Algeria

ABSTRACT: The California Bearing Ratio after immersion CBR_{imm} is a fundamental parameter used in preliminary studies of design and health monitoring of linear structures. This parameter is determined from laboratory testing, which requires skilled labour and time. This test is more expensive than a simple identification tests. A CBR_{imm} prediction model is presented, based on a database of easily measurable geotechnical parameter. Four models were investigated: Simple regression, multiple regression, artificial neural networks and neuro-genetic networks. The use of four simultaneous variables gave better results for the prediction of CBR ($R^2 = 0.91$). Using a neuro-genetic approach gave the least error. The current works will show whether this approach is good or requires some improvements.

1 INTRODUCTION

The Californian Bearing Ratio after immersion (CBR_{imm}) is a parameter used to characterize the bearing capacity of a material. To conduct CBR_{imm} test on subgrade soil, a representative sample shall be collected, from which a remolded specimen is prepared, compacted at the predetermined Optimum Moisture Content (OMC) at the standard Proctor compaction effort. The specimen is soaked for four days under water cofter which a penetration test is performed. The CBR test is relatively expensive, time consuming and laborious. In this paper, an estimation of the CBR_{imm} based on four models: Simple regression, multiple regression, artificial neural network sand neuro-genetic networks is based on simple, fast and less onerous tests. An explanation of the data used and methodology is provided. The practical application of the models is also discussed.

2 REVIEW OF PREVIOUS WORKS

Geotechnical properties of soils are controlled by factors such as mineralogy, fabric, and pore water, and the interactions of these factors are difficult to establish solely by traditional statistical methods due to their interdependence. (Shahin et al. 2008) have shown that the prediction by Artificial Neural Networks (ANN) gives good results, some authors (Flood & Kartam 1994, Ripely 1996, Sarle 1994) indicate that the use of multiple hidden layers in the methodology of ANN, provide the necessary flexibility for modelling complex phenomena.

Patel & Desai (2010) presented some correlations for soaked an unsoaked CBR values for alluvial soils of south Gujarat from different values of LL, PI, OMC and MDD of the soils. They concluded that as PI increases, CBR value decreases and CBR increases when PI value decreases. Ramasubbarao & Sankar (2013) presented the developing regression-based models for predicting soaked CBR value for fine-grained subgrade soils in terms of LL, PL, MDD and OMC the results obtained show that better performance can be obtained from the model developed using Multiple Linear Regression Analysis by showing the highest R-value of 0.96 and R^2-value of 0.92, Tang et al. (1991) observed that the ANN model which use a large number of inputs parameters gives better prediction. Smith (1986) suggested the following guidelines for values of the correlation coefficient $|R|$ between 0.0 and 1.0: $|R| > 0.8$—strong correlation exists; $0.2 < |R| < 0.8$—correlation exists and $|R| < 0.2$ weak correlation exists. All those studies of Willmott & Matsuura (2005) indicates that the Mean Absolute Error (MAE) is the most natural measure of average error magnitude, and that unlike Root Mean Square Error (RMSE) it is an unambiguous measure of average error magnitude. The idea of application of neuro-genetic modelling first originated in the late 1980s (Miller et al. 1989), and the reafter, an intense research was performed during the 1990s (Kinato 1986; Schiffmann et al. 1993). The investigation made by De Jong (1975) showed the utility of the Genetic Algorithms (GA) for function optimization. Genetic algorithms combine selection, crossover, and mutation operators to recognize the best solution of a problem.

3 METHODOLOGY

Analysis tools used in this study include a statistical method (regression analysis and multiple regression) and artificial intelligence method (artificial neural network sand neuro-genetic networks). Multiple regression analysis is always accompanied by an analysis of variance table. This table will test whether all explanatory variables have a significant effect on the dependent variable.

The strategy for obtaining ANN hinges on the development of a Matlab program, with several loops which is varied learning algorithms, and the number of neurons in each layer to minimize the cost function under the constraint of a fixed relative error. Genetic algorithms were integrated with the neural network model for optimization in order to obtain the desired level. The number of neurons in each layer in this hybrid, neuro-genetic model to compare the two approaches is kept the same. In this paper, the database used is based on 112 measures granted to different specimens: CH (silty clay of high plasticity), CL (sandy clay of low plasticity), MH (sandy loam high plasticity), ML (sandy clay purposes) and CL-ML (Limon low plasticity) made by Varghese, et al. (2013).

4 PREDICTION OF CBR$_{IMM}$ OF FINE GRAINED SOILS USING FOUR MODELS

Optimum Moisture Content (OMC), Maximum Dry Density (MDD), Plastic Limit (PL), and the Liquid Limit (LL) are used as input in the development of predictive models of CBR$_{imm}$ ratio.

4.1 Simple regression analysis

The simple regression analysis was performed to identify useful input parameters and to avoid irrelevant/weakly correlated input parameters so as to reduce the chance of neural networks being caught in local minima. The results of this analysis are presented in Figure 1 for the four parameters PL (Figure 1a), LL (Figure 1b), MDD (Figure 1c) and OMC (Figure 1d). The correlation is even more important and representative in simple regression for compaction parameters, namely OMC and MDD with the ratio CBR$_{imm}$.

4.2 Multiple regression analysis (RM)

Based on the results of the simple regression analysis of four parameters (LL$^{-1.34}$) (PL$^{-1.4}$), (OMC$^{-2.74}$) and (e$^{5.072*(MDD)}$), the contribution of each parameter is necessary to predict the index CBR$_{imm}$ fine soils. Due to the combined effect of several parameters in the multiple regression model, a minor

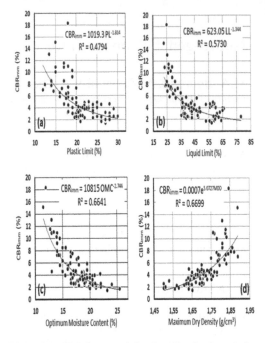

Figure 1.   CBR immersed simple regression correlation.

Table 1.   Output of multiple regression analysis.

Multiple regression analysis	
R Correlation	0.913
R^2	0.834
R^2 Adjusted	0.827
Standard Error	1.280
Observations	95

change in the parameter (PL$^{-1.81}$) obtained using the simple regression analysis was performed.

We have shown that the performance of multiple regression with parameter (PL$^{-1.4}$) is much better than that with the parameter (PL$^{-1.81}$). The results obtained are shown in Table 1.

The model gives; a correlation coefficient (R) equal to 0.913, determination coefficient (R^2) equal to 0.834, and determination coefficient equal to 0.827. If the value of the coefficient is close to (1) there is a strong correlation. The value of R^2 is close to (1), so the adjustment represents an excellent quality. But does not mean that there are no non-significant parameters within. Other checks are necessary as the significance test of the model coefficients. This test is performed by analysis of variance. The calculated F value is compared with a theoretical value F(α) obtained from Fisher tables for a significance level of $\alpha = 0.01$ and given two degrees of freedom k$_1$ and k$_2$. The null hypothesis is rejected if the value of F $\geq$ F(α).

Table 2. Analysis of variance (ANOVA).

	Regression	Residual	Total	
Df	$k_1 = 04$	$k_2 = 90$	94	
SS	740.68	147.34	888.02	
MS	185.17	1.64		
$F_{(calcul)}$	113.11			
$F_{(\alpha = 0,01)\ theoretical}$	3.04×10^{-34}			
	Constant	$LL^{-1.34}$	$PL^{-1.4}$	$OMC^{-2.74}$	$e^{5.072*MDD}$
Coefficients	−1.295	373.1	−128.4	6.7×10^3	3×10^{-4}
Standard error	0.474	92.122	62.6	$1. \times 10^3$	8.6×10^{-5}
Tstat	−2.731	4.05	−2.05	6.58	3.57
P-value	7.59×10^{-3}	1.1×10^{-4}	0.0433	3×10^{-9}	5.7×10^{-4}
Lower 95%	−2.24	190.08	−252.8	4.7×10^3	1.4×10^{-4}
Upper 95%	−0.353	556.12	−3.95	8.8×10^3	4.8×10^{-4}

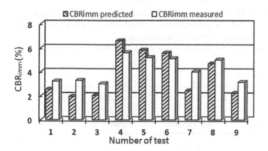

Figure 2. Comparison between predicted and measured.

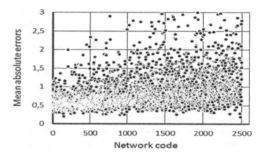

Figure 3. Variation in the MAE according to the neuron number of algorithm (trainoss).

The analysis of variance presented in Table 2 corresponding level of confidence is equal to 0.999. So at least one of the coefficients −1.295, 373.1, −128.4, 6.7×10^3 and 3×10^{-4} makes a significant contribution to the entire model.

The t static value calculated for all variables and constants are greater than the t critical value, with a significance level equal to 0.05, this implies that all variables and constants in the multiple regression models are statistically significant at the 0.05 confidence level.

The approach of the P value is also used to highlight significant differences between the values of coefficients and zero. The multiple regression model obtained is given by Equation 1, it is quite sufficient to establish the experimental data (Figure.2).

$$CBR_{imm} = -1.295 + 373.1 \times \left(LL^{-1.34} \right)$$
$$- 128.4 \times \left(PL^{-1.4} \right)$$
$$+ 6748.7 \times \left(OMC^{-2.74} \right) \quad (1)$$
$$+ 0.0003 \times e^{5.072 \times MDD}$$

4.3 Analysis of Artificial Neural Network (ANN)

For the prediction of the index CBR_{imm} using analysis of artificial neural networks the four input variables used are OMC, the PL and LL expressed as a percentage and MDD in g/cm^3.

The database for ANN was randomly divided into: training data, cross-validation and testing data. The cross-validation data set was used to test the performance of the network. The studied network is a multilayer perceptron type Feed-Forward with a learning algorithm of back-propagation type of error gradient. The learning rate set equal to 0.05 is considered suitable and adequate for good performance. To improve the behaviour of the ANN, both input and output data were normalized to the range of 0 and 1 by means of mini-max linear conversion function in order to obtain good network behaviour.

To determine the best network topology (the optimal number of neurons) we have fixed all parameters as (number of hidden layers, the type of activation function, learning algorithm) and varied the number of neurons in the hidden layers and then takes the test with the minimum mean absolute error (MAE), Figure 3 presents all

Table 3. The characteristics of the algorithms back propagation.

Algorithm	Description
Fletcher-Reeves conjugate gradient algorithm **(traincgf)**	Has smallest storage requirements of the conjugate gradient algorithms.
Polak-Ribiéreconjugate gradient algorithm **(traincgp)**	Slightly larger storage requirements than traincgf. Faster convergence on some problems.
Powell-Beale conjugate gradient algorithm **(traincgb)**	Slightly larger storage requirements than traincgp. Generally faster convergence.
Scaled conjugate gradient algorithm **(trainscg)**	The only conjugate gradient algorithm that requires no line search.
BFGS quasi-Newton method **(trainbfg)**	Requires storage of approximate Hessian matrix and has more computation in each iteration than conjugate gradient algorithms, but usually converges in fewer iterations.
One step secant method **(trainoss)**	Compromise between conjugate gradient methods and quasi-Newton methods.
Levenberg-Marquardt algorithm **(trainlm)**	Fastest training algorithm for networks of moderate size. Has memory reduction feature for use when the training set is large.
Bayesian regularization **(trainbr)**	Modification of the Levenberg-Marquardt training algorithm to produce networks which generalize well. Reduces the difficulty of determining the optimum network architecture.

Table 4. Absolute error test and validation for each algorithm.

Output	Input	Network structure	Algorithm	MAE	
				Testing	Validation
CBR	LL, PL, OMC, MDD	4-14-09-1	trainbfg	0.226	0.523
		4-44-11-1	traingdx	0.241	0.437
		4-12-24-1	traincgf	0.235	0.620
		4-34-27-1	trainscg	0.241	1.221
		4-13-03-1	traincgp	0.229	0.448
		4-42-29-1	trainoss	0.189	0.570
		4-12-24-1	trainlm	0.199	0.879
		4-29-03-1	trainrp	0.253	0.557

performed tests, the topology of the first test corresponds to 1-1-1 and the topology of the test 2500 is 50-50-1.

Table 3 summarizes the different learning algorithms used in this study with the advantages and disadvantages of each.

The use of validation set in the study is an important guard against this overtraining. To avoids overtraining; the error (maximum absolute error and/or mean absolute error) on the test set is monitored. Training is continued as long as the error on the test set decreased. It is terminated when the error on the test set started increasing again. Thus the training is halted when the error on testing dataset is lowest "Learning with early stop". We can conclude from the study of several algorithms that the best ANN model is a Multi-Layer Perceptron (MLP) neuronal network type that uses a step intersecting (trainoss) as a learning algorithm and functions (Logsig, Logsig and Purelin) as transfer functions with architecture [4-42-29-1]. The mean absolute errors of testing and model validation are 0.189 and 0.570 respectively.

Table 4 shows different prediction models obtained for the index CBR_{imm} using artificial neural network analysis.

4.4 Analysis of Neuro-Genetic (NG)

The work in NG gets started with the creation of a random generation composed of a chromosome collection. Normally, the size of initial population was considered to be 100.

For generating an initial population, a specified boundary should be defined for network parameters. The software randomly takes a value from the defined limits and then automatically produces the initial population. This particular population was then subjected to the genetic operators of selection, crossover, and mutation to produce a new evolved generation. The roulette Wheel method was used for the selection operator, whereas for crossover and mutation, probabilities of 0.9 and 0.01 were applied, respectively, the evolution process was repeated up to network with the lowest training error.

The fitness of each network is determined by calculating the mean square error (MAE) at the neural network with topology 12-42-1 was chosen. 649 synaptic weights was optimized by the GA, Figure 4 show best fitness.

4.5 Comparison of results

To evaluate the performances of the model of the multiple regression and the ANN simultaneously, the MAE parameter and the correlation coefficient are given in table 5. The study shows that ANN analysis gives more accurate results for predicting the for fine-grained CBR_{imm} index. The ANN correlation coefficient is higher than the multiple

regressions coefficient. Unlike the conventional statistical methods the use of parameters having an influence in the ANN templates provides soil properties prediction reliability (Figure 5).

ANN performances will improve with a wider and more representative database with a sufficient number of input variables. However, the analysis by ANN lack of transparency where the output is obtained as digital values. No information can be collected on the effect of each input on the output "black boxes". However, in the regression analysis, the output is obtained as equations or trend lines which give us an overall idea about the influence of each input variable on the prediction. Genetic algorithm was utilized to optimize synaptic weights. Recognition of the optimum model with this method as compared with the classic networks is faster and convenient.

The performance of the model was examined by statistical method in which absolutely higher efficiency of neuro-genetic modelling was realized. As such, a very low error mean (0.17) between predicted and measured was observed.

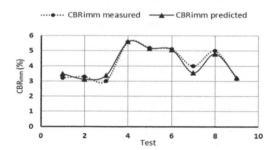

Figure 4. Comparison between predicted and measured.

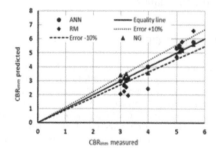

Figure 5. Comparison analysis of artificial neuron network, neuro genetic and multiple regression.

Table 5. Comparison between the neuron network, neuro-genetic and the multiple regression.

LL (%)	PL (%)	OMC (%)	MDD (g/cc)	CBR_{imm} (%)	NNA		NG		RM	
					P	E	P	E	P	E
37	20	21	1.63	3.21	2.99	0.22	3.48	0.22	2.53	0.68
42	21	19.89	1.52	3.28	3.14	0.14	3.12	0.14	1.94	1.34
42.5	17	19,7	1.66	3	2.96	0.04	3.38	0.04	2,04	0.96
52	26	14.6	1.81	5.6	5.77	0.17	5.61	0.17	6.58	0.98
56	24	14.8	1.79	5.2	5.23	0.03	5.17	0.03	5.8	0.6
59	25	14.9	1.78	5.1	5.41	0.31	5.10	0.31	5.56	0.46
60	25	19	1.67	4	3.99	0.01	3.56	0.01	2.42	1.58
26.5	15	18	1.71	5	5.34	0.34	4.81	0.34	4.68	0.32
37	21.5	21.4	1.55	3.14	2.67	0.47	3.23	0.47	2.24	0.9
Error absolute mean (MAE)					0.19		0.17		0.87	
Correlation coefficient (R)					0.99		0.97		0.96	

P The predicted values.
E The error between the predicted and measured values.

5 CONCLUSION

The objective of this study is to develop predictive models of the CBR index after immersion by operating enhanced data base easily measurable geotechnical parameters; the distribution was made by using the method of multiple regressions and the method of neural networks. The main conclusions from this study are:

- The prediction for CBR_{imm} ratio using ANN templates is better than the use of the regression models.
- The use of a large number of parameters makes a more correct learning and therefore increases the information available to the network, afterwards the performance is automatically improved.
- It is known that soil behaviour presents a spatial variability, for this it is always better to increase the number of influential parameters in the model, which provides predictive security.
- The neuro-genetic network reports reduced training time and the performance of the model can be improved with the inclusion of more training data sets.

REFERENCES

De Jong, K. 1975. *An analysis of the behavior of a class of genetic adaptive systems*. Ph.D. Dissertation.

Flood, I. & Kartam, N. 1994. Neural networks in civil engineering I: principles and understanding. *Journal of Computing in Civil Engineering* 8(2): 131–148.

Jong, YH. & Lee, CI. (2004). Influence of geological conditions on the powder factor for tunnel blasting. *Int J Rock Mech Min Sci* 41:533–538.

Kinato, H. 1990. Empirical studies on the speed of convergence of Neural Network Training using Genetic Algorithms. *In 8th national conference on artificial intelligence*: 798–795.

Lapedes, A. & Farber, R. 1988. How neural networks work, in neural information processing systems. *American Institute of Physics*: 442–456. New York.

Miller, G., Todd, P. & Hedge, S. 1989. Designing neural networks using genetic algorithms. *Proceeding of the 3rd International Joint Conference on Genetic Algorithms*: 379–384.

Patel, R. & Desai, M. 2010. CBR Predicted by index properties of soil for alluvial soils of South Gujarat. *Indian Geotechnical Conference* I: 79–82.

Ramasubbarao, G. & Sankar, G. 2013. Predicting soaked CBR value of fine grained soils using index and compaction characteristics. *Jordan Journal of Civil Engineering* 7(3).

Ripley, B. 1996. *Pattern recognition and neural networks*. Cambridge University Press.

Roy, T., Chattopadhyay, B. & Roy, S. 2007. Prediction of CBR for subgrade of different materials from simple test. *Proceedings of the International Conference on Civil Engineering in the new millennium-opportunities and challenges* III: 2091–2098.

Sarle, W. 1994. Neural networks and statistical models. *Proceedings of the 19th Annual SAS Users Group International Conference, Cary, NC: SAS Institute*: 1538–1550.

Schiffmann, W., Joost, M. & Werner, R. 1993. Application of genetic algorithms to the construction of topologies for multilayer perceptrons. *Proceeding of the international joint conference on neural networks and genetic algorithms*: 675–682. Innsbruk.

Shahin, M., Jaksa, M. & Maier, H. 2008. State of the art of artificial neural networks in geotechnical Engineering. *Journal of Geotechnical Engineering* 8: 1–26.

Smith, G. 1986. *Probability and statistics in civil engineering: an introduction*. London.

Tang, Z., Almeida, C. & Fishwick, P. 1991. Time series forecasting using neural networks versus Box-Jenkins methodology. *Simulation* 57(5): 303–310.

Varghese, V., Babu, S., Bijukumar, R., Cyrus, S. & abraham, B. 2013. Artificial neural networks: A solution to the ambiguity in prediction of engineering properties of fine-grained soils. *Geotechnical and geological engineering* 31(4): 1187–1205.

Willmott, C. & Matsuura, K. 2005. Advantages of the Mean Absolute Error (MAE) over the Root Mean Square Error (RMSE) in assessing average model performance. *Climateresearch 30*: 79–82.

Proceedings of the first Southern African Geotechnical Conference – Jacobsz (Ed.)
© 2016 Taylor & Francis Group, London, ISBN 978-1-138-02971-2

Numerical modelling rapid drawdown in riverbanks

C.J. MacRobert & M. van der Haar
University of the Witwatersrand, Johannesburg, South Africa

ABSTRACT: Flooding can result in high water levels in rivers, which can drop rapidly as floodwaters recede. There is a possibility of this leading to an elevated phreatic surface within the riverbank. Results from transient finite element seepage analyses show that bank geometry and material permeability play a key role. Flatter slopes provide a greater seepage surface, however they are also more stable. For low permeability clays, steeper slopes are more susceptible, whereas soils of intermediary permeability such as silty loam and loamy sands, flatter slopes are more susceptible. However, high permeability gravels are unaffected by floods as the phreatic surface recedes at the same rate as the flood. A table is presented to provide scoping level guidance on the risk of instability for various soil types, bank angles, and bank heights for different flood events.

1 INTRODUCTION

Riverbank failure can result in a loss of agricultural and natural lands, and potential damage to property. One mechanism of riverbank failure is rapid drawdown of bank phreatic surfaces following flooding. Such failure mechanisms have been reported by Duncan et al. (1990), Chen & Huang (2011) and Oya et al. (2015). During a flood event, water can seep into the riverbank, such that when floodwaters recede an elevated phreatic surface is left in the bank. This elevated phreatic surface can lead to a reduction in effective stresses, potentially leading to instability. Factors influencing this behaviour are the geometry of the riverbank, initial location of the phreatic surface, permeability of the material, duration of the flood and the rate of drawdown (Budhu & Gobin 1995; Green 1999; Garcia 2004).

The aim of this paper was to model how these various factors affect the stability of riverbanks. Thereby providing some guidance to the potential risks of instability following flooding and rapid drawdown.

2 METHODOLOGY

To establish how these factors affect riverbank stability the commercially available computer programs SEEP/W (2007) and SLOPE/W (2007) were used. A transient seepage analysis using SEEP/W was used to determine how the phreatic surface would change during flood events. The stability of the bank was then assessed using the Morgenstern-Price limit-equilibrium formulation within SLOPE/W. The critical slip surface at each stage of the flood event was obtained using the auto-locate function in SLOPE/W.

2.1 Material parameters

Although riverbanks may consist of various soil types, homogenous banks were considered in this study. Four general soil types were considered, a clay soil, a silt loam soil, a loamy sand and a gravel. The strength of these soils was defined by Mohr-Coulomb criteria (Table 1).

SEEP/W requires the definition of a Volumetric Water Content function (VWC) and a hydraulic conductivity function (KF) to model transient seepage. The VWC for the silt loam and loamy sand were estimated using the modified Kovacs formulation within SEEP/W using the parameters given in Table 2. The VWC for the clay and gravel soils were estimated using the sample functions in SEEP/W and the parameters in Table 2. To obtain the KF the van-Genuchten formulation in SEEP/W

Table 1. Soil strength criteria.

	Soils			
Property	Clay	Silt loam	Loamy sand	Gravel
Cohesion (kPa)	15	7.5	1	0
Friction Angle (°)	23	30	25.5	32
Unit weight (kN/m³)	21	16	19	25.5

Table 2. Soil permeability criteria.

| Property | Soils | | | |
	Clay	Silt loam	Loamy sand	Gravel
Saturated Water Content (m³/m³)	0.5	0.5	0.4	0.4
Residual Water Content (m³/m³)	0.07	0.02	0.04	0.04
Saturated Permeability (m/s)	1.7e–7	1.9e–6	1.7e–5	0.01
D_{10} (mm)	–	0.0017	0.041	–
D_{60} (mm)	–	0.028	0.47	–
Liquid limit (%)	–	25	27	–

was used. The coefficient of volume compressibility (m_v) was assumed to be 1×10^{-5} kPa^{-1} for all soils. A better definition of this value is only necessary for coupled stress and pore-pressure analysis (Geo-Slope, 2007).

2.2 Riverbank geometries

A range of geometries was analysed, to cover various typical riverbanks. Bank angles of 1:1 and 1:2 (vertical:horizontal) were analysed for all soil types. In addition, a 2:1 slope was analysed for the clay and silt loam soils, and a 1:3 slope was analysed for the loamy sand and gravel. Various bank heights between 1.0 and 10 m were investigated.

2.3 Flood events

For each riverbank, the initial river water level was assumed to be one-third from the bottom of the bank height. The initial phreatic surface within the bank was assumed level with the river. Flood events were simulated by raising the water level from the initial level to full bank height, maintaining it at this level and then lowering it to the initial level.

Three flood durations and three drawdown rates were considered. A short flood was defined as an increase of stage of 0.890 m/h, a medium flood by an increase of stage of 0.224 m/h and a long flood by an increase of stage of 0.023 m/h. The river was then kept at the full bank level for a period equal to that required to reach this level. Drawdown rates considered were 0.380 m/h, 0.224 m/h and 0.023 m/h for rapid, medium and slow drawdowns rates respectively. These flood durations and drawdown rates were based on a review of flood data (CSRA 1994; SANRAL 2006; DWAF 2007). This resulted in nine different combinations of flood duration and drawdown rate.

2.4 Stability analysis

A factor of safety of less than 1.3 was used as an indicator of an unstable slope (Hubble 2010). For each slope, the factor of safety was determined for initial conditions. At each stage of the flood event, a factor of safety was determined and the lowest factor of safety determined recorded.

3 RESULTS

Figure 1 shows an example of a 6 m high, loamy sand riverbank with a bank angle of 1:3 under the initial conditions. Figure 2 shows the same slope following a long flood followed by rapid drawdown. Under the initial conditions, the factor of safety was 1.5 and following the flood event, the factor of safety reduced to 1.0. Thus, this riverbank moves from a stable condition to an unstable condition due to the flooding event. Figure 3 shows the other factors of safety obtained for this bank but at other bank heights and for other flood events.

A similar assessment was carried out for the other soil types and bank geometries. Results from this analysis are condensed in Table 3. In this table, the left most column gives the soil types, the next column indicates the bank angle, and the next gives the height of the slope considered. Results from the stability analysis are summarised in the right hand columns. If a slope was unstable under the initial conditions this is stated, likewise if the slope

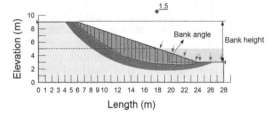

Figure 1. Example of initial slope stability analysis for a 6 m high 1:3 loamy sand riverbank.

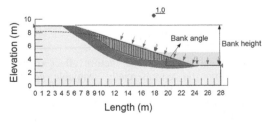

Figure 2. Impact of a long flood followed by rapid drawdown on the stability of a 6 m high 1:3 loamy sand riverbank.

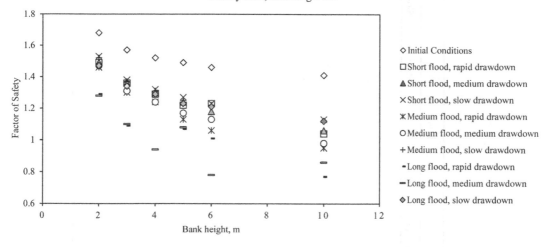

Figure 3. Factor of safety for various riverbank heights and flood scenarios.

was stable under all scenarios this is also indicated. An asterisk is used to indicate the flood conditions under which an initially stable slope becomes unstable.

For 1:1 clay riverbanks, bank heights less than 4 m were not affected by flood events. Bank heights between 4 and 6 m became unstable after long floods, followed by either rapid or medium drawdown rates. However, bank heights between 6 and 10 m all flood events resulted in unstable slopes. 1:2 clay riverbanks were stable under all scenarios. Steep 2:1 clay slopes were stable under all scenarios when the bank height was less than 4 m, but became unstable under all scenarios for bank heights between 4 and 10 m.

For 1:1 silty loam riverbanks, bank heights less than 4 m were not affected by flood events. Bank heights between 4 and 6 m became unstable for all flood durations followed by rapid drawdown. They also became unstable following medium and long duration floods followed by medium drawdown rates. Bank heights between 6 and 10 m were unstable under the initial conditions. 1:2 slopes were stable under all scenarios, except for the 10 m slope, which was unstable following long duration floods followed by either rapid or medium drawdown rates. Bank height was the main determining factor for 2:1 slopes; bank heights below 3 m being stable under all flood events but unstable under initial conditions when above 4 m.

For the loamy sand riverbanks, all 1:1 slopes were found to be unstable under the initial conditions. The 1:2 slopes were stable under all scenarios for bank heights below 1 m, but all bank heights above 2 m were unstable under the initial conditions. Loam sand riverbanks with a 1:3 slope were

affected the most by flood events. For bank heights less than 2 m, long duration floods followed by either rapid or medium drawdown rates became unstable. Bank heights between 2 and 3 m became susceptible to more flood events and between 3 and 10 m all flood events, resulted in instability.

Stability of gravel slopes was dependent on bank angle, as flood events did not result in instability. Slopes steeper than 1:2 were unstable under initial conditions and 1:3 slopes were stable under all scenarios.

4 ANALYSIS

From the above assessment of rapid drawdown in riverbanks, it is evident that slope geometry and material permeability play a key role. Generally, flatter slopes are at greater risk, as a greater seepage surface is available. However, flatter slopes are more stable and so these factors compete. For very low permeability soils, such as clay, steeper angles were more susceptible. Whereas for the silty loam and loamy sand the flatter slopes were at greater risk from flood events. However, the stability of steep silty loam and loamy sands was dependent on bank height. For the gravel riverbanks, which have a high permeability, flood events had little impact, as the phreatic surface receded at the same rate as the flood. Here the stability was then governed by bank angle and bank height.

Table 3 can be used as a scoping level design-aid, to indicate riverbanks susceptible to instability following flood events. The table can be used to determine which bank types may require stabilization.

151

Table 3. Synthesis of results—Design-aid.

Soils	Bank angle	Height m	Short — Rapid	Short — Medium	Short — Slow	Medium — Rapid	Medium — Medium	Medium — Slow	Long — Rapid	Long — Medium	Long — Slow
Clay	1:1	1–4		Stable under all scenarios							
		4–6			*		*	*	*	*	*
		6–10	*			*			*	*	
	1:2	1–10		Stable under all scenarios							
	2:1	1–4		Stable under all scenarios							
		4–10		Unstable under initial conditions							
Silty loam	1:1	1–4	*	Stable under all scenarios		*			*		
		5–6		Unstable under initial conditions			*			*	
		6–10		Stable under all scenarios							
	1:2	1–8		Stable under all scenarios					*	*	
		8–10		Unstable under initial conditions							
	2:1	1–3		Stable under all scenarios							
		3–10		Unstable under initial conditions							
Loamy sand	1:1	≤1		Unstable under initial conditions							
	1:2	≤1		Stable under all scenarios							
		1–10		Unstable under initial conditions							
	1:3	1–2				*	*		*	*	
		2–3		*		*	*		*	*	
		3–10			*			*		*	*
Gravel	1:1	1–10		Unstable under initial conditions							
	1:2	1–10		Unstable under initial conditions							*
	1:3	1–10		Stable under all scenarios							*

152

5 CONCLUSIONS

This paper reports findings from a study into whether riverbanks can fail following flood events due to rapid drawdown of phreatic surfaces. Factors affecting this are riverbank geometry, bank material strength, initial location of the phreatic surface, permeability of the material, duration of the flood and the rate of drawdown. Of these factors bank geometry and material permeability were found to play a key role.

Flatter slopes, due to longer seepage surfaces, are at a greater risk to flooding events. However, flatter slopes are more stable and these two factors compete. Soils of intermediary permeability, such as silt loam and loamy sands are most susceptible to flooding events. Low permeability soils, such as clays, can be susceptible if bank angles are steep. High permeability soils, such as gravels, are not susceptible to rapid drawdown following floods. The high permeability permits the phreatic surface to recede with the flood under all drawdown rates considered.

A preliminary design-aid is presented to guide designers as to which flood conditions may cause riverbanks to become unstable. For an improved design-aid, it is recommended that the effect of drawdown together with other bank erosion processes should be studied. It would also be beneficial to survey actual flood conditions for South African rivers and develop guidelines that reflect which rivers would be expected to experience stability problems because of those flood conditions.

ACKNOWLEDGEMENTS

The authors acknowledge the Water Research Commission for funding part of the study No.KSA2:K5/2270 from which this paper emanated.

REFERENCES

Budhu, M. & Gobin, R. 1995. Seepage-induced slope failures on sandbars in Grand Canyon, *Journal of Geotechnical Engineering*, 121(8): 601–609.

Chen, X. & Huang, J. 2011. Stability analysis of bank slope under conditions of reservoir impounding and rapid drawdown, *Journal of Rock Mechanics and Geotechnical Engineering*, 3:429–437.

CSRA (Committee of State Road Authorities). 1994. *Guidelines for the hydraulic design and maintenance of river crossings, Volume 1, Hydraulics, Hydrology and Ecology*. Committee of State Road Authorities, Pretoria.

Duncan, J.M., Wright, S.G. & Wong, K.S. 1990. Slope stability during rapid drawdown, *Proceedings of the H. Bolton Seed Memorial Symposium*, BiTech Publishers Ltd, Vancouver, British Columbia, Canada, Vol. 2.

DWAF (Department of Water Affairs and Forestry). 2007. *Berg River Baseline Monitoring Programme*. G. Ractliffe (Eds.), DWAF, South Africa.

Garcia, A.C.L. 2004. *Bank instability resulting from rapid flood recession along the Licking River*, Kentucky, MSc research thesis, University of Cincinnati, Ohio.

GEO-SLOPE. 2007. SEEP/W Version 7 Users Manual, GEO-SLOPE International, Calgary, Alta.

Green, S.J. 1999. *Drawdown and river bank stability*. MSc research thesis, Department of Civil and Environmental Engineering, University of Melbourne.

Hubble, T.C.T. 2010. Improving the stream of consciousness: A nomenclature for describing the factor of safety in river bank stability analysis, *Ecological Engineering*, 36: 1765–1768.

Oya, A., Bui, H.H., Hiraoka, N., Fujimonto, M. & Fukagawa, R. 2015. Seepage flow-stability analysis of the riverbank of Saigon river due to river water level fluctuation, Graduate School of Science and Engineering, Ritsumeikan University, Japan.

SANRAL (The South African National Roads Agency). 2006. *Drainage Manual*. The South African National Roads Agency, Pretoria.

SEEP/W. 2007. A software package for groundwater seepage analysis, Ver. 7. GEO-SLOPE International, Calgary, Alta.

SLOPE/W. 2007. A software package for slope stability analysis, Ver. 7. GEO-SLOPE International, Calgary, Alta.

Proceedings of the first Southern African Geotechnical Conference – Jacobsz (Ed.)
© *2016 Taylor & Francis Group, London, ISBN 978-1-138-02971-2*

FEM analysis of the behaviour of a flexible retaining wall

L. Hazout & A. Bouafia
University Saad Dahleb at Blida, Blida, Algeria

ABSTRACT: The results presented in this paper are part of a research program undertaken at the University of Blida focusing on the numerical and physical modelling of the behaviour of retaining walls. In order to investigate the effect of the proximity of a foundation on the displacement behaviour of a flexible retaining wall, a parametric study was carried out by using the finite elements software CRISP. The soil behind the wall was assumed following an elastic perfectly plastic constitutive law and obeying the Drucker-Prager failure criterion. A realistic behaviour of the soil/wall boundary was made possible by including nonlinear interface elements. After a description of the FEM model, the parametric study showed the distance between the wall head and the foundation is characterized by a threshold value beyond which the head deflections do not depend on this distance and the wall behaves independently of the presence of the foundation. The results were fitted and a practical formula for this threshold distance was suggested.

1 INTRODUCTION

In many parts of the world, the accelerated urban development of cities necessitates a maximum exploitation of the space by creating deep excavations supported by embedded flexible retaining walls.

Serviceability limit state design of such walls should account for the wall deflections. The latter might be caused by the active/passive pressures mobilized along the soil/wall interface by the proximity of a foundation as illustrated by the Figure 1. It is obvious an increase of the distance d may reduce the pressures mobilized along the wall and hence the head deflections y_0. It may be interesting to investigate the trend of such dependency

and the possibility of existence of a threshold distance beyond which the wall deflections are not influenced by the distance d. The finite elements method is a powerful tool which allows a displacement analysis of such a problem of soil/wall interaction.

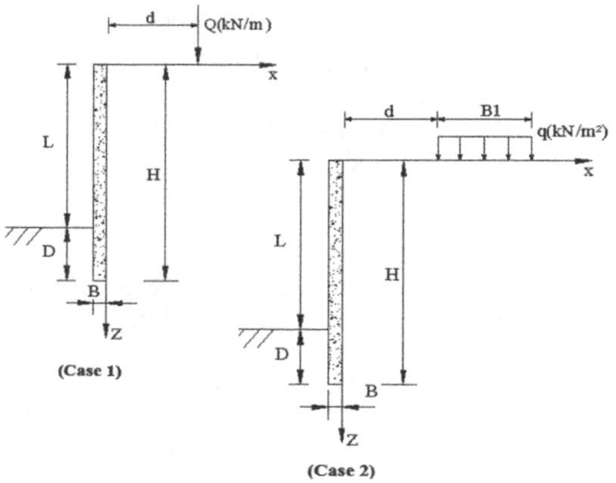

By using the FEM software CRISP, a parametric study was carried out by analysing the deflections of a flexible retaining wall supporting a homogeneous clayey soil exhibiting a drained behaviour. A shallow foundation (strip footing or a mat foundation) is assumed acting on the soil surface at various distances d from the wall head.

2 FINITE ELEMENT MODELLING

2.1 *Dimensional analysis of the problem*

According to the explanatory scheme illustrated in figure 1, the key parameters involved in such a problem may be compiled in the following general equation:

$$f\,(\varphi', C', \gamma, \nu_s, \nu_w, E_s, E_w, B, H, D, d, Q, y_o) = 0 \quad (1)$$

Dimensional analysis may be performed by using the Vashy-Buckingham theorem which leads to transform an equation of N physical parameters into another one of (N-k) dimensionless parameters called π parameters, k being the number of fundamental units (k = 3 for such a problem).

Equation (1) may then be replaced by the equivalent dimensionless equation as follows:

Figure 1. Explanatory scheme of the problem studied.

$$g(\pi_1, \pi_2, \ldots\ldots, \pi_{10}) = 0 \qquad (2)$$

$\pi_1 = \varphi'$ (drained angle of internal friction),
$\pi_2 = H/B$ (Slenderness ratio of the wall),
$\pi_3 = D/B$ (Embedment ratio of the wall),
$\pi_4 = d/B$ (distance ratio of the foundation),
$\pi_5 = v_w$ (Poisson's ratio of the wall material),
$\pi_6 = v_s$ (Poisson's ratio of the soil),
$\pi_7 = Q/\gamma H^2$ (Normalized foundation load),
$\pi_8 = y_0/B$ (Normalized wall head deflection),
$\pi_9 = E_w I_w/E_s H^4$ (Wall/soil stiffness ratio), and finally
$\pi_{10} = \gamma H/C'$ (Stability number for a general stability of the excavation).

2.2 Parameters of the soil/wall system

The retaining wall is a flexible infinitely long sheet characterized by a Young's modulus E_w of 4000 MPa, a Poisson's ratio of 0.15 (π_5), a width B taken equal to 0.50 m, an embedded depth D equal to 1 m, and a height H of 5 or 8 m, which correspond respectively to slenderness ratio (π_2) of 10 and 16. The wall/soil stiffness ratio (π_9) varies accordingly between 2.5×10^{-4} and 6.6×10^{-3}, corresponding to a flexible retaining wall.

The soil retained by the wall is a homogeneous saturated clay exhibiting a drained behaviour corresponding to a long term response after primary consolidation. Three types of soils were studied with the following geotechnical properties:

- Soft clay (Submerged unit weight $\gamma' = 10$ kN/m³, Drained Elastic modulus $E_s = 5$ MPa, Drained cohesion C' = 50 kPa, $\varphi' = 10°$ (π_1), $v_s = 0.33$ (π_6));
- Medium compact clay ($\gamma' = 10$ kN/m³, $E_s = 10$ MPa, C' = 100 kPa, $\varphi' = 20°$ and $v_s = 0.33$);
- Stiff clay ($\gamma' = 10$ kN/m³, $E_s = 20$ MPa, C' = 200 kPa, $\varphi' = 20°$ and $v_s = 0.33$).

It is to be noticed that the values of the stability number (π_{10}) vary between 0.25 and 1.6, to which correspond factors of safety against a general landslide of at least 2.87. The studied excavations are therefore stable without the presence of the retaining wall.

The foundation is either an infinitely long strip footing subjected to a distributed load Q (case 1 in Figure 1), or a long mat foundation having a width B_1 and subjected to a vertical pressure q (case 2 in Figure 2). Foundation loads are applied by increments in such a way the maximum values are less than the bearing capacity of the soil for each foundation configuration.

Since some dimensionless parameters are already fixed, equation (2) can be reduced to:

$$\pi_7 = h(\pi_8) \qquad (3)$$

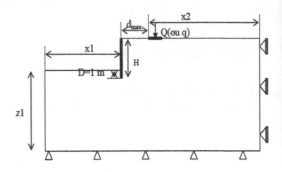

Figure 2. Scheme of design of the FEM mesh.

Function h describes a normalized loading curve, that is to say $Q/\gamma H^2 = h(y_0/B)$, for a given distance d of the foundation. A series of positions of the foundation was studied corresponding to distance ratios of the foundation (π_4) varying as follows: d/B = 0.5, 2, 4, 6, 8, 10, 12, 14, 16, 18 and 20.

2.3 Design of the FEM model

In order to guarantee representative results, the FEM mesh was designed with the objective to determine the minimum dimensions X_1, X_2 and Z_1 of the model, as illustrated in Figure 2, beyond which the computed displacements do not depend on these dimensions. The procedure consists of a regular increase of these dimensions as function of the height H. The lower horizontal boundary must be embedded (u = 0, v = 0) and the two lateral borders must be blocked horizontally (u = 0) (Nougier 1983).

The minimum sizes of the mesh are found as follows: $X_1 = 3H$, $Z_1 = 3H + D$ and $X_2 = 3H + 1$ (Hazout 2007).

2.4 Presentation of the FEM model

The soil mass was modelled by quadrangular finite elements of eight nodes, the behaviour being described by an elastic perfectly plastic constitutive law associated to the Drucker-Prager failure criterion. The retaining wall was also modelled by quadrangular finite elements of eight nodes and its response was assumed elastic.

The soil/wall interface was modelled by interface elements that can exhibit sliding, rubbing, or detachment and recollement. Their behaviour was described by an elastic perfectly plastic law characterized by an axial rigidity of 10000 kN/m/m, a tangent rigidity of 1000 kN/m/m, an adhesion equal to 2C'/3 and an angle of wall/soil friction equal to 2φ'/3, C' and φ' being the soil shear strength properties.

In order to increase the precision of the results, a mesh refinement was made within the zone of the soil/wall contact (around the wall). While moving away of this zone it is necessary to assure a progressive increase of the elements size (Nougier (1983), Delattre (2003)).

2.5 Presentation of the software CRISP

CRISP (Critical State Soil Mechanics Program) is a two dimensional software of finite elements applied in geotechnical engineering. CRISP was established in 1975 by a team of researchers from the Department of Engineering at Cambridge University.

The Pre-Processor provides an intuitive, interactive environment in which analysis can be quickly and easily created or revised. It is used to generate the input data for the analysis program. It also allows graphically to create the finite elements mesh, to define the soil properties, to specify the conditions in situ analysis, and finally to launch the analysis).

Figure 3 is an example of a deformed mesh in case of a strip footing in a stiff clay located to a distance of 4B from the wall head.

2.6 Presentation and interpretation of the results

Figure 4 illustrates typical normalized load-displacement curves obtained in case of a strip loading in a soft clayey soil. It is to be noticed that for a given load, the displacement y_0 at the wall

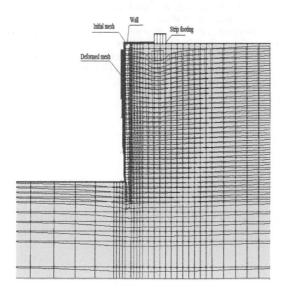

Figure 3. Illustration of a typical deformed mesh under a strip footing.

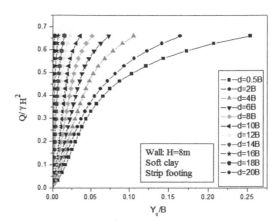

Figure 4. Dimensionless load-displacement curves in soft clay.

head decreases when the distance d between the wall and the strip increases.

In Figure 5 are compiled the curves of the wall head displacements as function of the distance between the strip footing and the wall in case of a wall high of 5 m. Whatever the type of clay, the curves stabilize as from a distance $d = 12B$, whereas in case of the wall of 8 m of height, the curves begin to stabilize as from 16B.

In case of a mat foundation, Figure 6 shows also the curves of the wall head displacements as function of the distance between the mat foundation and the wall in case of a wall high of 5 m. Independently of the type of clay, it can be seen that the curves stabilize as from a distance $d = 14B$, whereas in case of the wall of 8 m of height, the curves begin to stabilize as from 18B.

As shown in Figure 7, the dimensionless threshold distance d_{lim}/B is practically proportional to the wall slenderness ratio H/B in case of a strip footing as well as for a mat foundation. Simple relationship may be suggested as follows (Hazout 2007):

$$\frac{d_{lim}}{B} \approx (1 \ to 1.2)\frac{H}{B} \qquad (4)$$

3 CONCLUSION

A parametric study was undertaken on the basis of a finite elements analysis of the displacement behaviour of a flexible retaining wall supporting a saturated clayey mass and subjected to a nearby foundation loading. It has been shown the existence of a threshold distance beyond which the effect of the proximity of a foundation on the wall

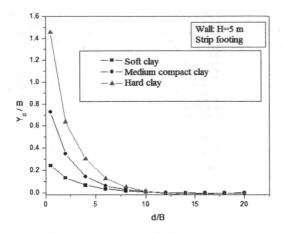

Figure 5. Variation of the normalized wall head displacement with the distance d.

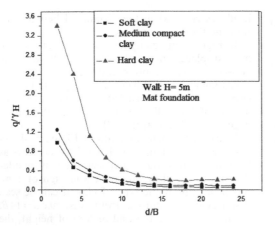

Figure 6. Variation of the normalized wall head displacement with the distance d.

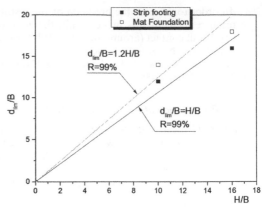

Figure 7. Variation of the threshold distance d_{lim} versus the wall slenderness ratio.

deflections vanishes. This distance is approximately equal to the height H of the retaining wall.

Further analyses of other soil conditions and foundations configurations may lead to generalize such a relationship.

REFERENCES

Ameur, Y. 2005. *Analyse Expérimentale et Numérique de l'interaction Sol/Soutènement Rigide* (in French), MSc Dissertation. University of Blida.

CRISP Consortium. 2006. Sage Crisp Geotechnical Finite Element Analysis Software. User's Manuel Ver 5.1b.

Delattre, L. 2003. Un siècle de méthodes de calcul d'écrans de soutènement-Les approches empiriques et semi-empiriques (in French). *Bulletin des laboratoires des ponts et Chaussées*, Vol. 244–245, Ref 4457, pp: 31–51, France.

Hazout, L.2007. *Modélisation numérique de l'effet de la proximité d'un ouvrage sur un mur de soutènement-Etude paramétrique* (in French), M.Sc Dissertation, University of Blida.

Nougier, J.P. 1983. *Méthodes de Calcul Numérique* (in French), Eds. Masson, Paris.

Potts, D.M. & Zdradcovitch, L. *Finite Element analysis in Geo technical Engineering—Theory*. Thomas Telford (ed), UK.

Proceedings of the first Southern African Geotechnical Conference – Jacobsz (Ed.)
© *2016 Taylor & Francis Group, London, ISBN 978-1-138-02971-2*

Trapdoor experiments studying cavity propagation

S.W. Jacobsz
University of Pretoria, Pretoria, South Africa

ABSTRACT: It is often assumed when analysing potential sinkhole size that larger diameter sinkholes are likely to form in deeper soil profiles because it is assumed that the potential development space expands from bedrock towards the surface in a funnel-shaped fashion (Buttrick & Van Schalkwyk 1995). A series of large-displacement trapdoor experiments were carried out in the geotechnical centrifuge to investigate the propagation of the zone of influence above the trapdoor towards the surface. Experiments were carried out in homogeneous fine and medium grained sand under respectively dry and moist and loose and dense conditions. It was found that the zone of influence propagated towards the surface in a vertical, chimney-like fashion and that it did not expand in width towards the surface, with the exception of tests in very loose sand. The width of the zone of influence was found to be equal to the width of the trapdoor. The results of trapdoor experiments may imply that the way in which maximum potential sinkhole diameter is assessed requires reconsideration.

1 INTRODUCTION

Sinkholes regularly occur in dolomitic ground in South Africa. In order to develop on dolomitic land it is often necessary to estimate the size of a sinkhole that should be designed for. Guidelines for the estimation of sinkhole size from literature are limited. The South African National Standard for the development on dolomitic land SANS 1936 (2012) advocates that potential sinkhole size be assessed using a "rational approach" and presents a "deemed to satisfy" method taken from Buttrick & Van Schalkwyk (1995).

Buttrick & Van Schalkwyk (1995) proposed that the maximum potential development space for a sinkhole be estimated by extrapolating a funnel-shaped zone from a receptacle in or near bedrock towards the surface at different "angles of draw". These angles (α) vary and depend on the properties of the overburden as shown conceptually in Figure 1. Where the funnel intersects the surface its diameter is usually taken as the maximum sinkhole diameter that should be designed against. An important consequence of this method is that the maximum sinkhole diameter is directly proportional to the thickness of the overburden above the receptacle that gave rise to the formation of the sinkhole. This method is known to provide conservative results and actual sinkholes are generally found to be smaller than predicted in this way.

The formation of sinkholes in the context of the South African dolomite is generally accepted to be associated with the presence of cavities and/or pockets of low density residuum in or just

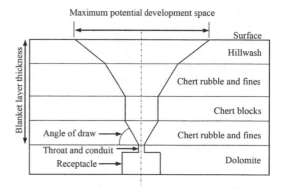

Figure 1. Estimation of the maximum potential development space of a sinkhole (modified from Buttrick & Van Schalkwyk 1995).

above the dolomite bedrock. Competent material would generally span or arch over these cavities or pockets to support the residuum and transported materials constituting the overlying soil profile. The material arching over these cavities may be extremely heterogeneous and may comprise boulders, gravels and slabs or blocks of chert, dolomite rock or transported materials, generally in a matrix of unsaturated soils which may range from clays to sands or gravels. Due to the presence of matric suctions, these soils possess a certain amount of "cohesive" strength providing the "mortar" keeping the larger fragments in position, increasing their ability to arch. Left undistributed, such arches may remain stable nearly indefinitely so that the risk of sinkholes development on undisturbed land is very low (Buttrick & Van Schalkwyk 1995).

Jennings et al. (1965) listed five requirements necessary for a sinkhole to form.

- Rigid material must be present along the perimeter of the soil arching over a cavity to provide abutments capable of supporting the load associated with an arching soil dome.
- Arching must develop within the residuum.
- A void must develop below the arch.
- A reservoir must be present below the arch to accommodate material removed to form the void and a means of transport of this material into the reservoir must be available.
- Once a stable void has formed, a disturbing agency is required to disrupt the equilibrium of the arching soil, resulting in a sinkhole eventually appearing atthe surface. Jennings et al. (1965) mentioned that the disturbing agency is most often infiltrating water, resulting in loss of strength of the arching soils, leading to collapse.

In a deep soil profile it may take a long time for a cavity at depth to gradually propagate towards the surface. Jennings et al. (1965) reported that collapse of the cavity roof occurs in an union-skin peeling fashion, with material falling from the roof to the floor so that the cavity migrates upwards. When the cavity approaches the soil surface the material arching over the cavity may eventually become too weak to support the load, resulting in the arch collapsing and a sinkhole appearing at the surface. This collapse of soil horizons leading to the formation of sinkholes has been studied by various researchers using numerical and physical models e.g. Augarde et al. (2003), Drumm et al. (1990), Craig (1990) and Abdullah & Goodings (1996).

The work reported here aimed to investigate the process of cavity propagation. The point of departure was the situation of a soil profile in equilibrium in which support was removed under a small plan area at the base of the profile in order to observe how the zone of influence would propagate towards the soil surface. The removal of support from the base of a soil profile can be studied by means of physical models using trapdoor experiments as has been done by researchers such as Terzaghi (1936), Santichaianaint (2002) and Costa et al. (2009). To ensure realistic stress-strain behaviour of the soil in a physical model in which soil self-weight is important, it is necessary to raise the stress level in the model to that acting in the full-scale prototype. The model study was therefore carried out in the geotechnical centrifuge at the University of Pretoria (Jacobsz et al. 2014).

2 PREVIOUS TRAPDOOR STUDIES

Several researchers have carried out model trapdoor experiments at normal gravity and in the centrifuge, with some of the first probably being the trapdoor experiments by Terzaghi (1936) and the latest that could be found in the literature the deep trapdoor experiments by Costa et al. (2009). Researchers found that, as soon as the trapdoor was lowered, the vertical stress acting on the trapdoor immediately reduced rapidly. To maintain equilibrium stress redistribution occurred, resulting in a significant increase in horizontal stress, i.e. the development of so-called arching conditions. As soon as the shear strength of the soil was exceeded, failure occurred along curved shear surfaces, the shape of which depends on the dilation properties of the sand (Costa et al. 2009).

As far as could be determined, published trapdoor experiments were carried out in dry cohesionless sands. For a fine grained soil to span a cavity, some tensile or cohesive strength is required. Cohesionless material does not have the ability to span as it possesses no tensile or cohesive strength and the development of a stable arch can therefore not occur.

3 CENTRIFUGE MODEL

A number of large-displacement trapdoor experiments were carried out in a fine grained and medium grained sand in a loading frame that allowed the upward propagation of the zone of influence to be studied in two-dimensional plane-strain conditions. What distinguished these tests from those in the literature was that moisture was added to the sand resulting in matric suction providing a small amount of tensile or cohesive strength.

The model set-up is illustrated in Figure 2. The model comprised a frame built from aluminium alloy channel sections that contained a volume of sand measuring 500 mm wide up to 450 mm high. The thickness of the sand body was 80mm. The front of the sand sample was contained by a 20 mm thick safety glass panel which allowed effect of the lowering of the trapdoor to be observed throughout the sand mass using a high resolution digital camera. A trapdoor that could be lowered during experiments were located in the base of the frame. The frame could accommodate trapdoors of 50mm, 75 mm and 100 mm width respectively. The trapdoor was supported by a water-filled piston. The lowering of the trapdoor was accomplished by extracting water from the piston using a second piston attached to a linear drive powered by

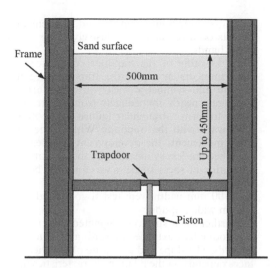

Figure 2. Illustration of test frame used for trapdoor experiments.

a stepper motor. The stepper motor was controlled from the centrifuge control room and allowed precise control over the lowering of the trapdoor during tests.

During model preparation dry sand was poured into the model using a flexible hose attached to a sand hopper. The sand density was controlled by maintaining a constant the drop height and flow rate. After sand pouring, the sand surface was levelled and five settlement transducers (LVDTs) were mounted to record the settlement of the sand surface. The settlement of the trapdoor was also monitored using a LVDT.

Tests were first carried out on dry cohesionless sand. In order to study the effect of a small amount of cohesive strength, a number of tests were carried out using moist sand with water providing a small amount of matric suction binding soil grains together. In test on moist sand, the sand was placed dry and the model was placed on the centrifuge. The soil surface was covered with a geotextile and water was introduced by means of a pipe with spigot holes along its length. The downward propagation of the wetting front was observed until the entire sand mass had been wetted to the base. The centrifuge was then started and allowed to accelerate slowly in steps of 5G to allow excess water to drain from the model. If the centrifuge was accelerated too rapidly, piping failure occurred and an uncontrolled sinkhole formed before the desired acceleration could be reached.

The model scale was 1:50 and therefore all tests were carried out at an acceleration of 50G.

4 MATERIALS TESTED

Tests were carried out on two grades of silica sand, a medium grained and fine sand respectively referred to as Consol sand and Cullinan sand. Consol sand is from a commercial source near Cape Town. It has a D_{50} particle size of 0.71 mm and the grains are rounded. Cullinan sand comes from a commercial source near Pretoria. It has a D_{50} particle diameter of 0.17 mm with sub-angular grains. The grading curves for both sands are presented in Figure 3. From the grading curves it is evident that both sands are quite uniformly graded. Other sand properties are tabulated in Table 1.

5 TESTS CARRIED OUT

Trapdoor experiments were first carried out on loose Consol sand in the dry and moist states, followed by test on dense sand also dry and moist. The relative density in the loose state was approximately 30% and in the dense state approximately 75%. This was followed by tests on dense Cullinan sand at a relative density of approximately 80% in the dry and moist states. These tests were all car-

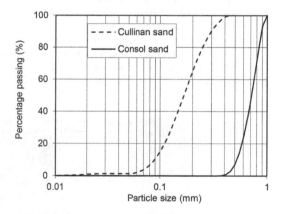

Figure 3. Grading curves for the sand tested in the trapdoor experiments.

Table 1. Selected properties of sand tested.

Property	Consol sand	Cullinan sand*
Minimum dry density (kg/m3)	1585	1392
Maximum dry density (kg/m3)	1735	1669
Specific gravity of grains (g/cm3)	2.652	2.666
Constant volume friction angle	34°	37°

*Measured by Archer (2014).

Table 2. Summary of tests carried out.

Sand type	Density	Moisture content	Trapdoor width
Coarse	Loose	Dry	50 mm
	Loose	Moist	50 mm
	Dense	Dry	50 mm
	Dense	Moist	50 mm
Fine	Dense	Dry	50 mm
	Dense	Moist	50 mm
	Dense	Moist	75 mm
	Dense	Moist	100 mm

ried out using a 50 mm wide trapdoor. Two further tests were carried out using a 75 mm and 100 mm wide trapdoor. In all tests the depth of sand above the trapdoor amounted to approximately 400mm. The tests reported on in this paper are summarised in Table 2.

6 RESULTS

The photographic record from the experiments was analysed using Particle Image Velocimetry (PIV) (White et al. 2003) to generate plots of the maximum shear strain distribution which vividly revealed the deformation mechanism during the course of the experiments. The observed development of maximum shear strain distributions in response to trapdoor settlement is illustrated in Figure 4. Figure 5 presents shear band development more generically. With the exception of the

dry, loose, medium grained sand, movement in al tests occurred almost exclusively along district shear bands.

As lowering of the trapdoor commenced, the sand immediately above the trapdoor began to expand and symmetric shear surfaces started to propagate nearly immediately from the trapdoor edges to form a triangular failure wedge moving downward with the trapdoor. With further trapdoor movement, the development of shear surfaces became less symmetric and more erratic. The formation of shear surfaces was found to be more symmetric when the wider trapdoors of 75 mm and 100 mm width were tested compared to the 50 mm wide trapdoor.

Further shear surfaces originated from near the trapdoor edges, extended upwards and then curved inwards. The slope and curvature of the shear surfaces depends on the dilation characteristics of the sand which is controlled by the relative density and the stress level (Bolton 1986; Costa et al. 2009).

As a shear band intensified, the wedge of sand below it eventually dropped down, followed by the formation of another shear band on the other side of the trapdoor. After the initial symmetrical behaviour, shear zone development alternated between the left and right hand sides of the trapdoor as the trapdoor was lowered. The sequence is illustrated in Figure 5.

With propagation, the apex of the advancing zone of influence gradually narrowed, becoming approximately parabolically shaped. As the material below the advancing front settled, the sides of the parabolic arch lost support and another shear

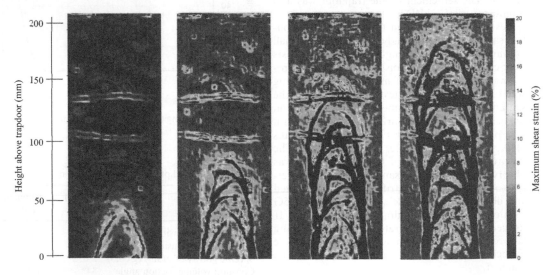

Figure 4. Evolution of maximum shear strain distribution illustrating zone of influence propagation above trapdoor.

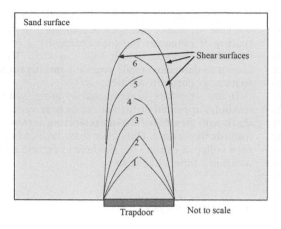

Figure 5. General illustration of the progressive development of shear surfaces in response to lowering of the trapdoor.

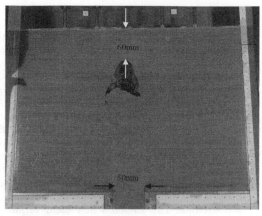

Figure 6. The formation of a stable cavity in moist dense Cullinan sand.

surface soon shot upward from above the trapdoor edges, widening the zone of influence to that of the trapdoor. As the zone of influence advanced, the sides of the shaft of sand moving downward with the trapdoor became vertical.

In tests with moist sand the development of a crack along a shear band was preceded by a change in colour of the sand along the developing shear band in response to dilation. In all cases, after sufficient trapdoor movement, the zone of influence propagated vertically upwards as also reported by other researchers (e.g. Costa et al. 2009).

In the case of the test on moist dense Cullinan sand lowering of the trapdoor resulted in the formation of a stable cavity after the apex of the zone of influence had propagated to approximately 60 mm from the surface and no visible subsidence appeared at the surface (see Figure 6). As the trapdoor movement commenced, the displacement transducers immediately registered a small amount of surface settlement which increased approximately linearly with increasing trapdoor settlement. However, the rate of surface settlement slowed considerably once a stable arch had formed.

The matric suction provided by the pore water was sufficient to allow the sand to span the cavity in the case of a 50 mm wide trapdoor. In the tests with wider trapdoors (75 mm and 100mm) a stable cavity could not form and the shaft of downward moving sand eventually breached the surface, manifesting as a shaft with vertical walls. The width of the shaft at the surface was similar to that of the trapdoor. As the trapdoor was lowered further, the sides of the shaft at the surface began to slump inward, causing the width of the

"sinkhole" at the surface to increase, taking on a funnel shape.

Only in the case of dry loose medium grain sand did the width of the zone of influence expand somewhat as it grew towards the surface. The addition of moisture to the loose medium grained sand caused the zone of influence also remain constant and equal to the width of the zone of the trapdoor.

7 DISCUSSION

In all trapdoor experiments, with the exception of that in loose medium grained sand, the propagation of the zone of influence occurred vertically and did not widen towards the surface. The addition of moisture to this material resulted in a small amount of matric suction and that was sufficient to stop the widening of the zone of influence, causing it to propagate vertically.

In sand the application of a horizontal stress equal or greater than the active soil pressure against the vertical sides of the zone of influence should be sufficient to support these sides. This will prevent soil failure into the zone of influence and therefore the widening thereof. Lowering of a trapdoor reduces the vertical stress in the sand above it (Terzaghi 1936, Handy 1985), requiring an increase in horizontal stress, i.e. the development of arching conditions to maintain equilibrium. These increased horizontal stresses will significantly exceed the active soil pressure and will therefore be even more efficient than the original horizontal stress to prevent widening of the zone of influence.

Cavity growth towards the surface was well illustrated in the test on moist dense Cullinan sand in which a stable arch eventually formed above the trapdoor. Initially a stable arch could not form due to the magnitude of the overburden. However, once the zone of influenced had propagated sufficiently, the overburden above it was small enough to be supported by the soil and a cavity formed. As the trapdoor was lowered, the system constantly strived to establish equilibrium by reducing the radius at the apex of the arch that formed, causing the cavity to take on a parabolic shape. However, as soil settled below the arch in response to the lowering trapdoor, the span of the arch at its base became too wide to be supported by the available soil strength. This caused a new shear surface to develop from near the base of the arch, propagating upwards and curving inwards, causing a wedge of soil to fall down and the cavity to propagate upwards. Cavity propagation eventually stopped when the overburden was insufficient to cause the arch to fail given its span and the cavity remained stable.

8 IMPLICATIONS FOR SINKHOLE DEVELOPMENT IN PRACTICE

In reality sinkholes do not form due to the lowering of a trapdoor. However, in the test on moist dense Cullinan sand, a cavity did eventually form, which allowed its propagation to be studied. It is believed that some of the mechanisms observed do resemble actual processes at play in the formation of dolomitic sinkholes.

Jennings et al. (1965) reported that cavities propagate to the surface due to deterioration of the arch supporting the overhead material in an union-skin peeling fashion. The maximum shear strain evolution presented in Figure 5 resembles this description.

Soil arching over a cavity would exert a high horizontal thrust, supporting the adjacent material, preventing it from failing into the cavity, so that it would not widen as it grows. In fact, the apex of the arch influence tended to narrow as the system tried to establish equilibrium by reducing the arch radius. Actual sinkholes are often observed to become narrower near the surface shortly after they appears.

Results of the trapdoor experiments indicated that the width of the sinkhole appearing at the surface in moist sand was approximately equal to the width of the trapdoor. The trapdoor width also controlled the width of the arch that formed (in the test in which an arch did form). It therefore appears that in order to predict sinkhole size, it is necessary to determine the span of the arch that

could progressively collapse to eventually form a sinkhole. In reality this span is controlled by the spacing of pinnacles and rock abutments at bedrock depth between which soils can arch. This cannot to determine easily given site investigation technology currently available.

In the trapdoor experiments on moist sand, sinkholes appeared at the surface as near vertical shafts with the sides of the shafts tapering narrower towards the surface. Only after some time did the sides collapse, causing the sinkhole to expand and widen in a funnel-like fashion.

9 CONCLUSIONS

A series of trapdoor experiments enabled the propagation of the zone of influence towards the surface resulting from the lowering of a trapdoor to be studied. Moist and dry, medium and fine grained sands of loose and dense consistencies were tested. It is believed that the mode of propagation could be indicative of the way in which full-scale cavities propagate towards the surface to form sinkholes.

With the exception of dry, loose medium sand, the zone of influence propagated to the surface in a vertical chimney-like fashion and did not widen as it grew. The thickness of overburden did not seem to influence the width of the zone of influence nor the sinkhole that eventually appeared at the surface.

The width of the sinkhole appearing at the surface was equal to the width over which support was removed at depth, i.e. the width of the trapdoor. Extrapolating to full scale, this may imply that in a homogeneous material, the diameter of a sinkhole would be similar to the diameter of the arch that was caused to deteriorate and propagate towards the surface. The dimensions of such an arch is controlled by the spacing of the abutments (e.g. rock pinnacles) at depth against which the arch can thrust. This is at present difficult to determine in sufficient resolution given current site investigation technology.

The presence of stronger horizons overlying the material through which a cavity grows was not studied and is a subject of ongoing research.

ACKNOWLEDGEMENTS

The author wishes to thank Mr Johan Scholtz, chief technician of the Civil Engineering laboratories at the University of Pretoria, who constructed the test frame and Mr Benjamin Oberholzer who prepared the centrifuge models.

REFERENCES

Abdullah, W.A. & Goodings, D.J. 1996. Modelling of sinkholes in weakly cemented sand. *Journal of Geotechnical Engineering* 122(12): 998–1005.

Archer, A. 2014. *Using small-strain stiffness to predict the settlement of shallow foundations on sand.* MEng dissertation. University of Pretoria.

Augarde, C.E., Lyamin, A.V. & Sloan, S.W. 2003. Prediction of undrained sinkhole collapse. *ASCE Journal of Geotechnical and Geoenvirmental Engineering* 129(3): 197–205.

Bolton, M.D. 1986. The strength and dilatancy of sands. *Geotechnique* 36(1): 65–78.

Buttrick, D. & Van Schalkwyk, A. 1995. The method of scenario supposition for stability evaluation of sites on dolomitic land in South Africa. *Journal of the South African Institution of Civil Engineering.* Fourth quarter 1995: 9–14.

Costa, Y.D., Zornberg, J.G., Benedito, S.B. & Costa, L. 2009. Failure mechanisms in sand over a deep active trapdoor. *ASCE Journal of Geotechnical and Geoenvironmental Engineering* 135(11): 1741–1753.

Craig, W.H. 1990. Collapse of cohesive overburden following removal of support. *Canadian Geotechnical Journal*, 27: 355–364.

Drumm, E.C., Kane, W.F. & Yoon, C.J. 1990. Application of limit plasticity to the stability of sinkholes. *Engineering Geology* (Amsterdam), 29: 213–225.

Handy, R.L. 1985. The arch in soil arching. *ASCE Journal of Geotechnical Engineering* 111(3): 302–318.

Jacobsz, S.W., Kearsley, E.P. & Kock, J.H.L. 2014. The geotechnical centrifuge facility at the University of Pretoria. *Proc 8th International Conference on Physical Modelling in Geotechnics*, Perth. CRC Press: 169–174.

Jennings, J. E., Brink, A.B.A., Louw, A. & Gowan, G.D. 1965. Sinkholes and subsidence in the Transvaal dolomites of South Africa: 51–54. *Proc 6th International Conference on Soil Mechanics.* Montreal, Canada: 51–54.

Santichaianaint, K. 2002. Centrifuge modelling and analysis of active trapdoor in sand. PhD thesis. University of Colorado at Boulder.

South African Bureau of Standards. 2012. *Development of dolomite land–Part 2: Geotechnical investigations and determinations.* SANS 1936–2.

Terzaghi, K. 1936. Stress distribution in dry and saturated sand above a yielding trap-door. *Proc 1st International Conference of Soil Mechanics*, Harvard University, Cambridge (USA), 1: 307–311.

White, D., Take, W. & Bolton, M. 2003. Soil deformation measurement using Particle Image Velocimetry (PIV) and photogrammetry. *Geotechnique*, 53(7): 619–631.

Site investigation

Proceedings of the first Southern African Geotechnical Conference – Jacobsz (Ed.)
© *2016 Taylor & Francis Group, London, ISBN 978-1-138-02971-2*

Discrete element analysis of granular soil recovery in a vibrocore

S.B. Wegener & D. Kalumba
University of Cape Town, Cape Town, South Africa

ABSTRACT: Vibrocoring consists of a vertical, tubular core barrel with a sharp cutting edge at its lower end vibrated into the seabed by a high-frequency, low-amplitude vibratory motor. Through a review of literature and the calibration of a discrete element model to physical vibrocore test results, this study investigated the soil mechanics phenomena influencing the recovery of granular soil in a tubular vibrocore. The accuracy of the numerical results was confirmed through comparison to the physical results using statistical methods.

1 INTRODUCTION

1.1 *Background and context*

Seabed soil sampling is crucial in deep-water engineering projects or geological studies in which a detailed knowledge of the seabed geology is required. Vibrocoring is a relatively new technique used to sample soil in offshore geotechnical practice. It is a form of drive sampling where a hollow, cylindrical core barrel, with a sharp cutting edge at its lower end, is driven into the ground by vibratory means. The sampling method is described in more detail by Smith (1998) and Weaver & Schultheiss (1990).

In the past, the success of a coring operation has often been judged primarily by the length of the recovered core. More recently, geotechnical studies have given focus to the problems associated with achieving a higher quality soil specimen (e.g. Parker & Sills 1990, Skinner & McCave 2003) When assessing the quality of a soil sample obtained through coring, there are two aspects that need to be considered: a) The influence of the sampling process on the laboratory measured mechanical properties of the soil specimen—i.e. the disturbance of the core sample, and b) Whether or not the sample is representative of the depth from which it is thought to have been recovered—i.e. the Recovery Ratio (RR) of the core sample (Lunne & Long 2006). The RR is defined as the ratio between the recovered length of core sediment and the length of core barrel penetrated into the soil. Therefore, an ideal specimen, in which the stratigraphy of the sampled ground is preserved, would have a recovery ratio of unity. However, due to the physical processes governing barrel-soil interaction during vibrocore penetration, ideal recovery is very rarely achieved in practice.

1.2 *Objectives*

The main objective of this study was to—through reviewing relevant literature and conducting physical testing and numerical modelling—identify and investigate the soil mechanics phenomena controlling the recovery of granular seabed soils in vibrocores. The following sub-objectives were developed: a) The calibration of a 3-Dimensional (3D) discrete element model to the physical results of vibrocoring tests conducted in saturated soil, and b) The assessment of the physical and numerical results, offering insight into the observed response of the vibrocoring system, including soil recovery and forces required for barrel penetration.

1.3 *Scope and delimitation*

All physical testing was conducted at the De Beers Marine Pty (Ltd)—henceforth referred to as DBM—Research and Development Test Facility in Cape Town. A large-scale vibrocore test rig was used. All numerical modelling was completed using ROCKY Discrete Element Method (DEM) software—henceforth referred to as Rocky. The soil samples investigated were poorly-graded gravels. There were two reasons for the selection of gravel as the tested and modelled soil type: a) Vibrocore testing completed by DBM engineers prior to this study showed that a very coarse, granular geology had a more problematic and less predictable effect on the mechanical system than that of finer-grained sands and clays (Raubenheimer 2015), and b) The discrete element modelling of particle sizes smaller than that of the gravels tested would result in simulations of impractically large computational expense.

2 REVIEW OF LITERATURE

2.1 The soil plug mechanism

During penetration of a corer into the seabed, under-sampling often occurs due to soil immediately below the barrel flowing around, rather than entering, the sample tube—discussed extensively by Parker & Sills (1990). Such a phenomenon is the result of a 'soil plug' forming within the barrel. The cause of plugging is explained by Skinner & McCave (2003): A descending core barrel accumulates an increasing sediment column inside it, subject to an increasing downward frictional drag. This friction is transferred through the soil column and imposed over the cross-sectional area of the core aperture as a vertical stress σ_v on the soil immediately below the corer at that instant in its descent.

From static equilibrium, it follows that plugging occurs when this vertical stress σ_v exceeds the mobilised bearing capacity q_{ub} of the soil about to be sampled. In simpler terms, plug formation is a result of punching shear failure of the soil immediately below the barrel, causing the sampler to penetrate as if it were close-ended. This leads to a decrease in recovery ratio.

2.2 The three modes of sampler penetration

The mechanism of plugging described above suggests that a sampler has only two modes of penetration: plugged or unplugged. However, there is a third: partially plugged. Once σ_v reaches equilibrium with q_{ub}, the two pressures must remain in equilibrium. An increase in bearing capacity due to greater penetration depth leads to sediment continuing to enter the corer to supply sufficient friction. Thus, partially plugged behaviour represents the squeezing of the soil into the sampling tube as σ_v and q_{ub} maintain equilibrium (Skinner & McCave 2003).

The Incremental Filling Ratio (IFR) is a parameter used to describe the mode of barrel penetration. The IFR is the RR over a given depth increment. Therefore, plugged behaviour yields an IFR of zero. A sampler penetrating without plugging has an IFR of unity. And partial plugging corresponds to $0 < \text{IFR} < 1$. Parker & Sills (1990) and DeNicola & Randolph (1997) showed that samplers penetrate primarily in a partially plugged mode. More specifically, these studies, which both assessed the incremental flow of soil into a descending tube, noted marked local variability in IFRs with penetration depth. These rapid and significant changes in IFR were attributed to the continual process of plugging—which causes soil densification around the core barrel head—followed by plug 'failure',

due to increase in bearing resistance from soil densification, ultimately allowing fresh soil to move into the tube.

2.3 Non-linearity of the soil plug mechanism

Randolph et al. (1991) derived an analytical formula for the imposed vertical stress σ_v for drained soil conditions—given by the equation:

$$\sigma_v = \frac{\gamma' D}{4K \tan \delta} \left(e^{\frac{4K \tan \delta}{D} L} - 1 \right) \quad (1)$$

where γ' = effective unit weight of the soil; D = inner diameter of the core barrel; K = ratio of horizontal to vertical effective stress at the barrel wall; δ = interface friction angle; and L = length of soil sampled within the barrel. The derivation of Equation 1 assumed that L is equal to penetration depth of the core barrel base d.

Through critical review the equation, it can be noted that a) The imposed stress increases exponentially with penetration depth, and b) The soil plug mechanism is a highly non-linear phenomenon. Only the barrel diameter D can be considered constant with penetration depth. According to Skinner & McCave (2003), it is expected that the interface friction and effective unit weight will evolve with penetration. Furthermore, K is not constant—both in its distribution along the length of the soil plug, and with penetration depth (De Nicola & Randolph 1997).

In addition to the properties of the sampled soil, other factors that influence the recovery of the core specimen include various physical characteristics of the core barrel, including cutting edge angle and wall thickness—discussed in more detail by Clayton et al. (1998), Lunne & Long (2006), and Wegener (2015).

2.4 Analogy between samplers and piles

In developing an understanding of how the soil plug mechanism is influenced by the cyclic action of vibrocores, a clear lack of geotechnical literature on the subject was noted. As such, a direct geometric analogy between samplers and cylindrical, open-ended, continuously hollow piles was assumed—as recommended by Paikowsky et al. (1989) and Skinner & McCave (2003). This allowed geotechnical study on the topic of plugging in samplers to be supported by the relative wealth of published literature analysing soil plugging in offshore piles. In completing this research, the terms 'pile', 'sampler tube', and 'core barrel' were considered synonymous. Nonetheless, it is worth noting that to date, no literature has been found that investigates the recovery of gravels in offshore vibrocores.

2.5 *The discrete element method*

The DEM is a numerical technique used to approximate soil behaviour. Unlike the finite element approach, the DEM does not assume the soil to be a continuum defined by some constitutive law. Rather, the DEM, developed by Cundall & Strack (1979), is a particle-based approach, in which the soil is modelled as an assembly of distinct particles. Discrete element calculations capture the complex behaviour of soil through simple assumptions and physical laws at interparticle contact, simulating the motion of every particle within a soil mass. However, with current computing capabilities, it is difficult to model the number of particles within soil volumes large enough to be of practical interest. Both Cundall (2001) and Ng & Meyers (2015) explain that full-scale 3D DEM simulations would take an excessively long time to compute due to the need to model billions, even trillions, of discrete particles.

Nonetheless, Jiang & Hu (2006) have shown the method to be a powerful tool for analysis in geomechanics, used to examine several aspects of soil behaviour—e.g. creep theory, anisotropy, particle crushing, strain localization, and liquefaction. A more detailed review of the history, theory and application of the DEM is provided in Cundall & Strack (1979) and Wegener (2015).

3 PHYSICAL TESTING EQUIPMENT AND PROCEDURE

3.1 *Soil material*

The gravels used for testing consisted of predominantly well-rounded, very hard quartzitic sandstone, sourced from the Orange River in South Africa. Particle sizes ranged from 16 mm to 64 mm.

The material was chosen for use at the facility based on its similarity to the transported, resilient seabed diamond gravels off the coast of Namibia, where DBM have conducted mining operations (Burger 2015). The Particle Size Distribution (PSD) of the soil is shown in Figure 1.

3.2 *Sampling equipment*

The components of the vibrocore system are summarised in Table 1 and illustrated in Figure 2.

3.3 *Monitoring devices*

Four monitoring devices were used during physical testing to measure the response of the soil

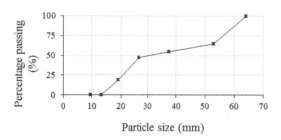

Figure 1. Particle size distribution of gravel used for vibrocore testing.

Table 1. Components of the vibrocore system.

Component	Description
Main frame	A 6.4 m high, 2 m wide truss structure, consisting of 250 × 250 mm hollow square steel cross-sections; Bolted to a specifically designed foundation; Supports the vibratory motor and core barrel
Sample bin	A 2.5 m high, hollow, cylindrical steel bin; 621 mm internal diameter; 0.75 m^3 volume capacity; Contains the sample soil during testing; Weighs approximately 1000 kg when empty
Vibratory motor	Motor consisting of rotating asymmetrical mass; Rotating mass creates high-frequency, low-amplitude oscillatory motion of the motor; Frequencies up to 200 Hz, Amplitude a function of the frequency; Motion limited to vertical direction
Core barrel	Hollow, high tensile steel tube; 150.4 mm inside diameter; 177.8 mm outside diameter; 2 m long; Has 100 mm thread at top and bottom for connection to the vibratory motor and cutting head, respectively
Cutting head	Hollow, cylindrical, steel cutter head; Total height = 344 mm; Wall thickness at base = 17.3 mm; Cutting shoe angle = 60°; Cutting edge is on the inside diameter wall; Screwed onto core barrel

and sampling equipment during or immediately after coring (Table 2). The measured data served two purposes: 1) Provided insight into the behaviour of the vibrocore system during coring, and 2) Provided the basis for calibration of the numerical model to the physical system. Due to practical limitations, strain gauges could not be placed on the core barrel during vibrocore

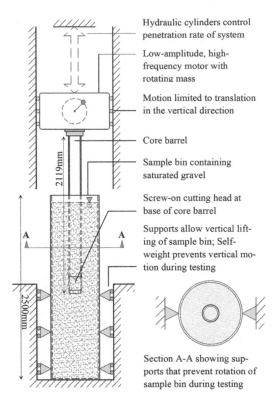

Hydraulic cylinders control penetration rate of system

Low-amplitude, high-frequency motor with rotating mass

Motion limited to translation in the vertical direction

Core barrel

Sample bin containing saturated gravel

Screw-on cutting head at base of core barrel

Supports allow vertical lifting of sample bin; Self-weight prevents vertical motion during testing

Section A-A showing supports that prevent rotation of sample bin during testing

2119mm

2500mm

A — A

Figure 2. Schematic elevation of vibrocore test rig sampling equipment and soil materials (Adapted from Wegener 2015).

Table 2. Summary of data provided by monitoring devices.

Monitoring device	Data provided
Existing monitoring system	Rotation frequency of the vibratory motor f_r; Penetration depth of the core barrel d; Hydraulic pressure in the cylinders controlling the vertical movement of the motor-barrel system P
Recovery measurement device	Recovery ratio at final penetration depth (The device consisted of a weight attached to a cable, which was lowered within the core barrel once at final penetration, allowing the measurement of sampled soil column height)
Accelerometer	Acceleration of core barrel a; Harmonic displacement of core barrel x
Strain gauges	Strain of metal plate supporting motor-barrel system ε; (Used to calculate vertical force on core barrel F_B)

testing. Thus, a series of preliminary coring tests were conducted to determine an empirical relationship between the vertical force acting on the core barrel during its penetration and the strain of the metal plates supporting the motor-barrel system. These tests are fully detailed in Wegener (2015).

3.4 Physical test methodology

Four vibrocore tests were conducted. In all tests, the height of the gravel within the sample bin was ~2.374 m and final barrel penetration depth was ~1.975 m. In tests one, two and three, all parameters specified in Table 2 were measured—except for core barrel acceleration. Due to signal interference from wireless transmitters within the monitoring devices, acceleration and strain could not be measured simultaneously. Only in test four was acceleration monitored. For each of the four tests, gravel samples were prepared using identical procedures. This was done to achieve, as close as was practically possible, constant soil conditions between successive tests. In preparing the gravel

samples before coring, the mass of the sample bin was measured a) when empty, b) when containing the dry gravel, and c) when containing the saturated gravel. By assuming the gravel had zero air voids when the sample bin was filled with fresh water, the average void ratio and specific gravity of the soil prior to sampling was calculated for each test. Of the samples prepared, the average void ratio was 0.510; the average specific gravity was 2.474. Slight deviations in void ratio between samples were observed. This was attributed to unavoidable losses of particles during sample preparation in practice. The differences between calculated specific gravities were negligible. During coring, the vibratory motor frequency f_r and core barrel penetration rate $\Delta d/\Delta t$ was manually controlled to be 157 Hz and 24 mm/s, respectively. If penetration rate or frequency was chosen too high, the vibrocore system would reach its operational limits, potentially leading to damage from repeated testing. Conversely, if $\Delta d/\Delta t$ was too low, the time required to be simulated in the calibrated DEM model would result in impractically long computation periods. If f_r was too low, the motor-barrel equipment would operate close to its natural frequency. The subsequent effects of resonance would result in large and unpredictable variations in core barrel amplitudes.

4 NUMERICAL MODEL ASSEMBLY

4.1 *Definition of geometries*

The Rocky model was a dynamic analysis. The simulation started with the input of soil particles into the sample bin and ended with the stationary core barrel at final penetration depth of 1.975 m. In assembling the model, the position and movements of the core barrel and sample bin could be specified at any point during the simulation. The bin and barrel were modelled as two separate 3D objects with dimensions the same as their physical geometry. The base of the barrel included the detailed geometry of the cutting head. To model the gravel, a Particle Shape Matrix (PSM) was developed to aid the quantification of the various particle shape types within the physical soil. The PSM is a 4×3 matrix with each matrix entry representing the approximate percentage of a particular particle shape within a total soil mass. The sum of all entries within a PSM is 100%. Columns are sorted according to the 'sphericity' of the particles. Rows are sorted according to the angularity of the particles. Tables 3a and 3b show the PSM of the physical gravel, as estimated from visual inspection, and the PSM of the modelled gravel. The computational expense of a DEM simulation increases significantly for increasingly complex particle geometries. In other words, DEM calculations are generally simpler and quicker for particles towards the bottom-right corner of the PSM. It was assumed that rounded-high-sphericity (RDH) particles could be modelled as perfect spheres. From Table 3, it is apparent that a) The physical gravel consisted mostly of sub-rounded and rounded particles, the majority of which were of medium sphericity, and b) The modelled gravel was selected

to have a markedly higher percentage of spherical (RDH) particles. A relatively high proportion of spherical particles were included in the modelled soil to simply the simulation. In evaluating the PSM shown in Table 4, matrix algebra was used to ensure that the PSD of the bulk gravel (Figure 1) was preserved. To model the more angular shape types towards the top left of the PSM, the simulated soil consisted of rounded polygons and polyhedrons—fully detailed in Wegener (2015).

4.2 *Selection of material properties*

DEM calculations required the density, stiffness and frictional properties of the materials within the model to be specified—summarised in Tables 4 and 5. It was assumed that the gravel was a cohesionless material. The software used did not allow the simulation of water in the model. Therefore to

Table 4. Density and stiffness parameters.

Material	Density kg/m³	Stiffness GPa
Gravel particles	1474	28
Core barrel	7800	200
Sample bin	7800	200

Table 5. Material interaction friction parameters.

Material interaction	Static friction factor	Dynamic friction factor
Gravel-gravel	0.725	0.725
Gravel-barrel	0.725	0.675
Gravel-bin	0.750	0.700

Table 3. Particle shape matrices of physical and modelled gravel.

a) Physical gravel		Sphericity			
		Low (L)	Medium (M)	High (H)	
Angularity	Angular (AR)	2.75	5.50	1.75	10.0
	Sub-Angular (SA)	3.44	6.88	2.19	12.5
	Sub-Rounded (SR)	7.56	15.13	4.81	27.5
	Rounded (RD)	13.75	27.50	8.75	50.0
		27.5	55.0	17.5	100
b) Modelled gravel		Sphericity			
		Low (L)	Medium (M)	High (H)	
Angularity	Angular (AR)	1.18	2.48	0.85	4.50
	Sub-Angular (SA)	1.71	3.51	1.17	6.39
	Sub-Rounded (SR)	6.27	13.30	4.61	24.18
	Rounded (RD)	7.69	16.12	41.12	64.93
		16.84	35.41	47.75	100

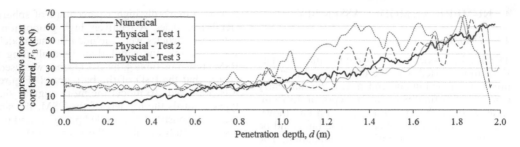

Figure 3. Comparison of numerical and physical force-penetration plots.

account for the effect of buoyancy, the density of water 1000 kg/m³ was subtracted from the physical gravel particle density 2474 kg/m³ to determine the density of the modelled gravel particles: 1474 kg/m³. The average void ratio of the modelled gravel was 0.585. The simulated soil had a slightly higher void ratio than that of the physical soil due to the differences in particle geometries between the two soils.

4.3 Core barrel penetration and dynamics

The modelled core barrel was specified to penetrate the simulated gravel at a constant penetration rate of 24 mm/s, vibrating at 157 Hz. The accelerometer readings from physical test four showed that the barrel amplitude increased from ~0.35 mm at zero penetration to ~0.50 mm at final penetration. This rise in harmonic displacement was attributed to the increasing stiffness offered by the soil as the barrel descends. In the Rocky model, only a constant barrel amplitude could be specified. To simulate the increase in amplitude with penetration depth, the descent of the sample tube in the model consisted of five successive stages. Each stage had a barrel amplitude greater than the previous stage, the value of which was the average of the physical amplitudes over the corresponding penetration interval.

5 RESULTS AND DISCUSSION

The core recovery ratio yielded by the numerical model was 54.1%. The RR values measured in physical tests one, two, and three were 50.7%, 46.6%, and 59.0%, respectively—giving an average of 52.1%. The low RR values suggest a clear difficulty in practically achieving high qualilty samples in gravel soils. To deteremine whether the test and model RR output was significantly different, a two sample t-test was conducted. Calculations showed that there was no statistical difference, at

95% confidence, between the physical and numerical recovery ratio results. The variability in physical results corresponded to variations in initial void ratios of the gravels sampled. This is in accordance with theory described in the literature: The extent of soil plugging is a function of soil bearing capacity, which in turn is dependent on soil density—i.e. void ratio.

Figure 3 shows plots of F_B as a function of penetration depth for both the physical and numerical results. According to reviewed literature, the force on the core barrel during sampling is a function of all vibrocore system variables—most importantly, the soil properties and core barrel depth and penetration rate. For this reason, force-penetration plots were considered an appropriate means to compare the micro-scale behaviour of the real and modelled soil during coring.

The strains measured by the monitoring gauges during the initial stages of barrel descent were primarily a function of the motor-barrel system weight—not the force acting on the barrel. Only upon reaching a penetration depth such that F_B exceeded the weight of the vibrocore system would the physical plots be representative of the force on the barrel. This is the reason for significant differences between the physical and numerical results during the first ~0.6 m of penetration.

The sharp decrease in physical plots during the last ~0.05 m of penetration was due to the decrease in core barrel penetration rate immediately prior to the end of sampling.

It is clear that no two force-penetration plots are the same. There is marked local variability in F_B with penetration depth. This was expected. The rapid local variations of the F_B-d distributions correspond to the micro-scale response and rearrangement of the particles as they enter the sample tube. Since it is statistically impossible that particle arrangements between two gravel samples would be the same, it followed that the resultant F_B-d plots were not indistinguishable. It was of greater importance to determine whether the difference between the numerical and force plots was significant. Notwithstanding the unavoidable local variability in F_B, the distribution trends for $d > {\sim}0.6$ m suggest that the numerical results approximate the physical result with reasonable accuracy. A Kolmogorov-Smirnov test—a non-parametric statistical method which effectively compares the shape of two bivariate data distributions—was conducted to confirm that the tests were of negligible difference. Thus, the numerical plot can be considered a good representation of the physical development of F_B with penetration depth. Results indicate that force on the core barrel increases approximately linearly from zero to a maximum value of 60kN at final penetration depth.

6 CONCLUSIONS

The soil mechanics phenomena controlling the recovery of granular seabed soils in vibrocores were identified and investigated through a review of relevant literature. Core shortening, or a decrease in recovery, often occurs in practice due to the plugging of soil inside the sample tube. The plugging mechanism is highly non-linear and a function a number of geometric and material factors—including core barrel cutting head design, rate and method of core barrel penetration, penetration depth, soil properties, and friction between barrel and soil.

Through assessment of physical and numerical results, the assumptions made in defining the DEM simulations were shown to be valid and the model was considered calibrated. It is recommended that the model serve as a basis for further research into the improvement of soil recovery in offshore tubular vibrocores.

ACKNOWLEDGEMENTS

The authors wish to sincerely thank De Beers Marine Pty (Ltd) for providing the funding, resources, equipment, and software that facilitated this research. Special thanks to Mr Johnny Lai Sang and Mr Gert Raubenheimer for their expertise, guidance and feedback.

REFERENCES

Burger, U. 2015. Personal communication. [2015, July 22].

Clayton, C.R.I., Siddique, A. & Hopper, R.J. 1998. Effects of sampler design on tube sampling disturbance—numerical and analytical investigations. *Geotechnique* 48(6):847–867.

Cundall, P.A. & Strack, O.D.L. 1979. A discrete numerical model for granular assemblies. *Geotechnique* 29(1):47–65.

Cundall, P.A. 2001. A discontinuous future for numerical modelling in geomechanics? Proceedings of the ICE—Geotechnical Engineering 49(1):41–47.

De Nicola, A. & Randolph, M.F. 1997. The plugging behaviour of driven and jacked piles in sand. *Geotechnique* 47(4):841–856.

Jiang, M. & Yu, H.S. 2006. Application of Discrete Element Method to Geomechanics. In Wu, W. & Yu, H.S., (eds). *Modern Trends in Geomechanics*. The Netherlands: Springer.

Lunne, T. & Long, M. 2006. Review of long seabed samplers and criteria for new sampler design. *Marine Geology* 226(1–2):145–165.

Ng, T. & Meyers, R. 2015. Side resistance of drilled shafts in granular soils investigated by DEM. Computers and Geotechnics 68(7):161–168.

Paikowsky, S.G., Whitman, R.V., Baligh, M.M. 1989. A new look at the phenomenon of offshore pile plugging. *Marine Geotechnology* 8(3):213–230.

Parker, W.R. & Sills, G.C. 1990. Observation of Corer Penetration and Sample Entry During Gravity Coring. In E.A. Hailwood et al. (eds). *Marine Geological Surveying and Sampling*: 101–107. Dordrecht, The Netherlands: Kluwer Academic Publishers.

Randolph, M.F., Leong, E.C. & Houlsby, G.T. 1991. One-dimensional analysis of soil plugs in pipe piles. *Geotechnique* 41(4):587–598.

Raubenheimer, G. 2015. Personal communication [2015, July 16].

Skinner, L.C. & McCave, I.N. 2003. Analysis and modelling of gravity-and piston coring based on soil mechanics. *Marine Geology* 199(1):181–204.

Smith, D.G. 1998. Vibracoring: A new method for coring deep lakes. *Palaeogeography, Palaeoclimatology, Palaeoecology* 140(1):433–440.

Weaver, P.P.E. & Schultheiss, P.J. 1990. Current methods for Obtaining, Logging and Splitting Marine Sediment Cores. In E.A. Hailwood et al. (eds), *Marine Geological Surveying and Sampling*: 85–100. Dordrecht, The Netherlands: Kluwer Academic Publishers.

Wegener, S.B. 2015. *Application of a discrete element model to the analysis of granular soil recovery in an offshore tubular vibrocore*. MSc dissertation, University of Cape Town, Cape Town.

Proceedings of the first Southern African Geotechnical Conference – Jacobsz (Ed.)
© 2016 Taylor & Francis Group, London, ISBN 978-1-138-02971-2

Continuous surface wave testing in UK practice

J. Rigby-Jones & C.A. Milne
Ground Stiffness Surveys Limited, UK

ABSTRACT: The attractive proposition posed by Continuous Surface Wave (CSW) testing in providing rapid, economic and accurate ground stiffness profiles has gained widespread use in Southern Africa over the last 15 years. In the UK commercial uptake has been slower and CSW testing has yet to become widely accepted as a 'mainstream' test. However, by building on the experience gained in Southern Africa, whilst developing this relatively new ground investigation technique and with suitable adjustments for UK geotechnical practice, demand in the UK is rapidly growing and it appears that the time when CSW testing in the UK is routinely specified is near. This paper describes, with case histories, the development of the CSW technique in the UK and explores the history behind its low commercial use there when compared to the relatively rapid acceptance of the technique in Southern African geotechnical engineering practice.

1 INTRODUCTION

Continuous Surface Wave (CSW) testing involves the use of a vibrating source (or 'shaker') and a series of surface geophones to measure Rayleigh Wave velocities at a range of frequencies. The resultant dispersion curve is then inverted to provide an in situ small strain stiffness (Go) profile, which can be strain softened to give values of E at known strains (Heymann 2007). The technique therefore addresses many of the limitations of alternative stiffness measurement testing, in that stiffness is determined at a known strain without disturbance and over a relatively large tested volume. The theoretical basis for Continuous Surface Wave (CSW) testing was developed in the UK in the 20th century, however CSW testing is only now beginning to be used in the UK for routine geotechnical design.

2 DEVELOPMENT OF CSW IN THE UK

The theoretical potential for determining ground stiffness through the measurement of seismic wave velocity was described in the early part of the 20th Century (Love, 1927). However the small strain levels at which such seismic techniques measure ground stiffness (typically <0.001%) were not considered relevant to the 'real world' strains operating around typical geotechnical structures leading to them being largely ignored for much of the 20th Century. It was only with the growing understanding that operational strain levels (typically <0.1%, Jardine *et al.*, 1986) are much closer

to seismically measured strain levels and that the strain softening response of soils is remarkably predictable (Clayton & Heymann, 2001), that interest in the CSW technique grew. The advent of increasing computer processing power, work done at BRE in the 1980s (Butcher & Powell, 1996) and University of Surrey in the 1990s (Matthews *et al.*, 1996), enabled the development of commercial systems for determining in situ ground stiffness profiles from surface waves (Heymann 2007).

Whilst in the UK there is acceptance of the need for accurate ground stiffness data for more complex geotechnical schemes, there has been reluctance to move away from traditional means of determining ground stiffness, typically empirical measures with SPT N values and/or laboratory testing, for more routine design. This resistance is, on the one hand, driven by the perception that reliable stiffness measurement is necessarily expensive, such as when undertaken by pressuremeter or large scale load tests. On the other hand there is concern from practitioners at the reliability of geophysical based testing based on experience of the limitations of geophysical surveying. By contrast, in South Africa, from 2001 practitioners began to specify CSW testing for stiffness profiling (Heymann 2007). This has been assisted by better integration between surface wave testing researchers and engineering practitioners and has resulted in CSW testing becoming commonplace on schemes where accurate prediction of ground movement is required.

Increases in CSW adoption in the UK are however finally being seen, partly due to an increasing amount of construction in settlement sensitive urban sites requiring accurate settlement prediction and increasing demand of clients to reduce costs through economical design. Learning from the lessons learnt in introducing CSW testing to the South African market the speed of uptake in the UK has been enhanced through the involvement of practicing engineers capable of demonstrating the practical application of CSW data and the benefits of design optimisation from reliable stiffness data. Considerable expansion in UK CSW testing is anticipated over the next few years with the potential for CSW to become the benchmark for most schemes where accurate ground movement prediction is required. Growth sectors include railways and renewable energy (particularly windfarms) where significant infrastructure spending is occurring and accurate prediction of ground movements is fundamental to design optimisation.

3 RAIL APPLICATIONS OF CSW

The strongest adoption of the CSW technique has been seen in the UK rail sector. Due to the sensitivity of the busy rail network and the low tolerances for movements, investigations for ground stiffness are frequently required to address movements of permanent and temporary structures, signalling, electrification infrastructure, embankments and track. However railway sites typically provide a number of challenges to conventional ground investigation including:

- difficult access to many locations requiring access both along the track and through operational stations, potentially requiring extensive night time possession working;
- short duration and infrequent track possession availability;
- typically narrow, linear test locations with difficult topography restricting orientation of the test set-up;
- limited options to avoid testing on ballast at in areas of dense services(including high voltage and fibre optic cables) and buried obstructions;
- high risk from any operations which may damage services or cause obstructions to the track.

For rail working CSW equipment has been modified to enable transportation using rail trolleys or skates and gained acceptance by the UK's national rail asset holder Network Rail. The non-intrusive nature of CSW, coupled with its portability, have allowed high quality stiffness rail investigation in circumstance which would have been impractical for conventional in situ or laboratory testing.

Figure 1. Wheelbarrow cradle for transporting 80 kg shaker.

Figure 2. Typical rail CSW testing environment.

Figure 3. CSW equipment being transported by rail trolley.

4 CASE HISTORIES

4.1 *Second forth crossing*

CSW testing was undertaken on behalf of the successful design & build contracting consortium for

178

Figure 4. CSW testing in quarry backfill at second forth crossing approach embankment site.

Figure 5. CSW testing at wind turbine location (peat stripped prior to testing).

this major UK project. Accurate stiffness profiles were required to demonstrate the suitability of replacing a piled approach viaduct with a shallow founded approach embankment. However the presence of quarry fill containing large boulders prevented sampling or probing and made the use of cross hole seismic techniques expensive.

A total of 12 ground stiffness profiles were obtained during 2 days fieldwork despite data acquisition having to be undertaken between nearby plant movements. High quality data was successfully obtained to a typical depth of 8 m including upon benches excavated into sloping ground. Good correlation was demonstrated between stiffness profiles and borehole data and the successful measurement of quarry fill stiffness allowed embankment settlements to be demonstrated as acceptable.

4.2 Wind farm development

Development of onshore windfarms remains a key part of the UK renewable energy strategy. In the UK onshore windfarms are predominantly located in remote upland areas, many of which are covered by relatively compressible soils and therefore large turbine bases are required. Accurate stiffness profiles are required to assess serviceability settlements of turbine bases and dynamic soil properties are needed assess dynamic response of turbine bases.

CSW testing has been undertaken at a number of windfarm locations for turbine base design. Very soft soils & peat (below which bases are founded) must be stripped to enable testing to be undertaken but the portable nature of the CSW equipment has enabled even sites with the most difficult access to be economically investigated.

4.3 Sheffield tram train

Upgrade of a little used railway line in Sheffield required installation of a large number of new structures, including Overhead Line Electrification (OLE) stanchions. The generally soft shallow ground conditions meant that the widespread use of piled foundations was anticipated. However access onto the railway corridor was severely constrained by limited cess clearances and the urban setting of the site and meant that little conventional intrusive ground investigation was practical.

A total of 22 CSW profiles were obtained along the railway corridor over a three day period to a typical depth of 10 m, with equipment being transported along the track using a portable rail trolley. This productivity was achieved despite bad weather requiring a temporary shelter to be erected at each test location to protect electronic equipment from the rain.

Use of the accurate ground stiffness values provided by the CSW testing allowed pile lengths to be refined provided significant scheme savings. The profiles also proved valuable in assessing stratigraphy between the widely spaced intrusive investigation boreholes which had been possible.

4.4 SEW eurodrive, Normanton

Accurate stiffness profiles were required to assess serviceability settlements of foundations for a new

Figure 6. CSW equipment being transported by rail skate.

Figure 7. Portable shelter for CSW testing.

settlement sensitive assembly facility where a varied thickness of fill potentially required the use of piled foundations A total of 12 CSW ground stiffness profiles were obtained during 2 days fieldwork. At the clients request 600 mm diameter Plate Load testing (PLT) was undertaken in parallel at each CSW test location to provide a control on CSW data.

High quality CSW data was successfully obtained to a typical depth of 10 m, far exceeding the depth investigated by the small diameter PLT data which would otherwise have been used for design. Whilst there was good correlation demonstrated between shallow stiffness determined from both the PLT and CSW testing, the CSW data demonstrated an increase in stiffness with depth which avoided the need for expensive piled foundations.

4.5 London OLE investigations

CSW investigations were undertaken in connection with the upgrade and modification to the existing overhead rail electrification infrastructure in London required as part of the Crossrail project. Due to the busy nature of the line and requirement to test at or close to stations, only investigation of short duration by portable non-intrusive means was possible along much of the site. As a result

Figure 8. Busy rail environment for London OLE investigations.

Soil Description	Legend	Depth (m)
Made Ground		1.05
Soft to firm sandy organic CLAY (LANGLEY SILT)		3.10
Dense to very dense slightly silty very sandy GRAVEL with a low cobble content (LYNCH HILL GRAVEL)		5.60
Medium dense slightly silty very gravelly SAND (LYNCH HILL GRAVEL)		7.10
Stiff closely fissured silty clay with occasional shell fragments and selenite crystals (LONDON CLAY)		10.45

Figure 9. Typical stratigraphy at London OLE site 1.

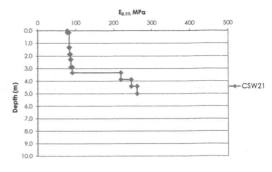

Figure 10. Typical CSW data at London OLE site 1 (note depth correlation with stratigraphy in Figure 9).

180

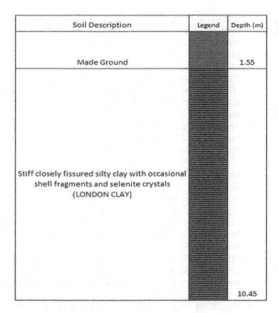

Figure 11. Typical stratigraphy at London OLE site 2.

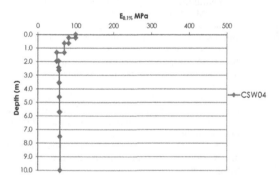

Figure 12. Typical CSW data at London OLE site 2 (note depth correlation with stratigraphy in Figure 11).

Table 1. Comparison of CSW and empirically derived design stiffness values for London OLE foundation design (site 1).

Stratum	Base depth (m)	Emp. $E'_{0.1\%}$ (MPa)	CSW $E'_{0.1\%}$ (MPa)	Emp $E'_{0.1\%}$/ CSW $E'_{0.1\%}$ (%)
Made ground	1.05	20	79	25
Langley silt	3.10	30	86	35
Lynch hill gravel	7.10	193	221	87
London clay	>10.45	97	58	168

Emp. $E'_{0.1\%}$ = empirically derived value of drained Young's Modulus at 0.1% strain CSW $E'_{0.1\%}$ = CSW derived value of drained Young's Modulus at 0.1% strain.

Table 2. Comparison of resultant design settlements using CSW and empirically derived design stiffness values for London OLE foundation design (site 1).

	Emp. $E'_{0.1\%}$	CSW $E'_{0.1\%}$	Emp $E'_{0.1\%}$/ CSW $E'_{0.1\%}$ (%)
Settlement of 2.5 m square pad under	6.6	3.6	183
250 kPa (mm)	9.3	5.2	179
Square pad dimension for 5 mm settlement (m)	2.8	1.3	215

of these constraints CSW testing was specified to enable stiffness investigations to be undertaken at locations otherwise unsuitable for other forms of stiffness testing.

The expected geological sequence across the sites comprised Made Ground associated with the railway trackbed and formation, overlying compressible Langley Silt, over the dense Lynch Hill Gravel with the bedrock geology comprising the very stiff London Clay Formation. Accurate ground stiffness profiles were required at each of the electrification structures to determine whether piling was required and to accurately dimension piled or spread foundations for limiting serviceability criteria.

CSW testing provided ground stiffness profiles to depths of 6–10 m in these difficult to investigate locations. Use of the CSW data resulted in a 2 to 3 times reduction in predicted settlements when compared to initial foundation analyses undertaken using stiffness values derived from conventional empirical relationships with undrained shear strength (Burland et al., 1979; Wroth 1972) and SPT N60 values (Clayton 1995).

5 CONCLUSIONS

Routine use of CSW testing in the UK remains in its infancy but is increasing rapidly in response to a more geotechnical, rather than geophysical, approach to its application. The portable, non-intrusive nature of testing has led to uptake within the rail industry in particular.

Recent UK CSW projects have emphasised the practical benefits of high quality stiffness data outside of very large or complex schemes, emphasising that design optimisation for ground movements relies on representative ground stiffness profiles. Experience on a wide range of schemes has demonstrated that the use of small strain stiffness values softened to typical operational strain levels in conjunction with simple analysis techniques provides

181

far better results than poor quality, empirically derived stiffness data used in conjunction with more complex analysis methods.

REFERENCES

Atkinson, J.H., Coop, M.R., Stallebrass, S.E. & Viggiani, G. 1993. Measurements of ground stiffness of soils and weak rocks in laboratory tests. In: *The Engineering Geology of Weak Rocks*, Engineering Geology Special Publication No. 8, Balkema, Rotterdam, pp. 21–27.

Aung, A. 2013. Detection of underground objects using surface wave methods. PhD thesis, Nanyang Technological University, p. 231.

Burland, J.B, Simpson, B. & St John, H.D. 1979. Movements around excavations in London Clay. Proc. Eur. Conf. Soil Mech. Found. Engng, Brighton 1, 13–29.

Butcherm, A.P. & Powell, J.J.M. 1996. Practical considerations for field geophysical techniques used to assess ground stiffness. Proceedings of the international conference on advances in site investigation practice, London: pp. 701–714.

Clayton, C.R.I. 1995. The Standard Penetration Test (SPT): Methods and Use. CIRIA report 143.

Clayton, C.R.I. & Heymann, G. 2001. The stiffness of geomaterials at very small strains. Géotechnique, 51(3): 245–256.

Heymann, G. 2007. Ground stiffness measurement by the continuous surface wave test. Journal of the South African Institution of Civil Engineering. Vol.49, No.1, pp. 25–31.

Heymann, G., Jacobsz, S.W., Vorster, T.E.B. & Storry, R.B. (2008). Comparison of ground stiffness from seismic surface wave and large scale load tests. Proc. 3rd Int. Conf. on Site Characterisation (ISC'3), Taipei, 843–848.

Heymann, G., Rigby-Jones, J., & Milne, C.A. (in print) The application of Continuous Surface Wave testing for settlement analysis with reference to a full scale load test for a bridge at Pont Melin Rûg, Wales. Journal of the South African Institution of Civil Engineering.

Jardine, R.J., Potts, D.M., Fourie, A.B. & Burland, J.B. 1986. Studies of the influence of non linear stress-strain characteristics in soil-structure interaction. Geotechnique, Vol.36, no.3, pp. 377–396.

Lai, C. & Rix, G.J. 1999. Inversion of multi-mode effective dispersion curves. In M. Jamiolkowski, R. Lancellotta, & D. Lo Presti (eds.), Pre-failure Deformation Characteristics of Geomaterials, 411–418. Rotterdam: Balkema.

Leong, E. & Aung, A. 2013. Global Inversion of Surface Waves Dispersion Curves Based on Improved Weighted Average Velocity (WAVe) Method. Journal of Geotechnical and Geoenvironmental Engineering, 10.1061/(AS CE)GT.1943–5606.0000939 Apr. 8, 2013.

Mair, R.J. 1993. Developments in geotechnical engineering research: application to tunnels and deep excavation. Unwin Memorial Lecture. Proceedings of the Institution of Civil Engineers, Civil Engineering, Vol.93, pp. 27–41.

Matthews, M.C., Hope, V.S. & Clayton, C.R.I. 1996. The use of surface waves in the determination of ground stiffness profiles. Proc. Instn Civ. Engrs Geotech. Engng, 119, Apr., 84–95.

Rollins, K.M, Evans, M.D. Diehl, N.B. & Daily, W.D. III 1998. Shear modulus and damping relationships for gravels. ASCE, Journal of Geotechnical and Geoenvironmental Engineering, 124(5), pp. 396–405.

Wathelet, M., Jongmans, D. & Ohrnberger, M. 2004. Surface wave inversion using a direct search algorithm and its application to ambient vibration measurements. Near surface geophysics, pp. 211–221.

Wathelet, M. 2005. Array recordings of ambient vibrations: surface wave inversion. PhD Thesis, University of Liege.

Wathelet, M. 2008. An improved neighborhood algorithm: parameter conditions and dynamic scaling. Geophysical Research Letters, Vol.35, L09301, DOI:10.1029/2008GL033256.

Wroth, C.P. 1972. Some aspects of the elastic behaviour of overconsolidated clay, Proc. Roscoe Memorial Symposium.

Proceedings of the first Southern African Geotechnical Conference – Jacobsz (Ed.)
© 2016 Taylor & Francis Group, London, ISBN 978-1-138-02971-2

A method to derive hydrogeological information from geotechnical data

G.J. Du Toit & A. Huisamen
Geo Pollution Technologies—Gauteng (Pty) Ltd., Pretoria, South Africa

ABSTRACT: Groundwater flux estimates require reliable hydraulic conductivity values. These are normally acquired by pumping or packer tests. However, these are not obtained upon commencement of engineering projects and often lead to additional project costs. This study illustrates that results obtained from these methods vary by orders of magnitude. In contrast, geotechnical data is normally collected on project commencement as part of project feasibility studies in core logs and/or geophysical data. This data is readily available and normally more abundant than pumping or packer test data. Analytical formulas were developed to convert this data into hydrogeological parameters, which are commonly site specific and disregard heterogeneity. This study describes methodologies to calculate hydraulic conductivity from core logs and calibration of results with piezometric levels. Calibration is performed by groundwater modelling, ensuring data which represents measured groundwater levels. The procedure is illustrated by computing inflow into a tunnel in Karoo Supergroup lithologies.

1 INTRODUCTION

1.1 *Civil engineering and groundwater*

Estimation of groundwater flux is essential for planning and contracting in most civil and mining engineering projects. It is also a requirement set by authorities before operations can commence, to estimate the impact of groundwater abstraction. Therefore, hydrogeological data is essential. However, it is seldom obtained at the early stages of projects. In contrast, geotechnical data is normally collected as part of the project feasibility study, such as core logs and/or geophysical data. If the geotechnical data can be used to reliably estimate hydrogeological parameters, it will result in cost and time saving as well as improving the reliability of flow calculations, the hydraulic parameters of which are often derived from literature. Analytical formulas have been developed to convert geotechnical data into hydrogeological parameters.

Unfortunately, most of these methods are calibrated and applicable to site specific materials only, often also ignoring heterogeneity. In Southern Africa`s fractured rock formations, deformational structures such as fractures, faults and joints are the main controls on groundwater flow. These structures are highly heterogeneous and calculations can be inaccurate if these non-uniformities are not considered.

Core borehole logs can represent readily available data of heterogeneous geotechnical characteristics of bedrock. Apart from the lithological and geotechnical strength data, many parameters of hydrogeological interest are also logged, e.g. core recovery, material recovery, Rock Quality Designation (RQD) and several joint parameters such as joint spacing, infill and weathering. This data is not only well represented laterally, but vertical data resolution could be on a meter scale, also representing vertical heterogeneity. In this paper, a methodology and case study will be described to calculate hydraulic conductivity from core borehole data and also to calibrate the parameters with site specific groundwater level data through groundwater modelling. This procedure ensures that the calculated parameters are a plausible deterministic data set which could result in site measured groundwater levels. This is illustrated by calculating groundwater inflow into a tunnel in Karoo Supergroup lithologies.

1.2 *Common inflow calculation methods*

Analytical formulas are widely used to predict groundwater inflow into tunnels due to their simplicity and ease of application (Farhadian et al., 2012). Several formulas have been proposed (Butscher, 2012, Katibeh and Aalianvari, 2012), varying from basic to elaborate equations. These formulas rely, almost without exception, on the hydraulic conductivity of the aquifer, the depth of the tunnel below the groundwater level, and the dimensions of the tunnel. While the latter variables are readily calculated or measured, the hydraulic conductivity presents a challenge.

In fractured consolidated rock, the primary porosity of the rock is practically zero, while hydraulic conductivity is dependent on fracturing of the rock. Fractures are highly heterogeneous and will vary both in vertical and horizontal extent. Pumping and packer tests conducted are frequently the only available data to acquire hydraulic conductivity and calculate inflow. However, the data is relatively expensive, especially lengthy pumping tests and are thus sparse by definition. It is thus not generally feasible to quantify a hydraulic conductivity field as a function of the length or the depth of the tunnel. Therefore, analytical equations are often used to calculate inflow, based on averaged hydraulic conductivity. In contrast, numerical models have been used to calculate groundwater inflow into tunnels (Molinero et al., 2002, Masset, 2011, Piccinini and Vincenzi, 2010, Dunning et al., 2004).

Although these methods are capable of accepting variable hydraulic conductivity in three dimensions as input, the available data is seldom sufficient to calculate such a detailed hydraulic conductivity field. Consequently, hydraulic conductivity is also mostly used as a constant along the length of the tunnel. In fractured rock where primary porosity is negligible, neither of these methods is satisfactory and thus suffers from the same shortcomings as the analytical formulations. The problem that arises, whether an analytical or numerical technique is being used, is the ability to calculate a reliable hydraulic conductivity for the rock above the tunnel.

1.3 Calculation of hydraulic conductivity from core log parameters

During core logging, several useful joint and fracture parameters are recorded (Bieniawski, 1989). Joint parameters are also sometimes recorded and include parameters such as inclination, spacing, roughness and filling. Several attempts have been made to create useful hydraulic rock mass classifications systems, such as the Hydro-Potential (HP) value (Gates, 1997) and the HC system (Hsu et al., 2011). While the HP value is based on parameters such as RQD, joint spacing, joint roughness, joint hydraulic conductivity, joint aperture factor and joint water factor, the HC system is based on RQD, a depth index, gouge content designation and lithology permeability index. Several attempts have been made to link these hydraulic rock mass classifications to hydraulic parameters (El-Naqa, 2001). The most basic of these attempted formulas relate only RQD to K.

$$K = 177.45 \times e^{-0.0361 \times RQD} \qquad (1)$$

A more elaborate approach was developed for the empiric formulation between hydraulic conductivity (K) and the HC system.

$$HC = \left(1 - \frac{RQD}{100}\right)(DI)(1 - GCD)(LPI)$$
$$K = 2.93 \times 10^{-6} \times (HC)^{1.342} \qquad (2)$$

In this study, the constants were calibrated with packer tests, rendering the resultant constants site specific and not readily applicable to alternative sites.

2 DESCRIPTION OF THE STUDY SITE

2.1 Geology and hydrogeology

A tunnel with inner dimensions of about 10 m by 10 m and a length of nearly 4 km is planned in the Vryheid Formation of the Ecca Group, Karoo Supergroup. It consists of layers of grit, sandstone, shale and several coal seams, intruded by Jurassic dolerite dykes and sills. The tunnel will be fully contained in such a sill with a thickness of at least 60 m. Core drilling indicated that the sill material is highly heterogeneous, consisting of fine to course grained plagioclase feldspar and pyroxene in diverse states of weathering. Fracturing and faulting was frequently observed during drilling, with displacements of up to 40 metres. The position, frequency and dimensions of these structures are thus a direct indication of the water bearing capacity of the host rock.

2.2 Geotechnical investigation

Ten percussion holes and thirty core boreholes were drilled at more or less regular intervals along the length of the tunnel. Pumping tests were conducted in six of the percussion holes. Eleven packer tests were done in six of the core holes at varying depths. Values for transmissivity were derived from the pumping tests and Lugeon values from the packer tests. Transmissivity values ranged from 0.3 to 70 m²/day for the pumping tests (based on late T), and 0.001 to 1 Lugeon for the permeability of the rock mass. Due to the sparsity of these values, they were considered only adequate to calculate mean values. With this data, an average inflow into the tunnel can be calculated at best. However, detailed borehole logs were completed for all core holes, presenting the best available information of bedrock fracturing at this site.

Core recovery, RQD and joint characteristics were recorded at intervals of one to three meters, resulting in data availability at every two meters on average. In contrast to the hydrogeological tests, this amount of this geotechnical data lends itself to the calculation of a field of hydraulic conductivity along the length and depth of the tunnel. This data opens the theoretical opportunity to calculate inflow as a function of position along the tunnel, as well as cumulative inflow while tunnelling. This approach could therefore add value to the planning and execution of the tunnel construction.

3 METHODOLOGY AND RESULTS

3.1 Calibration of hydraulic conductivity to pumping and packer tests

RQD was the most functional geotechnical data available for this site. In principle, core recovery and material recovery should also correlate with hydraulic conductivity. The measured RQD values above the tunnel were sorted to chainage in the horizontal and elevation in the vertical (Figure 1). The heterogeneity is evident, but there is also a clear tendency for low RQD values (higher hydraulic conductivity) to dominate the upper elevation closer to surface. This can be expected as these shallower sections are more exposed to stress release fracturing and subsequent weathering (Wang et al., 2009).

The relationships between the hydraulic conductivity and RQD relate through a generalised exponential relationship (equation 3).

$$K = C1 \times e^{-C2 \times RQD} \qquad (3)$$

The other parameters of the HP system were deemed to be non-applicable in this study, or were not recorded in the logs. This will be explained later in the text. Thus, the parameters C1 and C2 must be calibrated against available data to create an applicable site specific K value. Percussion

boreholes drilled at the site were constructed to only allow a selected portion of the casing open to the aquifer.

Transmissivities were thus obtained by pumping testing of these sections and could thus be used to calibrate the constants C1 and C2 in equation 3. Due to the nonlinear relationship between RQD and hydraulic conductivity, the Solver function in Excel was used to calculate a best fit. Similarly, the hydraulic conductivity calculated from RQD could also be calibrated to the packer tests. At least 12 packer tests could be used for this purpose, and the K values calculated from RQD could be compared with the K values calculated from Lugeon values, and calibrated to fit these results. The relationship 1 Lugeon is equivalent to 10^{-7} m/s, or about 10^{-2} m/d, was used. Although alternative, more complex relationships are available this simpler relationship was considered adequate due to the intended calibration.

Although a very good fit to the data was obtained (especially for the packer tests), the values are widely divergent. Values obtained from the pumping tests are unrealistically high, while those obtained from packer tests were considered too low. From the pumping tests, the average hydraulic conductivity value over the tunnel is in the order of 0.5 m/d, which is far too high, judged from experience in similar fractured rock. It is thus evident that the pumping test results are not necessarily representative of horizon specific hydraulic conductivity of the aquifer. The reason could be that percussion boreholes are normally sited at most favourable positions, judged from either geophysical and/or geological data. Structures such as these may have been targeted to obtain the highest potential hydraulic conductivity values during hydraulic testing, to represent a conservative inflow scenario. For the packer tests the values are much more realistic for a fractured aquifer and thus seem to show more reliable data. But the values seem to be too low in this case, and orders of magnitude less than

Figure 1. RQD in the vertical section above the tunnel; data bars represent RQD percentage. Higher percentage = larger bar.

the pumping test results. This is possibly explained by packers that seal better over intact sections and the best results are thus obtained at sections with low water take, thus favouring lower values.

These differences of results make it difficult to arrive at reliable values of the groundwater inflow to the tunnel, whether it is by analytical calculation or numerical modelling. It thus demanded further investigation. The measured groundwater levels must also be satisfied by the hydraulic conductivity field and could be verified by numerical groundwater model. Numerical modelling thus presents an alternative calculation to ensure that the calculated hydraulic conductivity field is representative.

3.2 *Calibration of hydraulic conductivity with a numerical model*

The advantage of using the numerical model is that the specified hydraulic conductivity must be of reasonable similarity to the actual subsurface hydraulic conditions to fit the measured groundwater levels, thereby providing an additional mechanism to calibrate the data. Should the hydraulic conductivity be too high, the groundwater levels as modelled will be too low and vice versa.

To implement this assumption, a 3D finite difference numerical model was created using the US Department of Defence Groundwater Modelling System (GMS10) as Graphical User Interface (GUI) for the well-established MODFLOW numerical code. Natural topographical water divides have been used as no-flow boundaries and terminated with constant head boundaries where needed. The 2D hydraulic conductivity field directly above the tunnel was transferred to the 3D numerical model, using the data calibrated to the packer tests as the initial conditions. For the lateral extent of the model, the mean horizontal hydraulic conductivity for each depth was applied over the full extent of the model. The assumption is that the distribution of hydraulic conductivity above the tunnel is fairly representative of the whole model area. This might not be correct everywhere, but the trend with depth is reasonable.

However, the flow to the tunnel is mostly vertical and the lateral extent of the model is optional; even a 2D model would suffice. While the measured variability directly above the tunnel is preserved (Figure 2), the hydraulic conductivity only decreases with depth over the lateral extent of the model. With the hydraulic parameters for the

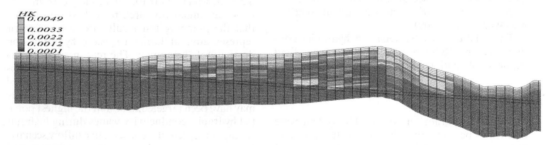

Figure 2.   Distribution of hydraulic conductivity.

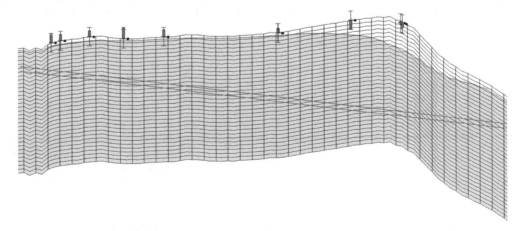

Figure 3.   Calibration of the hydraulic conductivity distribution.

Table 1. Calibration of hydraulic conductivity against pumping-and packer tests (each bar represent 10 meter).

Parameters	Pump test calibration	Packer test calibration	
C1	0.765066537	0.01897	
C2	0.203195082	1.92834	
Measured T (m2/day)	13.9662	0.00940	
Calculated T (m2/day)		21.0833	0.00931
RMS Difference (m2/day)		7.1171	0.00009

Table 2. Calibration statistics.

Mean Residual (Head in meters)	−0.65
Mean Absolute Residual (Head in meters)	7.71
Root Mean Squared Residual (Head in meters)	8.44

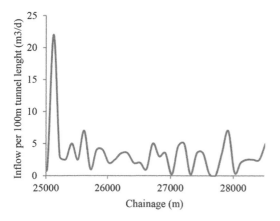

Figure 4. Inflow to tunnel.

model thus calculated, the model could be run in the natural (pre-tunnel) state. The predicted groundwater levels in this scenario could be compared to the current measured groundwater levels. The hydraulic conductivity could then be adjusted to fit the measured groundwater levels, thus calibrating water levels with the aid of the model. To fit the observed groundwater levels, the hydraulic conductivity field was multiplied by a factor determined through automated parameter estimation software.

Parameter estimation showed that the values were required to increase by about an order of magnitude to fit observed groundwater levels. This adjustment is considered reasonable, especially considering that the packer tests were suspected to underestimate hydraulic conductivity by about a similar amount. Additionally, this increase in hydraulic conductivity showed values well within the upper boundary, set by the pumping test results.

4 DISCUSSION

The final results of the calibration (Figure 3) show the finite difference cells of the model. The error bars in the model represent the difference between the calculated groundwater levels and the measured values. The bar depicts errors of 10 meters or less (green), or yellow if exceeding the ideal of 10 meters. The direction of the coloured bar indicates whether the calculated water level elevations are too high (upwards) or too low (downwards). The calibration statistics (Table 2) show that the mean errors are small in comparison to the depth of the tunnel below the groundwater level. The results calculated should thus also be reasonably correct.

With the hydraulic conductivity field calibrated to the measured groundwater levels, the inflow in the tunnel could be modelled with more confidence.

Except for the inflow (Figure 4) at chainage 25100, the inflow is low (less than 5 m³/day or 0.05 l/s). The high inflow at 25100 (22 m³/day or 0.2 l/s) is due to the poor rock conditions in that specific section as seen in the figure below. This is still a small volume and should pose no problem. Total inflow into the tunnel is predicted to be 120 m³/day, if all sections are open simultaneously.

5 CONCLUSIONS

It was shown that the use of hydraulic conductivity values obtained by either pumping tests or packer tests, are not always sufficient to calculate groundwater flux. Without having had two sets of data available, the differences could even have gone unnoticed resulting in an order of magnitude error in groundwater inflow calculated. However, the use of groundwater levels as a final calibration mechanism is not only inexpensive, but also a safeguard against biased data.

REFERENCES

Bieniawski, Z.T. (1989). *Engineering rock mass classifications: a complete manual for engineers and geologists in mining, civil, and petroleum engineering*, John Wiley & Sons.
Butscher, C. (2012). Steady-state groundwater inflow into a circular tunnel. *Tunnelling and Underground Space Technology*, 32, 158–167.

Dunning, C., Feinstein, D., Hunt, R. & Krohelski, J. (2004). *Simulation of ground-water flow, surface-water flow, and a deep sewer tunnel system in the Menomonee Valley, Milwaukee, Wisconsin*, US Department of the Interior, US Geological Survey.

El-Naqa, A. (2001). The hydraulic conductivity of the fractures intersecting Cambrian sandstone rock masses, central Jordan. *Environmental Geology*, 40, 973–982.

Farhadian, H., Aalianvari, A. & Katibeh, H. (2012). Optimization of analytical equations of groundwater seepage into tunnels: A case study of Amirkabir tunnel. *Journal of the Geological Society of India*, 80, 96–100.

Gates, W.C. (1997). The Hydro-Potential (HP) value: a rock classification technique for evaluation of the ground-water potential in fractured bedrock. *Environmental & Engineering Geoscience*, 3, 251–267.

Hsu, S.-M., Ku, C.-Y., Lo, H.-C. & Chi, S.-Y. (2011). *Rock mass hydraulic conductivity estimated by two empirical models*, INTECH Open Access Publisher.

Katibeh, H. & Aalianvari, A. (2012). *Common Approximations to the Water Inflow into Tunnels*, INTECH Open Access Publisher.

Masset, O. (2011). Transient tunnel inflow and hydraulic conductivity of fractured crystalline rocks in the Central Alps (Switzerland). Diss., Eidgenössische Technische Hochschule ETH Zürich, Nr. 19532, 2011.

Molinero, J., Samper, J. & Juanes, R. (2002). Numerical modeling of the transient hydrogeological response produced by tunnel construction in fractured bedrocks. *Engineering Geology*, 64, 369–386.

Piccinini, L. & Vincenzi, V. (2010). Impacts of a railway tunnel on the streams baseflow verified by means of numerical modelling. *Aqua Mundi*, 1, 123–134.

Wang, X.-S., Jiang, X.-W., Wan, L., Song, G. & Xia, Q. (2009). Evaluation of depth-dependent porosity and bulk modulus of a shear using permeability–depth trends. *International Journal of Rock Mechanics and Mining Sciences*, 46, 1175–1181.

Proceedings of the first Southern African Geotechnical Conference – Jacobsz (Ed.)
© 2016 Taylor & Francis Group, London, ISBN 978-1-138-02971-2

Aiding engineering and environmental problem solving using airborne geophysics

R. van Buren, O. Dingoko
CGG, Johannesburg, South Africa

ABSTRACT: When engineering the built environment or managing environmental problems, tools which improve the effectiveness and efficiency of projects are vital in achieving success. Airborne geophysical surveys have been used to solve many engineering and environmental questions. Decisions need to be made throughout engineering and other project life cycles, often relating to ground conditions; knowledge of ground conditions is often more readily determined by the application and integration of remote methods such as potential field (gravity and magnetic) or more contemporary electromagnetics with traditional engineering and environmental tools.

1 INTRODUCTION

Engineering of the built environment along with the management of a growing list of environmental problems which South Africa now faces, require new tools to improve the effectiveness and efficiency of projects. We aim to illustrate, through the presentation of a number of case studies, how airborne geophysical surveys may be used as just such a tool, to solve many engineering and environmental questions. Both frequency and time domain electromagnetic surveys, as well as more traditional gravity and magnetic surveys provide invaluable information about ground conditions, which can be easily integrated into existing engineering workflows, allowing informed, cost and time saving decisions to be made.

2 WHAT GEOPHYSICS OFFERS ENGINEERING

Although the use of geophysical surveys to solve engineering and environmental problems is not new, the adoption of airborne surveys within these disciplines is relatively new. This adoption is predominantly due to the improvement in resolution achieved in recent years from airborne systems through hardware and processing developments. The ability to acquire 5000–10000 acres of data per day (dependent upon the resolution) is common, facilitating rapid coverage of areas of interest, thereby allowing critical decision making to take place earlier in the project life cycle.

Geophysical methods indirectly measure resistivity, density, susceptibility and other rock properties. Civil and environmental engineers wish to determine factors such as formation thickness, salinity, friction ratio, rippability, or porosity. Often, the translation from geophysical properties to engineering quantities is achievable through well accepted transforms, making use of calibration coefficients readily derivable from in situ or laboratory measurements.

3 INTEGRATING GEOPHYSICS

Most subsurface investigation is performed via drilling. This historically preferred method of sampling the sub-surface for engineering applications is due to the relatively high level of confidence associated with the drill-derived information. However, drill holes sample small volumes of the subsurface and may pierce regions with conditions vastly different from those just meters away from the hole. Geophysical surveys, conversely, have lower resolution at any single point but provide near complete coverage over the area of interest.

The two methods—drilling and geophysics—are highly complementary and when integrated, return more than the sum of the two. Drilling provides the precise measurements at discrete points, while the geophysics provides the full area coverage, imaging between holes and detecting discontinuity.

Collection of airborne geophysical data generally does not depend on detailed knowledge of the ground, allowing acquisition to take place prior to drilling. The results have immediate value in indicating where drilling should be concentrated in order to maximize the information obtained via drilling, and to ensure that it is representative of the subsurface. Drilling hazards may also be avoided if a priori knowledge of the subsurface is available.

Further interpretation of the geophysical data is then possible, integrating the information obtained via the initial drilling. This process may be iterated, and maximizes the information contained in both data types to progressively build a superior model of the subsurface in the most efficient manner.

4 APPLICATIONS OF AIRBORNE GEOPHYSICS

When large construction projects are undertaken, subsurface characterization in terms of one or more of the following parameters is invariably required: compaction, stability, rippability, groundwater, salinity, electrical conductivity, and other factors. A number of airborne geophysical methods are commonly applied to solve these engineering and environmental problems. These include electromagnetics (to detect conductivity changes), magnetics (to detect magnetically susceptible minerals and materials), gamma-ray spectrometry (to detect anthropogenic radioactive materials and map geology via radioelement concentration) and gravity gradiometry (to detect density changes).

4.1 Mapping

Airborne geophysical systems are able to cover large areas very quickly, removing the need for ground access or exposure of personnel and equipment to subsurface hazards. Coverage can be achieved across all kinds of terrain, including continuous coverage from land across shorelines and over water. Figure 1 shows a conductivity section with continuous coverage from marine (left) across exposed land and into tidal flats (right) where the water table, seawater interface and sediment/bedrock contact were all able to be mapped. High resolution, helicopter-borne RESOLVE EM surveys acquire full coverage at a rate of 500–1000 Ha per hour at a typical line spacing of 100 m.

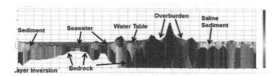

Figure 1. Continuous conductivity section from seawater (left) to land and tidal zone.

4.2 Construction and infrastructure

Airborne electromagnetic surveys detect ground conductivity and are sensitive to changes in three dimensions. A common example of this is the application to mapping ground conductivity for the design of infrastructure lightning protection grounding (Chisholm et al. 2003). Figure 2 shows a resistivity cross-section from New Brunswick, Canada, from the presentation by Chisholm et al. (2003). The resistivity distribution with depth can be used to design custom grounding arrays to take maximum advantage of the depth distribution of conductive soil and rock. Two layer resistivity models can be generated and used with standard ground electrode resistance equations to design electrode arrays to provide the necessary protection from lightning.

Pipeline corridors are well suited to airborne methods as the areas and distances to be investigated are extensive. The example in Figure 3 and Figure 4 (Hodges et al, 2000) shows a planned pipeline corridor in Quebec, Canada, which was flown with helicopter-borne EM in order to map the presence and thickness of overburden (principally glacial till related sediments). Pipeline construction was planned for a depth of 2 m, so knowledge of the depth of overburden allowed for planning where trenching with a backhoe would suffice and where blasting of bedrock would be necessary (at a much higher cost and level of complexity).

Surveys of a similar nature have been conducted in a wide range of geological environments, looking for soil types, soil depths, rock type, permafrost, palaeochannel aquifers and karst features.

4.3 Karstic geology and other hazards

Karstic cavities, sinkholes, and zones of weakness pose hazards to engineering projects. Airborne

Figure 2. Conductivity section from airborne EM and grounding recommendations for a 345 kV transmission line.

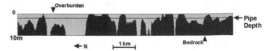

Figure 3. Overburden & bedrock fence section along the route of the pipeline showing pipe depth (From Hodges et al. 2000).

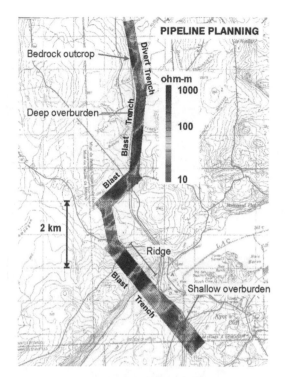

Figure 4. Resistivity map along a pipeline corridor used to interpret overburden thickness to estimate need for blasting (From Hodges et al. 2000).

geophysical data can be used to detect karsts where the geophysical properties of the cavity are distinct from their host.

The example shown in Figure 5 is from the Victor Diamond Mine in northern Ontario, Canada. The survey was conducted to detect and delineate a sinkhole known from drilling, as well as to search for other, unknown karstic hazards in the area before construction began. The limestone surface and sinkhole are covered by about 12 m of Quaternary glacial till, leaving no evidence of the sinkhole on surface. The location and 3D geometry of the sinkhole is clearly imaged with airborne EM due to the change in conductivity between the till filling the hole and the basal limestone. The airborne EM data were used in conjunction with seismic refraction data over the deepest parts of the sinkhole to construct a complete 3D image of the sinkhole. Aside from the sinkhole's location and geometry, the overlying cover was able to be classified into till and sand, further enriching the information at the engineers' disposal thereby reducing operating and construction hazard.

Conductivity contrasts can also be created by water filling cavities, and often by the higher porosity (and therefore saturation) of the extensive weak

zones along which cavities occur. This was the case in a survey described by Smith et al. (2003), which detected the previously-known Valdina Farm sinkhole in the Edward Aquifer in Texas, USA (Figure 6).

An air-filled cavity, a difficult target for EM, may be detectable with Airborne Gravity Gradiometry (AGG) due to the density contrast. Helicopter-borne AGG provides the lowest, slowest flight and therefore the highest resolution of an airborne method.

A prime example of what can be achieved with low altitude helicopter acquisition of gravity gradient data is from a project flown for Defense Advanced Research Projects Agency (DARPA), over the Low Energy Booster (LEB) and the Medium Energy Booster (MEB) at the Superconducting Super Collider Laboratory. The system was flown at approximately 2 m above ground (requiring special permissions) allowing the small 4 m x 4 m tunnels to be accurately imaged. Synthetic modelling was performed prior to the acquisition

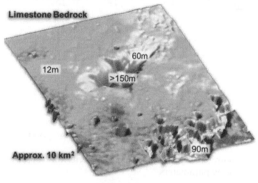

Figure 5. Bedrock surface (showing sinkhole) derived from airborne EM survey, constrained with seismic data (courtesy of De Beers).

Figure 6. Resistivity section and sinkhole section at Valdina Farms Sinkhole, Texas (Smith et al. 2003).

to ensure that the survey parameters would meet with the objectives (Figure 7). The survey results were as the modelling had predicted and facilitated mapping of the subterranean excavations to a high degree of accuracy (Figure 8).

4.4 Oil field applications

There are many demands for engineering and environmental work in the oil fields. The magnetic data in Figure 9, from Quarantine Bay in the Mississippi Delta, shows anomalies from surface infrastructure, wellheads and pipelines. Records of wellhead locations are often inaccurate, and old wells were often poorly sealed, if at all (Veloski et al. 2008). If a field is to be pressurized for enhanced oil recovery, all the wells ever drilled have to be located and the seal verified.

Figure 9. Magnetic data over Quarantine Bay, Louisiana, showing anomalies of wellheads, surface infrastructure, and pipelines.

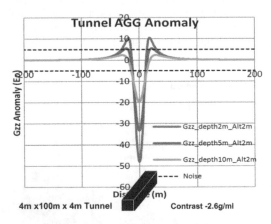

4m x100m x 4m Tunnel   Contrast -2.6g/ml

Figure 7. Synthetic modelling to determine appropriate survey parameters.

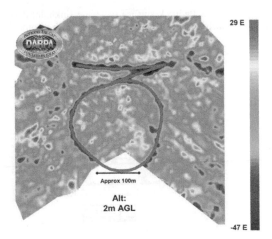

Figure 8. FALCON G_{DD} data collected over the SSC LEB & MEB subterranean tunnels.

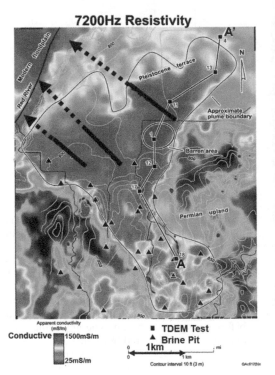

Figure 10. Resistivity map showing near-surface brine produced from oil wells (Paine et al. 2003).

Pipeline location surveys are often conducted in oilfield areas before new seismic surveys are shot, to prevent risk to the pipeline and interference in the data collected.

Oil well produced brine poses an environmental problem, and past practice was often to discharge this into surface brine pits (Paine et al. 2003). It is possible to map the subsequent distribution of this highly conductive saline water using airborne EM surveys. Figure 10 shows a map of 7200 Hz (mid-range) resistivity from a DIGHEM survey over the Permian Uplands and Red River Basin including part of the Nocona Oil Field. The brine pits are shown as black triangles.

Both the airborne EM and ground TDEM surveys show the brine is at the surface near the pits, travels below the surface under the Permian Upland (following ground water flow) to re-emerge in the Red River valley.

4.5 Contaminants

One could argue whether oil field produced brine (Figure 10) is a "contaminant", but there is no dispute that abandoned coal and hard rock mines can produce highly contaminated water, particularly those with elevated sulfide content.

The Sulphur Bank Mercury mine was intermittently worked from late in the 19th century until it was closed in 1957 (Hammack et al. 2002). It was known that the mine was leaking acidic, metal-contaminated water into Clear Lake, but the pathway was not known. The high-frequency 56 kHz resistivity map (Figure 11) shows the near-surface resistivity surrounding the water-filled open pit (the Herman Impoundment), and the conductive plume of contaminated water leading into the lake to the west. Lower frequency EM data (not shown) shows several other, possible contaminant pathways along geologic faults trending northwest to northeast.

Hammack et al. (2005) showed the results of an airborne EM survey to map contaminated water leaking from a coal mine impoundment (Figure 12). The survey detected strong conductivity in the decant basin, as well as the leakage pathway through the embankment, downhill to emerge at surface near the base of the impoundment.

Figure 13 shows a map of a different kind of contaminant—airborne Cs^{134} and Cs^{137}

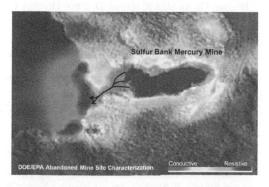

Figure 11. Resistivity map over sulphur bank mercury mine showing conductive contaminant plume from open pit flowing west to clear lake. (Hammack et al. 2002).

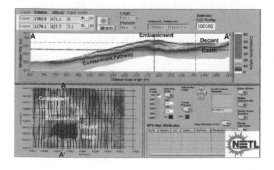

Figure 12. Resistivity section through coal tailings impoundment showing high conductivity of tailings and leakage path for contaminated waters (Hammack et al. 2005).

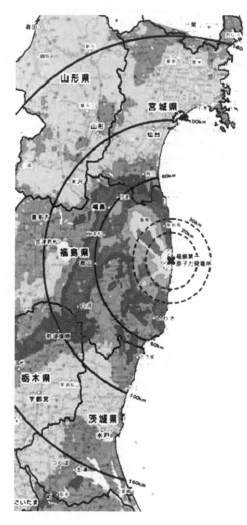

Figure 13. Airborne gamma ray spectrometry map showing Cs distribution from Fukushima nuclear plant disaster. (MEXT).

contamination carried by winds from the Fukushima nuclear power plant leak, mapped using airborne gamma-ray spectrometry. Gamma ray surveys have also been conducted to locate lost military and industrial radioactive sources—in one case found in a scrap metal yard.

5 DISCUSSION

There are many applications of airborne geophysics to engineering and environmental projects. Through the use of one or more of electromagnetic, gravity gradiometry, magnetic, or gamma-ray spectrometry, any feature which causes a significant change in resistivity, density, magnetic susceptibility or radioelement concentration can be detected and mapped. Airborne surveys offer fast coverage of large areas, provide continuous measurement of data sets, highly complementary to conventional engineering and environmental tools such as drilling. These data sets can and should be interpreted together; making use of the geophysics to guide the drilling and the drill-derived information used to refine and constrain the interpretation of the airborne geophysics. This joint process improves project efficiency and increases confidence in interpretations.

6 CONCLUSIONS

Engineering of the built environment along with the management of a growing list of environmental problems which South Africa now faces, require new tools to improve the effectiveness and efficiency of projects. We have illustrated, through the presentation of a number of case studies, how airborne geophysical surveys may be used as just such a tool, to solve many engineering and environmental questions. Frequency and time domain electromagnetic, gravity gradient, magnetic and radiometric surveys provide invaluable information about ground conditions, which can be easily integrated into existing engineering workflows, allowing informed, cost and time saving decisions to be made.

ACKNOWLEDGEMENTS

The authors thank CGG for permission to publish this paper.

REFERENCES

Chisholm, W. 2003. Recent Progress in Design and Test Methods for Transmission Line Ground Electrodes: Presented at the *International Symposium on Lightning Protection*, Curitiba. IEEE.

Hammack, R., Veloski, G., Sams III, J. & Mabie, J. 2002. The Use of Airborne EM Conductivity to Locate Groundwater Flow Paths at the Sulphur Bank Mercury Mine Superfund Site: *National Meeting of the American Society of Mining and Reclamation*, Lexington KY. American Society of Mining and Reclamation.

Hammack, R. 2005. Using Helicopter Electromagnetic Surveys to Identify Potential Hazards at Coal Waste Impoundments: *Proc 18th Symposium on the Application of Geophysics to Engineering and Environmental Problems (SAGEEP)*: 306–314, Atlanta GA. Environmental and Engineering Geophysical Society (EEGS).

Hodges, G., Rudd, J. & Boitier, D. 2000. Mapping Conductivity with Helicopter Electromagnetic Surveys as an Aid to Planning and Monitoring Pipeline Construction: *Proc 13th Symposium on the Application of Geophysics to Engineering and Environmental Problems (SAGEEP)*: 47–56, Arlington VA. EEGS.

Paine, J. G. 2003. Determining salinization extent, identifying salinity sources, and estimating chloride mass using surface, borehole, and airborne electromagnetic induction methods: *Water Resources Research*, 39(3): 3.1–3.10. American Geophysical Union.

Smith, B., Paine, J., Smith, D., Johnson, S., Waugh, J., Abraham, J., Blome, C. & Schindel, G. 2003. *Geophysical Characterization of Geologic Features in the Area of the Valdina Farms Sinkhole, Texas*: Geological Society of America Annual Meeting. Seattle WA. Geological Society of America.

Veloski, G., Hammack, R., Stamp, V., Hall, R. & Colina, K. 2008. Helicopter Magnetic Survey at the Teapot Dome Oilfield (Naval Petroleum Reserve No. 3)—A Case History: *Proc 21th Symposium on the Application of Geophysics to Engineering and Environmental Problems (SAGEEP)*: 264–276, Philadelphia PA. EEGS.

Proceedings of the first Southern African Geotechnical Conference – Jacobsz (Ed.)
© 2016 Taylor & Francis Group, London, ISBN 978-1-138-02971-2

Geotechnical investigations for low volume rural access roads

P. Paige-Green
Tshwane University of Technology, Pretoria, South Africa

ABSTRACT: The provision of access roads in rural areas is becoming increasingly important to fulfil basic human rights requirements. These roads need to be constructed at the lowest cost possible in order to allow provision of greater lengths of the roads for the available funds. In areas with problem subgrades and in difficult terrain (mountainous or wet areas), geotechnical investigations and design for foundations and structures can comprise a significant proportion of the total road provision costs and need to be minimised as far as possible. However, in certain situations conventional structures are required and they need to be designed and constructed to traditional standards using conventional geotechnical investigations. This paper describes various decision-making processes to be followed such that sufficient information is provided for different categories of road without excessive geotechnical investigations. The processes identify those situations related to the specific problems on different road categories that require specialist input and conventional investigation and design processes. Those decisions identified with less severe consequences for the specific road category can lead to significant savings through reduced geotechnical investigations.

1 INTRODUCTION

The provision of paved access roads in rural areas is becoming increasingly important to fulfil basic human rights requirements. These roads need to be constructed at the lowest possible cost in order to allow provision of greater lengths of the roads for the available funds. In areas with problem subgrades and in difficult terrain (mountainous or wet areas), geotechnical investigations and design for foundations and structures can comprise a significant proportion of the total road provision costs and need to be minimised as far as possible.

However, in certain situations conventional structures are required and they need to be designed and constructed to traditional standards using conventional geotechnical investigations. Deep cuttings, high fills, bridges and other structures that cannot be avoided or replaced with other options (e.g. lowered geometric standards, low water crossings, etc.) will usually have design lives considerably longer than the initial low volume road and should be investigated and designed to conventional standards such that the integrity of the structures are not compromised in the long term.

This paper describes various decision-making processes that can be followed, which will provide sufficient information for different categories of road without excessive geotechnical investigations. The processes identify those situations related to

the specific problems on different road categories that can be designed by road engineers and those that require specialist geotechnical input and conventional investigation and design processes.

Those decisions identified with less severe consequences for the specific road category can lead to significant savings in the costs of their geotechnical investigations.

2 LOW VOLUME ROADS

There are currently no international standard definitions for low volume roads. Two roads could have the same daily traffic counts but one could include mostly heavy commercial vehicles, while the other may have only light vehicles. Although the total traffic counts could be similar, the effects on the road pavement structures and the road capacity would differ significantly.

It has been shown (Figure 1) that traffic actually contributes considerably less than the environment to the performance of roads up to traffic counts of about 1 million Equivalent Standard Axles (ESAs) (Gourley & Greening (1999)).

For the purpose of this paper, low volume roads will be considered as those carrying not more than 300 vehicles per day (vpd) at the middle of their design life and with a cumulative axle count over their design life of not more than 1 million ESAs.

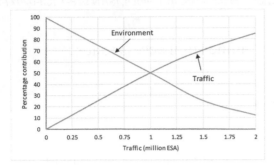

Figure 1. Impact of traffic and environment on performance of roads (Gourley & Greening, 1999).

3 GEOTECHNICAL PROBLEMS

3.1 *Subgrades*

Numerous potential problem soils can occur in road subgrades. Some of these may have a severe effect on the performance and integrity of the overlying pavement structures. The main subgrade problems encountered in the southern African region include:

- Expansive/heaving clays
- Collapsible soils
- Dispersive soil
- Erodible soils
- Slaking rock/soil
- Saline soils
- Soft clays
- Wet areas
- Karst areas.

Each of these problems requires its own investigation and test protocol as discussed by Paige-Green (2008).

The consequences of each of these problems on the road pavement structures differ between the actual soil problem and also between different road categories. Heaving clays, for instance, can result in significant deformation of the road surface with an undesirable impact on the safety and riding quality of the road and consequently the road user costs. The impact, however, on a road carrying 300 vpd is far more severe than on one carrying only 25 vpd and this should be taken into account in the design.

Conversely, highly saline material will often result in the total loss of a bituminous surfacing, which would have a similar impact on any road, irrespective of the traffic category.

3.2 *Earthworks*

Deep cuts and high fills are usually required to improve the geometric alignment of roads, and for many low volume roads, the costs of these expensive earthworks can seldom be justified. For most rural access roads, it is more important to provide all-weather access with variable speed regimes (indicated by the appropriate safety warning precautions and/or traffic calming measures), than to provide a road with a uniform higher design speed, requiring flat grades and large radii of horizontal and vertical curvature.

However, there may be isolated situations where deep cuts or high embankments are necessary. Obtaining the correct balance between the investigation and the requirements for these structures is important to minimise costs.

3.3 *Foundations*

Most foundations associated with low volume roads would be limited to water-crossing structures. Like the earthworks, the first choice is normally to make use of lower cost design solutions such as low-level structures, fords and larger culverts. These typically require a nominal subgrade investigation only. However, in certain situations, the need to provide conventional bridges may be unavoidable and the necessary site investigation should be carried out. It is not recommended that savings on site investigations for piers and abutments on concrete or steel bridge structures be considered to reduce costs—these can be costly structures and every care should be taken in their design.

4 INVESTIGATION SOLUTIONS

4.1 *Subgrades*

Figure 2 shows the relevance of the various potential geotechnical problems on different traffic classes of low volume road and provides an indication of when specialist geotechnical input should be sought.

The cells highlighted with vertical hatching indicate high importance/impact, those with dotted backgrounds have moderate impact and those with no background only have minor impact on the relevant road categories. The cells marked with an X indicate that the pavement engineer can in most cases make the necessary geotechnical design decision, based on experience, while those marked with √ indicate that a geotechnical or relevant specialist should be consulted.

It is clear that for the higher trafficked roads, subgrade problems, if not recognised and catered for will invariably result in expensive rehabilitation requirements. Only the erodible and slaking rocks/soils are unlikely to cause significant problems, but

Geotechnical problem	Traffic (vpd)			
	200-300	50-200	20-50	< 50
Subgrade:				
Expansive clay	√	√	x	x
Collapsible soil	√	x	x	x
Dispersive soil	√	√	x	x
Erodible soil	x	x	x	x
Slaking soil	x	x	x	x
Saline soil	√	√	√	x
Soft clay	√	√	x	x
Wet area	√	√	√	x
Cut slope	√	√	√	√
Embankment	√	√	x	x
Foundations	√	√	x	x

Figure 2. Importance and impact of various subgrade problems related to traffic and degree of geotechnical investigation required.

are included as it is imperative that they are identified and differentiated from the much more disruptive dispersive soils (Paige-Green, 2008). Problems such as potential sinkholes in karst areas need to be assessed in terms of the risk of a sinkhole actually causing an accident more than the effect for instance of a temporary road closure.

An extension of this for expansive soils is shown as an example. Typically, the potential expansiveness can be estimated using one of the traditional techniques, e.g. Van der Merwe, (1964). This can then be related to possible countermeasures including the following:

1. Retain the road over the clay as an unpaved section,
2. Flattening of side slopes,

3. Remove expansive soil and replace with inert material,
4. Pre-wetting prior to construction of the fill or formation,
5. Placing of un-compacted pioneer layers of sand, gravel or rockfill over the clay and wetting up, either naturally by precipitation or by irrigation,
6. Lime stabilization of the clay to change its properties (expensive),
7. The blending of fine sand with the clay to change its activity,
8. Sealing of shoulders,
9. The compaction of thin layers of lower plasticity clay over the expansive clay to isolate the underlying active clays from significant moisture changes, and
10. Limited success has been achieved using waterproofing membranes and/or vertical moisture barriers, which are generally geosynthetics.

Figure 5 provides a preliminary indication of possible counter-measure options (numbered as above) as a function of potential expansiveness.

It should be noted that usually a combination of these is most effective and all should go together with careful design and construction of side-drains, which should preferably be sealed.

This flow chart is specifically relevant to the higher trafficked roads, but it can also be related to the countermeasures required for other traffic volumes. By moving the solutions (right column) down one step for each other traffic class, i.e. for traffic of 50–200 vpd each of the solutions shown would be for a potential expansiveness class one higher. This would result in options 2, 4, 5 and 8 being applicable for a road carrying < 20 vpd with a potential expansiveness of > 8%.

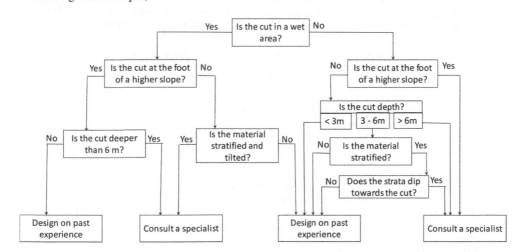

Figure 3. Decision chart for design of road cuttings.

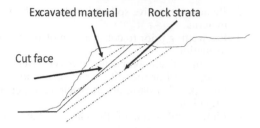

Figure 4a. Effect of cut at toe of a low slope.

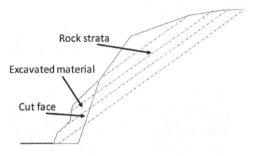

Figure 4b. Effect of cut at toe of a high slope.

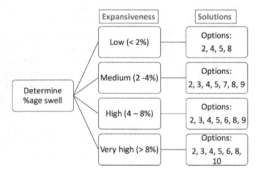

Figure 5. Possible solutions for roads on active clays related to potential expansiveness.

In many cases for roads with < 200 vpd, it may be better to retain the road as a gravel road over the expansive clay sections and apply the necessary maintenance (Option 1).

Similar systems can be developed for other types of problem subgrade but are not included here.

4.2 Earthworks

Large earthwork operations should be minimized, but as indicated previously, there will always be instances where they are essential. The construction of even the lowest class access road in mountainous areas could require either significant cut or fill, or both.

Figure 3 shows a decision tree that can be used to decide whether specialist geotechnical input is required or whether the project engineer can design the cuttings based on experience.

The question as to whether the cut is at the foot of a higher slope is extremely important. If the cut does not undercut a higher slope, failure will have minimal consequences (Fig. 4a). However, where the slope continues above a cut, failure of the cut will usually result in vast quantities of material higher up the slope becoming unstable (Fig. 4b).

If the designer goes through the decision process outlined in Figure 3 and ends up with a design on past experience option, the batters shown in Tables 1 & 2 can be used as a first estimate of possible slope angles for cuts.

Similarly, the slopes shown in Table 3 can be used as a first estimate for embankment fills.

It cannot, however, be overemphasised that significant engineering judgement and a good visual appraisal of the situation is required before any of the above batters are implemented.

Table 1. Suggested slope angles for cuttings in various materials (Modified after ERA, 2011).

Soil—rock classification		Slopes (V:H) for various cut heights		
			<6 m	>6 m
Hard rock (without adverse structure)			1:0.3–1:0.8	
Hard rock with joints and fractures			1:1–1:3 (depending on orientation of fractures)	
Soft rock			1:0.5–1:1.2	
Sand	Loose, poorly graded		1:1.5	
Sandy soil	Dense or well graded	1:0.8–1:1.0		1:1.0–1:1.2
	Loose	1:1.0–1:1.2		1:1.2–1:1.5
Sandy soil, mixed with gravel or rock	Dense, well graded		1:0.8–1:1.2	
	Loose, poorly graded		1:1.0–1:1.2	
Cohesive soil			1:0.8–1:1.2	
Cohesive soil, mixed with rock or cobbles		1:1.0–1:1.2		1:1.2–1:1.5

Table 2. Suggested slope angles for embankments constructed of various rock materials (Modified after NITRR, 1987).

Rock type	Moderately to highly weathered	Fresh to slightly weathered
Metamorphic (structured)	1:1.5	1:1.0
Metamorphic (massive)	1:0.5	1:0.25
Acid igneous	1:0.5	1:0.25
Basic igneous	1:0.5	1:0.25
Arenaceous and rudaceous	1:1.5	1:0.5*
Argillaceous	1:2.0	1:2.0**
Diamictites	1:1.5	1:0.5
Carbonate	1:1.5	1:0.25
Other sedimentary	1:0.5	1:0.25

* Subject to bedding not dipping into the cut.
** Subject to the material not being highly slaking and not dipping into the cut.

Table 3. Suggested slope angles for embankments constructed of various materials (Modified after ERA, 2011).

Fill materials	Embankment Side-slope (V:H) for various heights	
	<5 m	>5 m
Well graded sand, gravels, sandy or silty gravels	1:1.5	1:1.8
Poorly graded sand	1:1.8	1:2.0
Weathered rock spoil	1:1.5	1:1.8
Sandy soils, hard clayey soil and hard clay	1:1.5	1:1.8
Expansive clayey soils	1:1.8–1:4.0	1:2.0–1:4.0

4.3 Foundations

As discussed previously, no compromise should be allowed on the investigations for water crossing structures. Compromises can be allowed in the selection of such structures in that a limited period of lack of access may be accepted, e.g. fords or drifts subjected to periodic flooding may be constructed instead of a bridge. However, if a bridge is proposed a conventional geotechnical investigation appropriate for the specific structure must be carried out.

5 CONCLUSIONS

Geotechnical considerations can severely affect the cost of the provision of rural access roads. Even low volume roads such as these must be constructed taking into account the possibility that poor subgrades and local materials may affect the pavement structures and performance. It is, however, not always economical to carry out detailed geotechnical investigations.

Because of this, simple decision-making techniques have been provided to facilitate when specialist geotechnical advice should be sought and to assist with preliminary solutions for routine geotechnical problems. These include problem subgrade materials with some solutions provided as well as suggested slope angles for cuttings and embankments based on a simple flow diagram indicating when specialist input is necessary.

REFERENCES

Ethiopian Roads Authority (ERA). 2011. *Design manual for low volume roads: Part D.* ERA, Addis Ababa.
Gourley, C.S. & Greening, A. 1999. *Performance of low volume sealed roads: results and recommendations from studies in southern Africa.* TRL Ltd, Crowthorne, UK.
National Institute for Transport and Road Research (NITRR). 1987. *The investigation, design, construction and maintenance of road cuttings.* (Draft TRH 18). NITRR, Pretoria.
Paige-Green, P. 2008. Dealing with road subgrade problems in southern Africa. *Proceedings 12th International Conference of International Association for Computer Methods and Advances in Geomechanics (IACMAG),* Goa, India, 1–6 October, 2008, 4345–4353.
Paige-Green, P. 2008. Dispersive and Erodible Soils—Fundamental differences. *SAIEG/SAICE Problem Soils Conference,* Midrand, Nov 2008.
Van der Merwe, DH. 1964. The prediction of heave from the plasticity index and the percentage clay fraction of soils. *Trans. SAfr. Instn Civ. Engnrs,* 6(6), 103–107.

Proceedings of the first Southern African Geotechnical Conference – Jacobsz (Ed.)
© 2016 Taylor & Francis Group, London, ISBN 978-1-138-02971-2

Applicable station spacing for gravity surveys in dolomitic terrain

A.G. A'Bear
Bear GeoConsultants, Johannesburg, South Africa

R.W. Day
Engineering and Exploration Geophysical Services, Johannesburg, South Africa

L.R. Richer
LR Geotech, Johannesburg, South Africa

ABSTRACT: Gravity surveys are the most frequently applied geophysical method for investigating dolomitic terrain in South Africa. It has become the norm to apply a station spacing of 30 metres when carrying out gravity surveys, largely due to the competitive environment in which these surveys are conducted as well as lack of prior knowledge of site conditions. However, a closer spacing should be considered when rock head is shallow. This more detailed survey may have to be executed as a second phase of work when the extent of shallow rock head has been mapped, or after a decision on the footprint of a specific structure has been made. As the examples given here show, the results of a more detailed study can bring meaning to a diverse set of drilling results and perhaps a more appropriate classification or utilisation of the site. From this perspective, the guidelines provided in the National Department of Housing Specification, in which the station spacing is proportional to the depth to rock head, should be used as a means of determining the appropriate gravity station spacing used in investigations.

1 INTRODUCTION

Gravity surveys are carried out in the initial phases of dolomite stability investigations and the results used as an aid in selecting borehole sites and for determining hazard or geological zone boundaries. In the 1960's this method was employed on the Far West Rand, where sinkholes developed as a result of dewatering of the dolomite. After this, gravity surveys became the norm when conducting dolomite stability investigations.

The use of gravity is recognised in the standards governing the development of townships located in dolomitic areas within South Africa. However, there is some latitude in terms of the gravity spacing to be used. The National Department of Housing Specification (National Department of Housing 2002), for example, requires a station spacing equivalent to the average depth to bedrock and a maximum spacing of 30 metres for housing developments. Current standards, such as SANS 1936 (South African Bureau of Standards 2012a, b), do not specify the station spacing to be employed but rather leave it to the professional to decide on the appropriate density of data.

Financial considerations, often coupled to a lack of information about bedrock depth, frequently lead to a default choice of the largest station spacing possible; this approach may prove inadequate as this paper demonstrates. Here, case studies are used to illustrate the advantages of more appropriately spaced gravity surveys in areas where bedrock is shallow.

2 RELATIONSHIP BETWEEN GRAVITY AND BEDROCK MORPHOLOGY

Gravity surveys require the collection of gravity readings (observed gravity) along with determinations of differences between stations in elevation and latitude. The calculated relative Bouguer values are then separated into residual and regional components, where regional is defined as longer-wavelength changes in gravity that are of little interest to the study being undertaken. To a first approximation, the changes in residual gravity are attributed to variations in overburden thickness. This assumption is often sufficient because of a generally large density contrast between the dolomite bedrock and overburden which is commonly either a mix of, or singly, dolomite residuum, weathered Karoo Supergroup sediments and residual intrusive.

At some stage in the survey there is a reconciliation between residual gravity values and point samples of bedrock depths derived from drilling. This may result in the derivation of a new regional-residual separation of the Bouguer field and usually the acceptance that the particular gravity data set does not resolve all features of an assumed karstic bedrock topography. The overall bedrock depth usually influences the categorisation of the site stability.

There is a proportional relationship between the detail, or frequency, of bedrock variations recorded in a gravity map and the station spacing employed for the gravity survey; the closer the station spacing, the more detail that can be mapped. (This relationship is also dependent on the depth to bedrock compared to the magnitude of the changes in bedrock head, but here we are considering areas where variations in bedrock head are sufficiently large to noticeably affect the gravity field.) There will always be some practical limit to the amount of detail that can be picked up in a gravity survey and thus isolated pinnacles and very narrow or shallow solution features may not be detected. Drilling in dolomitic terrain is therefore always likely to find anomalous bedrock depths within what appears to be a uniform gravity feature.

Three case studies are presented below which illustrate the relationship between gravity station spacing and the detail that can be retrieved.

3 CASE STUDIES

3.1 Case study: Tembisa

The site is located in Tembisa, and is probably underlain by dolomite of the Oaktree Formation, Malmani Group, Transvaal Supergroup. The investigation commenced with a gravity survey on a 30 m grid. Once it became apparent that dolomite was very shallow over large portions of the site, a second gravity survey on a 5 m grid was commissioned. This revealed significant anomalies. that tied in with drilling results, although there were discrepancies with some deep bedrock in the gravity highs and shallow dolomite within the gravity lows.

In order to better define the exact boundaries of two significantly different hazard zones, the contractor was asked to remove the top 0,5 m of topsoil over the entire site. This exposed the area of predominantly shallow dolomite. The data from the two gravity surveys is shown on Figure 1. A photograph of the exposed dolomite pinnacles is shown in Figure 2. The mapped outcrop area ties in well with the gravity high areas as delineated by the 5 m grid survey.

It is apparent that the 30 m grid gravity survey and subsequent drilling could not have adequately

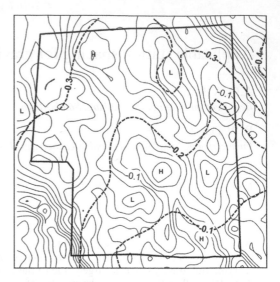

Figure 1.   Gravity survey comparison. Solid lines represent the 5 m grid and dashed lines the 30 m grid.

Figure 2.   Exposed dolomite pinnacles.

mapped the two distinctly different hazard zones with a reasonable degree of accuracy. The 5 m grid allowed the buildings which were the subject of the investigation to be located within an area of predominantly shallow bedrock.

3.2 Case study: Centurion

A large area was investigated for township development purposes in the southern part of Centurion. This is underlain by dolomite and chert of the Monte Christo Formation, Malmani Group, Transvaal Supergroup. A 30 m gravity grid was used to cover the 70 ha area as is the norm. One hundred and sixty-two (162) boreholes were drilled under the guidance of three different consultants in an attempt to delineate hazard zones within the site boundaries.

Although the final hazard zone plan appeared reasonable it was clear from the drilling results that numerous anomalous zones of poor conditions occurred within predominantly shallow dolomite bedrock areas. These were not associated with anomalies in the gravity plan making it very difficult to confidently proceed with development.

In an attempt to better understand the ground conditions and map a way forward, it was decided to cover a 60 m square area (3600 m²) with a gravity survey on a 5 m grid. This was done within a portion of the site where the 30 m grid showed reasonably uniform conditions but which drilling showed to be highly variable. The 5 m grid, overlain on the 30 m grid in Figure 3, shows distinctly different bedrock conditions which match the drilling results. Plans were subsequently drawn using the same data but reducing the data points as though a 10 m and 15 m grid survey had been carried out.

This recent study in Centurion was investigated with a 30 m spaced grid gravity and then subsequently investigated using 15, 10 and 5 m spacings. It is believed that the 15 m grid will still provide sufficient detail in this case for the bulk of the anomalous conditions to be delineated and a recommendation has been made to the client that the area be surveyed on this level. Not only will a survey of this nature allow poor zones to be delineated but it will allow better predictions to be made with respect to the width of solution features. The current requirement that footprint drilling be carried out to identify positions for individual structures could be revised on this basis so that a more economical second phase of drilling is implemented.

3.3   Case study: Kuruman

A 15 Ha stand in Kuruman was investigated using a 10 m grid as bedrock was expected to be close to surface. A single, narrow, linear feature formed by a gravity low was identified within an otherwise fairly uniform gravity high area. Drilling subsequently showed that this feature was at least 60 m deep. A borehole drilled 5 m to the side of the feature showed bedrock to be at 10 m which is the norm for that area. Subsequent appraisal by the geophysicist led him to estimate that the solution feature which this anomaly represented was probably no more than 2,5 m wide.

A broader gravity survey, on say a 30 m grid, may not have picked up this feature, although had it done so, it would not have allowed the geophysicist to deduce that the feature was no more than 2,5 m wide and a much larger area would have had to be excluded from development or subjected to a more intense drilling programme.

3.4   Case study: Lyttelton

This case study illustrates the reason for the disparities that can occur between borehole derived and expected bedrock depths, and the effect of sample spacing. The survey in question covered a plot of land underlain by rock of the Lyttelton Formation. The two-hectare area was first covered with a twenty-metre grid of gravity stations. The result is shown in the upper diagram of Figure 4,

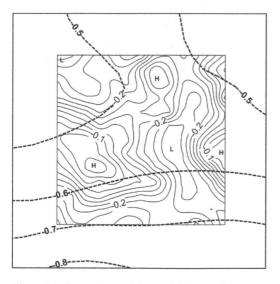

Figure 3.   Comparison of 5 m and 30 m gravity survey grids. Solid lines represent the 5 m grid and dashed lines the 30 m grid.

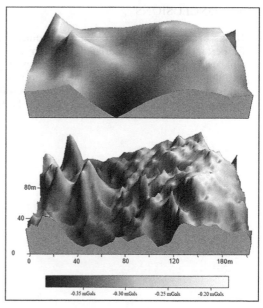

Figure 4.   Comparison of 5 m and 20 m gravity grids in three dimensions.

where the residual field varies from smooth in the east and has a somewhat pinnacled appearance in the west. Boreholes set out using the gravity results intersected rock from twelve to twenty-five metres below surface, depths that agree with the range in residual gravity values but, strangely, the shallowest rock was hit in a gravity low.

The area was resurveyed using a five metre grid with the result shown in the lower diagram of Figure 4. The trends are broadly the same but the gravity field is much more rugged, with the eastern high now dissected by linear troughs and overlain by spikes and ridges, and with the appearance of more and sharper pinnacles in the west. The newly-defined variations confirm the expectation of a pronounced karstic topography whilst improving the correlation between gravity and bedrock depth.

There were still discrepancies between actual and gravity-inferred depths which cannot be eliminated by decreasing the station spacing even further. For example, holes drilled on suspected pinnacles entered bedrock at depths less than those predicted from the gravity whilst the converse was sometimes found with the inferred slots. This result is typical of those obtained when drilling into pinnacles or narrow ridges. By their nature pinnacles, ridges and slots are restricted in lateral extent and thus have a limited effect on the amplitude of the gravity field, leading to inherent over and under estimates of bedrock depth.

4 CONCLUSIONS

It has been demonstrated using case studies that far better resolution of bedrock topography is possible using appropriate gravity survey grid spacings and that better resolution is required in certain circumstances to make sense of the drilling results. Appropriately spaced gravity grids allow for better zonation of sites as well as a better identification and prediction of the width of solution features within generally shallow bedrock areas. The additional cost of the gravity survey will be offset by reduced drilling requirements and the production of a more confident zonation and development plan.

The National Department of Housing Specification (National Department of Housing 2002) requiring that a gravity station spacing be used that is equivalent to the average depth to bedrock but never greater than 30 metres should be used as a guideline when deciding on a grid spacing. In areas where the depth to bedrock is unknown it is recommended that the site be covered using a 30 m grid which can subsequently be narrowed down in a second phase where warranted. For smaller sites and where individual structures, such as water towers, are to be investigated, grid spacings of as little as 5 m should be considered in shallow dolomite areas. It may not be feasible to investigate large areas at spacings of less than 15 m as the costs may become prohibitive. However, this study has shown that spacings of less than 30 m must be considered an imperative once areas of shallow bedrock are identified.

REFERENCES

National Department of Housing. 2002. *Generic Specification GFSH-2 Geotechnical site investigations for housing development*. Johannesburg: National Department of Housing.
South African Bureau of Standards. 2012a. *SANS 1936-1 Development of dolomite land: Part 1: General principles and requirements*. Pretoria: SABS Standards Division.
South African Bureau of Standards. 2012b. *SANS 1936-2 Development of dolomite land: Part 2: Geotechnical investigations and determinations*. Pretoria: SABS Standards Division.

Proceedings of the first Southern African Geotechnical Conference – Jacobsz (Ed.)
© 2016 Taylor & Francis Group, London, ISBN 978-1-138-02971-2

An account of tunnel support systems for soft rock mass conditions

J.E. Ongodia & D. Kalumba
University of Cape Town, Cape Town, South Africa

H.E. Mutikanga
Uganda Electricity Generation Company Limited, Kampala, Uganda

ABSTRACT: Subsurface works such as tunneling especially in soft rock require support, often immediately upon excavation. Support systems usually comprise of several individual components provided to suit the specific site's soft rock mass conditions. Rock mass conditions are classified according to strength, degree of weathering and discontinuities; based on empirical method applications for hard rock. The rock mass conditions hence rock class and hydrogeology are key inputs to tunneling and tunnel support systems. From the design process appropriate parameters of the support components, comprising a support system, are suggested. Research on tunneling in soft rock is evolving. Existing research and information highlight the need for further research on soft rock, including establishing its own rock mass classification system. It is upon this ground that this study discusses soft rock rather than hard rock. This paper reports on tunnel support systems in soft rock from related experiences and research. In summary, the need for further research on the subject of soft rock and constructing in it is apparent.

1 INTRODUCTION

Subsurface works test the peripheral of geotechnical engineering works. Tunnels are examples of subsurface works, categorized as rock engineering (Jones 1989). Tunnels were studied and presented because unlike open excavations, tunnels are bored into and through the ground. Besides internal loads, the weight of the overhang and surrounding material must be considered to avoid caving in and internal collapse of the tunnel. Thus the dynamics presented in tunneling are more than in open excavations which exclude overhang material. Based on this analogy, this paper studied tunneling. Typically rock is either soft or hard (He 2014). At its lowest limit range, soft rock is completely weathered therefore approximates soil. An overhang and surrounding of soil material presents several tunneling challenges when compared to hard rock. It was therefore chosen for this study as a weak material for tunneling projects.

Tunnel engineering and the concept of soft rock is gradually evolving (He 2014). The term 'soft' refers to mechanical strength. Soft rocks have low strength of below 25MPa amongst other characteristics (Kanji 2014 and Marinos 2014) which are generally undesirable for engineering purposes. Supports are therefore provided during construction to improve strength of excavations and make the structures more suitable for engineering purposes. In general, excavations especially in soft rock, are supported. Because of their low strength, once excavated they usually require immediate support to be provided (Marinos 2014). Immediate support is key to ensure stability and safety, of the surrounding rock mass. When a tunnel is excavated the ground around it, which is the surrounding rock mass, is destabilized.

Optimal resource utilization supports sustainability (Ghimire & Reddy 2013). Underground construction is an example of how the land resource is optimized. Such construction allows for multiple land use—both on and beneath the ground surface, concurrently. Engineering projects of such nature are complex. Both the surface and underground civil engineering infrastructures must be adequately supported. In other words, the surrounding ground must be safe guarded against failure. Both surface and underground structures must be constructed well to ensure stability, safety and soundness of the foundations or surrounding material. Tunnels are examples of underground construction projects, categorized under rock engineering (Jones 1989). They are constructed by excavating through ground (soil/ rock) for purposes of passage. Passage can be of wind, wildlife, traffic, or water (Yi 2006). Tunnels are usually constructed in rock mass and seldom in soil mass. Figure 1 shows tunnel project stages. Tunneling begins with routing, costing, detailed Geotechnical Investigations (GIs), boring method, and construction, (Blake 1989). This

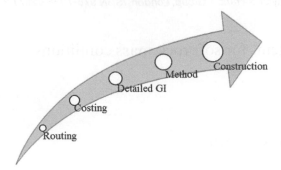

Figure 1. Tunnel project lifecycle.

paper focusses on the detailed GI stage and tunnel support systems.

2 DETAILED GEOTECHNICAL INVESTIGATIONS (GIS)

Investigation studies at this stage are more in depth than initial routing studies to establish site-specific rock parameters and profile. Key GI parameters are the rock mass condition and hydrogeology. Other important parameters include structure of the rock mass, void spaces, in_situ stresses, Poisson's ratio, elastic modulus, and construction stages. Material tests (in-situ and laboratory) to identify the parameters for soft rock are performed although with challenges. For instance, soft rock is brittle therefore procuring undisturbed samples for laboratory testing is hard and they easily crumble during in-situ testing thus affecting the quality of results. The challenges arise mainly because tests for soft rock are yet to be defined. Practice therefore improvises, borrowing mostly from similar tests for hard rock, in order to identify the requisite soft rock parameters. Besides sampling and testing challenges, their geomechancial classification is also indeterminate (He 2014 and Kanji 2014).

Rock mass condition is a function of rock strength, degree of weathering, and discontinuity sets. For engineering purposes, rock strength is most important. In terms of strength, rock can either be 'soft' or 'hard', depending on the expected engineering stresses (He 2014). The expected engineering stresses are the required bearing capacities to support the loads from the tunnel structures and vary according to the purpose and use of the structure. Rock strength factors include physico-mechanical properties and water content/humidity. The degree of rock weathering is a function of age and alteration processes. Weathering can be high, moderate, or complete. Alteration processes can be by cementing, transportation and deposition. Discontinuity sets, also referred to as discontinuities, include faults, joints, fractures, foliation, folding planes, among others. Strength, weathering and discontinuities determine the rock mass condition on which the rock class is based. Rock classification, distinguishing between the main categories, is a complex subject (Kanji 2014). Figure 2 shows rock classifications by several researchers. Each horizontal segment is a continuum (ranging from the lowest to the highest strength limits) of rock according to Unconfined Compressive Strength (UCS) by the attributed researcher, on the right hand side of the chart. The chart generally shows three classes of the ground material; soil, soft rock, and hard rock. Actual transitions between all three classes is indistinct; except for the common separator (the vertical broken line) distinguishing between soft and hard rock (Kanji 2014). Strength increases from soil to rock (*left to right*, across the continuum). Soft rock has intermediate strength between soil and rock.

Hydrogeology refers to the interaction between ground water and geology. Groundwater recharge sources include precipitation, surface runoff, percolating water, and nearby wells, among others. The hydrostatic head of water in the surrounding and knowledge of the pore water pressures, are important. Pore water pressure can contribute to further disintegration of the soft rock mass, as it flows through it. Water passage paths through the rock mass follow faults, joints, void spaces, and fractures. Hydrogeological studies for tunneling GI identify fault zones, open joints, aquicludes, and aquifers (Blake 1989). More specifically water bearing slightly-filled fault zones and aquicludes are sought. Furthermore, water can cause swelling depending on whether water comes from the surrounding rock mass or from an external source

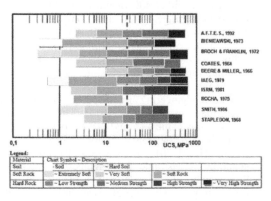

Figure 2. Rock classifications by various researchers (adopted from Kanji, 2014).

such as inside the tunnel (Jean et al. 2003). Soft rock is characterized by various discontinuities (faults, joints, fractures, foliation, and folding planes, among others) which are responsible for its poor engineering characteristics. Numerous discontinuities cause extensive internal deformations (Eberhardt 2008). Soft rock is discussed in the following subsection.

2.1 Soft rock

The term 'soft rock' remains a subject of debate in geotechnical engineering (Kanji 2014). However, numerous researchers suggest that soft rock has a maximum Unconfined Compressive Strength (UCS) of 25MPa. A study by the author found that soft rock had a porosity of less than 20% (Kanji 2014). Soft rock may therefore be defined as the ground material ranging in-between, soil and hard rock with an ultimate unconfined compressive strength of 25MPa and low porosity.

Rock strength is emphasized because the focus of geotechnical engineering is in the ability of a geological material to support stresses. Engineering rock underscores the rock's plastic deformation under stresses (He 2014). Generally, engineering soft rock descriptions should specify geology, lithology, complementary condition, structural defects, age of formation, and behavior and characteristics (He 2014 and Kanji 2014). Table 1 shows soft rock types and subclasses.

2.1.1 Classification of soft rock

Soft rock can be classified based on applied engineering stresses, mechanical properties and strength characteristics (He 2014). Based on

applied stresses, soft rock is categorized according to critical load for softening and critical depth for softening. Table 2 gives soft rock classes according to strength and table 3 shows the required tests to classify.

2.1.2 Failure systems in soft rock

The likelihood and magnitude of failure depends largely on inherent material properties. Soft rock has low strength, easily disaggregates, weathers fast, is plastic, slakes, and is very permeable (Kanji 2014). It also has various discontinuities (Panda et al. 2014), softens (He 2014), has a short stand-up time (Marinos 2014), deforms largely and fails (Wang et al. 2014). The shorter the rock stand-up time, the narrower the possible tunnel width to construct and the higher the need for immediate support upon excavation (Bieniawski 1992). Generally soft rocks deform by physical expansion, stress dilatancy and structural deformation (He 2014 and Wang et al. 2014). Since soft rock deformation is large, it translates into immediate failure. Knowledge of soft rock's specific large deformation at failure is important (He 2014); and both strain softening and nonlinear failure envelopes (Yoshinaka & Nishimaki 1981). Failure results from weak strata, discontinuities, in-situ stress, displacement, unsupported excavation, uncoupling of support systems, and deep-seated deformation. Table 4 shows failure mechanisms. Figure 3 shows soft rock failure mechanisms.

Based on the possible failure mechanisms, support systems are provided in order to arrest

Table 2. Classification of soft rock (Adjusted from He, 2014).

Soft rock class	Definitive characteristics
Expansive	High stress-strong expansion combined. Slips along clay minerals under external stresses. Water causes significant expansion.
High-stress	High stress-jointed. Slips along clay minerals under high stresses. Slightly inflates under water. Unconfined Compressive Strength (UCS) < 25MPa.
Jointed	High stress-jointed-strong expansion. Slip and expansion produced along jointed surface. UCS > 25MPa.
Combined	Combined soft rock. Complex mechanism (combines all characteristics).

Table 1. Types of soft rock (Adjusted from Kanji 2014).

Rock type	Sub classes
Igneous	Volcanic conglomerates, breccia, and lahar; Basaltic breccia; Piroclastic deposits, volcanic ash, tuff and ignimbrite; Weathering products of crystalline rocks such as granite, gneiss
Sedimentary	Clastic mudstones, shales, siltstones, cemented sandstones, conglomerates and breccia, and marl, Evaporites: salt rock, carnallite; Soluble limestone, dolomite, and gypsum; Coal
Metamorphic	Slate, phyllite, schists, quartzite lightly cemented, Metavolcanic deposits

Table 3. Soft rock tests (ASTM, 2015).

Test	Parameter
Brazilian Tensile Strength (BTS)	Strength and roughness
Acoustic velocities	Competency
Cerchar Abrasivity Index (CAI)	Resistance to wear
Tabor abrasion test	Abrasiveness
Schmidt hammer hardness	Soundness and hardness
Point Load (PL) index test	Classify strength
Rock Quality Designation (RQD)	Class and strength
Slake durability index	Disintegration resistance
Rock classification	Mineral identification
Permeability	Hydraulic conductivity

Table 4. Ultimate limit state failure in soft rock.

Type	Failure mechanism
Bearing capacity (Viggiani, 2012)	Shear
Tunnel collapse (Li et al., 2014)	Displacement
Sliding (Eberhardt, 2008)	Rock slope instability
Rock bursts (Viggiani, 2012)	Shear or splitting
Wall failure and Floating (Viggiani, 2012)	Shear
Heaving (Kovari et. al., 1988)	Expansion and strain softening

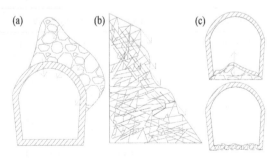

Figure 3. Failure mechanisms (left to right: Collapse, Sliding and Heaving).

the failures. The design, tunneling method and construction depend on good geotechnical investigation data and theory. The following section discusses tunnel support systems.

3 TUNNEL SUPPORT SYSTEMS

3.1 *The system*

Various factors contribute towards soft rock failure and critical loads/stresses acting on the tunnel from the surrounding rock mass (Blake 1989). The dynamics presented make tunneling in soft rock complex. Therefore in order to ensure stability, a combination of individual components collectively provide the necessary support as a system. Usual system components include reinforced concrete lining, bolts/anchors, reinforcement mesh, and shotcrete. Combination methods hinge on joint set orientation (Jean et al. 2003) and depend on rock strength, and whether support is required immediately/later or temporarily/permanently. For soft rock, immediate permanent support is required, usually comprising all components.

Bolt and anchor members strengthen and support the ground by extending beyond weak areas (slip circle). The mechanism of member support is by bolt strengthening alongside the coupling between the member-mesh and surrounding (He 2014). The coupling maximizes the bearing capacity of the surrounding rocks, thus ensuring stability. Individual members are evaluated and checked for adequacy using engineering software, finite element, numerical or empirical methods are used. Viggiani (2012) used empirical methods to find the tunnel-shaft's skin friction (NF). From Zeevaert's equation;

$$NF = \pi DK\phi \int \sigma dzi \qquad (1)$$

where D = external diameter; K_{ϕ} = lateral earth pressure; and $\int \sigma dzi$ = area of total vertical effective stresses. In contrast, however, recent research by Marinos (2014) suggested no related limit states for underground works were known. The different research opinions accentuate the need for further work on tunnel support systems in soft rock.

Overall, support systems constructed outside the tunnel to prevent inward movement of the surrounding rock (Hoek & David 1987), are required to ensure stability and safety of the structure. Figure 4 shows tunnel support components.

3.1.1 *Soft rock failure*

Failure in soft rock is complex in line with the suggested nonlinear large deformation theory (He 2014). A thorough understanding of the complex failure mechanisms of soft rocks and the expected engineering actions is important to design adequate support systems. To design, He (2014) hypothesizes that having established the complex large deformation of soft rocks; one general failure is put forth; and the method by which the complex deformation is transformed is used effectively. The generalized failure is comparable to the Ultimate Limit State (ULS) in concrete design (Bond & Harris 2008) and is critical to failure. The hypothesis forms the basis for tunnel support system design. The design process should firstly be strategic incorporating all actions and reactions including those likely to cause deformation. Secondly, the design process is optimized by providing all necessary countermeasures which are each evaluated and checked. Finally, individual system component parameters (diameter, length, spacing and tension in rod) are chosen such that they can adequately support all actions and resist reactions. Design is generally complex because it integrates several factors of which stability must be attained both of the structures and surrounding rock mass.

3.2 Rock classes and tunnel support systems

The Bieniawski (1989)'s Rock Mass Rating (RMR) method and Barton et al. (1974)'s Q system, developed for hard rock (Jean et al. 2003), are borrowed for soft rock and used to design tunnel support systems. The RMR classification defines five rock classes with their corresponding support systems. The Q system gives a quantitative indication of adequate tunnel supports to provide stability. The individual methods have some limitations, hence both methods are used concurrently to classify rock. The RMR method restricts the various excavation and support options while the Q system disregards the characteristic parameter of the rock mechanical strength (Jean et al. 2003). Table 5 illuminates both methods. The need of further research to classify (Kanji 2014) and design (Marinos 2014) soft rock, cannot be overstated.

Project-specific supports chosen depend on the rock class, construction method and the surrounding. For example, tunnel projects are aligned

Table 5. Common rock classification methods (Jean et al. 2003).

Parameter	RMR	Q system
Characterisation	Jointing patterns.	Mechanical properties of discontinuities. Natural stresses
Project purpose	Orientation with respect to axis structure.	Not relevant to orientation.
	Easily set bolt length.	Inaccurate bolt lengths.
	Stand-up time (conservative).	Easily choose roof Support.
	Not helpful to choose excavation method.	Instrumental to choose excavation method
Empirical relations to assess mechanical characteristics	RMR & strength & deformability parameters.	Q & physical mechanical parameters.

away from weaker areas (Kanji 2014) with the axis non-parallel to the orientation of discontinuities. Also where the project area has nearby structures, the flotation method is preferred to excavate (Viggiani 2012). The class and method are guided by the RMR and Q-system outputs. Relevant rock characteristics include Unconfined Compressive Strength (UCS), Rock Quality Designation (RQD), joint frequency, joint roughness, discontinuities and infilling, groundwater flow and pressure, and in_situ stresses (Kanji 2014). Also, the position of the structure with respect to tunnel driving and the Excavation Support Ratio (ESR) are unique to the RMR and Q method; respectively. ESR index is an indirect factor of safety. Recently, the Geological Strength Index (GSI) classification system by Hoek et al. (1998) incorporated jointing characteristics of the rock mass (Kanji 2014 and Marinos 2014). Panda et al. (2014) developed a design chart from which tunnel supports can be chosen based on the Q method. The chart relates rock classes, strength quality in terms of the Q-value, tunnel dimensions and bolt length. The latter two parameters were related to the Excavation Support Ratio (ESR).

From recent research, a new tunnel support system was suggested. He (2014)'s research team developed the *Continuous Resistance Large Deformation (CRLD) bolt/anchor* material to

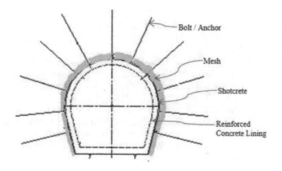

Figure 4. Tunnel support components.

Bolt / Anchor

Mesh

Shotcrete

Reinforced Concrete Lining

support surrounding rock masses, based on case studies and experiments in China. In their research, several theories were suggested, the most significant being the Systematic Coupling Support theory. Unlike ordinary bolts and/or anchors, the CRLD Bolt/Anchor's unique features include a negative Poisson's ratio effect, ability to resist large deformation, and endure impact resistance. In essence CRLD components thicken in tension, effectively ensure stability, and remain constant under repeated impact loading; respectively.

4 CONCLUSION

Research on soft rock engineering processes is young and evolving. Soft rock research was birthed from flaws and lessons learnt in practice as alternative sites became scarce. As a result, more engineering projects are constructed in unsuitable ground such as soft rock with poor engineering properties because of the increased demand for land. Tunnel engineering is evolving through observation and practice.

Tunnels in soft rock are complex and require specialized geotechnical engineering expertise and application (Blake 1989). Tunnel support systems in soft rock are provided immediately upon excavation. The systems comprise of several individual components which are integrated in the model and design to provide stability. Stability is ensured through adequate support of all loads, actions, and reactions in the engineering structures and surrounding rock mass.

Challenges in soft rock tunneling are evident. The methods to sample, test, and classify the soft rock are yet to be defined and hence practice borrows from the known hard rock principles. In other words, shortcomings are likely in design and construction tunneling in soft rock. Further research in the field of soft rock and tunneling in soft rock is necessary. Documented research (Kanji 2014) shows that artificial samples can be used to test in place of natural soft rock. Therefore the possibility of replicating and progressing research on soft rock may be easier without the need to designate for testing, actual natural rock.

ACKNOWLEDGEMENTS

The support of Julian Baring Scholarship Fund and Uganda Electricity Generation Company Limited is greatly acknowledged.

REFERENCES

ASTM 2015. American Standard of Testing Materials International, 1996–2015, West Conshohocken, PA, 19428–2959 USA.

Bieniawski, Z.T., 1992. Design Methodology in Rock Engineering. Rotterdam: Balkema.

Blake. L.S., 1989. *Tunneling. Civil Engineer's Reference Book*, Fourth Edition. © Elsevier Ltd ISBN-13: 978-0750619646.

Bond. A., and Harris. A., 2008. *Decoding Eurocode 7.* Taylor and Francis Group ISBN 0-203-93772-4.

Eberhardt. E., 2008. The Role of Advanced Numerical Methods and Geotechnical Field Measurements in Understanding Complex Deep-Seated Rock Slope Failure Mechanisms. *Twenty-ninth Canadian Geotechnical Colloquium.* DOI: 10.1139/T07–116.

Ghimire. N.S.B., and Reddy. M.J., 2013. Optimal Reservoir Operation for Hydropower Production Using Particle Swarm Optimization and Sustainability Analysis of Hydropower. *ISH Journal of Hydraulic Engineering*, Vol. 19(3): 196–210.

He. M., 2014. Latest Progress of Soft Rock Mechanics and Engineering in China. *Journal of Rock Mechanics and Geotechnical Engineering*: 165–179.

Hoek. E., 2001. Big Tunnels in Bad Rock. *Journal of the Geotechnical and Geo-environmental Engineering*, Vol 127, (9): 726–740.

Hoek. E., and Wood. F.D., 1987. Support in Underground Hard Rock Mines. *Underground Support Systems.* Edited by J. Udd. Special Volume 35: 1–6.

Jean. P., Amelot. A., Andre. D., Berbet. F., Bousquet-Jacq. F., Curtil. S., Durville. J., Fabre. D., Fleurisson. J., Gaudin. B., Goreych. M., Homand. F., Parais. G., Peraud. J., Robert. A., Vaskou. P., Vibert. C., and Wojtkowiak. F., 2003. *A.F.T.E.S Guidelines for Caracterisation of Rock Masses Useful For the Design and the Construction of Underground Structures.*

Jones. A.P., 1989. *Civil Engineer's Reference Book.* Edited by L S Blake © Reed Educational and Professional Publishing Ltd. The Bath Press, Bath ISBN O 7506 1964 3:10/3-10/39.

Kanji. A.M., 2014. Critical Issues in Soft Rocks. *Journal of Rock Mechanics and Geotechnical Engineering 6*: 186–195.

Kovari. K., Amstad. C., and Anagnostou. G., 1988. *Design/Construction Methods—Tunneling In Swelling Rocks. Balkema, Rotterdam.* ISBN 9061918359 pp. 17–32.

Li. L., Wang. Q., Li. S., and Huang, H., 2014. Cause Analysis of Soft and Hard Rock Tunnel Collapse and Information Management. *Pol. J. Environ. Stud.* Vol. 23(4): 1227–1233.

Marinos. V., 2014. Tunnel Behaviour and Support Associated With the Weak Rock Masses of Flysch. *Journal of Rock Mechanics and Geotechnical Engineering 6*: 227–239.

Panda. M.K., Mohanty. S., Pingua. B.M.P., and Mishra. A.K., 2014. Engineering Geological and Geotechnical Investigations along the Headrace Tunnel in Teesta Stage-III Hydroelectric Project, India. *Engineering Geology* 181: 297–308.

Tshering. K., 2012. *Stability Assessment of Headrace Tunnel System for Punatsangchhu II Hydropower Project*, Bhutan.

Viggiani. G., 2012. Geotechnical Aspects of Underground Construction in Soft Ground. Technology and Engineering *7th International Symposium on Geotechnical Aspects of Underground Construction in Soft Ground*, Rome, Italy.

Wang. S., Chunliu. L., Zhaowei. L., and Junbo. F., 2014. Optimization of construction scheme and supporting technology for HJS soft rock tunnel. *International Journal of Mining Science and Technology* 24: 847–852.

Yi, N. Tsing., 2006. *Tunneling*. Hong Kong Institute of Vocational Education. Lecture Notes. Accessed http://tycnw01.vtc.edu.hk/cbe2024/3-Tunnel.pdf on September 21, 2015 at 1020hours.

Yoshinaka. R., and Nishimaki. H., 1981. Bearing Capacity and Failure Mechanism of Soft Rock under Rigid Footing ISRM-IS-1981-109 *International Society for Rock Mechanics International Symposium, Tokyo, Japan*.

Soil properties

Proceedings of the first Southern African Geotechnical Conference – Jacobsz (Ed.)
© 2016 Taylor & Francis Group, London, ISBN 978-1-138-02971-2

Evaluation of dispersive soils in the Western Cape, South Africa

A. Mohamed
Aurecon, Cape Town, South Africa

G.N. Davis
Aurecon, Pretoria, South Africa

ABSTRACT: Dispersive soils were first identified in South Africa in the mid-1960's. Before then, several small earth embankment dams in the Free State and Northern Cape Provinces failed due to dispersive soils. A geotechnical investigation must include determination of the risk potential of dispersive soils, but no single test method can adequately confirm the risk potential of dispersive soils. The main objective of this study was to identify and rank dispersive soils in the Western Cape using the rating system developed by Jermy and Walker (1999) as a guide to help average out conflicting results that were obtained by the various test methods. In order to have achieved this, a series of both physical and chemical dispersity testing on 29 soil samples taken from new earth fill farm dams in the Western Cape, were conducted and results were combined into the rating system to allow the correct identification of the soils into non-dispersive to highly dispersive soils.

1 INTRODUCTION

1.1 Background

Dispersive soils were first identified in South Africa in the mid-1960's. Before then, several small earth embankment dams in the Free State and Northern Cape Provinces failed due to dispersive soils (Bell & Maud, 1994).

According to (Maharaj, 2013), the Senekal Dam in the Free State provides one example of the many earth embankment dams which failed due to dispersive soils. The earth dam was constructed of alluvial silty clay on a foundation of similar material overlaying sandstone and shale. Failure occurred after initial filling of the reservoir due to piping through the dam wall and floor. The remedial geotechnical investigation revealed that both the embankment and foundation soils had high Exchangeable Sodium Percentage (ESP) values (average 25). Soils with ESP of 10 or above are classified as dispersive.

(Maharaj, 2013), outlines another example of a 70 meter high earth fill embankment dam, known as Elandsjagt Dam, located on the Kromme River in the Eastern Cape. Double hydrometer tests were carried out which indicated that the soil was potentially dispersive. Thirty samples from the embankment core and basin was then taken through a comprehensive testing programme, and included the SCS dispersion test, pinhole, crumb and chemical tests. 19 of the 30 samples indicated dispersive

behaviour, therefore provisions were made for the reconstruction to allow for the addition of a chimney drain to act as a filter to prevent further piping erosion.

1.2 The research problem

The research problem in the literature is based on dam failures that were caused by dispersive soils. A geotechnical investigation must include determination of the risk potential of dispersive soils, but according to Bell & Maud (1994), no single test method can adequately confirm the risk potential of dispersive soils; each test method has different evaluation criteria and testing protocols that were developed by different researchers.

1.3 Research question

Are any of the new earth fill farm dams in the Western Cape constructed with dispersive soils; and if so, how do we characterize the soils dispersive potential?

1.4 Objectives

The aim of this research work was to identify and rank dispersive soils from new earth fill farm dams in the Western Cape by using the rating system developed for Kwazulu-Natal by Jermy & Walker (1999), as a guide.

2 TESTING FOR DISPERSIVE SOILS

As no single test method can adequately confirm the dispersivity potential of a soil, the testing should comprise a suite of specialized tests. These can be divided into physical and chemical tests. Physical tests show the effect of the dispersivity of the soil and chemical tests the cause. The tests used in this research included:

Physical tests: Double hydrometer, crumb and Pinhole.

Chemical tests: Exchangeable Sodium Percentage (ESP), Sodium Adsorption Ratio (SAR) and Total Dissolved Salts (TDS).

3 TEST METHODS AND EVALUATION CRITERIA

3.1 Physical tests

3.1.1 Double hydrometer

This test method was performed in conjunction with the American standard and is described in ASTM D 4221 (2006).

First a standard hydrometer is used in which the soil specimen is dispersed in distilled water with a strong mechanical agitation and a chemical dispersant. A parallel hydrometer test is then made on a duplicate soil specimen, but without mechanical agitation and without a chemical dispersant. The "percent dispersion" is the ratio of the dry mass of particles smaller than 0.005 mm diameter of the second test to the first expressed as a percentage.

The formula used to determine the percent dispersion is as follows:

$$\% \ Dispersion = \frac{P_B}{P_A} \times 100 \qquad (1)$$

where PB = the average percentage of soil smaller than 0.005 mm from the particle size distribution chart without adding chemical dispersant; and PA = the average percentage of soil smaller than 0.005 mm from the particle size distribution chart with chemical dispersant.

3.1.2 Crumb

Briefly, the test consists of selecting a soil crumb, or prepared cubical specimen, at natural water

Table 1. Dispersion evaluation criteria for double hydrometer (ASTM D 4221, 2006).

Percent dispersion	Degree of dispersion
<30	Non-dispersive
30–50	Intermediate
>50	Dispersive

content. The specimen is carefully placed in about 250 ml of distilled water or NaOH. As the soil crumb begins to hydrate, the tendency for colloidal-sized particles to deflocculate and go into suspension is observed.

3.1.3 Pinhole

The pinhole test was developed by Sherard *et al.* (1976), the test method directly measures dispersivity of compacted fine-grained soils in which water is made to flow through a small hole in a soil specimen, where water flow through the pinhole simulates water flow through a crack or other concentrated leakage channel in the impervious core of a dam or other structure. A 1.0 mm diameter hole is punched or drilled through a 25 mm long by 35 mm diameter cylindrical soil specimen. Distilled water is percolated through the pinhole under heads of 50 mm, 180 mm 380 mm and 1000 mm. The flow rate, effluent turbidity and size of hole are recorded.

Table 2. Dispersion evaluation criteria for crumb test (Maharaj, 2011).

Grade	Reaction	Description
1	No reaction	The crumb may slake and run out on the bottom of the beaker in a flat pile with no sign of cloudy water.
2	Slight reaction	There is a bare hint to easily recognizable cloud of colloids in suspension.
3	Moderate reaction	The colloids may be just at the surface of the crumb.
4	Strong reaction	The colloid cloud covers nearly the entire bottom of the beaker. In extreme cases, all the water in the beaker becomes cloudy.

Table 3. Dispersion evaluation criteria for pinhole test (Sherard et al., 1976).

Classification		Head	Effluent color	Hole size
Dispersive:	D1	50	Very distinct	>2 times
	D2	50	Distinct to slight	2 times
Intermediate:	ND4	50	Slight but visible	1.5 times
	ND3	180–360	Slight but visible	2 times
Non-dispersive:	ND2	1000	Barely visible	<2 times
	ND1	1000	Crystal clear	No erosion

3.2 Chemical tests

3.2.1 Exchangeable Sodium Percentage (ESP)

During the 1960's it was recognised that the presence of exchangeable sodium (Na) is a main contributing chemical factor to dispersive clay behaviour. The basic parameter to quantify this effect is ESP (exchangeable sodium percentage), where:

$$ESP = \frac{Exchangeable\ Sodium}{CEC(cation\ exchange\ capacity)} \times 100 \quad (2)$$

3.2.2 Sodium Adsorption Ration (SAR)

A further chemical test considered indicative of the degree of dispersion is the Sodium Adsorption Ratio (SAR), where Harmse (1980) indicated the dispersivity criteria illustrated in Table 5 were applicable in the South African context.

It must be noted, however, according to the rating system by Jermy & Walker (1999), SAR values greater than 2 are considered dispersive.

3.2.3 Exchangeable Sodium Percentage (ESP) and Cation Exchange Capacity (CEC) chart

A further criteria is to plot the ESP and CEC together on a chart with envelopes produced by Gerber & Harmse (1987) ranging from completely non-dispersive to very dispersive. Refer to Figure 1 taken from Bell & Maud (1994).

3.2.4 Total Dissolved Salts (TDS) and % sodium

Sherard et al. (1976) proposed a chart with percentage sodium (% Na) plotted against the Total Dissolved Salts (TDS) all in milli-equivalents per liter (meq/l) for assessing dispersivity.

If the percentage Na is greater than 60% and the TDS greater than 1 meq/l the soil is regarded as dispersive. If the % Na is less than 40% and the TDS less than 0.2 meq/l the soil is regarded as

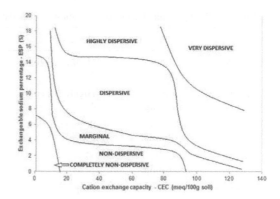

Figure 1.   ESP vs CEC chart by Bell & Maud (1994).

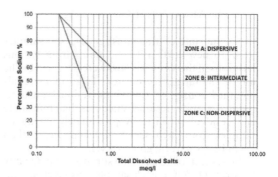

Figure 2.   TDS (1) vs % Na (2) by ICOLD Bulletin 77 (1990).

non-dispersive. An intermediate dispersive group is present between these two zones. Jermy and Walker (1999) found that this method identified the dispersive nature of soils with a fair degree of accuracy. The figure below taken from the ICOLD Bulletin 77 (1990) illustrates the above zones.

4   RATING SYSTEM OF JERMY & WALKER (1999)

Jermy & Walker (1999) and Bell & Walker (2000) combined the results of the above tests into a rating system, incorporating a weighting dependent on the reliability of each test, to identify dispersive soils with a greater level of certainty. Their rating system is presented in the table below.

According to the system a total rating score of less than 4 indicates the soil is non-dispersive, between 5 and 7 slightly dispersive, between 8 and 11 moderately dispersive and above 12 are highly dispersive. Furthermore, the rating system excluded the double hydrometer test. It is, however, still widely used elsewhere and found to be accurate by many researchers (Mellvil & Mackellar, 1980).

Table 4.   Dispersion evaluation criteria for ESP results (Harmse, 1980).

ESP	Degree of dispersion
<7	Non-dispersive
7 to 10	Intermediate
>10	Dispersive

Table 5.   Dispersion evaluation criteria for SAR results (Harmse, 1980).

SAR (%)	Degree of dispersion
<6	Non-dispersive
6 to 10	Intermediate
>10	Dispersive

Table 6. Rating system for the identification of dispersive soils (after Jermy & Walker, 1999).

Test	Class/rating	Dispersivity	Evaluation		
Pinhole	Class	Highly dispersive	Moderate	Slightly dispersive	Non-dispersive
	Rating	5	3	1	0
CEC vs ESP	Class	Highly dispersive	Dispersive	Marginal	Non-dispersive
	Rating	4	3	1	0
Crumb	Class	Strong reaction	Moderate	Slight reaction	No reaction
	Rating	3	2	1	0
SAR	Class	>2	1.5–2		<1.5
	Rating	2	1		0
TDS vs	Class	Dispersive	Intermediate		Non-dispersive
% Na	Rating	2	1		0

5 RESEARCH METHODOLOGY

From the start of this investigation, soil samples from 16 new earth fill farm dams from various locations in the Western Cape were obtained by the use of the farmer's excavator. Types of materials obtained consist of colluvium, residual siltstone, residual granite, residual shale, residual mudstone and residual phyllitte. Approximately 20 kg of samples from each site were taken deeper than 1.5 m below natural ground level.

In respect of Quality Assurance And Quality Control (QA/QC), each sample was labelled with reference number and tightly closed in bags to prevent drying out.

The samples were then taken to a storage facility where they were separated into two groups. The one group of samples was delivered to Geoscience laboratory in Airport Industria to perform all physical tests such as the double hydrometer, crumb and pinhole tests, and the second group of samples were delivered to Bemlab Laboratory in Strand to perform all chemical tests such as the ESP, SAR, TDS, % Na and CEC.

6 RESULTS

The results of this study summarized in Table 7 are based on the rating system by Jermy and Walker (1999).

7 DISCUSSION OF RESULTS

7.1 Crumb

Crumb test results in this study indicate no significant differences in soil reactions with either distilled water, 0.001N NaOH and 0.006N NaOH. Maharaj (2013), indicates that the use of 0.001N NaOH gives a good

Table 7. Summary of results and final classification.

Dam name	Reference	Total rating	Final classification
Avondson	Wall	8	Moderately dispersive
	Floor	5	Slightly dispersive
Poplar Grove	14	7	Slightly dispersive
	16	4	Non-dispersive
Kei Road	TP6A	5	Slightly dispersive
	TP7A	5	Slightly dispersive
	TP8A	7	Slightly dispersive
	TP8B	12	Highly dispersive
Soetmelksvlei	1	11	Moderately dispersive
	Green	11	Moderately dispersive
Nonna	TP5	11	Moderately dispersive
	TP12	4	Non-dispersive
Rietvlei	TP5	4	Non-dispersive
	TP6	6	Slightly dispersive
Oude Schuur	1	8	Moderately dispersive
Tienrivieren	Basin	5	Slightly dispersive
Riverside	River 1–4	9	Moderately dispersive
Skoongesig	TP4	7	Slightly dispersive
	TP9	3	Non-dispersive
Werda	TP2	13	Highly dispersive
	TP3	9	Moderately dispersive
Ruitevlei	TP1	1	Non-dispersive
Pieterse	East	8	Moderately dispersive
	West	8	Moderately dispersive
Bon Esperance	Core trench	3	Non-dispersive
	Borrow area	2	Non-dispersive
Vyeboom	Dam F	4	Non-dispersive
Eikendal	1	0	Non-dispersive
	2	0	Non-dispersive

indication of dispersive soils if the soils is dispersive. For all samples which indicated strong reaction with the use of 001N NaOH, such as Poplar grove (14), Kei road (TP6A + TP7A), Soetmelksvlei (1), Nonna

TP5 and Rietvlei (TP6), classify from slightly dispersive to highly dispersive in the rating system.

7.2 Pinhole

One of the problems encountered with the pinhole test procedure is the preparation of the sample in the cylinder. The in-situ sample is crushed and mixed with water until the sample has reached its plastic limit; the sample is then compacted with a hand compaction apparatus. If the sample has reached its plastic limit and is moist during compaction, then the soil particles at the bottom of the cylinder will squeeze through the mesh and into the voids of the pea gravel. Once the test starts the water will mix with the soil between the pea gravel and make the effluent highly turbid; leading to misleading results. For this research it was recommended to prepare the sample with 98% Proctor compaction effort to prevent such problem.

Kei Road sample, which gave a good indication of dispersive behaviour for the crumb test, indicates intermediate classification for residual mudstone and dispersive for colluvium soils. Both samples for Soetmelksvlei indicated dispersive behaviour at a minimum head of 50 mm and hole size to six times its initial size with the color of the effluent slightly dark to barely visible. Nonna Dam which one sample indicated dispersive and the other non-dispersive gives a good indication that dispersive soils varies within the same borrow area.

7.3 Exchangeable Sodium Percentage (ESP) and Cation Exchange Capacity (CEC) chart

For the graph in figure 3, some of the samples with ESP greater than 20, such as Nonna TP5, TP12 and Kei road TP6A, were estimated by extrapolation. The outcome of these results poses a significant problem, which could lead to incorrect classification of the soil.

7.4 SAR

From the SAR formula, the potentially dispersivity is clearly indicated if the amount of Na is significantly higher than Ca and Mg combined. The Na for all samples which classified as dispersive ranged from 7.4 to 23.4 meq/l of soil.

7.5 TDS vs % Na

Chemical results were also used for the analysis in the chart by Sherard et al (1976), that plots Total Dissolved Salts (TDS) against percentage sodium. The samples are well scattered on the graph and it clearly shows that most of the samples are classified as dispersive. Some of the samples which classified as dispersive on this graph are rated

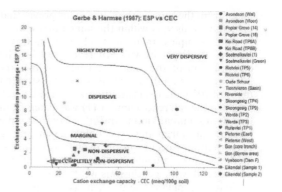

Figure 3.   ESP vs CEC (meq/100g Clay).

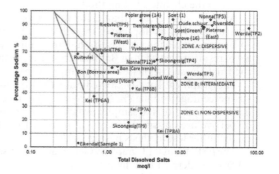

Figure 4.   TDS vs% Na.

non—dispersive for most of the other test methods in the rating system making this chart very unreliable.

8   CONCLUSIONS

The following conclusions are based on the final classification of the rating system:One sample each from Kei Road TP8B and Werda TP2 were classified as highly dispersive.

Soils rated moderately dispersive included samples from Oude Schuur, Riverside, Soetmelksvlei, Pieterse, including one sample from Werda, Avondson and Nonna dam sites.

Soils rated slightly dispersive included samples from Tienrivieren, three samples from Kei Road including one sample from Avondson, Rietvlei, Poplar Grove and Skoongesig dam sites.

Soils rated non-dispersive included samples from Ruitevlei, Bon Esperance, Vyeboom and Eikendal, including one sample from Poplar Grove, Nonna, Rietvlei and Skoongesig dam sites.

Both samples from Kei Road TP7A (residual mudstone) are classified as slightly dispersive and TP8B (colluvium) as highly dispersive. Both were taken from the same borrow area. This proves that dispersive soils vary within the same borrow area.

Both samples from Soetmelksvlei indicated moderately dispersive with equal ratings of 11. The construction of the dam commenced in January 2015 and was continued until work was stopped in April 2015 due to outstanding water use license. One sample reference number (1) was taken from the dam's embankment and the other sample (green) was taken from the dam's basin during construction. Specific problems with certain test methods and evaluation criteria's were identified and the following conclusions were drawn:

Only the physical tests such as the pinhole, crumb and double hydrometer test procedures and results seemed reasonable and very reliable compared to the chemical test methods.

A SAR value greater than 2 for the rating system is regarded as dispersive; these limits appear to be very unreasonable based on comparing the SAR results received from Bemlab laboratory with the final classification from the rating system.

The graph of Total Dissolved Salts (TDS) versus percentage sodium (% Na) seems to be unreliable as many of the samples which classified as dispersive are classified as non—dispersive for all of the other test methods besides SAR.

The ESP versus CEC graph by Gerber and Harmse (1980) was derived from their results and testing of 67 samples. Some of the samples plotted outside the graph, posing a significant problem, which could lead to incorrect classification of the soil.

9 RECOMMENDATIONS

Based on the research and conclusions of this study a follow-up investigation is required to propose a revised rating system, which could possibly include the double hydrometer test as the results seemed reasonable in this investigation. According to Mellvil & Mackellar (1980), the double hydrometer for dispersive identification, is found to be accurate by many researchers.

Further investigation for the variation in results should be assessed for the Total Dissolved Salts (TDS) versus percentage sodium (% Na) graph.

It is recommended that the Sodium Adsorption Ratio (SAR) limits from the rating system, origin and variation in the dispersive classification should also be assessed.

It is recommended to revise the ESP versus CEC graph by Gerbe and Harmse (1980), by taking the results and testing of many other researchers into consideration to derive a universal graph.

It is recommended to prepare the soils at 98% proctor compaction effort for the pinhole test.

REFERENCES

ASTM D 4221–99, reapproved 2006. Standard test method for dispersive characteristics of clay soil by double hydrometer.

Bell, F.G. & R.R. Maud (1994). Dispersive soils: a review from a South African perspective, *Quarterly journal of engineering geology*, 27, 195–210.

Bell, F.G. & D.J.H. Walker (2000). A further examination of the nature of dispersive soils in Natal, South Africa. *Quarterly journal of engineering geology and hydrogeology*.

Craft, C.D. & R.G. Acciardi (1984). Failure of pore water analysis for dispersion. *Proceedings American society civil engineers, Journal geotechnical engineering division*, 110, 459–472.

Elges, H.W.F.K. (1985). Dispersive soils. *The civil engineer of South Africa*.

Gerber, F.A. & H.J. Harmse (1987). *Proposed procedure for identification of dispersive soils by chemical testing.* The civil engineer in South Africa. 29, 397–399.

Harmse, H.J. von M. (1980). *Dispersive soils; their origin, identification and stabilisation* (in Afrikaans), ground profile no 22, April 1980.

Herrier, G., D. Lesueur, D. Puiatti, J.C. Auriol, C. Chevalier, I. Haghighi, O. Cruisinier, S. Bonelli & J.J. Fray (2012). *Lime treated materials for embankment and hardfilldam*. ICOLD Symposium: Dams for a changing world. Kyoto, Japan.

ICOLD, (1990). Dispersive soils in embankment dams. Bulletin 77. International committee on large dams.

Jermy, C.A. & D.J.H. Walker (1999). Assessing the dispersivity of soils, *proceedings of the African regional conference on geotechnics for developing Africa*, eds. Wardle, Blight and Fourie.

Maharaj, A. (2011). The use of the crumb test as a preliminary indicator of dispersive soils, *Proceedings 15th ARC on soil mechanics and geotechnical engineering*, Maputo, Mozambique.

Maharaj, A. (2013). The evaluation of test protocols for dispersive soil identification in Southern Africa. Unpublished MSc dissertation, university of Pretoria.

Mellvill, A.L. & D.C.R. Mackellar (1980). *The identification and use of dispersive soils at Elandsjagt Dam South Africa*. Seventh international conference for Africa on soil mechanics and foundation engineering. Accra. June 1980.

Mellvill, A.L. (1986). *Filters for dispersive soils*. Proc. SAICE symposium on filters. Johannesburg. October 1986.

Paige-Green, P. (2008). *Dispersive and erodible soils— fundamental differences*. SAIEG/SAICE problem soils conference, Midrand, November 2008.

Sherard, J., L. Dunnigan & R.S. Decker (1976). *Identification and nature of dispersive soils*. Journal of the geotechnical engineering division. April 1976.

Sherard, J., L. Dunnigan & J. Talbot (1984). *Filters for silts and clays*. Journal of the geotechnical engineering division, 110, 701–718.

Vorster, C.J. (2003). *Simplified geology South Africa, Lesotho and Swaziland (Map)*: Pretoria, South Africa, Council for Geoscience.

Proceedings of the first Southern African Geotechnical Conference – Jacobsz (Ed.)
© 2016 Taylor & Francis Group, London, ISBN 978-1-138-02971-2

A case study of heave due to flooding

G.A. Jones, H.A.C. Meintjes, P. Aucamp & J.R. Dutchman
SRK Consulting (Pty) Ltd., Johannesburg, South Africa

ABSTRACT: A townhouse complex in a low lying area on expansive clay suffered severe heave damage to the two storey structures after flooding by a burst pipeline. A geotechnical investigation was undertaken five years after the flooding, to determine the relevant material parameters so that the behavior history could be modelled. The traditional Van der Merwe (1964) approach seriously overestimated the heave, but the Weston (1980) method which allows for the change in moisture conditions, as well as pressure corrections, gave realistic heave predictions for the two stage history of initial to equilibrium moisture, followed by equilibrium to flooding induced saturation. It was found that little guidance appears to exist regarding the most appropriate remedial measures with the inevitable result that decisions may be based on a subjective assessment of cost, aesthetics and possible future risks as well as the assigned responsibility for the problem.

1 INTRODUCTION

A townhouse complex comprising fifty two storey buildings was constructed on expansive clays in 1995 and was flooded in 2009 by a burst major pipeline. The floodwater remained for about a month and the concrete block roads and some of the two storey structures showed significant distortion and cracking.

According to classification of damage with reference to masonry work in single storey units, NHBRC Home Building Manual, the damage classifies as minor (2), slight to significant (4), noting that the classification refers to single storey units but is used here as a convenient description. The variability of the damage between units in the flooded areas is most noticeable.

A geotechnical investigation was carried out to determine the cause of the damage. An important feature revealed by the investigation was that the piled foundations do not comply with standard recommendations for expansive clays.

The paper describes modelling of the before and after flooding behavior and discusses potential remedial measures.

2 SITE DESCRIPTION AND GEOLOGY

Published geology maps show the underlying profile to comprise the Platberg Group of the Ventersdorp Supergroup.

The present site investigation, which is in very close agreement with the original 1994 predevelopment investigation, comprised five hydraulic augured boreholes and showed the site is underlain by a superficial layer of made ground which is underlain by transported black cohesive clay to a depth of 0.7 m. This grades into a grey alluvium comprising silty clay to depths of 3.3 to 5.4 m which overlies residual greywacke, composed of clayey silt grading to very soft rock between 6.5 and 8.5 m depth.

Ground water was struck in all five auger holes at approximately 5m depth, which again is very similar to the 1994 results. Figure 1 shows a borehole log for the site.

Since the flooding only inundated about half the site the difference between the unflooded and flooded building zones could readily be observed. Generally the former was practically unchanged, with only very minor cracking, whereas the latter typically has 10 mm wide cracks predominantly at windows and doors. The concrete block roads were undulating with large gaps between blocks.

Trial pits immediately adjacent to the foundations showed no voids between the ground beams and the underlying slickensided clay and that the piles appeared to be simple end bearing. It is noted that the buildings had no impermeable surrounding slabs which are normally recommended in expansive clay areas.

3 LABORATORY TESTING

The laboratory testing consisted of moisture content, grain size analysis, hydrometer, specific gravity, Atterberg limits and swell pressure tests.

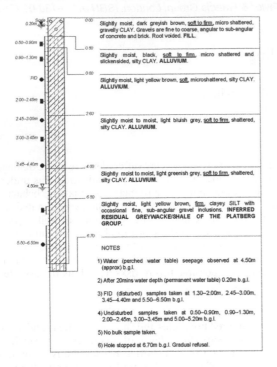

Figure 1.   Typical borehole log.

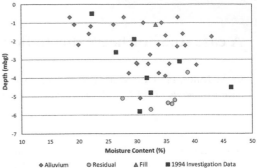

Figure 2.   Insitu moisture contents.

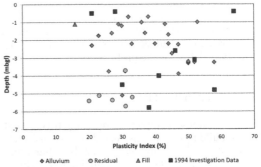

Figure 3.   Plasticity index.

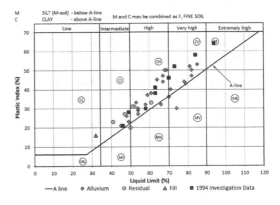

Figure 4.   Plasticity chart (BS 5930:1999).

The insitu moisture contents are shown in Figure 2 which distinguishes between the superficial made ground, alluvial and residual material and also shows data from the original 1994 investigation.

The alluvium has a moisture contents range from 19 to 43% with an average 31%; the lower moistures being closest to ground level.

The residual material has moistures varying from 28 to 39% with an average of 33%. It can be seen that both the alluvium and residual moisture are in agreement with the 1994 data.

Thirty five gradings, hydrometers and Atterberg limit tests and twenty five specific gravity tests were conducted.

The results are shown in Figures 3 and 4 respectively and summarized below.

The bulk of the alluvial samples fall in the CH group with some in the CV, whereas the residual material is divided between the CH and CI groups. The majority of samples classify above the A line.

The dominant profile is the alluvium and this has an average Liquid Limit (LL) of 66% and a Plasticity Index (PI) of 39%.

The underlying residual silty clay has an average LL of 48% and PI of 26%.

The alluvium clay fraction (<0.002) falls within a fairly wide range of 30 to 70% and the residual soil in a close range from 35 to 45%.

When plotted on the conventional plasticity chart: British system (BS 5930:1999) the majority of samples fall well within the Very High range with the remainder in the High to Medium ranges.

Swell Pressure tests were carried out on twenty two samples from a range of depth from 0.7 m to 5.2m on both alluvium and residual clays.

The data for the swell at 10 kPa is shown in Figure 5 as swell percentage versus moisture content.

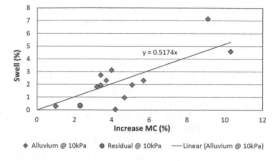

Figure 5. Swell pressure test results.

The tests also give swell at 50 kPa. The data is not shown on Figure 5 but although erratic indicates a trend of being approximately half the swell at 10 kPa. The Weston approach, illustrated later, confirms the relationship between swell at 10 kPa and at 50 kPa.

4 DISCUSSION

It is obvious that the flooding caused heaving but this observation alone is not sufficient to provide a comprehensive model of the soil behavior. There are two phases which require modelling, the first is the before flooding fifteen years ago, when despite inadequate foundations, little damage was observed and the second after the flooding period, when damage occurred very rapidly.

The explanation in general terms lies in the high initial moisture contents (see Figure 2 which shows the 1994 investigation data)—hence the relatively small increase to equilibrium before flooding.

In order to model this process, and the subsequent heave after flooding a series of moisture contents are required. All these, except a presumed equilibrium before flooding, are available from the test data and the equilibrium can be estimated from laboratory test data.

Jones & Els (1991) estimated the Equilibrium Moisture Content (EMC) at a large pre-wetting site at Vereeniging as:

$$EMC\% = 75 + 0.5PI \qquad (1)$$

where PI = Plasticity Index (%).

Weston (1980), quoting data by Haupt (1979) suggests:

$$EMC\% = 0.5LL \qquad (2)$$

where LL = Liquid Limit (%).

These estimated equilibrium moisture contents together with the laboratory data and calculated saturated moisture contents are given in Table 1.

A further estimation of the EMC is given by Emery (1988):

$$EMC(\%) = 0.45(LL^{0.7})((P_{0.425})^{0.3}) \\ + 5.29(Ln(100 + \text{Im})) - 29.5 \qquad (3)$$

where $P_{0.425}$ = Percentage material passing 0.425 sieve; Im = Thornthwaite's moisture value for the area.

Using the above results in EMC, is a little lower than those for Equation 2, but does not influence the conclusions. The initial period moisture change, to equilibrium, would be a little lower and the subsequent change, equilibrium to saturation, would be a little higher.

The table columns are designated i) to vii) and clarification of these is:

i. Initial moisture content from 1994 noting that there is one outlier and if this is omitted the average is 30%.
ii. Moisture contents from recent investigation, five years after flooding.
iii. Swell test data.
iv. Swell test data.
v. EMC using Equation 1.
vi. EMC using Equation 2 for swell test samples.
vii. Calculated saturated moisture contents from undisturbed laboratory samples.

A number of observations can be made from Table 1.

The first is that the insitu moisture contents for 1994 and the recent investigation are practically the same as the initial swell test data thus indicating that the swell test is representative of the actual conditions.

The twenty two swell tests show an average increase in moisture of 4.9% and a degree of saturation of 95%.

The calculated saturation moisture in column vii) is only 32.7% i.e. significantly less than those

Table 1. EMC, laboratory data & calculated moisture contents.

	Swell test				EMC		
	i)	ii)	iii)	iv)	v)	vi)	vii)
Moisture %	32.0*	30.6	30.8	35.7	28.0	U 33.4	32.7
						A 31.5	
Sample no:	9	33	22	22	33	33	22

*Only 9 samples one of which is 45.4%: if this is omitted the average is 30.0%. U is average at swell test undisturbed sample. A is average of all samples.

of the swell tests. There is no explanation for the discrepancy but it serves as a reminder that perfection in laboratory testing is difficult to achieve.

The discrepancy however has no relevance to the discussion since the final saturated moistures are taken as those actually measured in the laboratory swell tests.

It can be seen that there is a large difference between the two calculated equilibrium moistures. The former was derived from one site of residual clays from Ecca shale, whereas the latter is from numerous sites of geological variety and must be considered as more reliable.

A comparison of the swell test initial moistures, 30.8% and the estimated equilibrium for the same samples, 33.4%, suggests the reasonable outcome that the field equilibrium is say 2% to 3% lower than full saturation and that the increase from initial insitu moisture to equilibrium, i.e. before flooding is from 30.3 to 33.4%, say 3% and further that the increase in moisture from initial conditions to flooded conditions is close to 5.4%.

To summarize the moisture conditions at the site, the increase from natural insitu to equilibrium under covered area is about 3% and the increase from equilibrium to fully saturated after flooding is a further 3%.

It is now necessary to estimate the heaves that will result from these two moisture change phases.

The well-established Van der Merwe (1964) method has been used, despite its limitations, as a first step estimate with considerable success. In many cases it has resulted in foundation recommendations which effectively eliminate the problem, for example by piling and providing voids between ground beams and the potentially heaving soils. These and other foundation techniques often result in the reliability of prediction methods being untested. Nevertheless there are many examples in the literature, Meintjes (1991) and Williams (1980) where expansive clays have caused damage of buildings where the problem was not addressed and sufficient precautionary measures not taken and the Van der Merwe approach has given an acceptable correlation of measured and predicted heave. Often the reason for this is that the initial moisture contents were low and the unit heave assumed in the method based on experience with the Free State clays, corresponded with that for geologically similar materials.

Application of the Van der Merwe approach to the townhouse site is shown in Table 2 in which the soil profile is simplified as;

a. 0–4m: Silty clay; alluvium; very high expansiveness; 0.08m/m.
b. 4–5m: Silty clay; residual; highly expansiveness; 0.04m/m.

Since not all the alluvium is rated as very high expansiveness the estimated heave is reduced to 150 mm.

This is close to the estimate given in the 1994 site investigation report and hence the site rates as H3 using the NHBRC Home Building Manual (Part 1, Section 2, Table 5) and extreme construction precautions are necessary.

Figure 6 taken from the manual illustrates the recommended construction.

In addition to the voids below ground beams being necessary, it is recommended that buildings should have an impermeable surround of 1.5m width.

It has been found that the Weston (1980) method is very useful in that it provides a means whereby initial moisture contents and foundation pressures can be taken into account. It is interesting to note that Weston found that of the variables considered, namely; clayiness, density, initial moisture and vertical pressure, the swell was least sensitive to the density. In view of the work by Brackley (1973) and Pidgeon (1987), this may seem a surprising conclusion and the explanation is probably that Weston selected existing roadbed sites where the topography, geology and history were all fairly similar thus reducing the influence of variable densities.

Table 2. Van der Merwe method.

Depth m	Factor	Unit	Heave m
0–1	0.85	0.08	0.068
1–2	0.60	0.08	0.048
2–3	0.40	0.08	0.032
3–4	0.27	0.08	0.022
4–5	0.20	0.04	0.008
Total			0.178

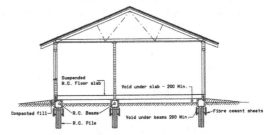

TYPICAL SECTION

Notes:
Site drainage, service and plumbing precautions should be adopted.
Piles to be designed in accordance with SABS 088 for:
– Compressive forces resulting from bearing pressures
– Net uplift (tensile) forces resulting from heave.
Reinforced concrete to be designed in accordance with SABS 0100.

Figure 6. Construction recommendations NHBRC.

Weston's method is shown in Figure 7 which shows that the relevant parameters are the initial moisture content, the liquid limit and the pressure.

Table 3 shows the application of Weston's method using the same profiles as those for Van der Merwe given above.

The result, total heave of 59 mm, illustrates the very large influence the initial moisture has. If the method is repeated using an initial moisture of 20%, which is a value often found at typical Free State sites, then the Weston method estimates a heave of 158 mm, i.e. similar to the Van der Merwe method. It may be noted that the realistic Weston value of 59 mm also places the site in the NHBRC Class H3 category.

A further method of swell estimation is given by El Sohby et al (1995). This is not so much a method in itself but a valuable amalgamation of many other methods combined into some charts. The chart given as Figure 8 is that for silty clay soils at 100 kPa foundation pressure. Using laboratory data a strain is estimated at 1.5% and if this is modified by using an average stress of 30 kPa for the 5 m thick layer then the heave is approximately doubled, using the right hand section of Weston's chart, to 3% or 150 mm, which corroborates the previous Van der Merwe and Weston's second estimates (i.e. from 20% MC). It is emphasized that these estimates are for

Table 3. Weston method.

Depth m	Pressure kPa	Swell %	Swell mm
0–1	7	1.9	19
1–2	21	1.3	13
2–3	35	1.0	10
3–4	49	0.9	9
4–5	63	0.8	8
Total			59

Initial Data:
Moisture content = 32.4% measured
Dry density = 1424 kg/m3
Liquid limit = 62%

moisture changes from an initial assumed dry state to an equilibrium moisture content, not to full saturation.

Summarizing these heave prediction methods is that they gave similar results from dry state to equilibrium. At this stage however the initial state, due to the high water table was already at 30% moisture, hence Weston's method taking account of this and the insitu pressures gives a heave estimate to equilibrium of approximately 60 mm. The process can be extended to assume the moisture content increases not to an equilibrium but to full saturation under flooding. i.e. a further about 2%

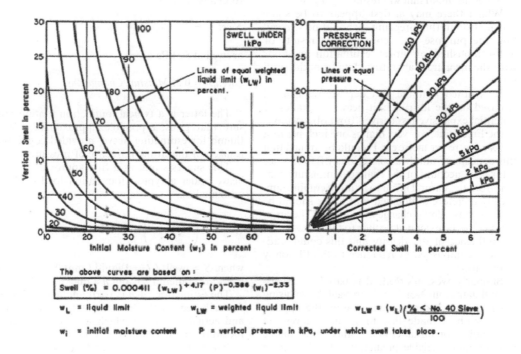

The above curves are based on :

Swell (%) = 0.000411 (w$_{LW}$)$^{+4.17}$ (P)$^{-0.386}$ (w$_i$)$^{-2.33}$

w$_L$ = liquid limit   w$_{LW}$ = weighted liquid limit   w$_{LW}$ = (w$_L$)($\frac{\% < No. 40 \ Sieve}{100}$)

w$_i$ = initial moisture content   P = vertical pressure in kPa, under which swell takes place.

Figure 7. Weston (1980) heave prediction method.

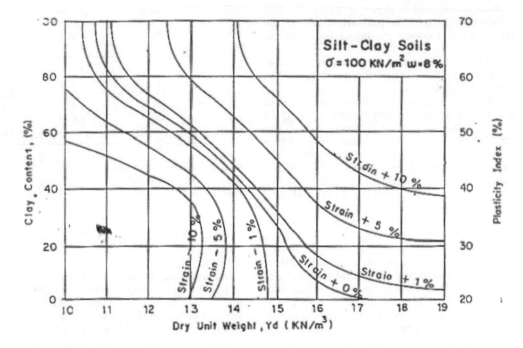

Figure 8. El Sohby (1995) method.

increase in moisture and this shows that a further heave of about 40 mm will result.

Whilst there may at first appear to be some speculation in the foregoing it should be stressed that the purpose is to compare the results of well-established empirical estimation methods with the measured laboratory data from the swell tests.

It is noted that the Van der Merwe method is based on the idea of unit heave, or grades of potential expansiveness derived from observations in the Free State, e.g. very high is one inch per foot or 0.08 metre per metre etc. It could be argued that this concept of unit heave is seriously flawed since it takes no account of the change in moisture content from the natural insitu to a future equilibrium. Ideally therefore a modified unit heave should be expressed as m/m/% change in moisture. The argument could be extended further to take account of the influence of dry density and this could readily be accomplished by reference to the El Sohby information.

Similarly Weston's method is based on experimental data from Northern Transvaal. When using either approach for another location the implicit assumption is made that these unit heaves are applicable. For initial estimates this is acceptable but for more reliable predictions the data from swell tests is invaluable.

The latter show that the measured swell at 10 kPa is given by:

$$Swell\% = 0.5 MoistureChange(\%) \qquad (4)$$

or swell at 50 kPa by:

$$Swell\% = 0.25 MoistureChange(\%) \qquad (5)$$

The latter is a typical foundation footing pressure that gives a possible approximate rule of thumb for estimating heave, provided cognizance is taken of the strong influence of pressure.

This suggests that as a first approximation the heave under a typical foundation load could be estimated as:

$$S = 0.25 \left(\frac{LL}{2} - w \right) h \qquad (6)$$

where S = swell, LL = Liquid Limit (%); w = insitu moisture content (%); and h = thickness of soil.

The estimation of h is partly judgmental and a factor based on the breadth of the covered area and actual soil depth which will be subjected to moisture changes should be applied.

For preliminary heave predictions a factor of unity is applicable, Equation 6 is based on data from this case study site and therefore incorporates the "unit heave" for the site at which the dry densities are approximately 1400 kg/m³. To apply this to other sites it may be necessary, as argued above, to correct for densities if they are significantly different from 1400 kg/m³ and the El Sohby chart should be used for this.

Since at the site two phases are represented by moisture changes:

Initial to equilibrium = 3.1%

Equilibrium to flooding = 2.3%

Then the swells can be derived for a 5 m thick layer as:

Initial to equilibrium = 40 mm

Equilibrium to flooding = 30 mm

Total = 70 mm

It may be concluded that both the laboratory swell test method and the Weston method give similar values for heave, i.e. 40 + 30 mm for swell tests and 60 + 40 mm for heave estimate. It could be argued that in a mathematical sense these are not in fact similar but from a pragmatic view, bearing in mind all the variables, the estimates are sufficiently similar so that identical engineering discussions regarding appropriate foundations would be made.

The question that remains is, how do these heave predictions relate to the building behavior?

The brief did not include a heave survey and since no initial level data was available such a survey would have been of limited reliability.

It is assumed that the building differential heaves would be about two thirds of the total heave. The differential heave, therefore for this first phase of wetting up to equilibrium would be about 30 to 40 mm and for the second phase to flooding a further 20 to 25 mm.

Since the building dimensions are approximately 10 m the angular distortion are about 1/200 for initial wetting and 1/120 for flooding.

Bjerrum (1963) and Burland & Wroth (1975) give tables of angular distortion and damage for various building types. From these it can be shown that at 1/150 structural damage would be expected.

The buildings are two storey, piled, with concrete slabs on ground and first floor levels, hence are relatively stiff. The initial wetting to equilibrium was therefore insufficient to cause noticeable damage as is exemplified by the buildings outside the flooded area. Heaving would have resulted in the build-up and storing of strain energy in the structures. Subsequent heaving pressures on flooding however, were sufficient to release the strain energy and cause cracking.

It is postulated that the preceding discussion effectively models the observed behavior of the structures.

Further questions remain and these are what remedial measures should be employed and inevitably where does the responsibility lie.

The view could be taken that since the maximum heave has already taken place in 2009, and five years have elapsed for the moisture contents to return to equilibrium, no further heave or shrinkage will occur. Remedial measures should therefore consist solely of repairing the obvious damage and constructing an impermeable surround. On the other hand the extreme approach could be adopted of reconstructing the foundations to NHBRC recommendations, i.e. creating voids under the ground beams and re-piling. The recommendation given in this case is the former, but inevitably remedial works will depend on who is responsible for the costs.

The literature has little to say on remedial measures in such cases and it would be useful if some guidelines could be published albeit without being over prescriptive.

Fortunately the brief did not include establishing responsibility in a legal sense. Conceptually it is a question of proximate and ultimate causes, an idea so well illustrated by Diamond (2006) who deals with the failures of civilizations, or is also demonstrated in official enquires into disasters where endless arguments ensue regarding ascribing faults, both immediate and pre-existing.

5 CONCLUSIONS

The modelling of the structure's behavior from the stages of construction to equilibrium and then flooding was accomplished using Weston's approach which allows changes in moisture to be readily incorporated in heave estimations together with site swell test data. In this case study the relatively high initial moisture contents due to the shallow water table strongly influenced both moisture increases to an equilibrium moisture content and the subsequent increase to saturation on flooding.

Evolving from this case study is the proposal that a rational approach to heave predictions is a two stage method using Equation 6, or a site specific equivalent derived from swell tests. The first stage is for the inevitable rise to equilibrium, assumed to be half liquid limit, and the second stage is to saturation conditions. The latter may occur through leaking pipes or inadequate storm water control and would be prudent to assume these for NHBRC classification and design purposes.

The recommended remedial works of simply repairing cracks and constructing an impermeable surround for the structures may not be the solution which will be adopted since responsibilities for costs lies somewhere in the range of inadequate foundation design for a highly expansive clay profile, to the proximate cause of a burst pipeline.

REFERENCES

Bjerrum, L. 1963. Discussion in Proceedings 5th International Conference SMFE, Paris, 2, 11–15.
Brackley, I.J.A. 1973. Swell pressure and free swell in compacted clays. *Proceedings 3rd International conference on expansive soils. Halifax*, 1, 169–176.
Burland, J.B. and Wroth, C.P. 1975. Settlements of buildings and associated damage. *Proceedings of Conference on Settlement of structures. British Geotechnical Society*, Pentech Press, London, 611–653.
Diamond, J. 2006. *Collapse*. Viking Press.
El Sohby, N.A., Rabbaa, S.A. & Aboutaha, M.M. (1995). On Prediction of Swelling and Collapsing Potential of Arid Soils. *Proceedings of the 11th Regional Conference SMFE, Cairo*, 2, 136–144.
Emery, S.J. 1988. *The prediction of moisture content change in untreated pavement layers and an application to design in Southern Africa*. DRTT 20, 644. CSIR Pretoria.
Johnson, M.R., Anhaeusser, C.R. and Thomas, R.J. (eds), 2006. Geological Society of South Africa. *Johannesburg Council for Geoscience*, Pretoria.
Jones, G.A. and Els, A.N. 1991. Pre-wetting expansive clay to minimize heave of structures. *Proceedings 10th Regional Conference SMFE, Maseru*, 1, 375–382.
Meintjes, H.A.C. 1991. A case history of structural distress on heaving clay: Colinda Primary School. *Proceedings 10th Regional Conference SMFE, Maseru*, 1, 99–104.
National Home Builders Registration Council. (1999). *Home Building Manual, Part 1 & 2. Rev 1*.
National Home Builders Registration Council. (1999). Foundation design, building procedures and precautionary measures for single story residential structures founded on expansive soil horizons. *Home Building Manual, Part 1, Section 2*, 76.
National Home Builders Registration Council. (1999). Classification of damage with reference to masonry walls in single storey units. *Home Building Manual, Section 2, Table 2*.
Pidgeon, J.T.A. (1987). *The prediction of differential heave for design of foundations in expansive soils areas*. NBRI Bou, 98. Pretoria.
Van der Merwe, D. (1964). The prediction of heave from the plasticity index and percentage clay fraction of soils. *The Civil Engineer in South Africa*. 6, 103.
Weston, D.J. (1980). Expansive roadbed treatment for Southern Africa. *Conference on expansive soils. ASCE, Denver*, 339–360.
Williams, A.A.B. (1980). Severe heaving of a block of flats near Kimberley. *Proceedings 7th Regional conference Africa. SMFE*, 1, 301–309.

Proceedings of the first Southern African Geotechnical Conference – Jacobsz (Ed.)
© 2016 Taylor & Francis Group, London, ISBN 978-1-138-02971-2

Use of a chilled-mirror dew-point potentiometer to determine suction in a railway formation material

L. Rorke
Aurecon SA, South Africa
University of Southampton, Southampton, UK

S. Tripathy
Cardiff University, Cardiff, UK

L. Otter, C.R.I. Clayton & W. Powrie
University of Southampton, Southampton, UK

ABSTRACT: Compacted soils are used in many civil engineering works, such as railway formations and highway pavements, earth dams, backfills, and soil covers. Compacted soils are invariably unsaturated and possess negative pore-water pressure or suction that can dominate their mechanical behaviour. A growing interest in unsaturated soil mechanics is increasing the need to routinely measure soil suction. The soil water retention curve, SWRC, is often used to relate the soil suction to water content or degree of saturation. A number of methods of measuring suction have become well established, including the filter paper technique, pressure plate apparatus, and tensiometers. In this study, the chilled-mirror dew-point technique has been used to measure suctions of a compacted material typical of South African railway foundations. The test allows for an evaluation of total suction greater than about 200 kPa but is of most value for measuring suctions greater than 500 kPa. The ability of the chilled-mirror dew-point potentiometer to measure high suctions, in the low water content range, means that it can be used to complement results from pressure plate tests at lower suctions and add to the results of filter paper tests at higher suctions. The paper compares suctions measured using the chilled-mirror dew-point technique with those obtained using more established techniques.

1 INTRODUCTION

The role of suction in the mechanical behaviour of unsaturated soils, such as compacted fill, has become of increasing interest especially in pavement design strategies (Lekarp et al. 2000, Liang et al. 2008, Cary and Zapata 2011, Salour and Erlingsson 2015). The influence of suction in shallow foundations with low confining stress can be significant (Fredlund et al. 2012). The benefit of good drainage to prevent saturation of the pavement layers is well recognised and is generally emphasized in railway foundation and road pavement designs.

A general inclusion of suction in effective stress theory was proposed by Bishop (1959) as:

$$\sigma' = \sigma - X(u_a - u_w) \qquad (1)$$

where: σ' = effective stress; σ = total stress; χ = Bishop's factor (function of degree of saturation); u_a = pore-air pressure; and u_w = pore-water pressure.

The difference between pore-air pressure (u_a) and pore-water pressure (u_w) is defined as matric suction, ssociated with the contractile skin or meniscus at the ir and water interface (Fredlund et al. 2012). Matric suction is one of two components of total suction, with osmotic suction being the other component.

Osmotic suction is dependent on the salt concentration in the soil and is mainly considered constant unless chemical changes in the soil occur. Environmental changes, such as rainfall or a rise in the water table, mainly result in changes in matric suction and will influence the behaviour of shallow foundations (Fredlund and Rahardjo 1993), such as pavements.

Measured suction is reported relative to the corresponding water content or degree of saturation. Once determined, the relationship between suction and water content, namely the Soil-Water

Retention curve (SWRC), can be used to obtain suction values for different water contents. The SWRC is not a unique curve as several hysteresis curves exist and is dependent on whether the soil is undergoing a wetting or a drying process (Mitchell and Soga 2005).

A range of laboratory methods exist for measuring otal suction, matric suction and osmotic suction. The advantages and limitations of different methods are summarised by Mitchell and Soga (2005). The use of different techniques to complement each other is commonplace. A study was conducted by Tarantino et al. (2011) to establish the repeatability of suction measurements using different techniques, including axis translation (pressure plate and suction-controlled oedometer), high-capacity tensiometer and the os-motic technique. The investigations were carried out at different laboratories in Europe. A chilled-mirror dew-point potentiometer was used to complete the measurements in the high suction range (2–100 MPa), but suctions in the lower range were not reported for the chilled-mirror dew-point technique and therefore comparison with other methods cannot be made.

This paper discusses the chilled-mirror dew-point technique, using a Chilled-Mirror Dew-point potentiometer (CMD), and reports suction measurements for a typical railway formation material. Comparison of results is made with suction measurements from other laboratory techniques including the filter paper technique (direct contact), pressure plate and highcapacity tensiometer (Ridley tensiometer) reported by Otter et al. (2015).

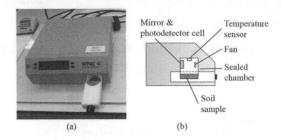

Figure 1. (a) Photograph of the WP4C dewpoint potentiameter (Decagon Devices, Inc.) used, and (b) a diagrammatic view of the CMD (after Leong et al. (2003)).

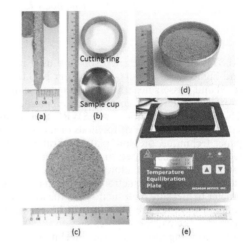

Figure 2. Preparation of a CMD sample (a) Trimming a suitable thickness (≤5 mm), (b) Sharp-edge cutting ring for trimming diameter, (c) Trimmed sample, (d) Sample lightly pressed into sample cup, (e) Sealed sample on a temperature equilibration plate.

2 CHILLED-MIRROR DEW-POINT TECHNIQUE

The Chilled-Mirror Dew-point potentiometer (CMD) shown in Figure 2a measures total suction. The device calculates total suction from the water activity, a_w, in a soil sample using the Kelvin equation (Equation 2) as given in the ASTM D6836–02 (2003) standard. Water activity, a_w, is determined from the dew-point temperature of the water vapour in equilibrium with the soil sample which is measured by the CMD.

$$\psi = \frac{RT}{M} ln(a_w) \qquad (2)$$

where: ψ = total suction; R = universal gas constant; T = temperature in Kelvin; M = molecular mass of water (R/M = 461 kPa/K); and a_w = water activity measured.

A schematic diagram of the CMD is shown in Figure 2b. A soil sample is placed in a sealed chamber and held at constant temperature. A small fan circulates the air within the sealed chamber. A small mirror with high thermal conductivity is located above the sample. The temperature of the mirror is reduced by means of a Peltier thermoelectric unit until condensation occurs. A Light Emitting Diode (LED) illuminates the mirror and the reflected light is received by a photodiode. The presence of condensation on the mirror is determined by measuring a change in reflected light from the mirror. The temperature of the mirror is controlled to achieve equilibrium of water mass on the mirror, thus a rate of condensation equal to the rate of evaporation. This equilibrium point is measured as the dew-point temperature of the ambient water vapour.

The claimed accuracy of measurement is ±50 kPa for the suction range of 0–5 MPa and 1% for the range of 5–300 MPa as per the manufacturer's specifications and discussed further by Leong et al. (2003). The test is thus of particular value for the high suction ranges. The device is calibrated, according to the manufacturer's instructions, using a potassium chloride (KCl) solution of known concentration.

In this research, CMD samples were trimmed in a temperature and humidity controlled laboratory from dynamically compacted specimens (target dry density of 2.1 Mg/m^3) at a range of water contents as outlined below.

- The compacted specimen (or part thereof) was trimmed, using a serrated knife, to a thickness of approximately 5 mm (Figure 2a).
- A sharp-edged ring (Figure 2b) was used to aid trimming of the sample diameter to fit in the sample cup of 37 mm. A trimmed sample is shown in Figure 2c.
- The sample was lightly pressed into the sample cup to ensure contact with the base of the cup (Figure 2d). Care was taken not to fill the sample cup more than half full as this could cause the mirror to become contaminated with soil resulting in measurement errors.
- The sample was sealed and placed on a temperature equilibration plate before testing (Figure 2e). The temperature of the specimen was brought within close range of the chamber temperature of the device to reduce testing time.

Each sample was placed and sealed in the sample chamber of the chilled-mirror dew-point potentiometer. Test commenced when the sample temperature was equal or less than the temperature of the cup base and sample chamber. The device was set to determine the dew-point temperature for several cooling cycles and produced a reading once the measurement was the same for three consecutive cycles. The sample cup was then removed and the combined cup and sample mass determined immediately. The sample (in the sample cup) was oven dried to determine the water content corresponding to the suction measured. The method described above was used to determine suction in a typical railway foundation material as discussed below.

3 SUCTION IN A RAILWAY FORMATION MATERIAL

Suction in a representative railway foundation material, based on a survey of the South African Coal Line (Gräbe and Clayton 2014), was measured in this study using the Chilled-Mirror Dew-point technique (CMD) and other techniques as previously reported by Otter et al. (2015).

3.1 Sample preparation

The reconstituted material tested, namely Material B, contains 11% clay and is considered to be representative of an engineered layer, Layer A, as in the Spoornet (2006) S410 railway formation design specification. Based on its particle size, Material B is a granular subgrade material according to the Flexible Pavement Design strategy (NCHRP Project 1–28 A) (Harrigan 2004). The quantities of the constituent materials of Material B are given in Table 3.1. The Atterberg limits of Material B determined according to BS1377:1990 (Part 2, Method 5) (2012) are given in Table 2. Activity, A, is calculated as the plasticity index, PI, divided by the percentage clay content.

Air dry aggregates were mixed together by hand according to the quantities given in Table 3.1 to obtain Material B. A specific amount of water was added to a known amount of dry material to create a mix of known gravimetric water content, following the approach by Sauer and Monismith (1968). A range of gravimetric water contents was used to create soil-water mixtures.

The required amount of dry material was transferred into sealable plastic bags containing no more than 800 g per bag. Water was added incrementally, and thoroughly mixed while the bag was sealed to prevent moisture and material loss. The wetted material was mixed by hand through the sealed bag for a minimum of 15 minutes, breaking up lumps in the material. During wetting, the fine particles (clay and silt) tended to lump together. Care was taken to reduce any lumps of fines as

Table 1. Particle size distribution and quantities of constituent materials for Material B.

Particle size % by mass	Sand 73			Silt 16	Clay 11
Material type	Leighton Buzzard sand—Fraction:			HPF5 silt	Hymod Prima clay
	B	C	D		
% by mass	47	10	10	21	13

Table 2. Atterberg limits of the representative railway subballast material, Material B.

	Liquid limit (%)	Plastic limit (%)	Plasticity index (%)	Clay content (%)	Activity A
Mat. B	27	14	13	11	1.2

much as possible as this could lead to variable fabric upon compaction (Bishop and Blight 1963, Otter et al. 2015). The mix was considered homogeneous when no dry lumps of fines (clay and silt) could be found in the mix.

Tarantino and Tombolato (2005) and Thom et al. (2008) passed wetted Kaolin clay through a sieve to reduce clay lumps or clay aggregates. Any remaining lumps after sieving were broken down and re-sieved. This method was not used as changes in the material composition might occur if wetted fines remain on the sieve. Moisture loss to the atmosphere from the clayey sand material tested occurs quickly, and moisture loss during sieving is difficult to control.

After mixing, the soil-water mixtures were left for a minimum of 24 hours in a temperature controlled room at 20°C. Care was taken to prevent moisture loss during this period by sealing the first bag inside a second. The amount of wet material needed to make a specimen was weighed out and put aside (in a sealed bag) once specimen preparation, by means of compaction, could commence.

Soil specimens of Material B were prepared at several water contents in the laboratory at the University of Southampton and transported to Cardiff University for testing. Soil water mixtures were dynamically compacted using adapted compaction energy to achieve a dry density of 2.10 Mg/m^3. From the BS1377–4:1990 (2012) Proctor compaction test the optimum water content (w_{opt}) and maximum dry density (ρ_{max}) were found to be 7% and 2.18 Mg/m^3, respectively. At the maximum dry density, the water content corresponding to 100% saturation was close to 10%. A single layer of material approximately 40 mm thick (the thickness of one layer in the Proctor compaction mould) was compacted in a 100 mm diameter mould, levelled and sealed in an air-tight bag to equilibrate before use.

3.2 Previous suction measurements

Otter et al. (2015) reported matric suctions of Material B. In their case the specimens were compacted in a 70 mm diameter by 140 mm high resonant column specimen mould in 10 layers. Otter et al. (2015) used the filter paper technique, pressure plates and a high capacity tensiometer (Ridley tensiometer) to establish the SWRC of Material B.

Otter et al. (2015) measured matric suctions of the soil using Whatman no.42 filter paper (ASTM D5298–4: 1994). For the filter paper tests, soil samples were cut from a compacted specimen as a set of two soil sample discs (each 20 mm thick and 70 mm in diameter). The matching surfaces

of the two discs were smoothed to ensure good contact when the filter paper was sandwiched between the two pieces. Once the discs were cut they were either left open to atmosphere to dry or wetted in a purpose-built tank containing a pond fogger. Once a sample was ready for testing three air dry filter papers were placed and sealed within the sample. Two of the filter papers acted as sacrificial layers against soil contamination of the third, central piece, used for the suction measurement. The combined sample was kept in a sealed, close-fitting glass container inside an insulated polystyrene box at 20°C to equilibrate. After 7 days the water content of the central filter paper (and soil sample, separately) was determined and related to matric suction using the ChandlerCrilly Montgomery-Smith wetting calibration curve (Chandler et al. 1992). The mass of the filter paper before and after drying was measured with an analytical balance, with 0.0001 g accuracy, to allow measurement of the small change in mass.

Two pressure plate systems, with a 500 kPa and a 1500 kPa high air-entry disc respectively, were used by Otter et al. (2015). The applied matric suction increments (equal to the applied air pressure) were 10 kPa for the 500 kPa system and 80 kPa for the 1500 kPa system. The pressure plate test samples (65 mm diameter and 20 mm high) were carefully trimmed from compacted specimens using cutting rings. The samples remained inside the rings during testing. A small weight (130 g) was placed on each sample to ensure good contact between the soil and the high air-entry disc. After an equilibrium period (1 day and 4 days for the 500 kPa and 1500 kPa systems, respectively), a sample was removed, weighed, and then oven-dried to determine the water content. This process was repeated for each increment of air pressure (matric suction) applied, until all samples were used.

For the purpose of comparison, Otter et al. (2015) carried out matric suction measurements using a high capacity suction probe (Ridley and Burland 1993). A number of specimens were compacted at a target dry density of 2.1 Mg/m^3 in plastic tube moulds of 77 mm diameter and 120 mm high. The suction probe was lightly spring mounted in a base platen on which the sample was placed. A soft kaolin paste was used to ensure good contact between the probe and the sample. The probe measurements were recorded for 24 hours and the maximum reading taken as the matric suction.

Matric suction measurements for Material B made by Otter et al. (2015) using the filter paper technique, pressure plate and high capacity tensiometer, are compared with the CMD results in Figure 3.

3.3 Suction measurement results

Figure 3 presents the total suction results (in log scale) of Material B. The total suctions of soil samples were measured using the CMD. The CMD test results are for two different test series, (i) as compacted samples at several water contents and (ii) the average test results of two compacted samples ($w_i = 7.1\%$) that underwent a drying process.

Matric suction measurements made by Otter et al. (2015) using the filter paper technique, pressure plate and high capacity tensiometer as described above are also shown in Figure 3. Filter paper and pressure plate measurements were made on specimens on a drying path, while the tensiometer specimens were tested at the compacted water contents. Figure 4 presents the water content versus suction (in normal scale) results of Material B for suctions less than 1000 kPa.

3.4 Discussion

It can be seen from Figures 3 that, in general, the total suctions of soil samples that underwent a drying process (symbol: filled square) remain greater than the total suctions of as compacted soil samples (symbol: filled circle). The differences in suctions of compacted samples and the suctions of a sample that underwent a drying process can be attributed due to the hysteresis effect. Tripathy et al. (2015) state that suction measurements of compacted samples in CMD correspond to the wetting SWRC of the soil.

Although repeatability of the results from filter paper tests was an issue (crosses in Figs. 3 and 4), the matric suctions measured by Otter et al. (2015) using filter paper and pressure plate techniques were found to be somewhat similar for suctions greater than about 200 kPa. Differences in the

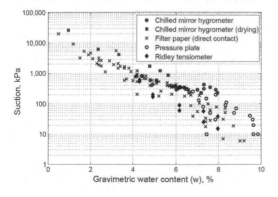

Figure 3. Suction in Material B at different water contents, measured using the CMD, filter paper method (direct contact), pressure plate and high capacity tensiometer.

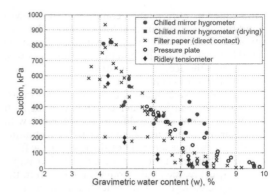

Figure 4. Suction in Material B at different water contents up to 1000 kPa, measured using the CMD, filter paper method (direct contact), pressure plate and high-capacity tensiometer plotted on linear scale.

results from the filter paper and pressure plate tests can be noted at suctions smaller than 200 kPa. The pressure plate, and also the high-capacity tensiometer, can only measure matric suction to the maximum air entry value of the high-air-entry disc (thus 500 kPa or 1500 kPa). The pressure plate systems were also limited by the capacity of the air-pressure supply and pressure plate readings therefore had a maximum of 400 kPa.

In general, the matric suction measurements from the high-capacity tensiometer remain lower than the measurements made using the other methods. This can be attributed due to the localised wetting at the measurement point where kaolin paste was used for creating a good contact between the soil samples and the ceramic disc of the tensiometer. Tarantino et al. (2011) also reported lower suctions measured by a high-capacity tensiometer as compared to pressure plate test results. They suggested that, in the pressure plate tests, occluded air in soil samples are expected to be under pressure resulting in higher pressure plate measurements than the test results from the high-capacity tensiometer (Marinho et al. 2008). Tripathy et al. (2012) have stated that the use of slurried kaolin paste as an interface between soil samples and the ceramic disc causes a reduction in suction of soil samples.

Different suction measurement techniques have different advantages and limitations, one of which is the range of suction that can be measured. One of the advantages of the CMD is the ability to measure high suctions (up to 300 MPa). Accuracy below 5000 kPa is ±50 kPa, making measurements below 200 kPa potentially erroneous. At high water content (w>7%) the total suction from CMD measurements were more than double the value of matric suction (Figure 4). This may be partly due to the added osmotic suction component (unknown) and the reduced accuracy at low suctions. The continuity of the water phase within the soil sample and between the sample and the high-air-entry disc influences the accuracy of matric suction measurement (Tarantino et al. 2011).

The filter paper method is often used for measuring suction at low water contents (i.e., at high suctions). One of the difculties associated with measuring suction using the filter paper technique is the precise measurement of the mass of the filter paper. Changes in the mass of a dried filter paper occur quickly due to absorption of water from the atmosphere, resulting in potential measurement errors (Chandler and Gutierrez 1986). Contamination of the filter paper with soil particles can introduce errors in suction measurements. Otter et al. (2015), therefore, placed the filter paper between two sacricial filter papers to reduce the potential soiling during suction measurements. The calibration curve used for converting filter paper water content to matric suction is also important and could cause some differences in the calculated matric suctions depending upon the calibration equation used (Leong et al. 2002).

The differences in suction measured using the different techniques are significant. At w = 5% for example (see Figure 4) the suction is measured as approximately 200 kPa with the tensiometer, 400 kPa to 600 kPa with the CMD and 200 kPa to 600 kPa with the filter paper technique. When assessing the result on log scale (Figure 3) the variation in results are less prominent. When fitting a Soil-Water Retention Curve (SWRC) the inflection point is around w = 5% at which the micropores start to dominate the suction response. The variation in the results from various tests may increase due to the bi-modal structure of the material resulting from clumping of fines when mixed with water. However, formation of the clay aggregates is expected to also occur in the field (Otter et al. 2015). The variation of results from various tests were found to be less significant at a water content of about 6%. In this case, pressure plate, filter paper and CMD results remain at about 350 kPa, whereas the tensiometer measured a suction of about 80 kPa.

Significant variations in the measured suctions can be noted from various tests for water contents of soil samples greater than about 8%. This may be attributed to low accuracies of suction measuring techniques at high water contents.

4 CONCLUSIONS

The measurement of suction is becoming increasingly important with the use of unsaturated soil mechanics for shallow foundation design such as pavements. The well-known importance of good drainage is partly attributed to the improved mechanical response of soils resulting from suction in the unsaturated material. The chilled-mirror dew-point technique has been discussed and measurements of suction in a typical railway foundation material (a clayey sand) have been compared with suction measurements from other techniques, including the pressure plate, filter paper technique (direct contact) and a high-capacity tensiometer.

The chilled-mirror dew-point technique enables a quick evaluation of soil suction. It is particularly useful for measuring high suctions (>1500 kPa) where other methods, such as the pressure plate and high-capacity tensiometer, cannot be used. The chilled-mirror dew-point technique also allows reasonable estimates of suction between 200 kPa and 1500 kPa and can provide an indicative range of suctions, to inform further refined measurements using other techniques.

The relative change and the potential impact of suction on the mechanical behaviour of soil is part of continued research and the ability to accurately measure suction is becoming increasingly important.

ACKNOWLEDGEMENTS

Funding provided under the EPSRC Track21 Project grant EP/H044949/1 and the Network Rail—University of Southampton Strategic Research Partnership.

REFERENCES

ASTM D5298–4: 1994. Standard test method for measurement of soil potential (suction) using filter paper.
ASTM D6836–02: 2003. Standard test method for determination of the soil water characteristic curve for desorption using a hanging column, pressure extractor, chilled mirror hygrometer, and/or centrifuge.
Bishop, A. W. 1959. The principle of effective stress. Teknick Ukeblad 39: 859–863.
Bishop, A. W. & Blight, G. E. 1963. Some aspects of effective stress in saturated and partly saturated soils. Geotechnique 13(3)(No. 3): 177–197.
British Standards Institution 2012. BS 1377 Methods of test for Soils for civil engineering purposes.
Cary, C. E. & Zapata, C. E. 2011, jan. Resilient Modulus for Unsaturated Unbound Materials. Road Materials and Pavement Design 12(3): 615–638.
Chandler, R. J., Crilly, M., & Montgomery-Smith, G. 1992. A low-cost method of assessing clay desiccation for low-rise buildings. ICE Geotechnical Engineering 92(2): 82–89.

Chandler, R. J. & Gutierrez, C. I. 1986. Technical notes: The filter-paper method of suction measurement. *Géotechnique* 36(2): 265–268.

Fredlund, D. G. & Rahardjo, H. 1993. *Soil mechanics for unsaturated soils.* New York: John Wiley & Sons Inc.

Fredlund, D. G., Rahardjo, H., & Fredlund, M. D. 2012. *Unsaturated soil mechanics in engineering practice.* Somerset, NJ, USA: Wiley.

Gräbe, P. J. & Clayton, C. R. I. 2014. Effects of Principal Stress Rotation on Resilient Behavior in Rail Track Foundations. *Journal of Geotechnical and Geoenvironmental Engineering* 140(2): 1–10.

Harrigan, E. T. 2004. Laboratory determination of resilient modulus for flexible pavement design. *Research results digest Number 285*: 1–52.

Lekarp, F., Isacsson, U., & Dawson, A. 2000. State of the Art. I:Resilient response of unbound aggregates. *Journal of Transportation Engineering-ASCE 126*(1): 66–75.

Leong, E. C., He, L., & Rahardjo, H. 2002. Factors affecting the filter paper method for total and matric suction measurements. *Geotechnical Testing Journal* 25(3): 322–333.

Leong, E. C., Tripathy, S., & Rahardjo, H. 2003. Total suction measurement of unsaturated soils with a device using the chilled-mirror dew-point technique. *Géotechnique* 53(2): 173–182.

Liang, R. Y., Rabab'ah, S., & Khasawneh, M. 2008. Predicting moisture-dependent resilient modulus of cohesive soils using soil suction concept. *Journal of Transportation Engineering-ASCE 134*(1): 34–40.

Marinho, F. A. M., Take, A., & Tarantino, A. 2008. Tensiometeric and axis translation technique for suction measurement. *Geotechnical and Geological Engineering* 26(6): 615–631.

Mitchell, J. K. & Soga, K. 2005. *Fundamentals of soil behavior.* (3 ed.). New Jersey: John Wiley & Sons Inc.

Otter, L., Clayton, C. R. I., Priest, J. A., & Gräbe, P. J. 2015. The stiffness of unsaturated railway formations. *Proceedings of the Institution of Mechanical Engineers Part F-Journal of Rail and Rapid Transit* 0(0): 1–13.

Ridley, A. M. & Burland, J. B. 1993. Technical note: A new instrument for the measurement of soil moisture suction. *Géotechnique* 43(2): 321–324.

Salour, F. & Erlingsson, S. 2015, mar. Resilient modulus modelling of unsaturated subgrade soils: laboratory investigation of silty sand subgrade. *Road Materials and Pavement Design 16*(3): 553–568.

Sauer, E. K. & Monismith, C. L. 1968. Influence of soil suction on behavior of a glacial till subjected to repeated loading. *Highway research board 215*: 8–23.

Spoornet 2006. S410: Specification for railway earthworks. Transnet Limited, South Africa.

Tarantino, A., Gallipoli, D., Augarde, C. E., De Gennaro, V., Gomez, R., Laloui, L., Mancuso, C., El Mountassir, G., Munoz, J. J., Pereira, J. M., Peron, H., Pisoni, G., Romero, E., Raveendiraraj, A., Rojas, J. C., Toll, D. G., Tombolato, S., & Wheeler, S. J. 2011. Benchmark of experimental techniques for measuring and controlling suction. *Géotechnique* 61(4): 303–312.

Tarantino, A. & Tombolato, S. 2005. Coupling of hydraulic and mechanical behaviour in unsaturated compacted clay. *Géotechnique* 55(4): 307–317.

Thom, R., Sivakumar, V., Brown, J., & Hughes, D. 2008. A simple triaxial system for evaluating the performance of unsaturated soils under repeated loading. *Geotechnical Testing Journal 31*(2): 107–114.

Tripathy, S., Elgabu, H., & Thomas, H. R. 2012. Matric Suction Measurement of Unsaturated Soils With Null-Type Axis-Translation Technique. *ASTM Geotechnical Testing Journal 35*(1): 91–102.

Tripathy, S., Thomas, H. R., & Bag, R. 2015. Geoenvironmental Application of Bentonites in Underground Disposal of Nuclear Waste: Characterization and Laboratory Tests. *Journal of Hazardous, Toxic, and Radioactive Waste-ASCE ISSN 2153.*

Proceedings of the first Southern African Geotechnical Conference – Jacobsz (Ed.)
© *2016 Taylor & Francis Group, London, ISBN 978-1-138-02971-2*

Variations in cohesion/friction with strain and significance in design

K. O'Brien & A. Parrock
ARQ Consulting Engineers (Pty) Ltd., Pretoria, South Africa

ABSTRACT: This article details findings into variations of cohesion and friction angle throughout the duration of a triaxial test. It is proposed that these parameters are, in fact, not only functions of soil tested, but also of strain. Strains within a structure under serviceability (SLS) and Ultimate Limit State (ULS) are vastly different. It is thus postulated that cohesion and friction angle values of a soil will also differ at these 2 limits states. A method of determining, via reference to triaxial test results, appropriate values of cohesion and friction angle under serviceability and ultimate limit states, is discussed. Generally it was determined that, under characteristic SLS strain levels, high cohesion and low friction values are attained, while the reverse is true at high strains, characteristic of ULS conditions. These findings are very relevant to limit state designs, which are now the norm in geotechnical codes.

1 INTRODUCTION

A soil's shear strength is defined by variable components of cohesion and friction angle, which fluctuate with strain level.

It is commonly assumed that set values for cohesion and friction angle can effectively describe a soil's behaviour. These attributes of shear strength, are then used in design for both serviceability and ultimate limit states. The triaxial test, discussed herein, is an effective test in determining cohesion and friction angle, either via maximum deviator stress or maximum effective stress ratio. Presently cohesion and friction angle are determined at soil failure by Mohr circles or stress paths.

In this paper, the findings of an investigation into the variation of friction angle and cohesion of a soil during the shearing phase of a triaxial test are presented.

2 PREVIOUS WORK

Hvorslev (1937) conducted numerous direct shear tests on remoulded clays and determined that cohesion was mainly a function of moisture content. Bjerrum (1954) and Skempton (1954) noted that for normally-consolidated clays the relationship between water content and consolidation pressure was independent of major and intermediate principle stresses. This was however not the case for other soil types.

Osterberg and Schmertmann (1962) and Schmertmann (1962) investigated change in cohesion (c') and friction angle (φ') with strain.

Consider Figure 1 extracted from Schmertmann (1962) which was computed from a triaxial test on a cemented sand with <5% passing the 0.075 mm sieve.

Now conventional values at or near failure would be considered as representing that material. In the above case it may be observed that at that condition the c'-value, as expected, is tending to near zero (but probably not quite getting there) while the φ'-value is tending to arctan of 0.9 i.e. some 42 degrees. This is what would be expected of a well-graded, non-plastic (<5% passes the 0.075 mm sieve) sand.

However for the serviceability case the situation is vastly different. Assuming the SLS represents on average, an axial strain of some 0.5%. For this strain value the c-values span the range 60–80kPa while corresponding φ' parameters reduce from the failure value of near 42° to a much lower 31° (arctan of 0.6).

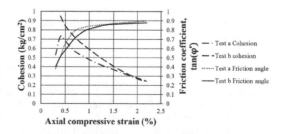

Figure 1. Triaxial test results showing cohesion and friction angle computations during shear extracted from Schmertmann (1962).

A summary of the findings of works done by Osterberg and Schmertmann (1962) is provided below.

- Cohesion and friction act independently of each other with increase in strain,
- Generally, the plot of strain vs. cohesion exhibits a peak at relatively low strains before decreasing to a fairly constant value,
- Cohesion is not only a function of moisture content, but also varies significantly with change in soil structure,
- For clays tested at the same pre-consolidation pressures, similar peak cohesion values resulted, and
- When peak cohesion values were compared for clays tested at different pre-consolidation pressures, no obvious trend was noted.

3 TRIAXIAL TESTING

Triaxial testing comprises testing of three soil samples, assumed to be identical.

Samples are generally, but not always, saturated and then consolidated to pre-defined pressures (representative of field conditions) in the triaxial apparatus. This all-round stress applied to undisturbed or reconstituted, cylindrical soil samples, is held constant, while axial pressure is applied to the sample via a constant rate of compressive strain. During testing, strain, axial pressure and pore water pressure, in a Saturated consolidated undrained (Scu) and volume change in a Saturated consolidated drained (Scd) test, within the sample are recorded. Various corrections for, *inter alia* membrane strength and area change during shear, are applied to produce accurate results.

Readings from the 3 tests are then used to determine stress conditions within the sample and subsequently shear strength properties.

4 SHEAR STRESS THEORY

Coulomb (1776) proposed shear strength behaviour as:

$$\tau = c + \sigma \tan \varphi \tag{1}$$

where τ = Shear strength (kPa); c = Cohesion (kPa); σ = Normal stress (kPa); φ = Internal angle of friction of the soil (°).

This formula was later modified to take into account effects of pore water pressure on shear strength as follows:

$$\tau' = c' + \sigma' \tan \varphi \tag{2}$$

where τ' = Effective shear strength of the soil (kPa); c' = Effective cohesion (kPa); σ' = Effective normal stress acting on a plane (kPa); φ' = Effective internal angle of friction of the soil (°).

Effective normal stress acting on a plane is defined as total stress on this plane minus pore water pressure:

$$\sigma' = \sigma - \mu \tag{3}$$

Furthermore, soil effective shear strength at failure can be defined as:

$$\tau'_f = c'_f + \sigma'_f \tan \varphi'_f \tag{4}$$

The term $\sigma'_f \tan \varphi'_f$ is the friction on the failure plane where normal stress acting on the plane is φ'_f. This equation defines the linear failure envelope of a soil, where cohesion (c'_f) is the y-intercept and the angle of friction (φ'_f) is the angle of inclination of the line at failure.

A stress state can be defined as a point with coordinates (σ';τ') or by a Mohr circle with major and minor effective principal stresses, σ_1' and σ_3' respectively. The effective principal stresses are defined as the normal stresses acting on an element which is rotated in such a way as to render shear stresses zero. As such, they are the x-intercepts of the Mohr circles as shown in Figure 2.

Mohr circles can be used to determine stress conditions not only at failure but also at any time during shear. Figure 3 provides Mohr circles and their corresponding tangent lines for various strain values. This data was obtained from a consolidated

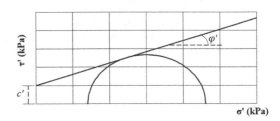

Figure 2. A Mohr circle with concomitant envelope.

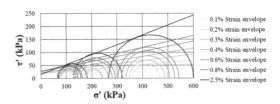

Figure 3. Mohr circle and envelope development with strain.

undrained triaxial test. As deviator stress increased and the sample strained, the Mohr circles for all three tests become larger (from light to dark in the graph). Envelopes were plotted for each set of circles, allowing determination of cohesion and friction with increasing strain. From this plot, the following is evident for this soil:

- Mobilised friction angle increases with strain, and
- Mobilised cohesion increases but then subsequently decreases with increasing strain.

Development of cohesion and friction angle with strain shown in the graph above is more effectively illustrated in the graph below.

This method was used to determine development of cohesion and friction angle with strain for various samples, as detailed below.

5 TEST DATA ANALYSIS

Data sets from saturated consolidated undrained triaxials deemed "high class" were analysed using the methods proposed by Osterberg and Schmertmann (1962) and Schmertmann (1962) described in the previous section. Duncan et al (1980) hyperbolic parameters of each soil type were also determined and provided in Table 1, along with the sample numbers and a brief soil description. The hyperbolic parameters are defined as:

- R_f–Failure ratio,
- K–Modulus number, and
- n–Modulus exponent.

Figure 5 to 9 provide cohesion and friction development with strain for each sample, grouped according to AASHTO classification.

The above graphs show cohesion peaking at some 35kPa at strains below 1% and then reducing again to fairly constant values. Friction angle typically increases until 4% strain and then remains fairly constant.

Table 1. Summary of samples incorporated into this article.

Sample number	AASHTO Classification	Soil description	R_f	K	n
1	A-2-4	Fine sand	0.89	646	1.3
2	A-2-4	Sand	0.93	61	2.4
3	A-2-4	Fine sand	0.71	867	0.6
4	A-2-4	Sand	0.33	31	2.1
5	A-2-4	Sand	0.81	632	1.57
6	A-2-4	Fine sand	0.81	901	0.64
7	A-2-4	Clayey sand	0.31	64	1.08
8	A-2-4	Clayey sand	--	755	--
9	A-2-6	G7 material	1.00	1 124	0.81
10	A-3	Sand	0.90	1 000	1.83
11	A-3	Clayey sand	0.50	511	0.51
12	A-3	Sand	0.83	698	0.79
13	A-7-5	Clayey silt	0.60	42	0.5
14	A-7-5	Clay	0.96	541	0.22
15	A-7-5	Clay	0.98	534	0.58
16	A-7-5	Clay	0.92	80	2
17	A-7-5	Clay	1.01	468	0.52
18	A-7-5	Clay	0.96	540	0.34
19	A-7-5	Sandy clay Gravelly clayey	0.88	404	0.41
20	A-7-5	Calcrete Sandy clayey fine grained	0.90	183	1.37
21	A-7-5	Calcrete Sandy	0.88	165	0.71
22	A-7-6	Clay	0.95	304	0.73
23	A-7-6	Clay	0.96	280	1.23
24	A-7-6	Clay Sandy clayey fine grained	0.95	427	0.35
25	A-7-6	Calcrete		388	0.77

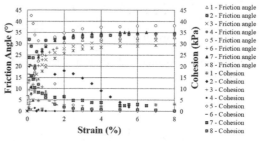

Figure 5. Cohesion and friction angle of A-2-4 materials.

6 SIGNIFICANCE OF THIS RESEARCH

Let us consider a slope, 6m high at a slope angle of 45°, comprising material with parameters derived from Triaxial Test 12 of this paper. This slope was modelled using RocScience's Phase^2 Finite Element software, using plane strain conditions as

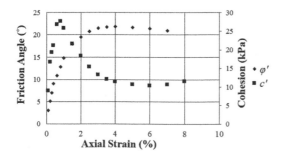

Figure 4. Example of variation of cohesion and friction angle with strain.

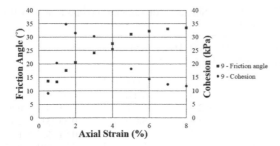

Figure 6. Cohesion and friction angle of A-2-6 materials.

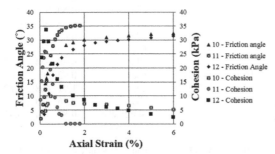

Figure 7. Cohesion and friction angle of A-3 materials.

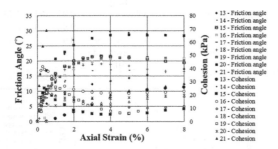

Figure 8. Cohesion and friction angle of A-7-5 materials.

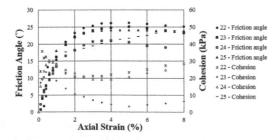

Figure 9. Cohesion and friction angle of A-7-6 materials.

shown in Figure 10. Six noded triangles were used in the model, with Mohr Coulomb theory used as the soil failure criterion under drained conditions and Duncan Chang hyperbolic parameters used to describe the stiffness properties of the soil.

Two scenarios were modelled, Scenario 1 representing SLS and Scenario 2 representing ULS conditions. SLS conditions are those experienced under normal operating conditions, for long periods of time, whereas ULS conditions seldom occur and represent the "worst case" loading scenario for the structure. Material properties used for each scenario are provided in Table 2.

It should be noted that strain dependency is not only applicable to shear strength properties but also to soil stiffness. Typically the stiffness of a soil decreases with increasing strain. The soil stiffness in this model is governed by the Duncan-Chang (1980) hyperbolic relation derived from the triaxial data (see Table 1). This nonlinear model calculates a stiffness for each finite element corresponding to the stain level present. With automated stiffness adjustment occurring in each iteration computed, the author is able to neglect stiffness-strain dependency, and instead focus on the subject strength-stiffness relation.

In addition to calculating strains, displacements and stresses within a model, Phase^2 is also able to determine a Strength Reduction Factor (SRF) of

Table 2. Summary of parameters used in analyses.

Scenario	Strain used for determination of cohesion and friction (%)	Cohesion (kPa)	Friction angle (°)	Pore pressure coefficient (r_u)	Seismic coefficient
Scenario 1	0.6	21.4	17.5	0.05	0.00
Scenario 2	1.2	11.9	25	0.15	0.08

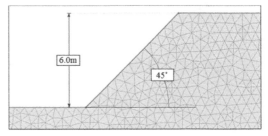

Figure 10. Dimensioned model of slope constructed in Phase^2.

240

a model. This is accomplished through an iterative process by lowering strength parameters until a point of non-convergence (failure) is reached. The factor by which shear strength parameters at point of failure are reduced is the SRF. SRF is therefore synonymous with a conventional Factor Of Safety (FOS), and will be considered as such henceforth.

The contour plot in Figure 11 provides the major principle strains within the model for Scenario 1, for which the maximum is 0.6%. The strain at which cohesion and friction angle were selected is therefore consistent with maximum strain in the model.

An SRF/FOS of 1.5 was calculated for this scenario, as shown in the analysis output in Figure 12.

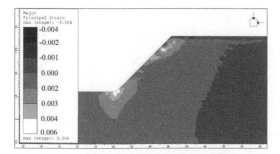

Figure 11. Contour plot of major principle strains under SLS.

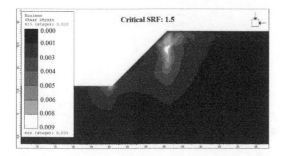

Figure 12. Analysis output for the model under SLS.

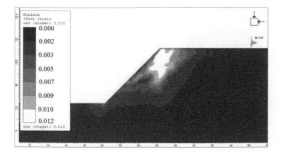

Figure 13. Contour plot of major principle strains for the model under ULS.

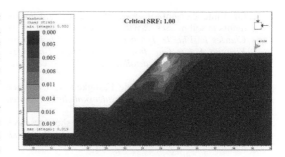

Figure 14. SRF output for the model under ULS.

A maximum strain of 1.2% was calculated for Scenario 2, as shown in Figure 13, which corresponds with the strain used in the material shear strength parameter selection.

An SRF of 1.00 was calculated for this scenario, such that the model was just able to withstand the "worst case" loading scenario. The SRF output for the model is provided in Figure 14.

7 CONCLUSIONS

The following conclusions may be deduced:

- A peak in cohesion is usually experienced at relatively low strains (<1%). These strain levels are indicative of a situation well before failure would occur, and
- Friction angle generally increases with strain until failure of the soil occurs.

8 RECOMMENDATIONS

For SLS conditions, appropriate expected insitu strain values shall be predetermined and concomitant c' and φ' values used if loci of expected failure points are to be realistically predicted and appropriate displacements calculated. For ULS conditions near-failure strains with their appropriate c' and φ' values shall be used.

REFERENCES

Bjerrum, L. 1954. Theoretical and experimental investigation on the shear strength of soils. *Norwegian Geotechnical Institute Publication No. 5.*

Coulomb, C. A. 1776. Essai sur une application des regles des maximis et minimis a quelquels problemesde statique relatifs, a la architecture. *Mem. Acad. Roy. Div. Sav., vol. 7, pp. 343–387.*

Duncan JM, Byrne P, Wong KS and Mabry P. 1980. Strength, stress-strain, and bulk modulus parameters for finite element analyses of stresses and movements in soil masses. *Report No UCB/GT/80-01 of the Charles E. Via, Jr. Department of Civil Engineering, Virginia Polytechnic Institute and State University. 70pp plus Appendix detailing FORTRAN computer printout listing.*

Hvorslev, M.J. 1937. Uber die Festigkeitseigenschaften Gestorter Boden. *Ingeniervidenskabelige Skrifter, Number 45.*

Rutledge, P.C. 1947. Cooperative triaxial shear research of the Corps of Engineers. *Soil Mechanics Fact Finding Survey.*

Schmertmann, J.H. and Osterberg, J.O. 1962. An experimental study of the development of cohesions and friction with axial strain in saturated cohesive soils. *Engineering Progress at the University of Florida, Volume 16, No 2.*

Schmertmann, J.H. 1962. Comparisons of one and two—specimen CFS tests. *Journal of the Soil Mechanics and Foundations Division, Proceedings of the American Society of Civil Engineers. Volume 888, Number SM 6.*

Skempton, A.W. 1954. The pore water coefficients A and B. *Geotechnique, Vol 4, No 4.*

Proceedings of the first Southern African Geotechnical Conference – Jacobsz (Ed.)
© *2016 Taylor & Francis Group, London, ISBN 978-1-138-02971-2*

Variability in soil properties and its consequences for design

P.R. Stott & E. Theron
Central University of Technology, Bloemfontein, South Africa

ABSTRACT: Samples from one particular layer of one particular test pit may give widely different results at different laboratories. Jacobsz and Day (2008) suggested possible slipshod testing. It appears that testing is sometimes not done as rigorously as it should be, but this does not convincingly explain all cases. Badenhorst et al (2015) suggested that preparation differences may be significant. Preparation can lead to different results (Blight 2012, Stott and Theron 2015a), but will also not convincingly explain some cases. Many soil properties are linked to suction. A procedure for assessing suction potential with little sample disturbance was described by Stott and Thereon (2015b). This technique allows reasonably rapid testing with little variation in preparation or testing procedure. Tests performed on a range of soils show large differences in variability. It seems possible that inconsistent laboratory results indicate that some soils cannot be reliably assessed with current procedures.

1 INTRODUCTION

The variability of soil properties has been noted by many observers, e.g. Singh and Lee (1970), Phoon and Kulhawy (1999). There appears to have been relatively little heed paid to this observation by many practicing engineers. Little attention is usually drawn to it in tertiary level geotechnical engineering courses. Little attention is drawn to it in many well-known soil mechanics text books. "Craig's Soil Mechanics" (Knappet & Craig 2012) mentions variability in three places, Das makes no mention of variability in either "Principles of Geotechnical Engineering" (Das 2006) or "Advanced Soil Mechanics" (Das 2008). Fredlund and Rehardjo mention it once in "Soil Mechanics for Unsaturated Soils" (Fredlundand and Rehardjo 1993) and once in "Unsaturated Soil Mechanics in Engineering Practice" (Fredlund et.al.2012). Blight's "Unsaturated Soil Mechanics in Geotechnical Practice" (Blight 2012) is one of few text books which not only stress the existence of variability but also point out the dangers of ignoring it.

Perhaps, therefore, it is not surprising that it is common practice, at least in the case of normal "bread and butter" engineering projects, to send one sample from each distinct horizon from a test pit to one laboratory for the performance of one set of a very limited selection of standard tests.

There have been clear indications that this method of working may be inadequate. Jacobsz (2013) described a situation resulting from this procedure at an electricity substation. Samples from a test pit were sent to a reputable laboratory

where the usual Technical Methods for Highways (TMH) (CSIR 1986) foundation indicator tests were performed. Results indicated that the soil had low expansive potential and no major precautions were taken against heave in the foundation design. Significant heave damage did, however, occur. Stott and Theron (2016) noted a case where samples from a housing development were analysed by the "foundation indicator" tests from TMH1. These tests indicated no risk of heave and the foundations and superstructures were designed accordingly. Heave did, in fact occur, and one house became structurally unsound and had to be demolished before its construction was even completed.

It is widely suspected that a prime cause of this situation is that engineering materials laboratories may be slipshod in their testing procedures e.g. Jacobsz and Day (2008). It has also been suggested that the tests may be critically dependent on details of sample preparation, which vary between laboratories (Badenhorst et al. 2015, Stott and Theron 2015a). The warning of intrinsic variability noted by, for example, Phoon (2008), however, suggests that it may rather be the normal practice of reliance on a single set of tests from each horizon which could be unsound.

It is almost certain that the main reason for relying on only one test is the expense of multiple testing. This paper outlines an investigation to assess the intrinsic variability of soils which requires relatively little time and input of skilled labour and little increase in these inputs for obtaining a significant number of results. It may therefore have the potential for indicating intrinsic variability in an economically feasible way.

2 TESTING OF SOILS FOR TYPICAL SOUTHERN AFRICAN PROJECTS

The majority of soils tests in Southern Africa deal with sites where moisture content experiences marked seasonal variation and unsaturated conditions are normal. Light structures like roads and low-rise buildings provide a significant fraction of the samples tested. It has long been realized that soil suction is the defining feature of unsaturated soils and unsaturated soil mechanics experts e.g. Fredlund et al. (2012) affirm that the correct way to precede with any unsaturated soils problem requires determining the Soil Suction Curve—also known as the Soil Water Retention Curve or the Soil Water Characteristic Curve. All soils property functions required for non-saturated soils analyses can then be derived by means of this curve. Using these soil property functions, differential equations can be set up, boundary conditions can be defined and fully automated solutions follow. Unfortunately the cost of producing the suction curve is quite high and the time required is considerable. Such a costly, time-consuming procedure is not feasible for most small Southern African engineering projects. Engineers continue to rely on simple, inexpensive tests and analyses which have been in existence for decades, but whose relevance and reliability may be questionable.

The Central University of Technology's Soil Mechanics Research Group has explored simple and potentially rapid suction tests as described by Stott and Theron (2015b). These tests use well known principles e.g. Blight (2013), maintaining samples at known temperature and humidity and using small sample size and high precision weighing to achieve significant reduction in time to reach moisture content equilibrium. One of the initial aims was to investigate the possibility of using a single suction value to assess a soil's expansive potential. This may seem a very unlikely possibility, but as can be seen in Figure 1, the relative values of water retention between various soils shows reasonable consistency over a considerable range of suctions. It can be seen that equilibrium is reached reasonably quickly for high suction/low water content conditions and much more slowly for low suction/high water content conditions. For most of this test, temperature was maintained at 20 degrees +/− 0.1 degree Celsius. For part of the test the samples were allowed to follow laboratory ambient temperature which varied between19 and 25 degrees. This demonstrates the feasibility of performing such tests in very economical circumstances.

Not all soils follow this pattern. Figure 2 shows curves for five specimens each of four different clayey soils, three of which follow the pattern of consistent relationship of water retention with suction, but the fourth breaks away from the pattern at low suction values and their curves cross over the curves of the other three.

This soil is not typical; it is an almost pure kaolinite from the Southern Cape. At high suction values the pattern of water content with suction variation is similar to the other clays, but is different for low suction values. The range of suction potential is also much larger than for the other soils at all suction values.

This break in the normal pattern has been noted in relatively few soils, and usually only for "pure" clays, which are not a common occurrence. An example can be seen in Figure 3, which shows curves of multiple samples for each of four different soils.

The five curves in Figure 3 with circular markers are for a clayey soil from Steelpoort in Limpopo. One of the curves (markers unfilled) does not maintain the usual pattern of constant relationship with the other samples. The sample showing the non-typical curve is not a natural clay sample,

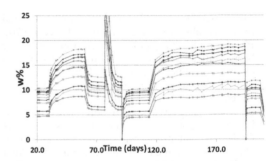

Figure 1. w%: time for 10 clays at various suction potentials:
Over NaCl, water and KCl at 20°C and over water at 19°C to 25°C with wetting and drying incidents. Time to reach equilibrium over water (low suction, high water content) is more than 30 days. Time to reach equilibrium over NaCl (38 MPa) and KCl (22 MPa) is about 5 days.

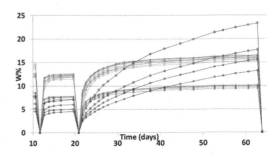

Figure 2. Five samples each of four soils over KCl and over water at 20°C.

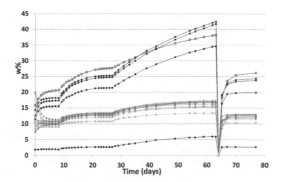

Figure 3.   20 samples (4 different clayey soils) at 37MPa, 10MPa, 1MPa, oven-dried and 22MPa.

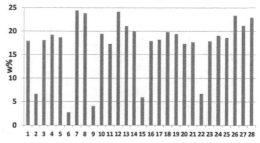

Figure 4.   Twenty eight samples of Steelpoort clay. Water retention at 28 MPa suction.

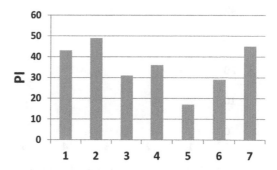

Figure 5.   Steelpoort clay PI values from 7 soils laboratories.

but was isolated from sediment in a settlement test of Steelpoort clay. It is the pure clay from the upper layer of sediment (probably montmorillonite). All of the other curves are for unprepared natural samples and all follow the normal pattern of maintaining a substantially consistent relationship.

3   ASSESSMENT OF VARIABILITY

A noticeable feature of Figure 3 is the variation indicated by the curves for the different clayey soils. Most show little variation, but the Steelpoort clay (circular markers) shows very large variation, from the highest suction by far to the lowest suction by far.

This observation of marked difference in variability led to a series of tests using multiple samples of several clayey soils and an assessment of the variability exhibited by these clays. Sample size was approximately 2–5 g. All samples were simply selected from appropriately sized pieces in the sample bags, or broken from larger lumps. They were then placed in glass weighing bottles with ground in lids after no further treatment. The samples were maintained at constant suction, either in a climate chamber or over solutions of various salts in a temperature-controlled chamber, until constant moisture content was achieved (usually from three to six days).

Figure 4 shows the results for 28 samples of Steelpoort clay at 28MPa suction. Samples were supplied by Prof SW Jacobsz on two separate occasions. Samples were taken randomly from both batches. Both batches gave similar results, which are combined in this graph.

Figure 4 tends to confirm the impression given by Figure 3 that there is huge variability in suction potential between individual samples of this

soil. The lowest of the values for water-retention (2.7%) suggests a non-plastic soil with a PI possibly less than10, the highest retention (24.4%) suggests extremely plastic clay with a PI possibly greater than 50. The average retention of 17.3% suggests highly plastic clay with PI probably in the region of 35 to 40. The Coefficient Of Variation (COV) is 35.6%. Samples of this clay were sent to seven reputable soils testing laboratories. Figure 5 shows values of PI from these laboratories.

The lowest PI is 17, the highest is 49 and the average is 34.2. The coefficient of variation is 33%. These values correspond well with the suction results. They raise the question whether the discrepancies between commercial laboratories—which have been noted for so long by so many people—may be due not to slipshod testing, as is often thought, but to intrinsic variability in soil properties. Such variability could lead to serious consequences if ignored.

Figure 6 shows suction values, again at 28MPa, for a typical clayey soil from Bloemfontein.

The retention values of Figure 6 range from 5.8 to 8.7 with mean of 6.8 and COV 13.1. Samples of this soil were sent to 5 commercial laboratories for testing. Figure 7 shows the values of PI obtained in these tests.

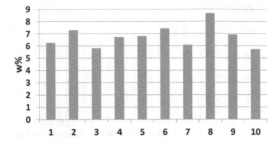

Figure 6. Ten samples of Brandwag clay. Water retention at 28 MPa.

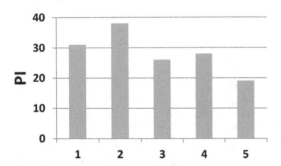

Figure 7. PI for Brandwag clay from 5 laboratories.

PIs range from 19 to 38 with mean 28 and COV 17.1. The variability here also corresponds well with that of the suction values.

It should be noted that although PI is the most commonly used heave indicator in Southern Africa it is not a direct measure of heave. Sridharan and Prakash (2000) noted many instances of poor correlation between PI and heave potential. Suction is directly related to heave potential and may be a better indicator than PI.

4 SAMPLE PREPARATION AND VALUE OF SUCTION FOR CONSISTENT COMPARISON

If the above way of assessing soil variability is to give consistent results and allow different soils to be compared meaningfully with each other, then two features need to be considered.

4.1 Hysteresis

Hysteresis has a significant effect on suction values in soils. Figure 8 shows the water retention at suction 28 MPa of ten different clayey Free State soils from both wet and dry condition. The retention for the initially dry samples is on average 18% less than for the initially wet samples. This is due to hysteresis effects.

On the principle of preferring in-situ conditions it might be logical to consider determining suction potential from natural moisture content. Unfortunately sample bags are often casually treated; they are often punctured, and stored and transported in the sun. Samples often reach the testing stage at well below their original moisture content. It appears that unless a concerted effort is made to improve sample treatment and storage, the most feasible consistent procedure would be to test dried samples. A reasonable procedure might be to dry at 40 or 45 degrees, since it is unlikely that any South African soil will be dried at a temperature higher than this under all but the most exceptional field conditions. A commercially more attractive procedure would be to dry at 105°C so that an additional drying step to establish water content would be eliminated. A disadvantage of this is that drying is known to affect the properties of residual soils more seriously than transported soils (Blight 2012). Investigations are being undertaken to assess this point.

4.2 Suction value

In general the higher the suction, the quicker a soil stabilizes to its equilibrium water-content. But as can be seen in Figures 1 and 3, the lower the suction, the greater the equilibrium water-content and hence the less sensitive the weighing procedure required to differentiate between soils. Tests were performed on a number of clays to assess the effect of suction value on COV. Table 1 shows values for ten samples each of ten clayey soils at suctions of 22MPa, 38MPa and 180MPa. The COV for each soil at any of the suctions measured is not far from the average of all of the values. A relatively quick test at reasonably high suction would probably give an adequate indication of variability over an appropriate range of water content.

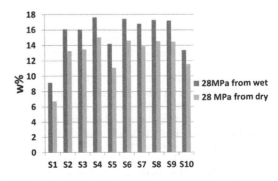

Figure 8. Differences in suction due to hysteresis.

Table 1. COV for 10 samples of 10 clayey soils.

Soil	22MPa	38MPa	180MPa	Average
Belcher 2	15.3	15.1	15.2	15.2
Lerato Park 1	10.5	10.9	10.7	10.7
Lerato Park 2	5.2	6.0	5.2	5.5
Fichardt Park	5.0	5.8	5.6	5.5
Botshabelo R	19.4	18.9	22.0	20.1
Botshabelo B	2.8	2.6	3.3	2.9
Dersley	7.1	7.0	7.1	7.1
BK 3270	2.3	2.3	2.4	2.3
Cecelia 5A	7.4	7.5	7.6	7.5
Cecelia 5B	14.2	14.0	14.4	14.2

5 CONCLUSIONS

Some soils appear to show little intrinsic variability others show very large variability. Conducting one set of tests on some soils would therefore probably lead to reasonable values for use in design. In other cases this procedure could lead to very unsound design.

The investigation described here gives a reasonably quick, easy and inexpensive way of assessing soil variability. The question remains of how to proceed to design. The best way of dealing with such uncertainty is undoubtedly Reliability Based Design.

This is not likely to be an attractive solution for a large number of engineering practitioners since it requires time-consuming and skills-intensive procedures like Monte-Carlo analysis. But in view of the apparent variability of some soils it might be preferable to invest the time and effort required rather than risk expensive failures. It may also be worth noting Phoon's comment *"probabilistic techniques do exist to calculate the probability of failure efficiently. The chief drawback is that these techniques are difficult to understand for the non-specialist, but they are not necessarily difficult to implement computationally."* (Phoon 2008 p 27). If a decision was made that such an approach was advisable, then specialists would probably be prepared to develop the required software for non-specialists at an acceptable price.

Another possibility might be to accept the most unfavourable values as indicated by a variability assessment (or some statistically acceptable compromise). This would probably lead to simpler, but less economic designs than the first alternative.

To do nothing, and to continue to base designs on isolated test results, is likely to perpetuate the occurrence of expensive failures—which are not at all uncommon in certain fields, such as low cost housing.

ACKNOWLEDGEMENTS

The authors would like to express their thanks to the following laboratories for their help and co-operation with parallel testing of samples: Matrolab, SNAlab, Soillab, Geostrada, Simlab, Letabalab and Roadlab.

REFERENCES

Badenhorst, W. Theron, E. & Stott, P. 2015. Duplicate testing conducted on the input parameters for the estimation of potential expansiveness of clay. *Proceedings of the 16th African Regional Conference on Soil Mechanics and Geotechnical Engineering. Tunisia 27–30 April 2015*. Tunis: ATMS.

Blight, G.E. 2012. Microstructure, mineralogy and classification of residual soils. In Blight G.E. & Leong E.C. (Eds), *Mechanics of Residual Soils*. London, New York, Leiden: CRC Press, Taylor and Francis.

Blight, G.E. 2013. *Unsaturated Soil Mechanics in Geotechnical Practice*. Boca Raton, London, New York, Leiden: CRC Press, Taylor and Francis.

CSIR 1986. Technical methods for highways, *TMH1, Standard Methods of Testing Road Construction Materials*. Pretoria: CSIR.

Das, B.M. 2006. *Principles of Geotechnical Engineering*. Toronto: Nelson.

Das, B.M. 2008. *Advanced Soil Mechanics*. New York: Taylor and Francis.

Fredlund, D.G. & Rehardjo, M. 1993. *Soil Mechanics for Unsaturated Soils*. New York: John Wiley and Sons.

Fredlund, D.G. Rehardjo, M. & Fredlund, M.D. 2012. *Unsaturated Soil Mechanics in Engineering Practice*. Heboken NJ: John Wiley and Sons.

Jacobsz, S.W. 2013. Site investigation on dry clayey soils, *SAICE Geotechnical Division Course on Site Investigation,19 March 2013*

Jacobsz, S.W. & Day, P.W. 2008. Are we getting what we pay for from geotechnical laboratories? *Civil Engineering*, 16(4): 8–11.

Knappett, J.A. & Craig, R.F. 2012. *Craig's Soil Mechanics*. Abingdon Oxon: Spon Press, Taylor and Francis.

Phoon, K.K. & Kulhawy, F.H. 1999. Evaluation of geotechnical property variability. *Canadian Geotechnical Journal*, 36: 625–639.

Phoon, K.K. 2008. Numerical recipes for reliability analysis—a primer. In K.K Phoon (Ed). *Reliability-Based Design in Geotechnical Engineering*. Oxford: Taylor and Francis.

Singh, A. & Lee, K.L. 1970. Variability in soil parameters. *Proceedings of the 8th Annual Engineering Geology and Soils Engineering Symposium*, Idaho, pp. 159–185.

Sridharan, A. & Prakash, K. 2000. Classification procedures for expansive soils. *Proceedings of the Institute of Civil Engineers, Geotechnical Engineering,143*.

Stott, P.R. & Theron, E. 2015a. Some shortcomings in the standard South African testing procedures for assessing heaving clay. *Journal of the South African Institution of Civil Engineering*. Vol 57 No 2, June 2015, pp. 36–44.

Stott, P.R. & Theron, E. 2015b. Assessment of clays by small-scale suction tests. *Proceedings of the 16th African Regional Conference on Soil Mechanics and Geotechnical Engineering*. Tunisia 27–30 April 2015. Tunis: ATMS.

Stott, P.R. & Theron, E. 2016. Shortcomings in the estimation of clay fraction by Hydrometer. *Journal of the South African Institution of Civil Engineering (in print)*.

Proceedings of the first Southern African Geotechnical Conference – Jacobsz (Ed.)
© 2016 Taylor & Francis Group, London, ISBN 978-1-138-02971-2

The effect of sample preparation and test procedure on the evaluation of tropical red soils

G. Brink
Exxaro Resources, Pretoria, South Africa

F. Hörtkorn
SRK Consulting, Johannesburg, South Africa

ABSTRACT: Tropical red soils occur as weathered insitu soils (residual soils) or transported soils. In many ways these soils have unique characteristics compared to soils found in temperate climatic regions. Testing and correlation of properties of tropical red soils must be questioned as the current test procedures are developed for temperate soils. This paper reviews the characteristics of tropical red soils summarized in the existing literature. This includes the variation in behaviour and test results when using "conventional" sampling and testing methods as typically done for the testing and evaluation of soils from temperate climatic regions. The findings are supported through the evaluation of test results obtained from selected study areas in Central Africa. The available results are discussed and recommendations given regarding the testing and assessment of the geotechnical characteristics and engineering behaviour of such soils.

1 INTRODUCTION

Tropical soils in many ways have unique characteristics which can be ascribed to the chemical compositions and micro-structures of a pedogenic material developed under hot, wet soil-forming conditions. These unique characteristics may, however, vary significantly as the factors which control pedogenesis (i.e. climate, time, geology, topographic setting, and drainage) vary. As such, the classification of soils formed under these conditions as distinct and well-defined soil types appears to have been avoided in the past. This has brought about the grouping and classification of materials of significantly varying chemical, geological and/ or geotechnical properties, as well as engineering behaviour, under a single, all-encompassing term of "tropical red soils". The unique conditions and soil characteristics further mean that conventional soil mechanics concepts and geotechnical investigative procedures, developed from work completed on temperate soils, do not apply to soils formed in tropical environments.

This paper briefly reviews the characteristics of tropical red soils summarized in the existing literature. This includes the variation in behaviour and test results when using "conventional" sampling and testing methods, as typically done for the testing and evaluation of soils from temperate climatic regions.

Only the findings pertaining to the determination of the particle size distribution and Atterberg Limits are further investigated and supported in this article. Test results obtained for such soils from selected case study areas in Central Africa are discussed. The paper concludes with recommendations in order to better understand the testing and future assessment of these specific properties for tropical soils.

2 BACKGROUND

The effect of the soil structure of residual tropical soils on its geotechnical characteristics and engineering behaviour has been discussed extensively in the literature (amongst others Terzaghi (1958), Gonzalez de Vallejo et al. (1981), Gidigasu (1974) and Vargas (1974)). It should be noted that the majority of results summarized in the literature are not recent in nature and pertain to residual tropical soils only.

These all attribute the soil structure as having the most important influence on the unique properties and behaviour reported for these soils. Continuous cementation of particles into aggregates, or "peds", through iron and aluminium oxides also further affects the characteristics and structure of the material. The resultant structure often results in behaviour and properties totally uncharacteristic for a "typical" material with the particular particle size distribution and Atterberg Limits. Structural breakdown and manipulation of the insitu soil structure (either during sampling or remoulding of test specimen) may therefore result in a significant change in the characteristics and engineering behaviour of such soils. It may also influence the classification of these materials (Mitchell and Sitar, 1982).

The type and amount of dispersing agent, used to determine the particle size distribution, may also have a significant effect on the reported particle size distribution. Terzaghi (1958) reported that when sodium hexametaphosphate is used as a dispersing agent the resultant clay-size fraction obtained is significantly higher than when using other dispersing agents. No difference in the effectiveness of two dispersing agents (sodium hexametaphosphate and sodium orthophosphate) were reported by Northmore et al. (1992). They however found that the particle size distribution and other index property values are extremely sensitive to pre-treatment of the samples prior to testing, particularly to pre-drying. Pre-drying of any sort was found to frequently result in inconsistent results during determination of grain size distribution and Atterberg Limits. Furthermore, the effect of drying on material properties has been found to be largely irreversible. This is largely ascribed to aggregate formation (through the cementation by iron and aluminium oxides) and the permanent loss of moisture from minerals that exist in a hydrated state (Mitchell and Sitar, 1982).

Values of effective cohesion (c') and effective angle of internal friction (φ') reported in the literature for tropical residual red soils were found to be highly variable, ranging from 0 to 345 kPa and between 11° and 57° respectively, see Table 1. These were generally reported to be established using either shearbox or triaxial test methods.

Table 1. Typical shear strength characteristics reported for residual tropical red soils.

Reference	Effective friction angle (degrees)		Effective cohesion (kPa)	
	Range	Average	Range	Average
Vargas (1974)	23–33	28	0–59	24
Tuncer and Lohnes (1977)	27–57	42	48–345	163
Foss (1973)	36–38	37	22–28	25
Northmore et al. (1992)	11–36	22	2–97	32
Northmore et al. (1992)	12–44	25	0–55	24

The large range of reported values for a "single" soil type therefore clearly indicates the need for a better terminology and classification when it comes to tropical red soils.

3 CASE STUDY: TRANSPORTED TROPICAL RED SOILS

3.1 Background

The Republic of the Congo is located in Central Africa between the parallels of 4° north and 5° south and the meridians of 11° and 19° east. The equatorial climate of this region is characterized by the absence of any seasonal winds, the lack of a well-defined dry season and consistently high humidity and temperatures. Mean Annual Precipitation levels of 1600 to 3200 mm/year have been recorded for the study area, with dual peak wet seasons.

During the completion of a number of geotechnical investigations in the study area, four distinct geotechnical horizons were identified and described from test pitting activities. Amongst these were two colluvial (transported) horizons; a dark red brown, soft, micro-shattered sandy to silty clay horizon and a yellow brown, soft, micro-shattered sandy to silty clay horizon. The thicknesses of the respective colluvial horizons varied between 0.70 metres and 1.50 metres throughout the study area.

Samples were collected from these colluvial horizons and submitted for laboratory test analysis in South Africa. All tests and sample preparation were completed according to the relevant international standard test methods (BS 1377–2), unless where otherwise indicated and discussed.

Test results of the red and yellow brown colluvial horizons reflected soils that have a similar range of particle size distribution and Atterberg Limit values. The results for both horizons are consequently discussed and represented as a single group (colluvial material) of results going forward.

3.2 Results

The collected colluvial samples were split into two groups of samples (i.e. different samples from the same geotechnical horizons) for hydrometer analysis purposes and different dispersing agents were specified for each group of samples. All other test and sample preparation procedures were kept identical.

A distinct and sudden decrease of between 10% and 20% in the particle distribution results occurs between the transition from sieve to hydrometer analysis (i.e. in the coarse silt fraction) when using a mix of 35 grams sodium hexametaphosphate and

7 grams sodium carbonate per litre of deionized water as dispersive agent. Thereafter, clay contents in the range of 52% (maximum) to 8% (minimum) can be observed (Figure 1).

Conversely, grading results obtained using a dispersant solution of 36 grams sodium pyrophosphate decahydrate per litre of deionized water continues from the final sieve aperture without any distinct drop-off in results between sieve and hydrometer analysis (Figure 2). Thereafter, similar clay contents as shown in Figure 1 can be observed.

Selected samples of colluvium material were further subdivided into three different specimens of which the particle size distribution and Atterberg Limits were established at the natural moisture content of the material, after partial air-drying of the test specimen and near-complete air-drying of tested specimens. The plastic limits of the material were found to remain fairly constant regardless of the exposure of the material to drying. The established liquid limits (and subsequent plasticity indices) of the material however decreased (on average) with increasing drying of the sample. Differences of up to 10% between tests conducted at the natural moisture content and at nearly completely air-dried samples were found.

Particle size distributions varied significantly with increasing exposure of the samples to drying. The percentage fines (silt- and clay-sized fraction) established at the natural moisture content of the sample decreased by between 10% (minimum) and 47% (maximum) on near-complete air drying of the test specimens, with corresponding increases in the sand and occasionally even gravel-sized fractions.

Additionally, the impact of oven drying temperature in the determination of the natural moisture content of samples was also investigated. Sub-samples of colluvium material dried for the same period of time, but at higher oven drying temperatures (105°C), report natural moisture contents of between 3.5% and 7.6% higher than the same sample material dried at a lower temperature of 65°C.

4 CASE STUDY: RESIDUAL TROPICAL RED SOILS

4.1 Background

The soil samples for the following case study were obtained from a site in the north-east region of the Democratic Republic of the Congo, which is located in Central Africa, close to the Ugandan border between parallels 6° north and 14°south, and the meridians between 12° and 32° east. The soil can be classified as a typical tropical residual soil.

Figure 3 shows a close up of the material investigated. The remnant rock structure of the completely weathered parent rock, in this case diorite, can clearly be seen. Based on a first impression on site one would describe the soil as a silt/clay type material.

A bulk sample of material was mixed and split into five identical sub-samples. The sub-samples

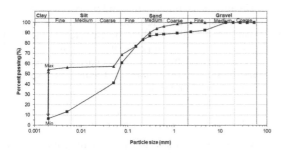

Figure 1. Particle size distribution (sodium hexametaphosphate and sodium carbonate agent).

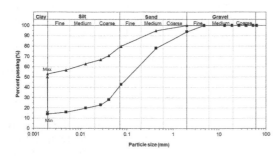

Figure 2. Particle size distribution (sodium pyrophosphate decahydrate agent).

Figure 3. Close-up of a residual diorite with visible remnant structure of the parent rock.

have been sent to five laboratories, three within South Africa and two to Europe. The laboratories were instructed to establish the soil grading and plastic indices. No specific testing recommendations were given to the laboratories so that the testing was done as per the methods typically used in the individual laboratories.

4.2 Results

As no clear instructions were given to the laboratories, it is important to link the methodology to the test result. The testing procedures used are therefore shown in Table 2. Further to the different procedures and dispersive agents different mixing techniques were used. The mixing times varied from 15 minutes to 24 hours.

The particle size distributions of the samples tested are shown below in Figure 4.

Subject to the procedures and technologies used a difference in the particle distribution results occurs between the transition from sieve to hydrometer analysis (i.e. in the coarse silt frac-

tion). The discrepancy is especially obvious when comparing the results of sample A1a with A4 and A6 respectively. Sample A1a and A6 used sodium hexametaphosphate as a dispersive agent but the mixing time and sample preparation were different. Sample A1a was oven dried and mixed for a short period only, sample A6 was air dried at 60 degrees and was mixed longer. Sodium pyrophosphate was used as dispersive agent for sample A4.

The discrepancies with regard to the clay content of the samples are even more distinct. Sample A6 is showing a clay content of 62% whereas sample A1a is only indicating a clay content of 12%.

When determining the plasticity indices, it was found that the plastic limits were fairly constant. The liquid limit however varied significantly (between 45% (A1, A4 and A5) and up to 80%. (A2, A3). The higher liquid limits were found for samples mechanically mixed thoroughly for a prolonged period and which were tested only air dried as much as required.

Table 2. Overview about the testing procedures used to determine the grading curve.

Sample	Testing code	Dispersive agent
A1a	ASTM D422	Sodium hexametaphosphate
A1b	ASTM D422	Sodium hexametaphosphate
A1c	ASTM D422	Sodium hexametaphosphate
A2	DIN 18213	Sodium pyrophosphat
A3	K.H. Head	Sodium hexametaphosphate
A4	TMH 1 (A5/6)	Sodium pyrophosphate
A5	TMH 1/ASTM422	Sodium pyrophosphate
A6	BS 1377 Part 2	Sodium hexametaphosphate

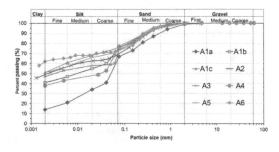

Figure 4. Particle size distribution of the residual soils tested.

5 DISCUSSION

5.1 Delineation of tropical soils

The unique characteristics and properties reported in the literature for residual tropical red soils were confirmed to also apply to transported tropical red soils. It therefore appears as if it is not justified to distinguish between and separately categorize residual soils and transported soils when specifying and assessing index test results.

5.2 Testing procedures and results

Sample preparation, drying and the mixing technique significantly impacted on the reported test results.

Different dispersing agents used during hydrometer analysis appear to have a significant effect on the resultant grain size distribution in the coarse silt-sized fraction of the tropical soils investigated during this study. The colluvium horizons investigated during the completion of this study proved to be highly sensitive to any changes in natural moisture content and mixing time. Significant variation in the liquid limit values and particle size distribution were reported when exposed to different periods of drying and mixing. The materials also further proved to be very sensitive to the aggressivity of the drying method (i.e. drying temperature), with higher percentage moisture contents reported for those test specimens exposed to higher drying temperatures.

6 SUMMARY AND CONCLUSION

The extremely wide range of results shows that great care must be applied when testing tropical soils and that clear instructions to the laboratories are required.

Based on the experience of the authors the following should be specified when submitting samples for testing:

- Drying method (advisable not to dry samples at all);
- Dispersive agent; and
- Mixing time and method.

However, it is not advised to draw any engineering conclusions based on such results alone, as variation in the overall percentage fines may still exist.

These results should at best only be used for comparative purposes due to the observed range of values in the fines.

When applying typical correlation methods on the results presented in this paper, very low shear strength parameters were found. Evidence from site and triaxial test results suggest shear strength and engineering performance better than the correlated values would suggest.

Until meaningful and conclusive testing and correlation methods have been established specifically for tropical soils, it is recommended to only conduct triaxial tests and/or suitable insitu testing methods to evaluate the shear strength properties of the material on site. If no triaxial or other insitu testing results are available, the authors want to encourage other engineers working in tropical areas rather to draw conclusions based on actual site observations (i.e. natural slope angles) than attempt producing results through existing correlation methods. Based on the personal experience of the authors, the engineering behavior and actual shear strength properties of tropical soils are often "better" than expected if only applying experience from temperate soils.

It is finally also worthwhile asking the question whether or not the minimum sieve size should be increased for such soils. This would expose larger grain size fractions, previously captured and recorded as fine sand in the small aperture sieves, to the dispersion process. This might lead to more reproducible results, especially in the transition from sieve to hydrometer.

The authors hope to inspire more research with this paper, especially considering the fact that a lot of the research and data reported in the available literature is not recent of nature.

REFERENCES

ASTM D422–63. 2007. e2. *Standard Test Method for Particle-Size Analysis of Soils.* ASTM International, West Conshohocken, PA, 2007, www.astm.org.

Blight, G. E. 1982. *Residual soils in South Africa, Engineering and Construction in Tropical Soils.* Proceedings of the ASCE Geotechnical Engineering Division Speciality Conference, Honolulu, Hawaii: 147–171.

BS 1377–2:1990 (replaced by BS EN ISO 17892–1:2014 and BS EN ISO 17892–2:2014). *Methods of test for soils for civil engineering purposes, Classification tests.* London, BSI.

DIN 18123:2011–04. *Soil, investigation and testing—Determination of grain-size distribution,* Beuth Verlag, Berlin.

Foss, I. 1973. Red soil from Kenya as a foundation material. *Proceedings of the 8th International Conference on Mechanics and Foundation Engineering,* Moscow, 2: 73–80.

Gidigasu, M. D. 1974. *Degree of weathering in the identification of laterite materials for engineering purposes—A review. Engineering Geology,* 8: 213–266.

Gonzalez de Vallejo, L.I., Jimenez Salas, J.A. and Leguey Jimenez, S. 1981. Engineering geology of the tropical volcanic soils of La Laguna, Tenerife. Engineering Geology, 17: 1–17.

Head, K.H. 2006. *Manual of Soil Laboratory Testing, Third Edition,* Dunbeath, Caithness, Whittles Publishing.

Mitchell, J.K. and Sitar, N. 1982. Engineering properties of tropical residual soils. Engineering and Construction in Tropical and Residual Soils. *Proceedings of the ASCE Geotechnical Engineering Division Speciality Conference, Honolulu,* Hawaii: 30–58.

Newill, D. A. 1961. Laboratory investigation of two red clays from Kenya. *Geotechnique,* 12: 302–318.

Northmore, K.J., Culshaw, M.G., Hobbs, P.R.N., Hallam, J.R. and Entwisle, D.C. 1992. *Engineering geology of tropical red soils: summary findings and their application for engineering purposes.* British Geological Survey Technical Report WN/93/15.

Saunders, M. K. and Fookes, P. G. 1970. A review of the relationship of rock weathering and climate and its significance to foundation engineering. *Engineering Geology,* 4: 289–325.

Terzaghi, K. 1958. The design and performance of Sasumua dam. *Proceedings of the British Institute of Civil Engineering,* 9: 369–394.

Technical Methods for Highways/TMH 1, 1986. *Standard Methods of Testing Road Construction Materials. Tests on soils and gravels,* Pretoria, Council for Scientific and Industrial Research (CSIR).

Tuncer, E. R. & Lohnes, R. A. 1977. An Engineering Classification of Certain Basalt-derived Lateritic Soils. *Engineering Geology,* Amsterdam, 11 (4): 319–339.

Vargas, M. 1974. Engineering properties of residual soils from the south-central region of Brazil. *Proceedings of the 2nd International Conference of International Association of Engineering Geology, Sao Paulo,* Brazil.

Proceedings of the first Southern African Geotechnical Conference – Jacobsz (Ed.)
© 2016 Taylor & Francis Group, London, ISBN 978-1-138-02971-2

Evaluation of methods to determine reference void ratios

C.J. MacRobert & L.A. Torres-Cruz
University of the Witwatersrand, Johannesburg, South Africa

ABSTRACT: Reference void ratios, that is minimum and maximum void ratios, are important values when investigating non-plastic soils. Reasons are two-fold: first, they reflect overall grain-size and particle shape characteristics of a soil, which largely dictate its physical nature, and secondly, the physical state of a soil is referenced to them. However, experimentally these values are not actual extreme values, but rather the densest and loosest packing possible under a given set of procedures. Various standard and non-standard techniques have been proposed to achieve these extremes. Variations to these methods are proposed, based on constraints set by available material and equipment in research situations. As repeatability is key, data illustrating this, for the proposed methods, are presented for a range of grain-size, particle shape and mineralogy. Results of the methods proposed herein are also compared with non-standardised methods proposed by other authors and with a standardised method.

1 INTRODUCTION

Soils consist of an assemblage of independent particles; the compactness of these particles influences their mechanical properties. The physical nature of the particles, that is overall grain-size and shape, dictates the loosest and densest possible states. The strength of a non-plastic soil will be a maximum at its densest packing and a minimum at its loosest packing, although this becomes less distinct at higher confining stress (Bolton 1986). Current compactness of a soil, or its physical state, can therefore be referenced to these limiting states of compactness. Common measures of compactness are dry density, ρ_d (mass of dry soil per unit volume), porosity, n (ratio of voids volume to total soil mass volume) or void ratio, e (ratio of void volume to solid volume in a unit total volume of soil). The selection of one measure of compactness over another is largely a matter of convenience or personal preference.

Theoretically, assuming spheres of uniform diameter, the minimum possible void ratio of 0.35 is obtained when the particles are arranged in a tetrahedral packing. Likewise, the maximum void ratio, in which all particles touch is 0.91, and is obtained when the particles are arranged in a cubic packing; however, this arrangement is not stable in a gravitational field. This makes a theoretical maximum void ratio impossible to model (Lade et al. 1998). Natural soils are not uniform spheres and therefore these limiting values are determined experimentally. Consequently, these values are not actual extreme values, but rather densest and loosest packing possible under a given set of procedures. They are therefore termed reference void ratios. Procedures used to determine loosest packing rely on assembling particles with the maximum possible hindrance. On the other hand, densest packing relies on assembling the particles with as little hindrance as possible, in this way particles can move about until the closest packing is achieved (Germaine & Germaine 2009).

The objective of this paper is to describe repeatable methods to determine reference void ratios, using equipment readily available within university research laboratories.

2 REVIEW OF METHODOLOGIES

Tavenas & La Rochelle (1972) reported more than thirty experimental techniques and advocated an adoption of standardised methods proposed by the American Society for Testing and Materials (ASTM). ASTM D4254 sets out three methods to determine maximum void ratio; funnel deposition or tube deposition into a standard mould and inverting a soil filled cylinder. ASTM D4253sets out the vibrating table method for determining minimum void ratio. Usual practice is first to determine maximum void ratio (e_{max}), by funnel deposition of an oven dry specimen into a standard mould. A collar is then fastened to the mould, a 13.8 ± 0.1 kPa surcharge placed on the specimen within the collar, the assembly secured to a vertically vibrating table and vibrated for a standard period, and the minimum void ratio (e_{min}) determined.

Of the three standard ASTM methods for determining e_{max}, the funnel method is reported to result in marginally higher void ratios (Tavenas & La Rochelle 1972). The only difficulty in carrying out the funnel method, in a research setting, is the size of the standard mould recommended, when insufficient material is available. However, special moulds are permitted, provided specimen height is 0.7 to 1.3 times the diameter. The funnel method outlined in ASTM D4254 was therefore adopted, although some minor variations, outlined below, were incorporated.

Criticism of the vibrating table method from literature are that many university laboratories do not have vibrating tables (Germaine & Germaine 2009), particle breakage is possible (Lade et al. 1998) and the stipulation that the method can only be used for soils with Fines Content (FC) less than 15% (Cubrinovski & Ishihara 2002); where fines are particles passing the 75 μm sieve.

Although a vibrating table was available to the authors excessive particle breakage was observed in some of the soils investigated. Vibration amplitude could not be altered and so this precluded any possible optimisation. Lade et al. (1998) proposed an alternative method to determine e_{min}, which is reported to result in minimal particle breakage. This method involved spooning material into a 2000 mL graduated cylinder, which is tapped 8 times, twice each on opposite sides, after three spoons. Such a method was also used by Cubrinovski & Ishihara (2002) for gravels.

When implementing this method, the current authors found that although the mass of material in the cylinder could be determined accurately, volume was difficult to determine accurately due to the resulting uneven surface within the cylinder. To determine volume accurately variations, outlined below, were therefore adopted.

3 PROPOSED METHODS

Different variations were found necessary to determine e_{max} and e_{min} for gravel-sand and sand-silt soils. These are now outlined.

3.1 Maximum and minimum void ratios for gravel-sand soils

A special mould, as per ASTM D4254, was fabricated for this study, relevant details of which are given in Table 1.

For a given material, an oven-dried specimen between 0.9 and 1.2 kg was prepared. The exact mass prepared was roughly 10% greater than the mass anticipated to fill the mould, based on the

Table 1. Dimensions of special mould.

Item	Value	Comment
Average	518	Direct measurement = 516 mL
Volume (mL)		Water filling = 519 mL (19.5°C)
Diameter (mm)	79.7	
Height (mm)	103.4	

expected density. As sandy gravels can segregate, the specimen was riffled into four equal portions, to minimise possible effects. After weighing the empty mould to an accuracy of 0.1 g, four equal portions were gently poured into a funnel with a 25 mm diameter spout and hose that rested at the bottom of the mould. Lifting the funnel gradually in a spiral allowed the material to be deposited with maximum hindrance, to roughly 25 mm above the rim. A straightedge was then used gently to level the material and a brush used to remove excess grains from outside of the mould. Mass of the specimen and mould was then determined, from which the mass of specimen could be determined, and subsequently the minimum dry density. Using particle specific gravity, G_s, e_{max} was determined.

The specimen used to determine e_{max} was then riffled into four equal portions. A collar was attached to the mould and material spooned into the mould from an average drop height of 300 mm. Following each approximately 15 g lift, the mould was tapped 8 times, twice each on opposite sides, with a rubber mallet. Once all material was rained into the mould, the standard surcharge (13.8 kPa) was lowered into the collar. The mould was tapped, as explained above, to level the surface of the specimen, so that an accurate volume measurement could be made.

Following careful removal of the collar and surcharge, a 5.0 mm thick steel plate was lowered into the mould. The distance between the edge of the mould and plate was measured at four points, using a digital height gauge with a 0.05 mm accuracy. From this, the average height of the specimen and hence specimen volume could be determined. Mass of mould and specimen was then determined. From these measurements, maximum dry density could be determined. Using the particle specific gravity, G_s, e_{min} was determined.

The specimen was then mixed with the material left over from the maximum void ratio determination and the procedure repeated. At least three individual determinations of maximum and minimum void ratio were carried out on each material.

3.2 *Maximum and minimum void ratios for sand-silt soils*

Riffling soils with high fines content generates significant amounts of dust, leading to a loss of fine particles. Instead, representability of test specimens was ensured by storing dry field sample in a sealed barrel and rotating the barrel on its side to mix material thoroughly before every e_{max} or e_{min} determination. The barrel was left in an upright position for at least 15 minutes before removing the lid to extract the test specimen. This allowed the dust inside the barrel to settle. Settling of the dust resulted in an excess of fines at the top of the sample, therefore the tailings were further mixed with a scoop when opening the barrel to collect each test specimen.

Although the ASTM D4254 procedures are only applicable to soils with FC smaller than 15%, the method was adopted because there is no standard procedure to determine e_{max} of soils with higher FC values. The mould used to find e_{max} for the sand-silt soils was different from the one described in Table 1 and had an average height and diameter of 150 mm and 152 mm, respectively. Using the water filling method volume of mould was calculated as 2753 mL.

The determination of e_{min} was conducted using a non-standardised method similar to the one described for gravel-sand soils. The two main differences consisted in the approach to control filling of the mould and determination of the volume occupied by the soil. In this case, the mould was filled in 40 layers of about 28 g each. The mass of each layer was calculated so that the total mass deposited extended into the collar by one or two centimetres. After depositing each layer, the mould was tapped eight times with a rubber mallet in identical fashion as was done for the gravel-sand soils. Once the last layer was deposited, the standard surcharge (13.8 kPa) was lowered into the collar and the mould was tapped eight more times. Negligible settlement of the surcharge mass was observed, which suggests that very little additional compaction took place during the last eight taps. The surcharge mass and collar were then removed and the test specimen trimmed level with the rim of the mould using a straightedge. Any grains on the outside of the mould were cleaned with a brush and mass of specimen was determined. Since volume of the mould was known, maximum dry density could be determined. Furthermore, determination of G_s enabled determination of e_{min}.

4 MATERIAL TESTED

4.1 *Gravel-sand soils*

Gravel-sand soils tested were mixed from a residual non-plastic granite, which was separated into various sizes. Grains can be described as sub-angular with moderate sphericity. Different composite gradations were prepared by adding increasing quantities of a fine sand, S, to a fine gravel, G. Figure 1 depicts these gradations along with an example of a composite gradation with the percentage of fine sand, F = 15%.

The small pyknometer method in accordance with clause 8.3 of BS1377:Part 2:1990 was used to determine G_s. Separate values (Table 2) were determined for sand and gravel constituents and were found to be consistent with expected minerals in residual granite. Repeatability is satisfactory as the standard deviations are close to or better than 0.007, which is the expected order of error for G_s (Germaine & Germaine 2009). For each composite gradation, G_s was interpolated based on proportion of components.

4.2 *Sand-silt soils*

Sand-silt soils were obtained from a platinum tailings dam close to Rustenburg, South Africa. Seven gradations (Figure 2 and Table 3) were prepared from separated particle size ranges. Five of these were designed to be representative of in-situ

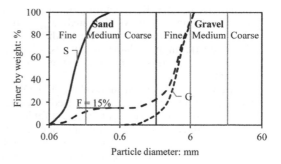

Figure 1. Gravel-sand gradations.

Table 2. Gravel-sand particle specific gravities.

Material	Particle size range mm	Particle specific gravity, G_s Average	Standard deviation
Gravel	1.18–0.710	2.66	0.008
Sand	0.150–0.105	2.69	0.001

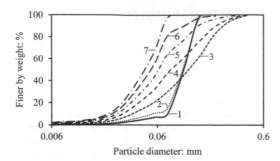

Figure 2.   Sand-silt gradations.

Table 3.   Sand-silt particle specific gravities.

Material	FC	Particle specific gravity, G_s	
	(%)	Average	Standard deviation
Mix 1	10	3.51	0.001
Mix 2	15	3.51	0.004
Mix 3	30	3.43	0.003
Mix 4	47	3.48	0.007
Mix 5	64	3.54	0.004
Mix 6	81	3.59	0.005
Mix 7	98	3.61	0.005

gradations, whereas the remaining two were made deliberately narrow to yield a high e_{min}. When observed under a microscope, grains exhibited the characteristic angular shape of tailings, which is due to the rock-crushing process from which they originate.

The water submersion method (ASTM D854) using a 500 mL pyknometer was used to determine G_s of the seven gradations (Table 3). Standard deviation values indicate that repeatability meets criterion suggested by Germaine & Germaine (2009). In Table 3, the FC value was chosen to characterise each mix, however, it should be noted that the entire shape of the Particle Size Distribution (PSD) curve, and not only FC, changes from one mix to the other (Figure 2). Accordingly, differences in properties of the mixes are a result of this overall change in the shape of the PSD curve, and not a result of changes in FC alone.

e_{min} was determined from three repeated tests for PSDs 1, 3, 5 and 6, two repeated tests for PSD 2 and only one test for PSDs 5 and 7. e_{max} was determined only for PSDs 3 to 7, from three repeated tests.

5   RESULTS

Due to difficulties in propagating error, error in reference states of compactness is typically reported as standard deviation of dry densities rather than of computed void ratios. Germaine & Germaine (2009) report these to be 0.01 g/cm³ for maximum density and 0.008 g/cm³ for minimum density based on tests on poorly-graded sands.

5.1   *Gravel-sand soils*

A decrease (Figure 3) in both e_{min} and e_{max} was observed up to approximately F = 30% after which both increased. This reflects sand initially sitting within gravel voids and then the sand pushing the gravel apart. Repeatability (Table 4) was on average satisfactory, although for a few tests standard deviation was greater than the values proposed by Germaine & Germaine (2009). Furthermore, it is apparent that the method to determine e_{min} was slightly less repeatable than the method to determine e_{max}. This is attributed to minor errors in determining the specimen height as the metal plate invariably had a slight tilt; on average 0.3 mm.

As e_{min} was determined with a non-standard method, a composite grading with F = 15% was tested according to the raining method outlined

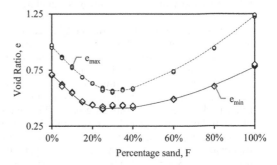

Figure 3.   Gravel-sand void ratios.

Table 4.   Gravel-sand soils dry density and void ratio errors.

State	Standard deviation	
	Dry density, g/cm³	Void ratio
	Average (Range)	Average (Range)
Loosest	0.005 (0.002–0.010)	0.006 (0.002–0.009)
Densest	0.007 (0.002–0.016)	0.006 (0.002–0.012)

in Germaine & Germaine (2009). These authors indicated that their method gives comparable results to the vibrating table method. Using this raining method e_{min} was determined as 0.51. However, with the method proposed herein a slightly denser state was achieved with, e_{min} determined as 0.47. The raining method was not adopted as it results in considerable material loss, which is problematic when material is limited.

5.2 Sand-silt soils

A trend of direct correlation between e_{max} and e_{min} is observed (Figure 4) which is in agreement with the results of the gravel-sand soils and with those reported by Cubrinovski & Ishihara (2002) for a wide variety of soil types. As the PSD of tailings changes and their FC increases from 10% to 98%, e_{min} values initially drop and appear to reach a minimum at an FC between 30% and 47%. At FC values higher than 47%, e_{min} continues to increase (Figure 4). Similar to the gravel-sand soils, repeatability (Table 5) of the procedures met, on average, the criteria suggested by Germaine & Germaine (2009).

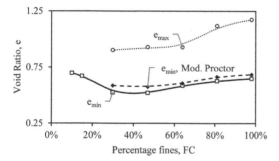

Figure 4. Sand-silt void ratios.

Table 5. Dry density and void ratio errors of the tailings.

State	Standard deviation	
	Dry density, g/cm³	Void ratio
	Average (Range)	Average (Range)
Loosest	0.005 (0.003–0.008)	0.005 (0.003–0.009)
Densest	0.008 (0.002–0.013)	0.006 (0.002–0.009)

As a way of validating the proposed e_{min} procedure, results of mixes 3 to 7 were compared to the e_{min} values obtained using the Modified Proctor Test (ASTM D1557). Both methods yielded the same trend (Figure 4). Such agreement can be interpreted as an indicator of suitability of the proposed e_{min} procedure.

6 CONCLUSIONS

A review of methods to determine e_{max} found that the standardised funnel method (ASTM D4254) could be implemented with minor variations within the constraints of a university laboratory. One variation proposed, which the standard provides scope for, was use of a smaller than standard mould when material constraints are present. The second variation proposed is to riffle the specimen into equal portions that are poured sequentially into the mould, if segregation is anticipated as would occur for sandy gravels. Testing on sand-silt soils without these variations and on gravel-sand soils with these variations resulted in similar repeatability.

Lack of access to an adequate vibrating table, that would prevent particle breakage, impeded the authors from using the standardised vibrating table method (ASTM D4253) to make e_{min} determinations. As noted by Germaine & Germaine (2009), this is a problem that many university laboratories are faced with. This situation encouraged the authors to review various methods proposed in literature to determine procedures to characterise non-plastic soils. Previous works (e.g. Cubrinovski & Ishihara (2002), Lade et al. (1998)) showed that the implementation of non-standard procedures is a viable alternative that can lead to repeatable and insightful results.

A method proposed by Lade et al. (1998), in which a specimen is incrementally spooned into a graduated glass cylinder and then tapped after each lift, was adopted. To improve accuracy with which volume could be determined, a steel mould of known volume was used. Two variations were proposed to determine specimen height. The first approach, applied to gravel-sand soils, involved carefully lowering a disk of known height unto the specimen within the mould. Then measuring the height between the mould rim and disk, to determine the specimen height. The second approach, applied to sand-silt soils, was to overfill the mould within a collar, which was then trimmed off. Different approaches to filling the mould were also tested. For gravel-sand soils, mass of each layer spooned into the mould was kept constant, whereas for the sand-silt soils, it was the total number of layers that was kept constant.

The methods yielded virtually identical repeatability (Tables 4 and 5) and both yielded results that were comparable to either non-standardised procedures proposed by other authors or to standardised methods. The consistency of results, good repeatability, and agreement with other methods lead the authors to conclude that the proposed methods are adequate for characterisation of non-plastic soils.

Finally, although it is known that e_{min} and e_{max} reflect overall shape of the PSD curve and particle shape characteristics of the soil, the authors opted, for convenience, to plot values of reference void ratios as a function of F (Figure 2) or FC (Figure 4). The authors wish to stress that this is just a means of presenting the data and that the plots are not intended to imply that e_{min} or e_{max} can be explained in terms of only the percentage of particles with a given size range that are present in a soil. To imply this would be to underestimate the number of factors that affect e_{min} and e_{max}.

REFERENCES

Bolton, M. D. 1986. The strength and dilatancy of sands. *Geotechnique*, 36(1): 65–78.

Cubrinovski, M. & Ishihara, K. 2002. Maximum and minimum void ratio characteristics of sands. *Soils and Found.*, 42(6): 65–78.

Germaine, J. T. & Germaine, A. V. 2009. *Geotechnical laboratory measurement for engineers*, New Jersey, John Wiley & Sons.

Lade, P. V., Liggio, C. D. & Yamamuro, J. A. 1998. Effects of non-plastic fines on minimum and maximum void ratios of sand. *Geotech. Testing J.*, 21: 336–347.

Tavenas, F. & La Rochelle, P. 1972. Accuracy of relative density measurements. *Geotechnique*, 22(4): 549–562.

Proceedings of the first Southern African Geotechnical Conference – Jacobsz (Ed.)
© *2016 Taylor & Francis Group, London, ISBN 978-1-138-02971-2*

Considerations for using soil-salt mixtures to model soil fabric changes

C.J. MacRobert
University of the Witwatersrand, Johannesburg, South Africa

P.W. Day
University of Stellenbosch, Stellenbosch, South Africa

ABSTRACT: Assessing changes in mechanical behaviour of soils due to changes in gradation, including particle crushing or internal erosion, frequently relies on computer modelling. Experimental validation is difficult, as the fabric changes usually occur over long periods. A novel approach is to use easily dissolvable salts to simulate these changes and assess their impact. This approach is currently being used to understand the mechanical consequences of internal erosion, by replacing erodible fines with sodium chloride. Although simple in concept, sodium chloride particles pack more efficiently than sand particles of an equivalent shape and size. It is therefore necessary to understand the fabric change that is to be modelled and then to use an appropriate soil-salt mix to model this. Although the current research is focused on internal erosion, the concept has the potential to be applied to various soil fabric studies, such as the propagation and collapse of underground voids.

1 INTRODUCTION

Traditionally geotechnical engineering has focussed on the design of structures on, in or of soil and rock. Very little attention is paid to the possibility that the properties of the soil or rock may change over time. However, both because of ageing infrastructure and requirements for long design lives, such considerations need to be addressed.

One such long-term effect is the potential for the gradation of a soil to change either through particle crushing, internal erosion, weathering or chemical dissolution of bonds. Muir Wood (2007) outlined the state of knowledge in this field highlighting how Discrete Element Modelling (DEM) and continuum modelling are predominantly being used.

The authors' current research is into the change in the gradation of a soil due to internal erosion particularly by suffusion (loss of finer particles from gap-graded or widely graded soils). In some such soils, the coarse particles dominate load transfer to the extent that the fine particles within the voids carry little or no load. These soils are termed internally unstable (ICOLD 2014). DEM and continuum modelling of this phenomenon (Muir Wood et al. 2010, Hicher 2013, Shire et al. 2014) has shown that fines carry low loads and can be eroded with the result that initially dilative soils become contractive. However, these remain numerical models; based on idealised Particle Size Distributions (PSDs) that require experimental validation.

For DEM studies, the number of particles needs to be limited to allow simulations to run in realistic periods. Within a given volume, as particles get smaller, their number increases substantially. This limits gradations that can be modelled with DEM to narrow PSDs, which do not adequately model the full range of internally unstable soils.

2 FABRIC CHANGES DUE TO INTERNAL EROSION

MacRobert (In Press) has suggested that in internally unstable soils, loads are transferred dominantly through the coarse particles. For low fines contents ($F < F_t$, F_t = approximately 15%) the void ratio of the coarse particles remains essentially unchanged and fines can be lost with very little change to the arrangement of coarse particles. Consequently, the ability of the overall fabric to resist either deviatoric or isotropic stresses is unlikely compromised.

However, for intermediate fines contents ($F_t < F < F_c$, F_c = approximately 26%) the coarse particles are less tightly packed, with fines partially filling voids to maintain an overall dense fabric. The impact of fines loss on such a soil's ability to resist either deviatoric or isotropic stresses is then dependent on the remaining coarse fabric. Thus, the fabric change to be modelled is the manner in which coarse particles rearrange as fine material within the coarse voids is removed.

For high fines contents ($F > F_c$) fine particles completely fill coarse voids, separating the coarse particles, resulting in an internally stable grading.

3 EXPERIMENTAL METHODS

Experimental approaches to date (Sterpi 2003, Shwiyhat & Xiao 2010, Chang & Zhang 2012, Ke & Takahashi 2012) have relied on reconstituted samples, modified triaxial tests in which suffusion is initiated under high seepage velocities, and cone penetration tests on an internally unstable material in a permeater subject to upward flow. These studies are somewhat inconclusive and provide only a general, high-level understanding of the behaviour. It is clear that a better method is needed to investigate the consequences of fines loss on mechanical behaviour.

Using seepage alone to erode fines from a soil is fraught with difficulties. Firstly, the loss of fines can take place over a significant period as is often the case with embankment dams (Fell et al. 2003). Secondly, scaling erosional processes is difficult (Bezuijen & Steedman 2010) as the scaled sample needs to be capable of reproducing single grain erosion processes (Marot et al. 2012). As such, modelling the process within realistic periods and under representative hydro-mechanical conditions is difficult in a laboratory setting.

One alternative considered was to place the unstable material in a hydraulic flume (900 mm wide, 800 mm high and 7200 mm long) and subject it to long-term flow. Deformations at the boundary could possibly be tracked using particle image velocimetry (White et al. 2003) and deformations within the soil mass tracked using fibre optic cables (Vorster et al. 2006). The practicalities involved and questions about whether any quantifiable data on the mechanical behaviour would be obtained, precluded further development of this idea.

X-ray computed tomography (CT scanning) is another non-destructive method of observing the interior of physical systems and this is increasingly finding application in geotechnical research. The development of ice lenses has been investigated by Torrance et al. (2008), development of shear zones in triaxial testing by Hall et al. (2010) and soil arching by Eskişar et al. (2012). The possibility was considered of using CT scanning to track developing strains in a soil system undergoing internal erosion from which changes in mechanical behaviour could be inferred.

The difficulties involved in conveying water through a specimen in an environment with sensitive electronic equipment were considered prohibitive. It was proposed that suitably crushed and sized solid carbon dioxide (dry ice) and chilled soil particles could respectively form the fine and coarse particles of an unstable material. This material formed around a thin heating element, which is then slowly heated, within a CT scanning unit, could shed light on the mechanical processes taking place. Practicalities around handling materials at temperatures of −78 °C, and uncertainty whether any quantifiable data would be produced, precluded further development of this idea.

The concept of replacing the erodible particles with an analogue material that could be eroded in shorter periods and under realistic hydraulic conditions was pursued further. It was proposed that sodium chloride salt would be the most practical material. This would allow controlled quantities of fines to be lost under atmospheric saturation; precluding the need for high seepage velocities and resulting high pore pressures. Differences in particle specific gravity, hardness and shape are possible limitations and these are explored further in the next section. Overall, it is felt that the observations made provide ample justification to overlook these limitations.

The approach of using soluble particles in a soil is not entirely new in geotechnical research. A similar approach was used by Shin & Santamarina (2009) to investigate the evolution of the coefficient of earth pressure at rest, k_0 during soil diagenesis. In their study, similar sized salt and glass beds were placed in a soft oedometer and the change in horizontal stress under a constant vertical stress was monitored as the salt was dissolved. Following dissolution, no attempt to determine the shear behaviour was undertaken.

To obtain quantitative data on the impact of fines loss on settlement and shear resistance, suitable soil-salt mixtures are placed in an appropriately altered direct shear box. At the time of writing testing is ongoing; however, considerations on how to prepare suitable soil-salt mixtures are discussed here.

4 SOIL-SALT MIXES

Mixtures of a fine Gravel (G) and fine Sand (S) that were found to result in internally unstable gradations in tests by Skempton & Brogan (1994), Li (2008) and Shire et al. (2014) were used to investigate the fabric changes as fines are lost. In this case, the gravel forms the coarse non-erodible component, to which a variable percentage (F) of sand, were added to represent the erodible fines. An example of a grading with F = 15% is indicated in Figure 1.

The soil used to prepare the various gradations was a non-plastic residual granite, which was separated into different size ranges. Light microscope images (Figure 2) of the various sand size ranges show that the particles are sub-angular with moderate sphericity. The shape of the gravel particles was similar to that of the sand.

As noted earlier, sodium chloride salt was used as an analogue for the sand particles, to facilitate particle loss under realistic seepage conditions. Ordinary table salt was crushed and separated into equivalent particle size ranges as were present in the sand. Figure 3 shows light microscope images of the various salt size ranges. Although the sand and salt particles are of similar size and sphericity, key differences are observed. Firstly, the salt particles are smoother, a consequence of cleavage planes, due to preferential planes of weakness,

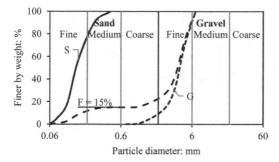

Figure 1.   Gradations investigated.

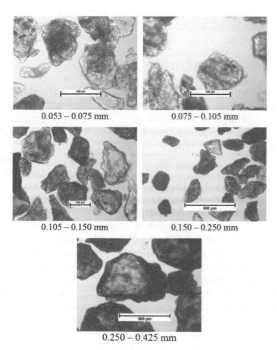

0.053 – 0.075 mm   0.075 – 0.105 mm

0.105 – 0.150 mm   0.150 – 0.250 mm

0.250 – 0.425 mm

Figure 2.   Light microscope images of sand particles.

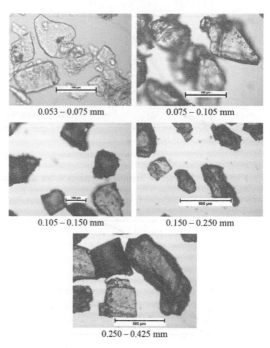

0.053 – 0.075 mm   0.075 – 0.105 mm

0.105 – 0.150 mm   0.150 – 0.250 mm

0.250 – 0.425 mm

Figure 3.   Light microscope images of salt particles.

which are absent in the sand particles. Secondly, the salt particles are more angular, again a consequence of cubic cleavage in sodium chloride.

Another difference is particle hardness. On the Mohs hardness scale, the minerals making up granite have a hardness of approximately 6, whereas sodium chloride has a hardness of 2.5 (Kehew 2006). To investigate whether sample preparation would cause particle breakdown, a sand sample and a salt sample were prepared. Each was then sieved to determine their PSDs. Two more samples were prepared and riffled a number of times to simulate the abrasion that may occur during sample preparation. These riffled samples were then sieved to determine whether their PSDs had altered.

It is clear (Figure 4) that sample preparation has minimal effect on the particle size of the sand. On the other hand, the salt sample (Figure 5) undergoes some particle breakdown, especially during sieving. This can be inferred from the fact that the added riffling step did not result in a significant change in PSD. Thus, some particle crushing can be expected, although it is unlikely it will influence sample preparation significantly.

Due to differences in particle specific gravity G_s, the theoretically correct approach would be to replace sand particles with an equivalent volume of salt. However, it was apparent from testing to determine reference void ratios with increas-

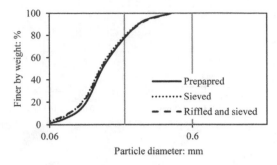

Figure 4. Impact of sample preparation on sand particle sizes.

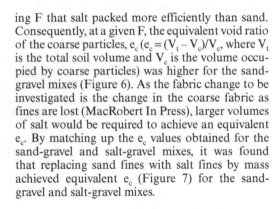

Figure 5. Impact of sample preparation on salt particle size.

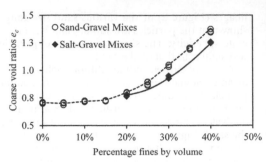

Figure 6. Coarse void ratios for mixes prepared by volume.

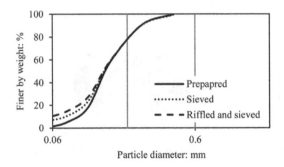

Figure 7. Coarse void ratios for mixes prepared by mass.

ing F that salt packed more efficiently than sand. Consequently, at a given F, the equivalent void ratio of the coarse particles, e_c ($e_c = (V_t - V_c)/V_c$, where V_t is the total soil volume and V_c is the volume occupied by coarse particles) was higher for the sand-gravel mixes (Figure 6). As the fabric change to be investigated is the change in the coarse fabric as fines are lost (MacRobert In Press), larger volumes of salt would be required to achieve an equivalent e_c. By matching up the e_c values obtained for the sand-gravel and salt-gravel mixes, it was found that replacing sand fines with salt fines by mass achieved equivalent e_c (Figure 7) for the sand-gravel and salt-gravel mixes.

5 FURTHER APPLICATIONS

A similar approach, as proposed in this paper, may be suitable for investigating the effect of weathering and chemical dissolution of particles on mechanical behaviour of soils. Various soil fabric interactions could be investigated using salts with different degrees of solubility and different solvents. Centrifuge modelling of the propagation and collapse of underground voids could be investigated by dissolving appropriately shaped salt pockets. This may be more analogous to the actual mechanism than deflating liquid filled bladders.

6 CONCLUSIONS

With demands for longer design lives and questions of how best to ascertain continued performance of ageing infrastructure, geotechnical engineers are increasingly required to look at how soil and rock performance may change over long periods. One such example is how mechanical behaviour of internally unstable soils change as fines are lost, altering the soil fabric. The long periods over which these changes take place makes experimental investigation difficult.

A novel approach is to use salt particles as an analogue for the erodible soil particles. Particle loss can then be simulated using realistic seepage velocities, but within shorter periods. Differences in particle shape, hardness and particle specific gravity have been explored in this paper, showing that consideration must be given to the fabric changes that are to be modelled rather than simply replacing sand with salt.

In the case of internal erosion, because the change to be modelled is the change in the fabric of the coarse particles as fines are lost, replacing fines on a mass basis rather than a volume basis was found to recreate the desired fabric changes better. This was because of salt packing more efficiently than sand.

REFERENCES

Bezuijen, A. & Steedman, R.S. 2010. Scaling of hydraulic processes. *Physical Modelling in Geotechnics, Two Volume Set: Proceedings of the 7th International Conference on Physical Modelling in Geotechnics (ICPMG 2010), 28th June-1st July, Zurich, Switzerland.* 93–98. CRC Press.

Chang, D.S. & Zhang, L.M. 2012. Critical hydraulic gradients of internal erosion under complex stress states. *J. Geotech. Geoenviron. Eng.*, 139(9): 1454–1467.

Eskişar, T., Otani, J. & Hironaka, J. 2012. Visualization of soil arching on reinforced embankment with rigid pile foundation using X-ray CT. *Geotextiles and Geomembranes*, 32: 44–54.

Fell, R., Wan, C.F., Cyganiewicz, J. & Foster, M. 2003. Time for development of internal erosion and piping in embankment dams. *J. Geotech. Geoenviron. Eng.*, 129(4): 307–314.

Hall, S.A., Bornert, M., Desrues, J., Pannier, Y., Lenoir, N., Viggiani, G. & Bésuelle, P. 2010. Discrete and continuum analysis of localised deformation in sand using X-ray μCT and volumetric digital image correlation. *Geotechnique*, 60(5): 315–322.

Hicher, P.-Y. 2013. Modelling the impact of particle removal on granular material behaviour. *Geotechnique*, 63(2): 118–128.

ICOLD 2014. *Bulletin 164 Internal erosion of exisiting dams, levees and dikes, and their foundations.*, Paris, International Commision on Large Dams (ICOLD).

Ke, L. & Takahashi, A. 2012. Influence of internal erosion on deformation and strength of gap-graded non-cohesive soil. *Sixth International Conference on Scour and Erosion*, Paris. 847–854.

Kehew, A.E. 2006. *Geology for engineers & environmental scientists*, New Jersey, Pearson Prentice Hall.

Li, M. 2008. Seepage induced instability in widely graded soils. Vancover, BC, Canada, Univ. of British Columbia.

MacRobert, C.J. In Press. A theoretical framework to understand the mechanical consequences of internal erosion. *84th ICOLD Annual Meeting.* Johannesburg, SANCOLD.

Marot, D., Le, V.D., Garnier, J., Thorel, L. & Audrain, P. 2012. Study of scale effect in an internal erosion mechanism: centrifuge model and energy analysis. *European Journal of Environmental and Civil Engineering*, 16(1): 1–19.

Muir Wood, D. 2007. The magic of sands. 20th Bjerrum Lecture, presented in Oslo 25 November 2005. *Can. Geotech. J.*, 44(11): 1329–1350.

Muir Wood, D., Maeda, K. & Nukudani, E. 2010. Modelling mechanical consequences of erosion. *Geotechnique*, 60(6): 447–457.

Shin, H. & Santamarina, J.C. 2009. Mineral dissolution and the evolution of k0. *J. Geotech. Geoenviron. Eng.*, 135(8): 1141–1147.

Shire, T., O'Sullivan, C., Hanley, K.J. & Fannin, R.J. 2014. Fabric and effective stress distribution in internally unstable soils. *J. Geotech. Geoenviron. Eng.*, 140(12): 04014072.

Shwiyhat, N. & Xiao, M. 2010. Effect of suffusion on mechanical characteristics of sand. *Scour and Erosion.* 378–386. ASCE.

Skempton, A.W. & Brogan, J.M. 1994. Experiments on piping in sandy gravels. *Geotechnique*, 44(3): 449–460.

Sterpi, D. 2003. Effects of the erosion and transport of fine particles due to seepage flow. *International Journal of Geomechanics*, 3(1): 111–122.

Torrance, J.K., Elliot, T., Martin, R. & Heck, R.J. 2008. X-ray computed tomography of frozen soil. *Cold regions science and technology*, 53(1): 75–82.

Vorster, T.E.B., Soga, K., Mair, R.J., Bennett, P.J., Klar, A. & Choy, C. K. 2006. The use of fibre optic sensors to monitor pipeline response to tunnelling. *GeoCongress 2006: Geotechnical Engineering in the Information Technology Age*, 2006: 33–33.

White, D.J., Take, W.A. & Bolton, M.D. 2003. Soil deformation measurement using Particle Image Velocimetry (PIV) and photogrammetry. *Geotechnique*, 53(7): 619–631.

Proceedings of the first Southern African Geotechnical Conference – Jacobsz (Ed.)
© 2016 Taylor & Francis Group, London, ISBN 978-1-138-02971-2

Interface shear: Towards understanding the significance in geotechnical structures

G.C. Howell & A.H. Kirsten
SRK Consulting, Johannesburg, South Africa

ABSTRACT: It is well known that low strength natural materials under foundations and in slopes can cause bearing capacity and slope failures. With the increased use of manufactured materials such as HDPE liners and geotextiles, similar conditions can inadvertently be built into the structure. These interface shear aspects occur between liner and geotextile, liner and soil, geotextile and soil and even within the confines of the geotextile. Injudicious use of low interface shear materials can lead to excessive deformation and even catastrophic failure. The many potential interface shear planes that can exist in geotechnical structures are considered with reference to 20 years of published research and actual recent shear box tests carried out for validation purposes for lined structures. The paper further considers a simplistic calculation method using block limit equilibrium and finite element simulations to understand the problems that exist in these structures. The concept of 'excess shear' which leads to overstressing of liners is also discussed.

1 INTRODUCTION

Recent failures of geotechnical structures have again highlighted the requirement of engineers to fundamentally understand the materials that they are working with. Unlike other spheres of engineering where material properties are known to a large extent within a defined band (concrete, steel, aluminum), geotechnical materials are variable and subject, amongst other issues, to changed conditions due to pore pressures and seismic excitation. In one such case (Independent Expert, 2015), an apparently stable structure subject to pore pressures and susceptible low strength foundation clay strata, lead to failure. The lessons learnt were that the geotechnical investigation failed to adequately identify the properties of the clay in the foundation and the design engineers failed to appreciate the significance of the location and strength of the material within their structure. Another case recently of a tailings dam failure in Brazil, lends weight to the issue of fundamental engineering appreciation for such structures.

Here is Southern Africa there have been numerous examples of slope and bearing capacity failures that show the same lack of appreciation of engineering principles that are evident in other incidents. In particular, the existence of weak subgrade materials, weak or gouge filled joints in otherwise strong rock and layered materials in stockpiles and tailings dams are examples of these issues. Even so, more and more, engineers are using 'new age'

materials in designs for practical and economic reasons, without possibly comprehending the associated risks. HDPE, LLDPE, PVC, geosynthetics and geotextiles in general, in combination with soils in structures, represents an instance of such risks unless judicious appreciation of their interaction is clearly understood.

The objective of this paper is mainly educational in terms of two themes:

- By presenting the interface friction properties of commonly used manufactured liners with soil and other geosynthetics in order to highlight the relative strength (rather weaknesses) inherent in these interfaces. This has been achieved by interrogating literature and own results from direct and ring shear testing done on such interfaces. Comments and cautions are provided to guide the design engineer in new applications; and
- By presenting an approach to conceptualizing the mechanics of a slope stability design problem that includes interfaces in a coherent, practically and understandable way by using a first principles approach. By so doing the physical importance of the design parameters can be visualized and appreciated. The method uses a limit equilibrium, or static equilibrium, approach by balancing forces in a slope, with the objective of understanding how the energy in structure is sustained by the parent material itself, by the interface and how liner tension is developed as a consequence.

The overarching objective is to foster an appreciation for the importance of clearly 'defining the problem' in fundamental terms before attempting a solution. Clear understanding of the problem from the outset is a prerequisite for a solution.

2 PROBLEM DEFINITION

Figure 1 has been drawn to indicate 5 (of many) possible low strength planes within a slope (natural or manufactured). 1 represents a foundation plane of weak material; 2 represents a weak interface at ground level, possibly as a result of the insitu strength of the material, or from a manufactured interface such as a geomembrane; 3a and b represent inclined planes of viable angle, either from natural jointing (in a rock slope), deposition of variable low strength materials in a tailing dams or a manufactured interface surface in a constructed embankment; 4 represents similar horizontal natural or manufactured interfaces; 5 represents a common circular failure; and 6 represents a piecewise linear failure plane resulting from failure along the base interface and inclined through the body of the material.

In this paper, we will be concerned mainly with interfaces 2 (a manufactured interface nominally along the base of a dam or embankment) and 6 the resulting failure plane along the base and through the parent material, but other combinations of interface layers and failure planes are also addressed intrinsically but not specifically in the discussion that follows.

From a design perspective, there are a number of issues that need to be addressed. These include the following:

1. The interface shear strength along the potential failure plane;
2. The development of deformation and strain along the failure plane; and
3. The consequent development of tension in the geomembrane.

These aspects have to be fundamentally understood for any design. Interface shear strength properties are sourced from soils laboratory testing and from experience with the use of these materials. Section 3 considers these properties for a range of geomembranes and textiles with interfaces to soils (granular and clay) and to other geomembranes. Section 4 considers fundamental concepts, while the approach covered in Sections 5 and 6 assists in defining the fundamental analysis aspects in a simple but effective limit equilibrium treatment.

3 INTERFACE SHEAR PROPERTIES

A potential weak zone (either natural or manufactured) consists of 2 or more interfaces that need to be assessment individually and collectively, since the interface with the minimum shear strength will dominate the overall behavior. In addition, a weak zone can be due to the nature of the material itself, for example a clay layer, or the bentonite clay in some GCL products that are not thermally locked within the geotextile carrier layers. The most common situations with manufactured materials occur on the upper and lower surfaces of the interface that is created between the geosynthetic, the soil and/or other geosynthetics.

The information that is given here is derived from literature and from physical testing on the interface properties carried out by the authors' colleagues during the course of projects. However, it should be noted that this information is purely indicative for use in preliminary design only. Once the interface materials have been chosen, then careful and thorough physical testing of the sandwich of materials to be used in the design is essential, since variations in actual behavior can be expected.

Only interfaces from commonly used material are reported. The current data base available to the authors includes 143 tests direct and ring shear tests on various interfaces. With time, this data will be augmented to include additional information from other physical testing sources.

Although subgrouping is difficult since (particularly for soil) the descriptions are not always clear, an attempt has been made to assess the data base in terms of the following materials:

- HDPE-S and HDPE-T: HDPE geomembranes (Smooth and Textured)
- LLDPE-S and HDPE-T: LLDPE geomembrane (Smooth and Textured)

  interfaced with:

- Granular Soil (USCS classification S and M)
- Cohesive Soil (USCS classification C)
- Geotextile (Needle punched)
- GCL (thermally welded needle punched).

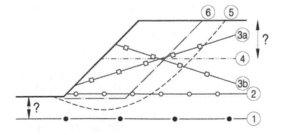

Figure 1.   Potential failure planes in a slope.

It should be noted that geotextiles are produced in various grades and types including needle punched and woven fabrics. Only needle punch geotextile of any grade is reported here. Similarly, GCLs are produced in various grades and types. Only thermally welded needle punched GCLs are reported here. It should also be noted that some GCLs are not thermally welded (meaning that the upper and lower geotextile carrier layers are not physically connected by needle punching strands through the clay (usually bentonite) and welding them in place. Thermal welding creates an internally stable 'lattice' structure within the GCL layer, otherwise internal interfaces exist that themselves can have very low frictional properties, sometimes as low as 1° for bentonite itself.

Tables 1 and 2 give the statistical representation from the database in terms of mean, (standard deviation) and [number] of tests available for peak and residual friction angles for various interfaces under saturated conditions.

The implications of Tables 1 and 2, are inter alia noted below. Textured geomembrane improves the interface shear by at least 10° in all cases. The variability of the results given by the standard deviation in ( ) shows that the mean value used in analyses cannot be directly justified and that a more realistic design value is the mean less 1.2 to 1.5 times the standard deviation. LLDPE appears to give more consistent results than HDPE. This is probably due to the lower density and softer modulus of the LLDPE which allows more mechanical frictional interaction between the soil (whether granular or finer grained) and the geomembrane. The peak and residual values are also interesting showing that the residual friction angle is some 2 to 5 degrees less than the peak in general for HDPE material. It is however larger (10 degrees) for the LLDPE textured material. Whilst an explanation

Table 2. Interface residual friction angle properties.

Interface	Soils		Geosynthetics	
	Granular	Cohesive	Geotextile	CL
HDPE-S	15.05	10.49	10.5	-
	(5.05)	(5.57)	(2.12)	(-)
	[4]	[4]	[2]	[-]
HDPE-T	26.42	19.47	17.08	10.1
	(6.85)	(11.56)	(1.61)	(-)
	[16]	[24]	[4]	[1]
LLDPE-S	21.7	9.07	9	-
	(6.29)	(2.70)	(-)	(-)
	[4]	[3]	[1]	[-]
LLDPE-T	26.10	28.08	17	-
	(5.50)	(11.17)	(-)	(-)
	[7]	[9]	[1]	[-]

is not immediately obvious, this could be due to the collapse of the textures under strain because of the softer modulus of the material.

Compare the friction angles from Tables 1 and 2 with of USCS graded materials for gravels (G) (φ = 33 to 40°), Sand (S) (φ = 31 to 38°), Silt (M) (φ = 30 to 33°) and Clay (C) (φ = 22 to 30°). It is apparent that interfaces are significantly inferior to clay in most cases.

The information indicates that extreme caution must be taken by the designers to fundamentally assess the interface effects on the structure. It is also indicative that the structure must be maintained well within the limits of the peak shear stress and strain parameters and 'elastic' range, since once the peak is reached, the residual nature of the interfaces will not sustain the loading. Alternatively put, interfaces of the type considered here exhibit significant strain softening behavior and can never be considered to be elastic perfectly plastic as would be the default in some numerical models.

4 FUNDAMENTAL STRESS CONCEPTS

Conceptually speaking, due to gravity, every structure is represented by a potential energy state that is counteracted internally by the strength (or strain energy) capacity of the system. Alternatively put, strain energy is induced in a structure by virtue of its size and shape which is sustained by the strength of the multiple 'elements' that make up the whole, provided that the fundamental principle of Capacity (C) always being greater than Demand (D) is never violated. If one element in the system (of many elements) is unable to sustain the demand placed on it, then, like a weak link in a chain, failure will occur. This sounds obvious, but it is still surprising how often this principle is forgotten.

Table 1. Interface Peak Friction Angle Properties.

Interface	Soils		Geosynthetics	
	Granular	Cohesive	Geotextile	GCL
HDPE-S	21.26	10.64	11.33	8
	(8.03)	(6.75)	(3.51)	(-)
	[5]	[5]	[3]	[1]
HDPE-T	30.51	20.65	25.14	24.2
	(6.52)	(6.75)	(4.87)	(13.0)
	[18]	[27]	[8]	[2]
LLDPE-S	26.63	8.53	10	-
	(0.99)	(2.55)	(-)	(-)
	[4]	[3]	[1]	[-]
LLDPE-T	35.89	33.53	26	-
	(5.04)	(12.92)	(-)	(-)
	[7]	[9]	[1]	[-]

Fortunately most structures exist in a state of redundancy, meaning that no one element alone is responsible for accepting and sustaining the full demand on it. Demand is 'shared' (in a complex way) between many elements in the structure. Should one element not be able to accept the full demand placed on it, then it will 'shed' the excess demand to other elements in the structure. The other elements can therefore 'accept' an increase in stress/strain within their own capacity, or 'shed' the excess demand to the next element in line. This is the phenomenon of system shakedown, or stress/strain relaxation/redistribution and is accompanied by plastic strain deformation.

4.1 Model representation

In the design of geotechnical structures, the development and positioning of a failure plane is an essential requirement in conceptualizing the failure mechanisms within the structure.

Modeling using finite element, finite difference or discrete element software programs are favoured by 'new age' engineers. But these methods are just 'black boxes' to some and injudicious use can lead to serious flaws in the design and must be used with caution. Yet they are essential to develop an appreciation for deformation and stress trajectories in the detailed design phase.

Limit Equilibrium (LE) methods, while not capable of providing detailed stress, strain and deformation output, fundamentally concentrate on block stability by ensuring that the forces are in equilibrium subject to the simplistic assessment of shear stress capacity usually modelled in terms of the Mohr-Coulomb failure criterion. LE is nothing else than the balance between demand, represented by mass and gravity, and capacity, represented by shear stress/force 'available' on a defined failure plane by virtue of its shear stress capacity integrated along the defined failure plane.

LE should not however be underestimated as a tool to fundamentally understand the nature and mechanisms of failure that prevail in a structure. The method requires that a failure mechanism or block model be defined inductively whereupon the full suite of applicable forces are assembled to represent the state of force equilibrium on the block. This is instructive in two aspects: i) it forces the user into inductive (as opposed to deductive) reasoning but asking the question "what could be the mechanism of failure" and ii) it requires the calculation of basic first principles forces and counter forces. The benefit of this thinking will be demonstrated in the section of Translational analysis below.

4.2 The 'excess shear' concept

'Excess shear' is not a physically reality, but a numerical construct/concept that greatly enhances the understanding of the mechanics of a problem particularly when the problem involves plastic strain and the redistribution of stress in the structure. The state of stress cannot physically exist outside of the failure surface (Mohr Coulomb for example) and in reality, as the failure surface is approached (from below), stress and strain redistribution occurs which ensures that the stress state is maintained at least at the failure surface. In numerical analysis, however, the stress state at every point is first calculated and then checked against the failure state. If it is found to be within (below) the failure surface, then calculation proceeds. If a stress point is outside (above) the failure surface, then the 'excess shear', that measure that is above the failure surface, is redistributed to other points in the vicinity so that the overall stress state is maintained at least at the failure surface.

To investigate this concept further, consider a structure built from homogeneous material which is in equilibrium under all applied loads and stresses. Introduce an inclined plane into the structure which has the same capacity as the parent material. The equilibrium condition is not altered and the structure is stable. Now, progressively decrease the strength capacity of the plane until failure is reached—failure plane. The difference between the equilibrium and failed state is defined here as the "excess shear" in the system. "Excess shear" in the system contributes to the lack of 'factor of safety', but if the capacity is not available and redistribution to maintain equilibrium is not possible, then numerical (and physical) failure is precipitated (Howell, 1992).

There are a number of options that the designer can choose to deal with the problem:

- Include reinforcement in the system to contribute to reducing the excess shear;
- Reduce the overall potential energy/strain energy in the system by changing the geometry (shallower angles, reduced height, for example).

In the former case, added reinforcement in the modern geotechnical context includes the addition of geomembranes and geogrids that provide a tension component to improve shear strength (effective cohesion) and to 'consume' the excess shear. This is an important aspect since the tension that is generated must be sustainable by the geomembrane, remembering that the strain deformation of geomembranes at maximum stress is of the order of 100%, far above the strain development in the structure. Therefore, strain compatibility and strain limitation is necessary in the design, which in turn means limiting the stress in the element. This potentially is a complex design procedure and is not easily conceptualized or understood in modelling (with FE/FD), but must be considered in order to adequately account for the ultimate and serviceability limit states. LE formulations can assist greatly with visualization of this issue as will be considered in the next section.

4.3 *Homogeneous materials*

To demonstrate the excess shear concept further, consider the states of stress in a structure (Figure 2). The M-C failure criterion is represented by the line $\tau = c_m + \sigma_n \tan(\varphi_m)$ where τ is the shear stress, σ_n is the normal stress and φ_m is the friction angle for the material (denoted by the subscript m). Note that any other relevant failure criterion other than M-C could be used as it does not change the fundamental concepts.

It is immediately obvious that the state of stress can either be inside the failure envelope (point 1) or outside (point 2). The stress state at point 2 is physically impossible as it lies outside of the failure envelope. But conceptually the vertical distance that the stress point is outside the failure envelope, is an indication of the state of 'excess shear' that exists and that needs to be accounted for, either by redistribution or by accommodation by other means, such as tension in the liner.

Integration of the effects over the continuum defines a failure zone that describes the failure surface in the structure and typically shown in Figure 3.

4.4 *Constructed interface*

Install/construct into the continuum an inclusion that has strength parameters (φ_i, c_i) as represented in Figure 4 as an additional sub-horizontal potential failure surface.

Clearly, whereas the stress state of the homogeneous slope was in equilibrium, the new constructed structure may have stress states that violate the failure conditions along the inclusion

that lead to excess shear. The net effect of this is that shear deformation (or block deformation in Figure 4) will occur sympathetic with the excess shear component that exists in the system.

There are 4 identifiable zones in the block model (Figure 6b):

- Zone 1: represents an area where excess shear stress is manifest along the inclusion, leading to deformation and potential failure;
- Zone 2: due to strain deformation, the stress conditions in this zone are altered sufficiently for them to exceed the failure condition in the parent material, leading to the formation of a failure plane;
- Zone 3: similar to Zone 2, but the shear stress development is sympathetic with the mass movement along the inclusion;
- Zone 4: represents the area behind the failure plane where the stresses do not violate either that of the parent material (φm, cm) or the inclusion (φi, ci).

The effect of the zonal nature of the stress field subdivides the continuum into failure blocks in a piecewise linear fashion, where the excess shear stress is translated into extension strain (or tension) at the bifurcation point (A). This clearly demonstrates the dilemma of geometry that the engineer faces in this design conceptualization. Numerical analysis can easily be used to calculate the stresses and forces that develop in the structure and thereby define the block geometry, but the fundamental behavior that the engineer needs to anticipate is best demonstrated by a limit equilibrium approach using a hand or spreadsheet calculated translation analysis.

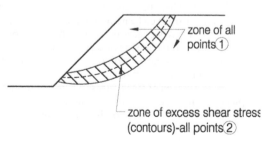

Figure 3. Failure zone in a continuum.

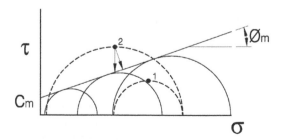

Figure 2. State of stress and M-C failure criterion.

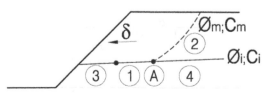

Figure 4. Zoned constructed continuum.

5 LIMIT EQUILIBRIUM: TRANSLATIONAL ANALYSIS

Qian et al (2002) defined a calculation procedure for translational failures using a piecewise linear limit equilibrium approach in landfill sites. This method has been adapted here to analyze block failures and the commensurate increase in liner tension in constructed structures. The translational (or two wedge) failure analysis was originally used to calculate the factor of safety against possible mass movement along a liner, but in this application, the objective is to reformulate the method to calculate tension in the liner directly. This method can, however, just as easily be used to analyze any generic slope stability problem including (and very specifically) low strength interface problems once the failure mechanism of the slope has been defined. For this purpose, the slope is divided into sectors/regions as shown in Figure 5 with reference to the zones defined in section 4.4 above.

In Figure 5, Sector 1 is called the passive wedge since it is acting to resist deformation, Sector 2 is the active wedge since it is subject to gravitational energy/movement in a lateral direction; and Sector 3 is the stationary wedge held in place by virtue of its location and force equilibrium.

Translational movement takes place when the active wedge (2) under gravity fails on the incipient planes bc and bd and drives the passive wedge laterally along ab. Along these planes, excess shear stress has been developed resulting in deformation which is counteracted by:

- Frictional strength on plane ab
- Internal friction on planes bc and bd
- Tension in the liner at b.

Conceptually, the passive wedge is akin to the commonly observed heaving portion of the toe of a slope, the active wedge is the slump that takes place and points c and d are the observable manifestation of shear strain deformation that occur between the wedges on surface. Physically, planes bc and bd are zones of high shear strain/stress where failure takes place. While it can be shown numerically (FE) that the plane bd is inclined as shown in Figure 5, the LE mathematics (below) becomes too complicated for hand or spreadsheet calculation and hence bd is considered to be vertical for this purpose. Line bd is the interface along which equilibrium between the active and passive blocks is calculated. Taking the plane as vertical introduces some minor error, but the principles are still valid and instructive. The resulting model is shown in Figure 6.

With the objective of calculating the minimum T (tension in the liner) to maintain equilibrium, the static force equilibrium equations $\Sigma F_y = 0$ and $\Sigma F_x = 0$ can be derived in terms of the following parameters:

Passive Wedge:
W_p = weight of the passive wedge
N_p = normal force acting on the bottom of the passive wedge
F_p = limiting frictional force acting on the bottom of the passive wedge (subject to MC limit for the interface)
E_{HP} = normal force from active wedge acting on the passive wedge
E_{VP} = frictional force from active wedge on the side of the passive wedge
Φ_P = interface friction angle under the passive wedge
Φ_M = friction angle of the parent material
α = outer angle of the parent material slope
θ = angle of the liner/subgrade to horizontal
Active Wedge:
W_A = weight of the active wedge
N_A = normal force acting on the bottom of the active wedge
F_A = limiting frictional force acting on the bottom of the active wedge (subject to MC limit for parent material)
E_{HA} = normal force from passive wedge acting on the active wedge
E_{VA} = frictional force from passive wedge on the side of the active wedge
Φ_A = interface friction angle under the active wedge
β = base angle of the active wedge to the horizontal
General:
W_T = total weight of the active and passive wedges
T = Tension in the liner

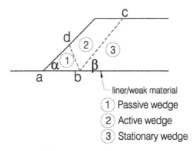

liner/weak material
(1) Passive wedge
(2) Active wedge
(3) Stationary wedge

Figure 5. Block zones in a structure.

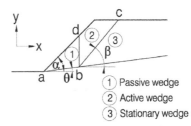

(1) Passive wedge
(2) Active wedge
(3) Stationary wedge

Figure 6. Dual wedge model.

Considering the force equilibrium of the passive wedge for $\Sigma F_y = 0$:

$$W_P + E_{VP} = N_P \cos\theta + F_P \sin\theta \qquad (1)$$

$$F_P = N_P \tan\varnothing_P + T \sin\theta \qquad (2)$$

$$E_{VP} = E_{HP} \tan\varnothing_M \qquad (3)$$

When $\Sigma Fx = 0$:

$$F_P \cos\theta + T \cos\theta = E_{HP} + N_P \sin\theta \qquad (4)$$

Considering the force equilibrium of the active wedge for $\Sigma F_y = 0$:

$$W_A = F_A \sin\beta + N_A \cos\beta + E_{VA} \qquad (5)$$

$$F_A = N_A \tan\varnothing_A \qquad (6)$$

$$E_{VA} = E_{HA} \tan\phi_M \qquad (7)$$

When $\Sigma Fx = 0$:

$$F_A \cos\beta + E_{HA} = N_A \sin\beta \qquad (8)$$

Equation (1) to (8) together with the equality $E_{HP} = E_{HA}$ produces a complex expression for T in terms of the known parameters above. The expression, however, for T is best solved using a spreadsheet. This allows the parameters to be tested and a coherent design achieved from the treatment. The fundamental treatment from very basic principles (in this case, static equilibrium) is very powerful in gaining an in-depth understanding of the engineering processes that are involved.

Cohesion and the size of a toe berm to provide additional passive pressure and so render the liner tension to zero can also be simply include by adding a stabilizing term to equation 4.

The tension in the liner is due to the inability of the interfaces to sustain the excess shear stress (or force) that the system, driven by gravity, imposes on it. The interaction between shear capacity along the failure planes and interfaces and the resultant need for reinforcement (in this example, the liner strength) is clearly demonstrated in this treatment.

The corollary is that other more esoteric design calculations can be carried out quickly and efficiently using the spreadsheet. These include, but are not limited to, the assessment of the height of the structure at the material's normal angle of repose (cascaded stockpile) subject to zero tension in the liner; the stacking angle for a given height for zero tension; the height and/or stacking angle for a predefined tension in the liner commensurate with an specified target strain, the relationship of the base/subgrade angle to liner tension; or the calculation of the required tensile reinforcement (geogrid) required for a given set of parameters.

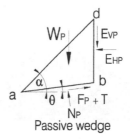

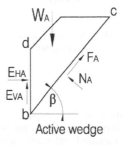

Figure 7. Components of force on active and passive Wedges.

One question that remains is the inclination of the base angle (failure plane, bc) under the active wedge (denoted by β) that gives rise the minimum release energy within the system. Intuitively, using Rankine theory, the angle should be in the range of $45 + \varphi_m/2$. To test this intuition, a simple finite element analysis has been performed as shown in the following section.

6 FINITE ELEMENT ASSESSMENT

Figure 8 shows the results of a finite element analysis using Phase[2].

The figure shows the zones of maximum shear strain (light coloured diagonal zones) that demarcate the development of failure planes within the stockpile parent material. The mass is shown divided into the passive wedge (Sector 1: lower left), the active wedge (Sector 2: upper material) and the stationary wedge (Sector 3: bottom centre). This example is slightly more complicated than the two wedge LE translational analysis given in Section 5 since the wedges form to the left and right in symmetrical format, but with a little imagination the applicability of the 2 wedge analogy is clearly evident.

The stockpile material friction angle is 37° and that of the interface is 10° in this example. In this case, again with reference to the annotations in Figure 6, the diagonal line bd (line of equilibrium) is inclined at 68° while the base of the active wedge above the stationary wedge is inclined at 47° to the horizontal. From a translational analysis calculation

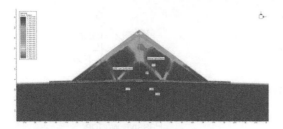

Figure 8.   Maximum shear strains for stockpile.

perspective then a good initial approximation for the base angle would be 50° or $45 + \varphi_i/2$.

7   CONCLUSIONS

The objective of this paper has been to highlight the intrinsic weakness properties of natural and manufactured interfaces and to develop a fundamental understanding of the nature of the physical mechanisms playing out within a structure supported on such interfaces. In particular, an understanding of the development of tension is a liner is shown by the judicious use of a simple limit equilibrium approach that provides a versatile method for calculation not only liner tension but also other aspects that are required in a design.

The lessons learnt from this treatise are the following:

- Fundamental understanding of the behavior of the interface materials are required to produce a competent design.
- Physical shear testing of the actual samples of the liner system including the soils should be carried out.
- Develop a conceptual model that interprets the essence of the problem physically, that is 'define the problem' both fundamentally and numerically.
- Develop a hand/spreadsheet calculation model that describes the mechanics of the problem and test the design variables accordingly.
- Only once the physical mechanics of the design are fundamentally understood, then resort to more sophisticated numerical modeling.

Moreover, it is vital that a coherent design strategy or philosophy is developed for any problem to be solved. This naturally develops from a fundamental familiarity with the mechanics thereof.

REFERENCES

Akber, S.Z., Hammamji, Y. and Lafleuer, J. (1985). Frictional Characteristics of Geomembranes and Geomembrane-Geotextile Composites, *Proc. Second*

Canadian Symposium on Geotextiles and Geomembranes, Edmonton, Alberta, 209–217.

Ali, F., Salman, F.A., Subramanaim (2012). Influence of surface texture on the Interfaces Shear Capacity of Landfill Liner *Electronic Journal of Geotechnical Engineering (EJGE)* Vol 17 2012.

Ali, J., Chen, J.F., Rotter, J.M. and Ooi, J.Y. (2009). Finite Element Prediction of Progressively Formed Conical Stockpiles *Simulia Customer Conference 2009.*

Bhatia, S. and Kasturi, G. (circa 1996). Comparison of PVC and HDPE Geomembranes (Interface Friction Performance), *Research Report, Dept of Civil and Environmental Engineering, Syracuse University*, www. geomembrane.com/userfiles/filemanager/451/

Blond, E. and Elie, G. (2012). Interface Shear-strength Properties of Textured Polyethylene Geomembranes http:// www.solmax.com/wp-content/uploads/2012/08/1_ ShearStrenghtproperties-PEmembrane.pdf.

Effendi, R. (2011). Interface Frciton of Smotth Geomembrane and Ottawa Sand *INFO TEKNIK* Vol 12 No 1 July 2011.

Gourc, J.P. Reyes-Ramirez, R. and Villard, P. (2004). Assessment of Geosynthetics Interface Friction for Slope Barriers of Landfill www.geosynthetica.net/ Uploads/GeoAsia04Gourc.pdf

Howell, G.C. and Kirsten, H.A.D. (1992). Kinematic Indeterminacy as basis for the determination of optimum mining procedures *Rock Support in Mining and Underground Construction*, Kaiser and McCreath (eds) 1992, Balkema.

Independent Expert Investigation and Review Report (2015), Report on Mount Polley tailings Storage Facility Breach, January 2015, www. mountpolleyreviewpanel.ca

Koerner, G.R. and Narejo, D (2005). Direct Shear Database of Geosynthetic-to-Geosynthetic and Geosynthetic-to-Soil Interfaces, *GRI Report # 30, 2005.*

Krahn, J. (2003). The 2001 R.M. Hardy Lecture: The Limits of Limit Equilibrium Analysis" *Can Geotech J* Vol 40 pp. 643–660 (2003).

Lupo, J.F., Morrison, K.F. (2008). Innovative Geosyntheric Liner Design Approaches ad Construction in the Mining Industry *Geo-Frontiers Congress* 2005.

Martin, J.P., Koerner, R.M. and Whitty, J.E. (1984). Experimental Friction Evaluation of Slippage Between Geomembranes, Geotextiles and Soils, Proceedings of the Intl. Conference on Geornernbranes, Denver, Colorado, pp. 191–196.

Narejo, and Allen (2009). Final Report Evaluation of Geosynthetic Clay Liners as a Substitute for Compacted Clay Liners in Large Impoundments, *Report for ConocoPhillips by TRI Environmental, Inc.*, 2009.

PHASE² Finite Element Software for the Analysis of 2D Continua, Rocscience, Canada.

Qian, X. Koerner, R.M. and Gray, D.H. (2002). Ch 13 *Landfill Stability Analysis, in Geotechnical Aspects of Landfill Design and Construction*, Prentice Hall, 2002.

SLIDE Limit Equilibrium Software, Rocscience, Canada.

Vaid, Y.P. and Rinne, N. (1995). Geomembrane Coefficients of Interface Friction *Geosynthetics International*, Vol2 No 1 pp. 309–325.

Williams, N.D. and Houlihan, M.F. (1987), Evaluation of Interface Frictional Properties Between Geosynthetics and Soil, *Proc. Geosynthetics '87*, New Orleans, USA, pp. 616–627.

Proceedings of the first Southern African Geotechnical Conference – Jacobsz (Ed.)
© *2016 Taylor & Francis Group, London, ISBN 978-1-138-02971-2*

Determination of steady state lines for non-plastic platinum tailings

J.C. Hemer, N.L.M. Mincione & L.A. Torres-Cruz
University of the Witwatersrand, Johannesburg, South Africa

ABSTRACT: The Steady State Line (SSL) is essential to characterizing the mechanical response of soils. Previous works have correlated the SSL to Grain Size Distribution (GSD) descriptors such as Fines Content (FC) or coefficient of uniformity (C_u). However, such correlations have limited applicability because GSD descriptors do not account for particle shape which is also known to significantly affect the SSL. Conversely, the maximum and minimum void ratios (e_{max} and e_{min}) reflect both the GSD of a soil and its particle shapes. Accordingly, the current paper proposes that e_{max} and e_{min} can be correlated to the SSL intercept (Γ_1) of non-plastics soils regardless of GSD (including FC) or particle angularity. The SSLs of three soil types covering the range of e_{max} and e_{min} of a platinum tailings deposit were defined using one drained and one undrained triaxial test. The results suggest an approximately linear correlation between e_{max}, e_{min} and Γ_1.

1 INTRODUCTION

It is generally acknowledged that the Steady State Line (SSL) represents the ultimate conditions of mean effective stress (p'), deviator stress (q) and void ratio (e) that a soil attains after being sheared to large deformations (Schofield & Wroth 1968). Given that it represents an ultimate state towards which a soil moves when sheared, the SSL is central to understanding mechanical response. Although some authors prefer to use the term Critical State Line (CSL), it will be assumed henceforth that the SSL and CSL are one and the same. This assumption is based on previously reported experimental results (e.g. Been et al. 1991, Verdugo & Ishihara 1996).

With the objective of gaining a better understanding of its current state, the current authors took on the task of characterising a platinum tailings dam from a SSL standpoint. It is known that the multiplicity of soil types typically present in a single tailings deposit means that it is not possible to use a single SSL to characterise the entire deposit (Fourie & Papageorgiou 2001). Accordingly, the variety of soil types present in the tailings dam studied by the current authors, means that multiple SSLs have to be experimentally determined. The time and cost resources needed for the rigorous determination of each SSL are significant. This is because usually, the testing programme required to determine one SSL, involves a series of Consolidated Drained (CD) and Consolidated Undrained (CU) triaxial tests (e.g. Been et al. 1991, Verdugo & Ishihara 1996). Each test, yields one set of steady state values namely, p' = (σ'_1 + 2σ'_3)/3,

q = (σ'_1–σ'_3) and e, which together define one point of the SSL. A reduction of these times and costs can be achieved with a clear understanding of how the SSL is correlated to easily measured soil index properties.

The purpose of the current paper is to present preliminary results that suggest that the location of the SSL in e-p' space can be correlated to the maximum and minimum, index void ratios (e_{max} and e_{min}, respectively) provided that the soil is non-plastic. Importantly, the correlations appear to be independent of non-plastic Fines Content (FC). This independence from FC contradicts previous works which have attempted to explain the location of the SSL in e–p' space in terms of FC.

2 SOIL PROPERTIES THAT AFFECT THE E—P' PROJECTION OF THE SSL

The three dimensional nature of the SSL is generally represented using its projections on e–p' and q–p' space. The scope of this work is limited to the e–p' projection. It is common to assume that if p' is plotted on a logarithmic scale, the SSL can be represented by a straight line modelled with Equation 1:

$$e = \Gamma_1 - \lambda \cdot \log_{10}(p')$$ (1)

where Γ_1 is the steady state void ratio at p' = 1 kPa, and λ is the slope of the SSL.

Both parameters, Γ_1 and λ, have been correlated to index soil properties. For instance, Been &

Jefferies (1985) and Bouckovalas et al. (2003) suggested a trend of direct correlation between FC and λ. However this trend has not been observed in other works (Zlatovic & Ishihara 1995, Thevanayagam et al. 2002, Carrera et al. 2011, among others) in which mixes of sands with different amounts of non-plastic fines have yielded SSLs that are roughly parallel, and with λ appearing unrelated to FC. Thevanayagam et al. (2002) and Rahman & Lo (2008) proposed that variations in Γ_1 could be explained by variations in FC. Specifically, they suggested that as the FC of a sand increases from zero, the SSL will initially shift downwards in e–p′ space Until a Threshold FC (TFC) value is reached. Thevanayagam et al. (2002) and Rahman & Lo (2008) hypothesised that the TFC represents the FC value at which the soil's behaviour transitions from being dominated by contacts between coarse particles, to being dominated by contacts between fine particles. Application of the TFC framework to tailings, however, is limited because the framework can only predict the SSL of soil types with FC < TFC. Rahman & Lo (2008) report that TFC is generally around 40%, however the FC of tailings can easily exceed this value (Fourie & Papageorgiou 2001).

Other authors have explored the correlation between other GSD descriptors (besides FC) and the SSL. Poulos et al. (1985) and Li (2013) reported inverse relationships between the coefficient of uniformity ($C_u = D_{60} / D_{10}$) and Γ_1. Although insightful, the use of this result is limited because the Γ_1–C_u correlation is dependent on particle shape (Li 2013) and also because, two soils with identical C_u can still have different GSDs and hence, different SSLs (Li 2011). In other words, it is not possible to establish a generally applicable correlation between Γ_1 and C_u. Similarly, Muir-Wood & Maeda (2008) proposed using the ratio between the maximum and minimum particle size to characterise GSD curves that exhibited a normal distribution around a mean particle size. They termed this ratio R_D and conducted 2D Discrete Element Method (DEM) simulations of assemblies of discs showing that there is an inverse relationship between R_D and Γ_1 and also between R_D and λ. Again, although these results are insightful, it is not clear how the Γ_1–R_D and λ–R_D correlations would be affected by changes in particle shape.

Cho et al. (2006) studied the effect of particle shape on the SSL of sands with little or no fines and with mostly narrow GSDs ($C_u \leq 4$). Their results show that particle regularity (similarity to a sphere) was inversely correlated to Γ_1 and that it also had a weak effect on λ. Results obtained by other authors (Poulos et al. 1985, Cubrinovski & Ishihara 2000, Li 2013, Yang & Luo 2015, among others) confirm that particle shape has a significant and systematic effect on the SSL.

The reviewed literature makes it clear that it is not possible to establish a unique correlation between the SSL and any GSD descriptor (e.g. FC, C_u) because such a correlation cannot account for particle shape effects. Conversely, it is well known that the e_{max} and e_{min} of a soil reflect both its GSD and its particle shapes (Biarez & Hicher 1994, Li 2013). Accordingly, attempts to correlate SSL parameters with e_{max} and e_{min} seem logically sound. Such attempts have been made by Cubrinovski & Ishihara (2000) and Cho et al. (2006). Both of these works found significant correlations between the SSL and limiting void ratios. Nonetheless, in both works the focus was on sands with limited amounts of fines.

The hypothesis explored in the current paper is that e_{max} and e_{min} can be correlated to Γ_1 regardless of FC. This hypothesis is based on the fact that the classification of a particle as fine or coarse is largely arbitrary. Most geotechnical practitioners use 75 μm or 63 μm as a coarse-fine boundary, but it is known that there are no significant changes in the nature of soil particles across this boundary (Vermeulen 2001, Been et al. 2012). This is particularly true of non-plastic soils in which the greater specific surface of the fine particles is not expected to significantly affect mechanical response. So, as noted by Been et al. (2012), it is a *"dubious assumption that fines content is a good proxy for material behaviour"*.

3 MATERIALS AND METHODS

3.1 *Materials tested*

The tailings tested were obtained from the Impala Platinum tailings dam located near Rustenburg, South Africa. Preliminary index testing indicated that the soil types present in the dam had values of e_{min} ranging from 0.52 to 0.65 and values of e_{max} ranging from 0.90 to 1.18. To determine e_{min}, the current research called for a non-standard procedure from which reliable results could be achieved at least expense of soil material. Therefore one of the two non-standard methods described in detail in MacRobert & Torres-Cruz (2016) was used. In this method, a mould with a known volume is fitted with an extension collar and filled in 40 layers of dry soil of about 28 g each. The mould is tapped 8 times with a rubber mallet after depositing each layer. After tapping the last layer, an overburden surcharge of 13.8 kPa is applied to the specimen which is tapped eight more times. The surcharge and extension collar are then removed and the specimen is trimmed even with the upper rim of the mould. The mass of the specimen, the volume of the mould and the particle density (ρ_s) are then used to determine e_{min}. Determination of e_{max} was done as per ASTM D4254.

With these in situ ranges of e_{min} and e_{max} in mind, three laboratory mixes of platinum tailings were prepared so that their values of e_{max} and e_{min} covered and exceeded the range of values observed in situ. The GSD curves of the three mixes are shown in Figure 1, and additional index parameters are presented in Table 1. The code assigned to each mix is composed of two numbers separated by a forward slash. The first is the FC ($< 75\ \mu m$) expressed as a percentage, and the second number is the e_{min} multiplied by 100. The code is intended to emphasise the view of the authors that a GSD descriptor alone (e.g. FC) is not a good predictor of mechanical response.

The fines were found to be non-plastic and as such it was anticipated that FC would not largely influence the location of the SSL. However, to maximise representability, it was found convenient by the authors that the FC of the tailings mixes tested in the laboratory covered a range similar to the range of in situ FC values. This range was estimated from the results of a site exploration programme that included GSD determinations and which was carried out by an engineering firm investigating the dam. Based on the GSD of 49 samples, collected mostly from different positions along the beach of the dam, FC was found to vary

between 33% and 95%. Accordingly, the range of FC values covered by the tested mixes (i.e. 10% to 81%, Table 1) appeared adequate to investigate the range of SSLs present in the dam. Although the mix 10/70 has a FC significantly lower than that of the coarsest material present in the dam, this mix was used because its high e_{min} value expanded the range of limiting void ratios covered by the testing programme. The wider range is desirable when trying to identify correlations between SSL parameters and e_{max} or e_{min}.

3.2 Triaxial testing programme

Noting that a unique SSL is defined by drained or undrained triaxial tests, Jefferies & Been (2006) pointed out the convenience of using both types of tests to determine the SSL. They recommended using drained tests to obtain SSL points beyond p′ values of 200 kPa, and undrained tests to obtain SSL points below this p′ value. Therefore, the triaxial programme presented herein relies on a combination of Consolidated Drained (CD) and Consolidated Undrained (CU) tests to make a preliminary determination of the SSLs for each soil type. As such, two triaxial tests, one CD and one CU, were carried out for each mix. The authors wish to emphasise the preliminary nature of these SSLs. The reader must be aware that rigorous SSL determinations require a greater number of triaxial tests for each soil type (e.g. Zlatovic & Ishihara 1995, Thevanayagam et al. 2002, Carrera et al. 2011, Li 2013). Nonetheless, the authors believe that these preliminary SSLs are adequate to explore how Γ_1 correlates to e_{max} and e_{min}.

All tests followed the general outline presented in ASTM D4767–04 for the CU triaxial compression testing of cohesive soils. A few modifications were introduced however and will be described herein. Specimens were formed using the moist tamping method. Each specimen was tamped into a split mould in six layers of equal mass and height to achieve a uniform void ratio. The moist tamping method is often adopted in triaxial testing for SSL determinations due to its good control over the preparation void ratio (Been & Jefferies 1985, Verdugo & Ishihara 1996, Fourie & Papageorgiou 2001). Each layer was pre-mixed with distilled water (at water contents of 7% to 11%) and allowed to cure for at least 4 hours. The membrane was placed after the specimen was formed in the split mould to prevent accidental damage of the membrane during tamping. The split mould had an internal diameter of 70 mm and a height of 141 mm.

Specimen saturation was aided with Carbon Dioxide (CO_2) treatment. After being flushed with CO_2, the specimens were flushed with 2 litres of

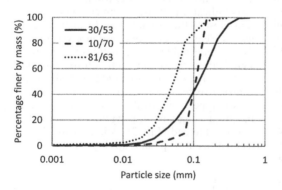

Figure 1. Grain size distribution curves of the soil types tested.

Table 1. Material parameters of soil types tested.

Soil types	D_{60}/D_{10}	FC %	e_{min}	e_{max}	ρ_s* g / cm³
10/70	1.6	10	0.70	1.08	3.51
30/53	4.5	30	0.53	0.90	3.43
81/63	2.7	81	0.63	1.12	3.59

*ASTM D854.

de-aired water and saturation proceeded using back-pressures until the Skempton B parameter was greater than 0.96. All specimens achieved B > 0.96 after 24 hours of being subjected to a back-pressure of 220 kPa. Once saturation was complete, specimens were consolidated isotropically. The excess pore water pressure (pwp) dissipated within the first few minutes, however secondary consolidation was allowed to continue for at least 24 hours.

To measure the post consolidated height of the specimen (H_{pc}), the outstanding height of the loading rod (i.e. the height outside of the cell when the rod is in imminent contact with the top-cap) was correlated to the height of the specimen. Measurement of this outstanding height after the consolidation of each specimen allowed for a simple way of estimating H_{pc}.

Accurate determination of the void ratio of the specimens at the end of the triaxial test is a key step towards accurately determining the SSL. This measurement was done herein following the general guidelines of the method described in detail by Verdugo & Ishihara (1996). In this method, the water content of the specimen at the end of shearing is used to calculate the volume of voids of the specimen. That is, given that the specimen is saturated, the volume of water is taken as equal to the volume of voids. The volume of solids can also be readily obtained from the mass of solids in the specimen and the particle density (ρ_s, Table 1). This method of void ratio determination was followed albeit with one modification. The original method recommends using a spatula to remove the particles attached to both the membrane and porous stones. However the current authors found it difficult to do this using only a spatula and no water. Instead, the bulk of the specimen (with its remaining water) was removed from the membrane into a first drying bowl. Next, the remaining particles that clung to the membrane and porous stones were washed out into a second and separate drying bowl. The remaining water content (ω_r) was calculated only from the portion of the specimen contained in the first bowl while the total mass of specimen solids (m_s) was calculated using the oven dried mass of solids contained in both the first and second bowl. The void ratio (e) could then be computed as:

$$e = \left(\frac{V_{sq} + \omega_r \cdot m_s}{m_s / \rho_s} \right) \qquad (2)$$

where V_{sq} is the volume of water squeezed out of the specimen before disassembling the triaxial cell.

To calculate the deviator stress, the load on the rod was divided by the cross sectional area (A) of the specimen. Calculation of A must account for the deformations that the specimen undergoes during shearing. As the specimens tended to deform in a barrel shape wherein the cross section was greater towards its mid-height, the correction suggested by the French Standard NF P 94-074 to account for this barrel-like geometry was adopted herein:

$$A = \left(\frac{V_c - \Delta V_i}{H_{pc} - \Delta H} \right) \left(1 + \frac{\Delta H}{2 H_{pc}} \right) \qquad (3)$$

where V_c is the post-consolidation volume of the specimen, ΔV_i is the specimen volume change during shearing ($\Delta V_i = 0$ for undrained tests), and ΔH is the change in specimen height during shearing.

The adopted shearing rate was of 4.4% axial strain per hour. All tests were conducted in a temperature controlled environment as preliminary tests showed that thermal expansion and contraction of water could affect the pwp readings of undrained tests. All data acquisition was automated using the suitable transducers connected to a data logger.

4 RESULTS

Figure 2 presents the six resulting p'–ε_a plots. Additionally, Figure 3 presents the e–ε_a plots for the CD tests and the pwp–ε_a plots for the CU tests. The stability of the readings towards the end of the tests indicates that the specimens had reached or were close to their steady state. In all but one of the tests a clearly contractive response was observed. This is reflected by the reductions in void ratio or significant increase in pwp that took place during the tests. The only test that did not exhibit a clearly contractive response was the CU test conducted on soil type 30/53. In this test, p' dropped during the initial stages of shearing but recovered slightly towards the end (Fig. 2). Additionally, there was no significant pwp build up (Fig. 3).

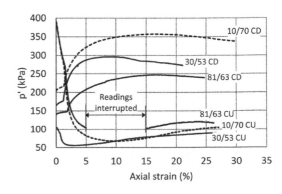

Figure 2. Resulting plots of p'–ε_a for each triaxial test.

The steady state values of e and p′ can be read off Figures 2–3 to plot the preliminary SSLs of each soil type. In the case of the CU tests, the void ratio remains constant throughout the test and the steady state void ratio is simply the post-consolidation void ratio which is measured at the end of the test. Table 2 presents the steady state values of e and p′ for each test and the preliminary estimation of Γ_1 for each soil type. Preliminary values of λ can also be readily calculated from the data in Table 2. However, given that λ is particularly sensitive to the results of additional triaxial tests, the authors have preferred not to explicitly report any λ values. The preliminary SSLs (Fig. 4) are in agreement with the observation made by Fourie & Papageorgiou (2001) regarding the variety of SSLs that can exist in a single tailings deposit. That is, it is not realistic to characterise the entire tailings dam with a single SSL. Furthermore, the preliminary SSLs appear to be roughly parallel, which is consistent with results reported by others (e.g. Zlatovic & Ishihara 1995, Thevanayagam et al. 2002, Carrera et al. 2011).

Values of Γ_1 (Table 2) are plotted against e_{max} and e_{min} in Figure 5. To allow comparison, the data from Cubrinovski & Ishihara (2000) and its best fit lines has also been included in Figure 5. The data of Cubrinovski & Ishihara (2000) corresponds to 52 sandy soils whose non-plastic FC varies from 0% to

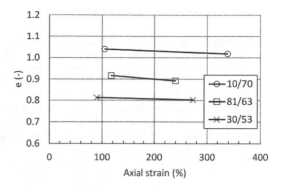

Figure 4. Preliminary SSLs of the three soil types tested.

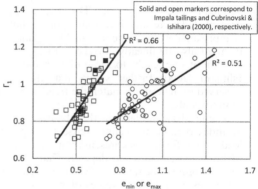

Figure 5. Values of Γ_1 plotted against e_{max} and e_{min} for each of the three soil types tested, and data from Cubrinovski & Ishihara (2000).

20%. The values of Γ_1 corresponding to the Impala tailings are approximately linearly correlated to e_{min}. This linear correlation is much weaker for the Γ_1 vs e_{max} data corresponding to Impala. However, with only three data points it is not possible to rule out the possibility that such a linear correlation indeed exists. Furthermore, the overall trend of the Impala tailings is the same trend of the sands reported by Cubrinovski & Ishihara (2000). This match in trends occurs even though two of the Impala soil types (i.e. 30/53 and 81/63) had FC values that were significantly higher than those used by Cubrinovski & Ishihara (2000). Accordingly, the results presented herein suggest that the correlation between Γ_1 and e_{max} or e_{min} is applicable to soil types with significant amounts of non-plastic fines. Additionally, although this is not evident from Figure 5, it is of relevance to mention that Cubrinovski & Ishihara (2000) noted that the correlation between the vertical position of the SSL in e-p′ space, and the limiting void ratios was the same regardless of whether the soils had angular or rounded particles.

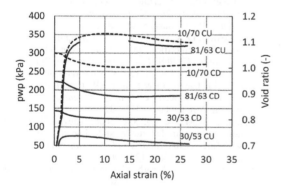

Figure 3. Plots of e–ε_a and pwp–ε_a.

Table 2. Results of triaxial testing programme.

Soil types	Test type	e*	p′* kPa	Γ_1
10/70	CD	1.02	338	1.13
10/70	CU	1.04	105	–
81/63	CD	0.89	239	1.08
81/63	CU	0.92	117	–
30/53	CD	0.80	273	0.86
30/53	CU	0.81	90	–

*Value at steady state.

5 CONCLUDING REMARKS

Preliminary results of a triaxial testing programme on platinum tailings have been presented. The testing programme was conducted to obtain the range of SSLs characteristic of a platinum tailings dam. The tested soil types were chosen in such a way that the results could be used to infer the range of SSLs occurring in the tailings dam. Based on a literature review, it was hypothesised that the limiting void ratios (e_{max} and e_{min}) of non-plastic soil types could be used as predictors of the y-intercept (Γ_1) of a straight SSL in e–log (p') space. The literature review also suggested that the SSLs of platinum tailings with different GSDs and non-plastic FC can be expected to be roughly parallel.

Three preliminary SSLs, each of which was defined by one CU and one CD triaxial test, were presented. The results show that Γ_1 of the tailings is approximately linearly correlated to e_{min} and that it might also be linearly correlated to e_{max}. Comparison of our results to those of Cubrinovski & Ishihara (2000) further shows that the trends observed in the tailings match the trends observed in sandy soils.

The results also highlight that the use of FC as an index parameter to characterise soil types can lead to significant misrepresentations of the range of SSLs present in a soil deposit. That is, notice that in Figure 4 the range of SSLs indicated by the soil types with extreme values of e_{min} (10/70 and 30/53) is wider than the range of SSLs indicated by the soil types with extreme values of FC (10/70 and 81/63).

The research group of the authors continues to conduct additional triaxial testing to obtain better defined SSLs and to analyse an expanded database to further explore the correlations between the SSL and the limiting void ratios.

ACKNOWLEDGEMENTS

The authors gratefully acknowledge Impala Platinum Ltd. and SRK Consulting for providing samples and information that made this work possible. We also sincerely thank Prof Heyman (University of Pretoria) and Soillab for their technical support with the triaxial testing. Lastly, the first and second authors gratefully acknowledge the guidance of Mr Torres-Cruz in the preparation of this paper and the use of his PhD findings.

REFERENCES

Been, K. & Jefferies, M.G. 1985. A state parameter for sands. *Géotechnique* 35(2): 99–112.

Been, K., Jefferies, M.G. & Hachey, J. 1991. The critical state of sands. *Géotechnique* 41(3): 365–381.

Been, K., Obermeyer, J., Parks, J. & Quinonez, A. 2012. *Post-liquefaction undrained shear strength of sandy silts and silty sand tailings.* Colorado: Proceedings tailings and mine waste.

Biarez, J. & Hicher P.Y. 1994. *Elementary mechanics of soil behaviour.* London: Balkema.

Bouckovalas, G.D., Andrianopoulos, K.I. & Papadimitriou, A.G. 2003. A critical state interpretation for the cyclic liquefaction resistance of silty sands. *Soil dynamics and earthquake engineering* 23(2): 115–125.

Carrera, A., Coop, M., & Lancellotta, R. 2011. Influence of grading on the mechanical behaviour of Stava tailings. *Géotechnique* 61(11): 935–946.

Cho, G.C., Dodds, J. & Santamarina, C.J. 2006. Particle shape effects on packing density, stiffness and strength: natural and crushed sands. *Journal of geotechnical and geoenvironmental engineering* 132(5): 591–600.

Cubrinovski, M. & Ishihara, K. 2000. Flow potential of sandy soils with different grain compositions. *Soils and foundations* 40(4): 103–119.

Fourie, A.B. & Papageorgiou, G. 2001. Defining an appropriate steady state line for Merriespruit gold tailings. *Canadian geotechnical journal* 38: 695–706.

Li, G. 2013. Étude de l'influence de l'é talement granulometrique sur le comportement mecanique des materiaux granulaires. PhD thesis. àl'Ecole Centrale de Nantes.

Li, S.L. 2011. *Effects of particle gradation on static liquefaction behaviour of sands:* Harbin Institute of Technology.

MacRobert, C.J & Torres-Cruz, L.A. (2016). Evaluation of methods to determine reference void ratios. *In Proceedings of the First South African Geotechnical Conference.* Rustenburg. South Africa.

Muir-Wood, D. & Maeda, K. 2008. Changing grading of soil: effect on critical states. *Acta geotech* 3(1): 3–14.

Papageorgiou, G. 2001. *Liquefaction of tailings.* PhD thesis. Johannesburg: The University of the Witwatersrand.

Poulos, S.J., Castro, G. & France, J.W. 1985. Liquefaction evaluation procedure. *Journal of geotechnical engineering* ASCE 111(6): 772–792.

Rahman, M.M., & Lo, S.R. 2008. The prediction of equivalent granular steady state line of loose sand with fines. *Geomechanics and geoengineering: An international journal* 3(3): 179–190.

Schofield, A.N. & Wroth, C.P. 1968. *Critical state soil mechanics.* London: McGraw-Hill.

Thevanayagam, S., Shenthan, T., Mohan, S. & Liang, J. 2002. Undrained fragility of clean sands, silty Sands, and sandy silts. *Journal of geotechnical and geoenvironmental engineering* 128(10): 849–859.

Verdugo, R. & Ishihara, K. 1996. The steady state of sandy soils. *Soils and foundations* 32(2): 81–91.

Vermeulen, NJ. 2001. *The composition and state of gold tailings.* PhD Thesis. Hatfield: University of Pretoria.

Yang, J., & Luo, X. 2015. Exploring the relationship between critical state and particle shape for granular materials. *Journal of the mechanics and physics of solids.*

Zlatovic, S. & Ishihara, K. 1995. On the influence of nonplastic fines on residual strength. *Proceedings of IS-TOKYO'95: The first international conference on earthquake geotechnical engineering:* 239–244.

Proceedings of the first Southern African Geotechnical Conference – Jacobsz (Ed.)
© *2016 Taylor & Francis Group, London, ISBN 978-1-138-02971-2*

Assessment of reliability of the hydrometer by examination of sediment

P.R. Stott, P.K. Monye & E. Theron
Central University of Technology, Bloemfontein, South Africa

ABSTRACT: A fundamental aspect of the characterization of any soil is the assessment of its particle size distribution. While this is relatively easy for the coarse fraction it remains problematic for soil fines particularly for the fraction less than 2µm. Hydrometer analysis has been the standard tool for fines assessment for many years but there may be serious shortcomings. Nettleship et al. (1997), Savage (2007), Rodrigues et al. (2011) and many others have pointed to a number of problems facing the hydrometer. Some of the questions have been addressed by laser scattering techniques e.g. Eshel et al. (2004), but others, including completeness of dispersion and the amount of clay carried down with coarser fractions remain problematic. This investigation assesses some aspects of the reliability of the hydrometer by isolating and testing the sand and silt fractions after settlement. Microscopic examination is used to compare the composition of sediment layers with that expected according to hydrometer theory.

1 INTRODUCTION

Hydrometer analysis is widely used internationally for determination of the clay and silt fraction of soils in both engineering and soil science practice. The standard procedure used in South Africa for Civil Engineering purposes is detailed in SANS 3001 GR3 (SABS 2012). The test has a number of theoretical weaknesses. It monitors the change in density of a settling suspension—theoretically at one level in the suspension. It has been pointed out, however, (e.g. Rolfe et al. 1960) that it averages the specific gravity over the submerged part of the instrument and therefore depends on the shape and depth of hydrometer submergence, which may bring errors into the analysis.

Others (e.g. Savage 2007) have noted that hydrometer analysis relies on Stokes Law (which assumes all particles to be spherical); it assumes that all soil particles have the same density; it assumes complete dispersion of clay particles at the time of testing; and it assumes that fine particles are not carried down by coarse particles—all of which assumptions are dubious. Nettleship et al. (1997) suggested that hydrolysis of polyphosphate dispersants may cause underestimation of clay fraction. Rodrigues et al. (2011) proposed that soil mineralogy should be taken into account and probably different treatments and different dispersants are required for different clay types. In addition Keller and Gee (2006) noted that hydraulic soil property estimates work best for coarse-textured, structureless soils of low clay content.

One of the most problematic aspects of South African soils (from an insurance claims point of view, from the point of view of frustration to the government's attempts to provide durable low-cost housing to the poor, and from the point of view of several aspects of the performance of roads, water supply conduits, etc.), is the behaviour of expansive clay.

Most of the methods for predicting the severity of heave which may be expected on a site underlain by fine-textured soil rely on an estimate of clay fraction. As noted above by Keller and Gee, clay is the soil material which is least amenable to hydraulic methods of analysis, and hydrometer results can be expected to be particularly unreliable for these most problematic soils.

Critical aspects of soil behavior depend on clay mineral content rather than particle size as given by the hydrometer test. It is assumed that clay particles range from 2 microns downward, and that particles larger than 2 microns are silt. There are, however, some clays whose particles can be considerably larger than 2 microns (e.g. kaolinite, illite and halloysite) also silt particle sizes may range down to 1 micron. The range of sizes between approximately 1 and 2 microns may therefore be mineralogically either clay or silt.

Stott and Theron (2016) examined suspensions after preparation for hydrometer testing following the procedures of SANS 3001 GR3. The suspensions were examined using a light microscope/digital camera combination. They found three aspects of incomplete dispersion which suggested that unreliable results might be obtained by hydrometer analysis. The first was adhesion of clay to larger-sized particles, the second was agglomeration of clay particles among themselves and the third was agglomeration of clay and silt particles into groups of appreciable size. They speculated that these agglomerations would not precipitate at the rate expected for clay and would therefore give misleading results in hydrometer analysis.

The objective of the investigation here described is to evaluate the hydrometer test for assessing typical South African clayey soils in light of the above objections.

2 METHODS AND MATERIALS

2.1 Soil samples and treatments

A range of plastic clays from typical roads and housing projects were tested. Samples were treated with Sodium Hexametaphosphate and Sodium Carbonate as per SANS 3001-GR3. For control purposes examination of untreated samples were considered to serve as a check on the effectiveness of the dispersant for each soil. Preliminary tests, however, showed that without treatment in dispersant, agglomerations are so large that meaningful information would not be obtained.

Samples were prepared in accordance with SANS 3001-GR3:2012. After being soaked in dispersant solution for not less than the prescribed minimum of 16 hours they were mechanically stirred at 1570 rpm for a minimum of 15 minutes and transferred to a settlement container specifically designed for this investigation. Distilled water was added to reach the dilution indicated in SANS 3001 GR3.

Samples were left to settle for approximately two days, excess water was carefully removed by a suction pipe. Containers were then transferred to an oven and dried at 45°C until the sediment was solid enough to be removed from the containers. The various fractions, which were usually visually distinct, were separated. SANS 3001 GR3 is concerned primarily with determining sand and silt fraction and the last reading which is given as normal is 12 minutes. The sub 2μm fraction can be deduced by subtraction. An optionl final reading at 12h is mentioned if a reading is required for the sub 2μm fraction. The clay topping was treated simply as a protective layer to shield the silt layers from disturbance while removing water from the container and was not examined. The separated layers were prepared for microscopic examination by the addition of de-ionised water.

2.2 Settlement container

A settlement container was designed to be waterproof while having one side removable for the extraction of settled samples. Figure 1 shows two of these containers, one before closure, the other with water seal and front cover in place.

2.3 Microscopic examination

An optical microscope with 10x, 40x, 60x and 100x objectives was used for this investigation. The microscope was equipped with a 9 mega-pixel digital camera.

The combined optical and digital magnification was assessed by measurements on a micro ruler. Magnification can be expressed in various ways. For example the spacing between 10 micron graduations on the micro ruler was 60 mm on the viewing screen when using the 60x objective, indicating a magnification of 6000 times. Alternatively the magnification can be expressed as the number of pixels per micron (22.48 for the same objective). The most useful way of indicating magnification is by the inclusion of a scale on photographic results. All photographs shown here have a scale rectangle 30μm × 2μm.

After removal of sediment from the container and separation of the visible layers, small samples of each layer were suspended in de-ionised water. A drop of that suspension was placed on a meticulously cleaned microscope slide and covered with a similarly cleaned cover slip. Photographs were taken of samples before and after addition of methylene blue.

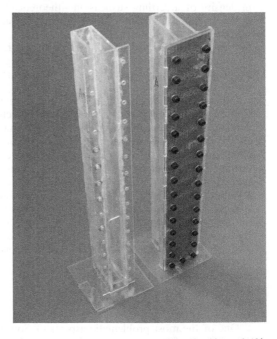

Figure 1.   Settlement containers: 430 × 50 × 20mm: 0.43ℓ.

2.4 Methylene blue

Methylene blue—C16H18 N3SCl—(MB) has been found to be effective in labeling clay minerals by replacement of exchangeable cations. Clay minerals have Cation Exchange Capacity (CEC) which depends on their atomic structure and Specific Surface Area (SSA). The more active clays, like the smectites, have high CEC and SSA. Such clays exchange cations with MB readily and become stained easily. Less active clays like kaolinite and halloysite have low CEC and SSA and require higher concentration of MB to become visibly coloured.

MB staining techniques have been found useful in evaluating possible clay activity and estimation of CEC. Procedures are specified by the *Association Française de Normalization* (AFNOR) and by the American Society of Testing Materials (ASTM). Both standards describe procedures for obtaining a semi-quantitative evaluation of the activity of a soil and indications of the probable type of clay minerals contained in the soil. AFNOR (1993) derives the *"valeur de bleu"* (V_B) and ASTM (1984) derives a comparable "methylene blue index" (MBI).

Chiappone et al. (2004) came to the conclusion that both procedures can give good estimates of clay content in certain circumstances.

For the purpose of this investigation a MB solution with a concentration of $0.5g/\ell$ was used for staining purposes. After examining unstained samples, small quantities of this MB solution were progressively added to the soil suspensions and samples were examined microscopically after each addition.

3 PHOTOGRAPHS AND FINDINGS

3.1 Sand layer

The sand layers were expected to contain small quantities of silt and clay present at the lower part of the containers at the start of precipitation. This expectation was met, as can be seen in Figure 2, a photograph of material from the sand layer from an active clayey soil from Bloemfontein. A small quantity of Methylene Blue has been added. A scale-rectangle of 30×2 microns has been included in all of the microscope images. Some photographs show conditions before, others after, addition of MB.

A few silt and clay particles are visible in Figure2. The silt-size particle within the white circle is completely covered by particles of approximately 2 microns which show very little Methylene Blue staining and are probably kaolinite. Almost all of the sand particles have a thin covering of extremely

small, deeply blue-stained particles which are probably montmorillonite. The two sand grains at upper left appear to be joined by a solid bridge of these clay particles (within the white rectangle) and at the lower left there is a silt-size arc (within the white pentagon) which appears to be entirely made up of the same minute clay particles.

Figure 3 is a view of a grain of coarse sand from the lowest deposition layer of a clayey soil from a roads project in Thaba Nchu.

The sand grains visible in this photograph show fuzzy blue-stained edges consisting of very small, deeply stained clay particles (probably montmorillonite). The larger sand grain has a marked projection of these particles near the upper right corner and a more tenuous projection on the upper left side.

Figure 4 is from the sand layer of a soil from a housing project in Kimberley. The hydrometer analysis performed as part of the geotechnical investigation for the project indicated very little clay. Heave damage, however, indicated significant clay.

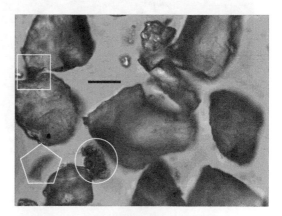

Figure 2. Sand layer from Bloemfontein soil.

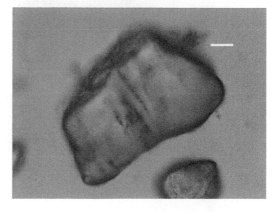

Figure 3. Sand layer from Thaba Nchu soil.

The agglomeration shown consists of silt particles of various sizes. Each silt particle has a coating of deeply blue-stained fine clay particles, and it appears to be these clay particles which hold the silt together in a large, sand-sized agglomeration. Since it is found in the sand layer it seems probable that it precipitated out of suspension at a rate suitable for a sand grain, rather than at the rate expected for silt.

Figure 5 is from the same sand layer of the Kimberley soil shown in figure 4. The structure visible appears to be an agglomeration, possibly of silt, (possibly also sand) or possibly part of a root-hair, within a mass of particles with size close to 2μm. These show only slight staining from the very small quantity of Methylene Blue added and are probably low-CEC clay. At top right and along the left side smaller, deeper stained particles are visible.

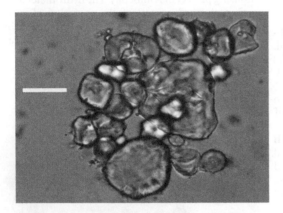

Figure 4.   Sand layer from Kimberley soil.

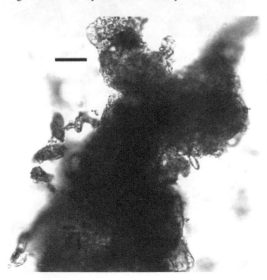

Figure 5.   Sand layer of Kimberley soil.

They are high-CEC / high SSA clay. The size of this agglomeration is that of a large sand grain, and it appears to have precipitated with the coarse sand particles. Little, if any of it appears to be sand. A significant fraction is clay.

One of the clays (from a Bloemfontein housing project) showed good dispersion of clay from the sand particles. Figure 6 shows grains from the coarse sand layer.

Very few clay particles appear to be attached to the sand grains at the right and centre. The sand grain on the left has a partial coating of very small clay particles visible as a fuzzy covering, particularly at the top. Dispersion for the sand grains of this soil appears to be more successful than for the other soils examined.

3.2   Coarse silt layer

Figure 7 shows a photograph from the coarse silt layer of the soil shown in Figure 6.

The structures visible are of coarse-silt size, but they appear to contain fine silt sized particles bound together and covered by clay particles. A similar situation is shown in Figure 8. Where the largest silt particle (which is only partially visible) could be coarse silt, but the agglomeration is of sand size.

The coating obscuring the details of the structure appears to be about half of fine, high CEC/SSA clay and half of larger, low CEC/SSA clay.

Figure 9 is from the coarse silt layer of the Thaba Nchu soil shown in Figure 3.

There are agglomerations of finer material together with the silt. Again the agglomerations tend to be larger than the clean members of the group. This tends to confirm the impression gained from most of the photographs; that

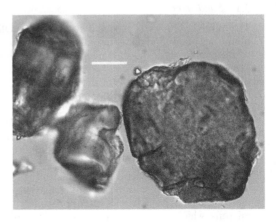

Figure 6.   Sand layer from a Bloemfontein housing project.

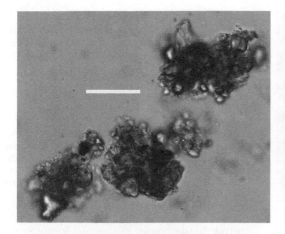

Figure 7.   Coarse silt layer, Bloemfontein soil.

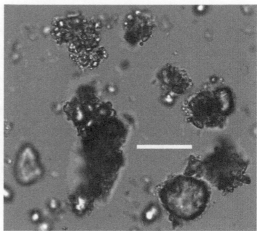

Figure 9.   Coarse silt layer of Thaba Nchu soil.

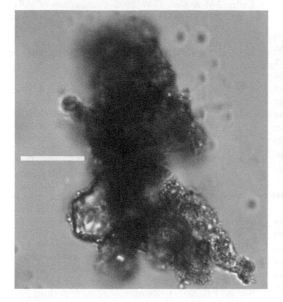

Figure 8.   Coarse slit layer of Bloemfontein soil.

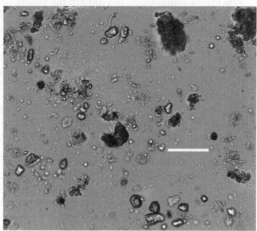

Figure 10.   Fine silt layer of Bloemfontein soil.

agglomerations precipitate more slowly than single particles of similar size.

3.3 *Fine silt layer*

Figure 10 shows the fine silt layer from the Bloemfontein soil of Figures 6,7 and 8. The agglomerations are again larger than the clean silt particles. Some fine silt particles appear to be fully coated with very fine, deeply stained clay particles. Some of the agglomerations appear to be clay particles with no silt core, the largest agglomerations probably do have silt cores. Genuine fine silt particles do not appear to make up the majority of the material present.

Figure 11 is a photograph from the fine silt layer of a soil from Wepener in the Free State. A large part of this layer appears to be made up of agglomerations of clay without silt cores.

Many of these agglomerations appear to be tenuous sting-like structures, in some cases up to $100\mu m$ long. Observation of their movement under the microscope suggests that their bonds, though flexible enough to allow them to bend easily, are strong enough to hold the structures together robustly.

The silt particles present are completely coated by fine, high CEC/SSA clay which is deeply stained by the small quantity of methylene blue added to the suspension. A large proportion of the material appears to be clay.

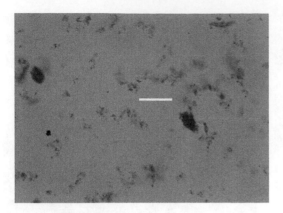

Figure 11.   Fine silt layer: Wepener soil.

4   CONCLUSIONS

The investigation tends to confirm the widely held view that hydrometer analysis is not always reliable, particularly for high clay-content soils. A number of the queries raised by Savage (2007) can be seen to be justified. The photographs generally indicate relatively few free clay particles in the sand layer but a noticeable increase in the silt layers. The fine silt layer has generally more free clay particles than the coarser layers (Figures 10 and 11), suggesting that smaller particles may be more effective in carrying clay particles down with them than large particles.

Almost all sand and silt particles were seen to have at least a few attached clay particles. The coating of clay noted in the sand-layer was generally quite thin in comparison with the size of the sand grains and therefore unlikely to make up a large fraction of the material in this layer. Hydrometer analysis might therefore be expected to give an acceptable assessment of the sand fraction in many cases.

The silt layers, on the other hand were frequently found to be made up of agglomerations of fine silt and clay, or in some cases, of clay alone. These agglomerations were frequently larger than the bare particles of the host layers, suggesting that the agglomerations settle more slowly than single particles of their aggregate size, but more quickly than the individual small particles of which they are made.

In the analysis of high clay content soils it seems likely that the hydrometer will over-estimate the amount of silt to a considerable extent. Consequently the clay content will almost always be under-estimated.

REFERENCES

Chiappone, A. Marello, S. Scavia, C. & Setti, M. 2004. Clay mineral characterization through the methylene blue test: comparison with other experimental techniques and application of the method. *Canadian Geotechnical Journal Dec. 2004; 41, 6; ProQuest Central pp. 1168–1178.*

Eshel, G. Levy, G.J. Mingelgrin, U. & Singer, M.J (2004). Critical evaluation of the use of laser diffraction for particle-size distribution analysis. *Soil Science Society of America Journal: 68:736–743(2004).*

Keller, J.M & Gee, G.W. 2006. Comparison of ASTM and SSA Methods for Particle-Size Analysis. *Soil Science Society of America Journal; Jul/Aug 2006; 70, 4; ProQuest Central pp. 1094–1100.*

Nettleship, I. Cisko, L. & Vallejo, L.E. 1997. Aggregation of clay in the hydrometer test. *Canadian Geotechnical Journal 34:621–626 (1997).*

Rodrigues, C. de Oliveira, V.A. da Silveira, P.M. & Santos, G.G. 2011. Chemical Dispersants and Pre-Treatments to Determine Clay in Soils with Different Mineralogy. *Journal of the Brasilian Society of Soil Science. Vol 35. No.5. 2011, pp.1589–1596, 2011 Brasil. Available online at http://www.redalyc.org.*

Rolfe, B.N. Miller, R.F. & McQueen, I.S. 1960. Dispersion Characteristics of Montmorillonite, Kaolinite, and Illite Clays in Waters of Varying Quality, and Their Control with Phosphate Dispersants. *Geological Survey Professional Paper 334-G. U.S. Government Printing Office In cooperation with Colorado State University, Washington, 1960.*

SABS. 2011. *Civil Engineering Test Methods. Part GR3. South African National Standard SANS 3001:2011 Edition 1.1.* Pretoria: South African Bureau of Standards.

Savage, P.F. 2007. Evaluation of possible swelling potential of soil. *Proceedings of the 26th South African Transport Conference (SATC 2007), 9–12 July 2007.* Pretoria: DTT.

Stott, P.R. & Theron, E. 2016. Shortcomings in the estimation of clay fraction by Hydrometer. *Journal of the South African Institution of Civil Engineering (in print).*

Proceedings of the first Southern African Geotechnical Conference – Jacobsz (Ed.)
© 2016 Taylor & Francis Group, London, ISBN 978-1-138-02971-2

Untreated fine sands and lateritic materials in low volume sealed roads

F. Netterberg
Frank Netterberg, Pretoria, South Africa

M.I. Pinard
Infra Africa Consultants, Gaborone, Botswana

P. Paige-Green
Tshwane University of Technology, Pretoria, South Africa

ABSTRACT: Fine sands and lateritic materials are widespread sources of road construction materials in southern Africa. Although the use of untreated fine sands is usually restricted to the lower layers and untreated lateritic materials seldom meet traditional requirements for base course, both a review of the literature and of experience showed that both such materials have successfully been used as base course. Greater acceptance of the use of such materials will enable the wider provision of more economic low volume sealed roads.

1 INTRODUCTION

The purpose of this note is to raise awareness of ongoing work in southern Africa aimed at the best use of untreated fine sands and untreated lateritic materials (laterite and ferricrete gravels and lateritic soils), especially in low volume sealed roads.

The practical importance of this work is that it will enable the confident and more economical provision of low volume sealed roads in areas where conventional materials are scarce.

These projects have so far included literature and local experience reviews of the distribution, origin, composition, engineering properties and specifications for both lateritic materials and fine sands.

2 LATERITES

In the case of lateritic materials the work has only proceeded as far as the review stage, but has included a comparison of the known material properties of over 30 laterite-based roads in Botswana, Ethiopia, Malawi, Mozambique, South Africa and Zimbabwe with the relaxed Brazilian national specification for laterite gravel bases for roads designed to carry up to five million equivalent standard 80 kN axles (5M E80). These southern African roads varied in age but had nearly all performed satisfactorily and had variously carried between about 13, 000 and 2M E80.

Preliminary findings indicated that, in many cases, lateritic material of even poorer quality in terms of grading, plasticity and/or CBR than that permitted by the Brazilian specification can be used.

Further work envisaged includes the pavement evaluation of and additional laboratory testing of lateritic materials from selected roads in southern Africa and the construction of experimental sections. This work will also establish the necessity or otherwise for certain special tests to establish whether or not a material is a lateritic material before any relaxations in material properties can be permitted.

3 SANDS

Vast areas of southern Africa such as the Kalahari and west and east coasts are covered by fine sands of aeolian and other origins—the Kalahari-type sands extending as far north as the Congo and into the eastern half of South Africa. Many of these sands are thick and hide any gravel and rock sources of conventional materials.

The initial project involved the testing of about 100 samples of sand from roads and borrow pits in Mozambique and Namibia.

Most of the sands tested were fine-grained (0.06–2.00 mm) and were either non- or slightly plastic when tested on the usual fraction passing 0.425 mm. However, Plasticity Indices (PIs) of between about 6 and 35 (PI_{075}) were obtained on the minus 0.075 mm fraction (P075) and appear to be an essential factor in their good performance by providing substantial apparent cohesion and high CBRs when unsaturated.

A comparison of the results of these tests with the Metcalf—Wylde specification for sand bases in Australia suggested that classification in terms of the mean and standard deviation of the particle size (using the phi scale) and the PI of the minus 0.075 mm fraction followed by direct strength testing using the DCP or CBR could be used as the basis of a subbase and base course specification for low volume sealed roads.

Two experimental sections of fine Kalahari-type sands in South Africa and in Botswana—now over 50 and 25 years old respectively and still carrying traffic—have proven the viability of such untreated sand bases for low volume sealed roads. Whilst full details of the Botswana section have not yet been released, it was still in good condition after 0.3M E80 in 20 years.

After 50 years and about 1.0M E80 the South African section was in poor condition, largely due to edge breaking of the narrow 6.0 m-wide seal caused by the six- and seven- axle trucks currently using it. However, shear failures and excessive rutting were absent and the in-situ DCP-CBRs were all in excess of 100. The results of a pavement evaluation including laboratory testing by two different laboratories indicated that a conventional type of specification including, among others, a PI_{075} of SP—6, a $PI_{075} \times P075$ of 20–120, a minimum soaked MAASHO CBR of 50, a minimum CBD-extractable Fe content of 0.3%, and compaction to at least 100% MAASHO should prove adequate as base course for at least 0,1M E80 over the first 20 years.

Critical factors required for the success of such untreated sand bases appear to be the presence of a small amount of clay and/or free iron and/or aluminium hydroxide in the material, together with a high degree of compaction and confinement, and a good seal and drainage.

Further work envisaged includes the construction of longer demonstration sections and the extension of the work to include cement,- lime- and bitumen-treated sands for higher levels of traffic.

ACKNOWLEDGEMENTS

These projects were funded by UKaid from the United Kingdom Department of International Development, carried out under the Africa Community Access Programme (AFCAP), under the auspices of the Association of Southern African National Road Authorities (ASANRA) and initially managed by the Crown Agents. However, the opinions expressed herein are those of the authors and do not necessarily reflect the opinions or policies of the organisations referred to.

REFERENCES

The project reports available to date can be found on www.afcap.org and only publications in the general literature are listed here.

Botswana Roads Department. 2010. *The use of Kgalagadi sands in road construction.* Gaborone: Botswana Roads Department.

Netterberg, F. & Elsmere, D. 2015. Untreated aeolian sand base course for low-volume road proven by 50-year old road experiment. *J.S. Afr. Instn Civil Engng* 57 (2): 50–68.

Paige-Green, P., Pinard, M.I. & Netterberg, F. 2015. A review of specifications for lateritic materials for low volume roads. *Transportation Geotechnics* 5: 86–98.

Paige-Green, P., Pinard, M.I. & Netterberg, F. 2015. Low-volume roads with neat sand bases. *Low-volume roads; Proc. internat. conf., Pittsburgh*, July 2015. Washington D.C: *Transportation Research Record* 2474: 56–62.

Proceedings of the first Southern African Geotechnical Conference – Jacobsz (Ed.)
© 2016 Taylor & Francis Group, London, ISBN 978-1-138-02971-2

Typical strength properties of South African soils

G. Heymann
University of Pretoria, Pretoria, South Africa

ABSTRACT: Design engineers require accurate soil parameters to design geotechnical structures. In recent years the demand for high quality strength parameters has prompted a number of South African commercial soils laboratories to improve and expand their triaxial testing capability both in terms of equipment and personnel. This paper reports typical soil parameters measured during triaxial testing for a large number of projects by the soils laboratory at the University of Pretoria over a period of 13 years. Results presented include natural materials such as gravels, sand, silts and clays as well as man made soils such as tailings from mine waste storage facilities. Guidance is given on the strategy used and interpretation of triaxial data and the typical results that can be expected for these South African soils.

1 INTRODUCTION

In South Africa commercial soil testing has traditionally been conducted by laboratories whose primary focus has been material testing for large pavement and earthworks construction projects. These tests included grading analysis, Atterberg limit tests and compaction tests. Geotechnical tests such as oedometer and shearbox tests have also been conducted by these laboratories but triaxial testing was viewed as a specialist test conducted by university laboratories and a few commercial laboratories. However, during the last two decades triaxial testing has become more widely available in South Africa after much investment in equipment and training of personnel by a number of commercial laboratories.

This paper reports the results form thirteen years of triaxial testing by the soils laboratory at the University of Pretoria. The aim of the laboratory has been to conduct research, teach post graduate geotechnical students, as well as doing limited commercial testing. Tests strategies are discussed and typical soil parameters are reported for a range of South African soils.

This paper reports on 395 triaxial tests conducted with a minimum of three specimens sheared per test. Of these, 272 tests were on disturbed soil and 123 on undisturbed soil. All of the soils were visually described prior to testing and for 105 tests sufficient information was available to classify the soil. For the purpose of this paper the soil is categorised in five categories as shown in Table 1. Tailings material was sampled form mine waste storage facilities from numerous mines which included gold, platinum, iron ore and coal mines.

Table 1. Soil categories.

Soil type	No. of tests
Gravel	21
Sand	128
Silt	56
Clay	105
Tailings	85

2 TEST STRATEGY

Unconsolidated undrained triaxial tests require an undisturbed sample to be taken of a fully saturated soil. This is only practical for clay samples below the water table. However, the geology for large parts of South Africa consists of unsaturated residual soils and therefore unconsolidated undrained triaxial tests are not performed often in South Africa. During the thirteen year period reported in this paper unconsolidated undrained triaxial tests were conducted for only three projects and the data set is deemed to small to include.

All consolidated triaxial tests were consolidated under isotropic stress conditions. For the test reported in this paper 93% were consolidated undrained tests and 7% were consolidated drained tests. Clearly undrained tests are much more popular. The reason is that for an undrained test a stress path can be plotted which contains valuable information with regard to the shear behaviour of the soil. The tendency for the material to contract or dilate may be easily observed from the direction of the stress path. In addition, yielding may be observed as a sharp turn of the stress path. In contrast stress paths for drained tests show little variation and information with regard to the contraction or dilation is contained in the volumetric strains that the specimen undergoes during the shear phase. However, drained tests are preferable for coarse material such as gravel or coarse sand, particularly if these materials are densely compacted. Such materials tend to dilate aggressively which may cause cavitation of the pore fluid leading to early termination of the test.

Critical State Soil Mechanics is a useful framework within which to interpret the behaviour of soil. Figure 1 shows the Normal Consolidation Line (NCL) and swell line which may be quantified during consolidation and swell stages in the triaxial test. Figure 2 shows typical contractive stress paths for a normally consolidated (A) and lightly overconsolidated clay (B) and a dilative stress path for heavily overconsolidated clay (C) sheared undrained.

A carefully planned test strategy may allow valuable information to be acquired without much additional testing effort. For high permeability soils little additional time is required to conduct multiple consolidation stages, swell stages and permeability stages. For the tests reported here multiple consolidation stages were conducted for 33%,

swell stages for 3% and permeability stages for 43% of the tests.

An important part of the test strategy is to select the initial mean effective stresses for each of the triaxial specimens. In general, the range of stresses is chosen to span the range of stresses that the soil currently experiences and will experience after completion of the project under consideration. Figure 3 shows the distribution of the initial mean effective stresses, prior to the onset of the shear stage, used for the tests. It is shown as box and whisker plots indicating the minimum, maximum and median values as well as the 25th and 75th percentile values. It may be seen from the figure that the minimum mean effective stresses for a set of tests ranged between 25 kPa and 200 kPa with a median value of 50 kPa, whereas the maximum mean effective stresses ranged between 100 kPa and 800 kPa. The range of mean effective stresses for a set of specimens varied between 65 kPa and 700 kPa.

3 NON-LINEAR FAILURE ENVELOPE

Figure 2 shows the critical state framework for soils. It is a useful reference within which to interpret the behaviour of soil but it makes some simplifications which may not be appropriate for some soils. It suggests that normally consolidated (A) or lightly overconsolidated clays (B) will reach the Roscoe surface before failing at the Critical State Line (CSL). The critical state line passes through the origin which implies that normally consolidated or lightly over consolidated soil has no strength at zero effective stress. Many engineers prefer to use the Mohr-Coulomb failure envelope as failure criterion which is the boundary between possible and impossible states (Equation 1). The

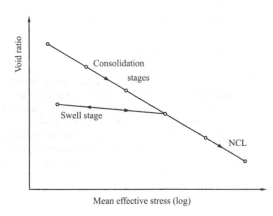

Figure 1. Consolidation and swell stages.

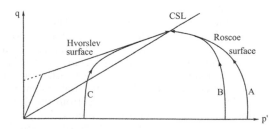

Figure 2. Critical state framework.

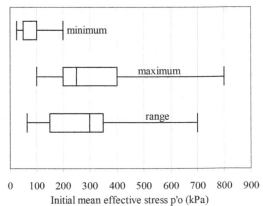

Figure 3. Initial mean effective stress.

290

equation assumes that the failure envelope is a straight line defined by two soil parameters c' and φ'. If c' = 0 the Mohr-Coulomb failure envelope passes through the origin, implying no strength at zero normal effective stress (σ').

$$\tau_f = c' + \sigma' \tan\phi' \qquad (1)$$

Heavily overconsolidated soils will yield at the Hvorslev surface which is often assumed to be a straight line and the extension of the Hvorslev surface (dashed line) has a positive intercept with the deviatoric stress axis (q) as shown in Figure 2. The Hvorslev surface is a state boundary and conveniently it may also be quantified with Equation 1 in which case c' will have a positive value. However, a positive c' value does not necessarily imply shear strength at zero normal effective stress. It may simply be a consequence of the assumption that the Mohr-Coulomb envelope is a straight line. It is well known that the failure envelope for rocks is non-linear (e.g. Hoek & Brown 1980). There is also clear evidence that the failure envelope for soils is non-linear. Sture et al. (1998) observed friction angles between 63° and 70° degrees for sands under micro gravity with normal effective stresses less than 0.1 kPa. At the same density the sand exhibited a friction angle of about 42° at a normal effective stress of 70 kPa, clearly demonstrating the non-linear nature of the failure envelope.

Charles and Watts (1980) suggested a power law non-linear failure envelope for soils with A and b as material parameters:

$$\tau_f = A\sigma'^b \qquad (2)$$

If plotted in terms of normal effective stress (σ') and shear stress (τ) the slope of the failure envelope can be calculated at any normal effective stress as:

$$\frac{d\tau}{d\sigma'} = Ab\sigma'^{(b-1)} \qquad (3)$$

and the intercept (c') as:

$$c' = (1-b)A\sigma'^b \qquad (4)$$

Figure 4 shows triaxial results from a consolidated drained triaxial test on gravel with particle sizes ranging between 6.7 mm and 1.18 mm. The specimens had a diameter of 75 mm and a height of 150 mm and were compacted to an initial void ratio of 0.82.

Two failure envelopes are shown in Figure 4. One is the non-linear failure envelope according to Equation 2 with A = 1.84 and b = 0.84. Figure 4 also shows a linear failure envelope with φ' = 31.2° and c' = 114 kPa. It was calculated using Equations 3 to find the slope of the non-linear envelope at a mean effective stress of 1000 kPa and Equation 4 to calculate the intercept of the straight line (c').

The linear failure envelope fits the experimental data well for the stress range (s') 500 kPa to 1500 kPa, and will give satisfactory results if applied in this stress range. However, it should be emphasised that the value of c' = 114 kPa does not imply that the gravel has 114 kPa shear strength at zero normal effective stress.

4 DISCUSSION OF TEST RESULTS

4.1 Gravels

Gravels often exhibit a non-linear failure envelope as shown in Figure 4. If a linear Mohr-Coulomb failure envelope is used (Eq. 1), careful consideration should be given to the stress range within which the equation is applied. Alternatively a non-linear failure envelope (Eq. 2) should be applied.

The dataset contained 21 triaxial results for gravel. For 12 of the tests the failure envelope was a straight line passing through the origin (c' = 0). The friction angle (φ') for these test are shown in Figure 5. The remaining 9 tests showed non-linear failure envelopes and the material parameters A and b are shown in Table 2.

For the sand, silt, clay and tailings samples a linear Mohr-Coulomb envelope was assumed. The distribution of the friction angles (φ') and cohesion (c') are shown in Figure 5 and Figure 6 as box and whisker plots with the number of test sets for each material shown in brackets. For the cohesion values distinction was made between specimens prepared from undisturbed samples and those prepared from disturbed samples. Undisturbed samples were typically block or Shelby tube samples and the specimens prepared from disturbed material were typically compacted or moist tamped to the required target density.

4.2 Friction angle (φ')

Figure 5 shows the friction angle measured for the soils. The median friction angle (φ') for the clay material was 28° The lowest friction angle for clay was 15°, but interestingly, of the 105 clay samples the friction angle was below 20° for only seven tests. The highest friction angle measured for clay was 38°, but for the 105 tests only six were above 35°. This agrees well with data for tests on clay by Mesri & Abdel-Ghaffar (1993), who reported maximum friction angle of 38° and Mitchell (1976) for who reported 39°.

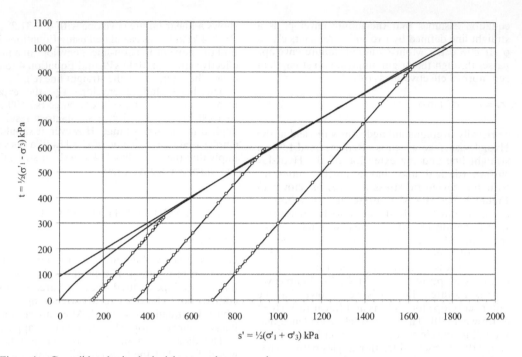

Figure 4.   Consolidated rained triaxial test results on gravel.

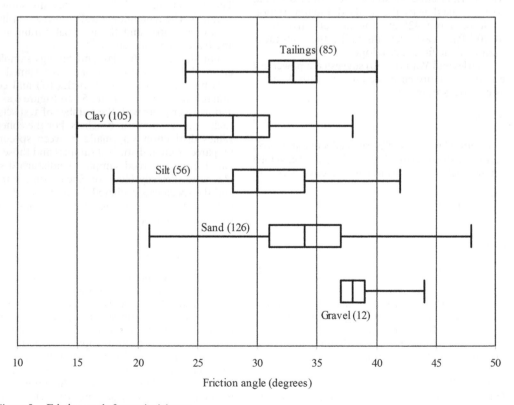

Figure 5.   Friction angle from triaxial tests.

292

The median friction angles observed for the tailings material was 33°, with the 25th percentile 31° and 75th percentile 35°. Therefore, for half of the 85 tests on tailings material the fiction angles varied in a small range between 31° and 35°.

4.3 Cohesion (c')

Figure 6 shows the cohesion values (c') measured for the soils. Distinction was made between tests on specimens prepared from Undisturbed (UD) block or tube samples and specimens prepared from disturbed (D) samples. The number of tests for each case is shown in brackets. The median cohesion values was zero for all material types except for undisturbed silt for which c' was 1 kPa and undisturbed clay for which c' was 4 kPa. Clearly, low cohesion values were observed in most cases and this has important implications for geotechnical design. High cohesion values were observed in some cases and, with the exception of tailings, cohesion values above 20 kPa was observed for all material types. The very high c' values, in excess of 50 kPa, observed for sand was as a result of assuming a linear failure envelope and does not imply that the sand exhibited shear strength at zero effective stress.

4.4 Permeability

The permeability of the soils tested is shown in Figure 7. The median permeability of gravel was roughly one order of magnitude larger than that of sand and two orders larger that natural silt. The median permeability of mine tailings was between those of sand and natural silt. The permeability of clay spanned more than six orders of magnitude,

Table 2. Parameters for gravel failure envelope.

Test	A	B
1	1.19	0.89
2	1.44	0.86
3	0.61	1.00
4	0.93	0.94
5	1.16	0.91
6	1.84	0.84
7	0.76	0.98
8	1.64	0.85
9	1.12	0.92

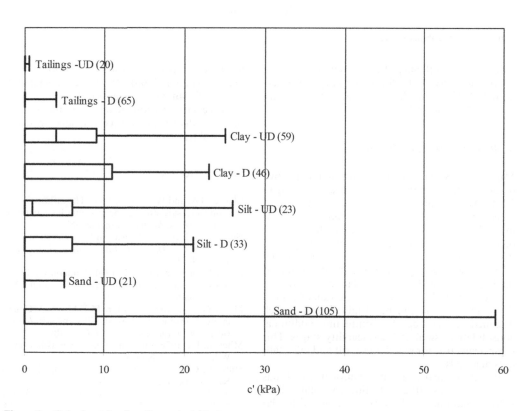

Figure 6. Cohesion (c') values from triaxial tests.

293

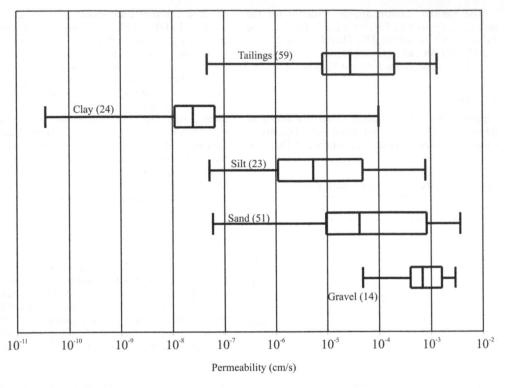

Figure 7.   Permeability test results.

indicating the difficulty of accurately determining the permeability of clay for design purposes. Interestingly half of the clay tests had a permeability between 10^{-7} and 10^{-8} cm/s.

5 RECOMMENDATIONS

Much progress has been made during the last two decades to improve the quality of triaxial testing in South Africa. This is primarily due to investment in equipment and personnel by numerous commercial soils laboratories. However, the majority of triaxial testing still focuses on measuring effective stress strength parameters of soil. This paper shows typical strength parameters measured at the University of Pretoria soils laboratory over an extended period. It is argued that valuable additional information can be obtained from a well designed test strategy by conducting additional consolidation, swell and permeability stages. This is especially true for gravel, sand, silt and mine tailings materials.

In resent years, the international trend has been to also use triaxial testing to obtain shear stiffness parameters. This can be done with relatively little effort and expense by incorporating local strain instrumentation and bender elements (e.g. Heymann 2000). It is recommended that South African commercial soils laboratories make the necessary investment to follow this trend and further enhance the service provided to the geotechnical fraternity.

REFERENCES

Charles, J.A. and Watts, K.S. 1980. The influence of confining pressure on the shear strength of compacted rockfill. *Géotechnique* 30(4): 353–367.

Heymann, G. 2000. Advances in triaxial testing. *Journal of the South African Institution of Civil Engineering* 42(1): 24–31.

Hoek, E. and Brown, E.T. 1980. Underground excavations in rock. *The Institution of Mining and Metallurgy*, London.

Mesri, G. and Abdel-Ghaffar, M.E.M. 1993. Cohesion intercept in effective stress stability analysis. *Journal of Geotechnical Engineering, ASCE*, 119(8).

Mitchell, J.K. 1976. Fundamentals of Soil Behavior, John Wiley and Sons, New York.

Sture, S., Costes, N., Batiste, S., Lankton, M., Alshibli, K., Jeremic, B., Swanson, R., and Frank, M. 1998. Mechanics of Granular Materials at Low Effective Stresses. *Journal of Aerospace Engineering, ASCE*, 11(3): 67–72.

Proceedings of the first Southern African Geotechnical Conference – Jacobsz (Ed.)
© *2016 Taylor & Francis Group, London, ISBN 978-1-138-02971-2*

Initial consumption of stabiliser testing—quo vadis?

P. Paige-Green
Tshwane University of Technology, Pretoria, South Africa

ABSTRACT: In order to improve the quality of many marginal materials for use in roads and foundation layers, chemical stabilisation is often resorted to. During the design of these stabilised layers using lime or cement, the result of the Initial Consumption of Lime (ICL) or more recently the Initial Consumption of Stabiliser (ICS) test is essential to ensure material durability. Experience has shown that by ensuring that the ICS is satisfied, a cementitiously stabilised material is almost guaranteed to be durable over the service life of a typical road. With the increasing use of common extended cements as stabilisers and a reduction in the use of lime and Ground Granulated Blast-furnace Slag (GGBS), the understanding of the processes measured in the ICS and its interpretation has been cast into doubt. It is not clear whether the basic principles on which the original ICL test were based are still valid for modern testing. This paper assesses some of these fundamental principles in terms of conventional chemical understanding and makes recommendations on improved ways of interpreting the results of ICS testing.

1 INTRODUCTION

The quality of many natural materials, and increasingly these days, recycled materials, requires improvement for use as structural layers in roads. This is frequently done by adding a cementitious stabiliser, to the material, mostly cement in current construction projects. The basis of any stabilisation design (DPW&T, 2004) using either lime or cement is the result of the Initial Consumption of Lime (ICL) or more recently the Initial Consumption of Cement (ICC) (or more generally Stabiliser) (ICS) test. Experience has shown that by ensuring that the ICS is met, a cementitiously stabilised material is almost guaranteed to be durable over the service life of a typical road (Sampson and Paige-Green, 1990). By ensuring that there is an excess of stabiliser remaining after the initial reactions, provision is made for the material to undergo the traditional cementation reactions.

With the increasing use of common extended cements as stabilisers and a reduction in the use of neat lime and lime blends with flyash (PFA) or Ground Granulated Blast-furnace Slag (GGBS), the understanding of the processes determining the ICS has been cast into doubt. It is not clear whether the basic fundamental principles on which the original ICL test were based are still valid for testing of the cements currently used for road stabilisation.

This paper assesses some of these fundamental principles in relation to basic chemical doctrines and makes recommendations on improved ways of interpreting the results of ICS testing.

2 HISTORY OF THE TEST

The original test was developed in the United States by Eades and Grim (1966) to assist with identifying the quantity of lime required to stabilise any soil. The test consisted of measuring the pH of slurries of 20 g of soil passing the 0.425 mm sieve in 100 ml water after adding different quantities of lime and finding the minimum lime percentage that gives a pH of 12.4 after 1 hour. They explained the loss of pH from the standard value of lime (12.4) within the first hour as being caused by a number of chemical reactions:

These include:

- The reaction of calcium in the lime with organic and inorganic compounds in the soil;
- The absorption of calcium by hydroxyl-quinones (oxidised derivatives or aromatic compounds) found in soil;
- Absorption of calcium ions onto the exchange sites of clay minerals (displacing sodium and hydrogen), and
- Reaction with unleached soluble silica, alumina, sulphur and phosphorous remaining in the soils.

Clauss and Loudon (1971) reviewed this document and made some modifications for South African use. These included:

- The use of a standard lime as well as that to be used on the project;
- The use of a standard lime solution to account for the ICL of the water and for comparison of the maximum pH values;

- The introduction of a correction for the percentage passing the 0.425 mm sieve, and
- The option to use a pH other than 12.4 by taking the lime content at which the pH becomes constant.

In the late 1970s it was realised that the ICL was not always only related to the soil fines but for many materials (basic crystalline rocks in particular), there were "accessible" components in the coarser fractions that also reacted with the lime. The so-called "gravel ICL" was thus developed in which the entire fraction of material finer than 19 mm was tested (at saturated moisture content) and no correction for the particle size was applied.

Ballantyne and Rossouw (1989) discussed the rapid early reactions and identified various causes. These included:

- Ion exchange—the large calcium ions in the lime replace the smaller sodium and hydrogen ions on the surface of the clay particles. The sodium cation can hold 79 molecules of water, whereas the calcium cation only holds 2, making the clay more hydrophobic.
- The "crowding" of additional cations onto the clay surface together with the cation exchange reactions, alter the electrical charge on the clay surface allowing a greater attraction between the particles (flocculation).

Additional work has been carried out investigating the longer term effects of lime on soils, much of this unpublished. The pH has been measured after 24 hours, 7 days and even 28 days with corrections being made to the conventional measured results. The wisdom of this can be questioned—the test is related to the "initial" consumption of lime—longer term reactions cannot really be classified as such. This is particularly the case when cement is being used as the stabiliser because of the confounding effect of the hydration reactions of the cement and the continual release of lime into the solutions during these reactions.

ASTM (2006) adopted the Eades and Grim test as a full standard in 2006. However, as all ASTM standards require updating by the end of the eighth year since the last approval date, the standard was withdrawn in 2015. It is interesting to note this as the method often gave incorrect results by taking the ICL as the lime content at a pH of 12.4 and not at the point of constant pH.

More recently, the South African Bureau of Standards (SABS) has published a local standard (SANS 3001 GR-54:2014) for the ICS test. This is a slight modification of the traditional gravel ICL test used in South Africa.

3 MATERIALS FOR STABILISATION

Materials that would normally be tested for possible stabilization fall into 5 groups:

- Crushed aggregates
- Natural residual gravels
- Transported materials
- Pedogenic materials
- In situ materials for recycling

3.1 Crushed aggregate

Aggregate derived from the crushing of unweathered rock materials is seldom stabilised as the properties of these materials are generally adequate for most construction purposes and do not require alteration. Occasionally, particularly with basic crystalline rocks that have been subjected to deuteric alteration and may contain a small quantity of smectite clay, a nominal amount (1 to 2%) of lime may be added to reduce possible degradation of the aggregate with time. ICS testing is not usually carried out in these cases.

The crushing of rock results in fracturing of individual crystals and disruption of the neutral chemical bonding, with the development of surface charges on aggregate particles. These can affect the cation exchange/absorption capacity, an important property in material usage and stabilization as discussed later.

3.2 Residual gravels

Materials originating from weathered and altered rock are probably those most commonly used for stabilized layers in roads. It is essential that these materials are treated with the correct quantity of stabilizer in order to optimize the durability and cost of the layers and these are thus the materials most commonly tested for their ICS.

Residual materials are derived from any type of rock with varying composition and thus contain a wide variety of minerals and chemical constituents. It is mostly the effect of these secondary minerals (clays) that affect the ICS and need to be assessed.

Materials derived from acidic rocks (e.g. granites) tend to have higher sodium contents compared with basic rocks (dolerite and gabbro), which tend to have higher calcium contents. It would thus be expected that, if the assumptions regarding the ICL test are correct, the acid materials would have higher ICS values than the basic materials. This is usually not the case.

3.3 Transported materials

Transported materials are usually residual materials, which have been moved by wind, water, ice or gravity to a new location. During this movement various actions affect the materials changing their mineralogical and physical properties. These mostly include a reduction in mean particle size, the reduction in quantity of softer and less stable minerals (e.g. feldspar and mafic minerals), a relative increase in more resistant materials (e.g. quartz) and a decrease in secondary minerals. Transported materials can therefore have almost any mineralogical composition and thus highly variable properties.

3.4 *Pedogenic materials*

Pedogenic materials consist of high percentages of secondary minerals (iron oxides, re-precipitated calcite and silica, gypsum, etc.) and tend to have unusual properties. Their properties thus vary widely and can be highly variable even within small areas. Their properties often affect normal stabilization reactions, with laterites and ferricretes, for instance, from within the same borrow pit reacting totally differently with stabilisers—some materials being almost unaffected by stabilisation.

3.5 *In situ materials for recycling*

In situ recycling and stabilization of existing pavement layers is increasing rapidly with the improvement in technology to carry this out quickly and economically. It is also introducing new problems, especially on old roads that have been extensively patched and repaired, where rehabilitation involves widening of the road or where existing layer thicknesses are highly variable.

In these cases, the material being recycled can vary significantly over short distances and the road could require constantly changing stabiliser contents (and possibly even stabiliser types) along its length.

This is, however, a technical problem for the designer more than an ICS problems and is not discussed further.

4 ICL TEST

Lime (Ca(OH)$_2$) is only slightly soluble in water. However, as soon as it is placed in water, it dissociates into the calcium cation (Ca^{++}) and the hydroxyl (OH$^-$) anion—if the equilibrium is upset by removing any of these ions, more lime will dissociate to restore equilibrium. The pH measurement determines the quantity of hydrogen ions in the solution. When lime is added to pure water, the concentration of OH$^-$ ions increases above the neutral value of 1×10^{-7} M for the water and the concentration of H$^+$ decreases below 1×10^{-7} M.

As the lime is "consumed", the concentration of OH$^-$ ions decreases and the concentration of H$^+$ increases. The pH measurement records this increasing H$^+$ concentration. A balance is then determined between this "acidity" and the lime required to neutralise it and provide an excess for cementation to occur.

However, the fundamentals of clay chemistry dictate that when lime is added to the water (and material) the high content of calcium ions released displace Na$^+$ (and probably H$^+$) ions from the clays. The sodium ions will remain dissociated in the solution, effectively forming a sodium hydroxide solution, whose pH is higher than that of calcium hydroxide. In this case the pH measured in the ICL test will increase as more sodium ions are released, and the point of stabilisation of the pH will be difficult to identify.

Conversely, if hydrogen ions are displaced from the soil, this would acidify the solution and only when sufficient lime has been added to the solution will the pH increase and stabilize at the normal pH of lime.

It is also known that organic materials in soils tend to be acidic. These "acids" would neutralize a certain amount of the lime added and reduce the pH, before neutrality is reached and the pH starts increasing again. However, in most borrow pits, the material being processed is beneath the depth of typical organic material (the occasional root or organic acid resulting from seepage along joints, may be present, but generally in low quantities).

The clays in many common construction materials, however, are effectively calcium saturated (dolerites, calcretes, limestones, etc) and yet they still have an ICL. This initial consumption of the lime cannot be ascribed to cation exchange capacity involving sodium. What then is consuming the lime?

5 ICS TEST

A typical fresh cement contains about 5% free lime that is available for reaction. However, shortly after the hydration reactions commence, additional lime is produced as a by-product of the formation of hydrated calcium silicates, up to a quantity of about 20% of the total mass of cement being hydrated. The lime (albeit less) goes into solution as is the case during the ICL test and the reactions essentially proceed in a similar manner.

One big difference, however, is that the cement starts to hydrate as soon as it comes into contact with water and the normal cement reactions take place. These form products (hydrated aluminium and calcium silicates) with pH values equivalent to that of lime or even higher and thus will tend to offset any lime consumption in the pH measurements.

It has become common to extend the test to longer periods in order to assess any longer term changes in pH. When cement is being assessed as the potential stabilizer, this results in more hydration occurring (more lime being released) and changes in the pH related to these reactions being measured in addition to any initial "consumption reactions".

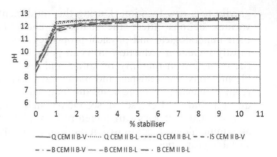

Figure 1. Typical ICL test results with a single material and different stabilisers.

6 PH MEASUREMENT

The measurement of pH, particularly small changes requires high quality equipment. The process essentially measures a small electrical potential (voltage) produced by a solution whose acidity is to be measured. This is compared with the voltage of a known solution (second electrode), and uses the "potential difference" between them to quantify the pH.

One electrode contains a silver-based electrical wire suspended in a solution of potassium chloride. This is contained in a thin bulb made from a special glass containing metal salts (typically compounds of sodium and calcium). The second (reference) electrode has a potassium chloride wire suspended in a solution of potassium chloride. The potential difference measured between the two electrodes indicates the pH.

It is essential that the glass electrode used for ICS testing is in good condition and is replaced regularly. Each time it is used, the salts in the electrode are modified and even the glass is "attacked" by the high pH of the lime/cement materials. It is thought that much of the "wander" of the pH measurements may be due to deterioration of the glass electrodes.

7 TEST RESULTS

Investigation of many test results recently has shown that significant difficulties are increasingly being encountered with the interpretation of the results. The biggest problem is the failure of the pH to stabilize at a uniform level. Typical ICS results using 3 different cements (CEM II B-V and two different CEM II B-L cements) are shown in Figure 1 and Table 1. These results are for a residual quartzitic sandstone, (Q), in situ material (IS) consisting of local soil with some manganiferous slag and a 2:1 blend of these two materials (B). The materials had low Plasticity Indices (4–6%) and between 8 and 13% finer than 0.075 mm.

It is clear that the ICS of the material tested is not easily determined from the typical graph with most of the plots appearing to have stabilised at a cement content of about 5%, a content probably too high to be economic. By "zooming in" on the results, a slightly different picture is seen (Fig. 2).

Each of the cements used has a different endpoint pH and in every case the pH appears to still be increasing at a cement content of 10%. However, the saturated (100%) pHs are generally similar to the 10% values. This type of result is invariably obtained during ICS testing using cement, indicating that there are problems encountered when extrapolating the original ICL test using lime to one using cement.

8 WHAT ARE WE ACTUALLY MEASURING?

The chemical process of measuring pH assumes that the initial reduction of pH is indicative of the OH^- (not the calcium) concentration in the solution being reduced. This is supposedly brought about by consumption of the lime or cement through the early reactions (ion exchange and flocculation) described in most stabilization texts (Ballantine and Rossouw, 1989) that occur rapidly. Once these reactions are complete all additional stabilizer added is available for cementation reactions.

It can be noted from the reactions with the materials discussed earlier that the hydroxide does not come into the equation. All of the theoretical causes discussed are related to the effects of the addition of stabiliser on the calcium ions, an aspect not measured by the pH. It is possible that the overall equilibrium changes in the lime as calcium ions are involved in "reactions" may affect the hydroxyl content (and thus pH), although chemical laws indicate that this equilibrium will be maintained as long as there is still undissolved lime in the solution. On this basis, we do not actually know what we are measuring.

Table 1. ICL test results.

Material	Q	Q	Q	IS	Bl	Bl	Bl
Cement type	CEM II B-V	CEM II B-L	CEM II B-L	CEM II B-V	CEM II B-V	CEM II B-L	CEM II B-L
Cementcontent (%) pH							
0	8.8	8.8	8.8	9.03	8.36	8.36	8.36
1	12.01	12.2	12.36	11.97	11.67	11.76	11.59
2	12.12	12.37	12.43	12.09	12.22	12.15	12.01
3	12.22	12.46	12.49	12.19	12.33	12.3	12.15
4	12.35	12.49	12.53	12.27	12.45	12.39	12.24
5	12.44	12.51	12.59	12.33	12.5	12.47	12.38
10	12.57	12.65	12.66	12.53	12.57	12.51	12.53
100	12.7	12.65	12.62	12.7	12.62	12.56	12.54
ICL[1]	5.0	5.0	5.0	4.0	5.0	5.0	5.0

[1]Interpreted according to SANS 3001 GR-57:2014. In fact 3% CEM II B-L cement was used on the project and this carbonated rapidly at the surface.

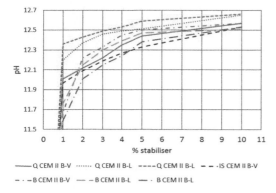

Figure 2. Enlargement of detail in Figure 1.

In the original ICL test, a quantity of lime equal to various percentages of the fine soil material fraction (minus 0.425 mm) was added in a solution of 100 mℓ of water. The solubility of $Ca(OH)_2$ is very low (1.85 g/ℓ). In the 100 mℓ of water used for the original ICL, only 0.185 g of lime would dissolve. As the sample size was 20 g, the quantity of lime added would be 0.2 g for 1% stabiliser up to 1.2 g for 6% stabiliser. Thus, other than for the 1% treatment, there is probably always sufficient lime remaining undissolved for additional lime to dissolve and restore equilibrium as either the Ca or hydroxide ions are removed from the solution. With the gravel ICL test, the quantity of material tested is increased to 200 g but the water content is decreased to just above saturation moisture content, probably resulting in only about 50 to 75 cc of water and 2 to 12 g of lime being used in the test. However, again there will always be sufficient undissolved lime in the mixture to retain equilibrium between the calcium and hydroxide ions during the test.

With cement stabilisation, a different scenario is observed. In the original ICL test of the 1 to 6% stabiliser, only 0.01 to 0.06 g of lime is available for initial reaction, this increasing with hydration of the cement to a maximum of 0.04 to 0.24 g after complete hydration, which could take many months or even years. The availability of lime to restore equilibrium is thus limited. The respective quantities in the gravel ICS test, however, are 0.1 to 0.6 g initially, increasing to between 0.4 and 2.4 g over time. Despite this, the pH increases to in excess of 12.4 in every test, probably more because the pH of the cementitious reaction products is expected to be greater than 12.4. A totally different scenario thus exists between the lime and cement quantities in the gravel ICL test.

9 CONCLUSIONS

The Initial Consumption of Stabiliser (ICS) test has evolved from the ICL test and is a necessary part of any road material stabilisation design. However, difficulties are regularly being encountered when interpreting the results of the standard test when using cement. The actual chemical processes measured in the test in relation to the changes in pH of the soil stabiliser mix are not clearly understood.

The fundamental assumption that cation exchange and flocculation involving the calcium ions in the stabiliser appears to be flawed when determined using the pH measurement, which is a measure of the hydroxyl or acid content of the mix. The initial loss in pH appears to be more the result of activity of hydrogen ions displaced by the calcium or acid organic matter/solutions in the material, possibly affected by equilibrium changes in the lime as calcium ions are involved in "reactions".

Despite this, exceeding the "ICL" or "ICS" by 1% of additional stabiliser has shown to allow a durable stabilised material to be created. Thus although the theory does not hold up to scrutiny, the end result seems to be correct.

It is recommended that this issue is urgently researched in order to improve the interpretation of the ICS test. The researcher carrying out this work should have a good grounding in both soil stabilisation and fundamental chemistry.

REFERENCES

ASTM. 2006. *Standard Test Method for using pH to estimate the soil-lime proportion requirement for soil stabilization*. Method D6276, ASTM, Philadelphia.

Ballantine, R.W. & Rossouw, A.J. 1989. *Stabilisation of soils*. PPC Lime, Johannesburg.

Clauss, K.A. & Loudon, P.A. 1971. The influence of initial consumption of lime on the stabilization of South African road materials. *Proc. 5th Reg. Conf. Africa Soil Mech. Fndn Engng*, Luanda 1 (5): 61–68.

DPWT&R. 2004. *Stabilization Manual*. Gauteng Department of Public Transport, Roads and Works, Pretoria.

Eades, J.L. & Grim, R.E. 1966. A quick test to determine lime requirements for lime stabilization. *Highway Research Record* 139: 61–72.

Sampson, L.R. & Paige-Green, P. 1990. *Recommendation for suitable durability limits for lime and cement stabilised materials*. Division of Roads and Transport Technology, Pretoria. (Report DPVT 130).

South African Bureau of Standards (SABS). 2014. *Determination of the initial stabilizer consumption of soils and gravels*. (SANS 3001 GR 57: 2014). SABS, Pretoria.

Proceedings of the first Southern African Geotechnical Conference – Jacobsz (Ed.)
© 2016 Taylor & Francis Group, London, ISBN 978-1-138-02971-2

Measuring the tensile strength of gold tailings

T.A.V. Gaspar & S.W. Jacobsz
University of Pretoria, Pretoria, South Africa

ABSTRACT: This paper discusses the modification of the Brazilian disk test so that it can be applied to measure the tensile strength of unsaturated soils. The study involved testing the tensile strength of unsaturated gold tailings prepared across a range of moisture contents. In order to monitor the initiation and propagation of cracking within each sample, Particle Image Velocimetry (PIV) was utilised. The study revealed that across the range of moisture contents considered the Brazilian disk test, performed with curved loading strips, consistently resulted in cracking initiating in the centre of the sample, thereby validating the measured failure load as the result of a tensile failure. An investigation of load-displacement relationships for all samples tested showed that in the pendular regime, the behaviour of the material was particularly erratic. However, at higher saturation levels, material behaviour was found to be significantly more stable.

1 INTRODUCTION

1.1 *Background*

Soils can possess a certain amount of tensile strength provided by cementation, particle interlock, pore water suction under partially saturated conditions, or a combination of these factors. Although the tensile strength is normally small, it is significant when studying problems such as soils arching across cavities in the formation of sinkholes, slope stability problems, the design of retaining structures, bearing capacity analyses and other stability problems.

The Brazilian disk test is a method commonly used to indirectly determine the tensile strength of brittle materials such as rocks, concrete and in more recent years, weakly cemented soils (Consoli et al. 2012). Despite its frequent use in the abovementioned fields, the test has rarely been performed on soils and, as a result, little development has been made regarding the applicability thereof. This deficit in research can possibly be attributed to the fact that in practice, soils are generally regarded as having very little to no tensile strength. Over the past few decades however, a greater amount of emphasis has been placed on the research of unsaturated soils. It has become commonly understood that as a soil becomes unsaturated, it develops some additional strength from pore water suction, the extent of which is related to the degree of saturation of the soil in relation to its particle size distribution (Lu et al. 2009). This additional strength can be described as being a result of two separate components, both of which have been highlighted in

Figure 1: (1) the surface tension which acts along the air-water interface and (2) the pressure difference acting across the air-water interface which is termed matric suction (u_a–u_w) (Kim & Changsoo 2003).

When a given soil is subject to negative pore water pressure, three distinct regimes can be identified. At low moisture contents, the water phase is discontinuous and exists only as thin layers surrounding isolated soil particles. This soil state is referred to as the pendular regime (Kim & Changsoo 2003). As the soil reaches higher levels of saturation, the water phase becomes more continuous and the soil transitions into the funicular regime. Finally, the capillary regime is a term used to describe the portion of soil above the phreatic surface which is kept saturated due to the presence of a negative pore water pressure (commonly referred to as the capillary fringe) (Lu & Likos 2004).

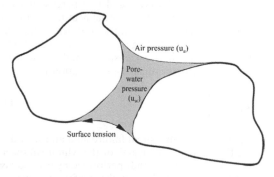

Figure 1. Components contributing to additional strength of unsaturated soil.

Although the transition between each of the phases described above is not distinct, each state will result in a different soil behaviour, all of which have been investigated in the current study. The additional strength present in unsaturated soils arising from matric suctions results in a non-zero value of tensile strength. Depending on the particle size, the strength magnitudes are generally small and difficult to measure accurately. The contribution of a certain amount of tensile soil strength may, however, be significant in several classes of problems. Examples could include slope stability and bearing capacity problems, as well as heterogenous soil masses arching over cavities, i.e. the sinkhole problem.

A modification of the Brazilian disk test, which is relatively simple in comparison to direct tension tests, to measure the tensile strength of soil samples was therefore undertaken in the present study to determine the tensile strength of unsaturated soil samples accurately and repeatably. The study will examine a range of material types, but commenced with the testing of silts. Gold tailings was selected as the first silty material tested and selected results are reported in this paper.

1.2 The Brazilian disk test

Perhaps the most extensive development of the Brazilian disk test has been in the field of rock mechanics and it is from this viewpoint that the present study departed (Li et al. 2013). The test involves diametrically loading a disk specimen, resulting in the development of tensile stress perpendicular to the direction of load application. It is obvious that at the portions of the sample where the load is applied and in the immediately adjacent zones close to the loading strips, the sample experiences compressive stresses. If the applied load resulted in tensile failure, cracking would initiate in the centre of the disk sample and propagate outwards towards the loading strips. If this failure mechanism is observed, and if it is seen that the sample is not crushed appreciably at the points of load application, it provides confidence that the sample is indeed failing in tension (Li et al. 2013). In this case the indirect tensile strength of the specimen can be determined using Equation 1:

$$\sigma_t = \frac{2P}{\pi DL} \tag{1}$$

where P is the measured failure load and D and L are the diameter and length of the cylindrical specimen respectively. A study performed by Erarslan & Liang (2012) highlighted the profound effect of loading conditions on the resulting failure of the disk specimen. The study revealed that by loading

samples of Brisbane tuff with the use of standard Brazilian jaws, a compressive failure was observed. In contrast, it was found that through the use of 30° loading arcs, a centrally located crack and thus a tensile failure could be achieved. One of the aims of this study was therefore to investigate whether the conclusions obtained by Erarslan & Liang (2012) are applicable to the testing of unsaturated soil samples across a range of moisture contents.

2 EXPERIMENTAL PROCEDURE

2.1 Sample preparation procedure

The Brazilian disk tests carried out for the current study were performed on a number of disk samples of gold tailings. The particle size distribution is included in Figure 2. In order to avoid unwanted stress concentrations within the samples as they were loaded, care was taken to ensure that the samples prepared were free of defects. Due to the low tensile strength of the soil tested, any defects present in a sample would likely have resulted in the failure mode being affected, possibly initiating at one of the sample defects rather than in the centre of the sample where the soil would be in tension.

Disk specimens, with a diameter and thickness of 50 mm and 25 mm respectively, were prepared to achieve gravimetric moisture contents of 5, 8, 12 and 17%. Due to the amount of moisture loss that would take place as each sample was being placed onto the testing equipment, final (more representative) measurements of moisture contents were taken immediately after each sample was tested. The samples were prepared from a uniform reconstituted slurry with a moisture content of approximately 30% (based on workability considerations). The slurry was used to fill up several cylindrical split moulds in five separate lifts, lightly tamping the mould after each lift to allow the slurry to distribute itself evenly and to allow air bubbles to escape. The slurry mixtures were subsequently

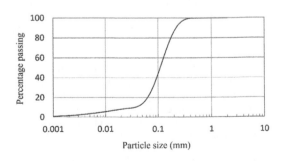

Figure 2. Particle size distribution.

placed in an oven at 65°C and allowed to lose some moisture. In order to monitor the amount of moisture lost throughout the drying process, periodic mass readings were taken until such point that the desired moisture content of each sample had been reached.

Once a sample achieved its desired moisture content, a final mass reading together with a series of height and diameter measurements were taken to determine the unit weight of each sample. Finally, before being stored, samples were wrapped with a combination of both clingfilm and aluminium foil, as suggested by Heymann & Clayton (1999), to minimize the amount of moisture loss prior to testing.

2.2 Test setup and procedure

The test setup, illustrated in Figure 3 consisted of the following equipment:

- two curved loading strips with 30° arcs as suggested by Erarslan & Liang (2012). The strips were manufactured to have a 27 mm radius (2 mm greater than the sample radius) to cater for initial deformation of the soil without the edges of the loading strips cutting into the sample;

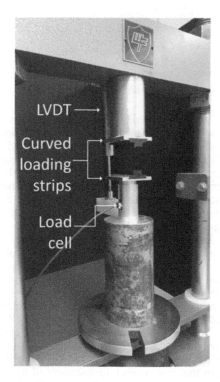

Figure 3. Test setup.

- a Linear Variable Differential Transformer (LVDT) to measure the vertical displacement of the sample throughout loading;
- a sensitive load cell with a maximum capacity of 200N;
- triaxial loading rig, allowing the diametric load to be applied at a constant rate of deformation;
- a high resolution digital camera (not visible in Figure 3) was mounted in front of the specimen so that strains within the sample could be monitored. The camera was set such that a photograph of the sample would be taken every five seconds.

3 RESULTS

The results of the testing performed were sorted into four separate groups according to water content. This was done to characterise the behaviour of the soil across varying degrees of saturation. The data for all tests performed have been included in Table 1. From both Table 1 and Figure 6 it can be seen that at lower moisture contents a higher degree of scatter is observed owing to the extremely brittle nature of the material in that state. As the saturation levels of the samples increased, it was however observed that the material behaviour became less erratic, as is evident by the corresponding lower standard deviations in Table 1.

The variation in the consistency of the results with moisture content is further illustrated when considering the load-displacement curves presented in Figure 4 & 5. Figure 4 illustrates the load-displacement curves for an average water content of 5.18%. The load-displacement curves initially rise steeply, then yield, after which the load increases more slowly to a peak. From this figure it can be seen that there is relatively poor consistency in the results for the samples tested.

In contrast, the results presented in Figure 5 illustrate considerably less variation. In general, the load-displacement curves rise more gradually, then plateau, exhibiting a more ductile response than the dryer, more brittle material. The consistency of results within this moisture regime is highlighted when considering the amount of displacement that took place before the yield point was reached, as well as the maximum tensile strength that was achieved for each sample.

One trend that is present across all samples tested is related to the double stress peak seen in both Figure 4 and 5. This is attributed to two failure mechanisms which mobilised. Initially, as the diametric load was applied, the soil experienced tensile stresses perpendicular to the direction of load application, resulting in the soil failing in

Table 1. Brazilian disk test results.

Sample number	Water content(%)	Average water content(%)	Max. recorded stress (kPa)	Standard deviation of max. stress (kPa)	Yield stress (kPa)	Standard deviation of yield stress (kPa)
1	4.29	5.18	3.34	0.97	3.31	1.31
2	4.58		2.51		1.78	
3	4.98		1.19		0.62	
4	6.89		3.15		0.51	
5	7.26	7.69	1.73	1.06	0.83	1.39
6	7.29		3.59		3.41	
7	7.69		2.69		2.53	
8	8.51		4.15		4.05	
9	11.85	12.49	4.97	0.52	4.15	0.81
10	12.54		4.15		3.97	
11	12.58		4.18		3.03	
12	12.99		3.71		2.44	
13	13.82	16.59	2.86	0.44	2.17	0.90
14	13.97		3.87		1.29	
15	18.83		3.68		3.33	
16	19.74		3.45		2.93	

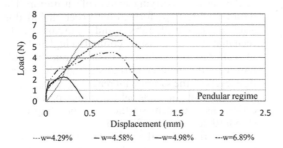

Figure 4. Load-displacement (w = 5.18%).

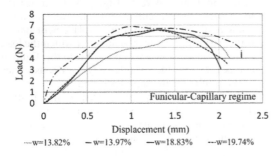

Figure 5. Load displacement (w = 16.59%).

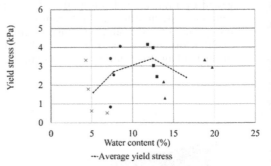

Figure 6. Yield stress vs water content.

Figure 6 plots the results of yield stress vs moisture content. Also displayed is a trend line connecting the average yield stress within each water content group. These results point towards a similar trend as observed by Lu et al. (2009) where the tensile strength of unsaturated sands was noted to increase in the pendular regime, achieve a maximum value in the funicular regime, and decrease again in the capillary regime.

In order to more closely monitor the opening of the crack, the coordinates of two, vertically centred patches on the extreme left and right of the sample were analysed. Using these coordinates, the distance between the two patches in successive photographs was tracked photogrammetrically using Particle Image Velocimetry (White et al. 2003). The value of horizontal crack extension for Sample 12 is plotted as a function of vertical displacement in Figure 7. From this figure it can be seen that at a vertical displacement of 0.0264 pixels, behaviour of the soil became linear and it is at

tension at the first stress peak. Thereafter, the sample was able to sustain higher compressive loads as the soil was gradually crushed. This behaviour was confirmed when examining the photographic records. Minimal sample deformation occurred as tensile stress mobilised. Then, at the first peak, a crack rapidly started to grow and then, with further deformation, crushing became evident at the points of load application.

this point that the crack in the centre of the sample became visible. At this point it can be stated that the maximum tensile strength of the soil had been mobilized and that any increase in load recorded after this point, was as a result of compression of the sample.

Figure 8 illustrates the load-displacement behaviour of Sample 12 with Figure 9 presenting the corresponding development of horizontal strain in

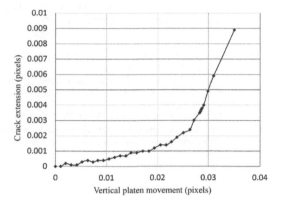

Figure 7. Crack extension vs vertical platen movement.

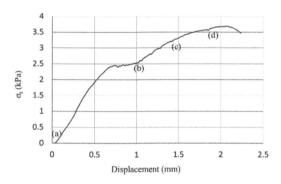

Figure 8. Load vs displacement (w = 12.99%).

the sample during load application. From Figure 9 the initiation and propagation of cracking with an increase in diametric load can be seen.

Upon first load application, no visible straining of the sample was detected. In fact, it can be seen that the first peak (Figure 8) was mobilised before measureable strains were detected. As is required for a valid Brazilian disk test result, the crack initiated in the centre of the sample and propagated outwards towards the loading strips until failure of the sample occurred.

Of the 16 samples tested, 14 were observed to have an initial crack positioned in the centre of the sample. This result highlights the effectiveness of the use of curved loading strips in achieving a tensile failure as was observed by Erarslan & Liang (2012) on samples of Brisbane tuff. In contrast, Hudson et al. (1972) found that the use of flat loading platens consistently resulted in failure originating directly below the loading points.

4 CONCLUSIONS

For the range of moisture contents considered, it was observed that the use of 30° curved loading strips consistently resulted in cracking initiating in the centre of the sample which is essential for carrying out a valid tensile test. In contrast, compressive failures originating directly below the point of load application were observed when flat loading strips were used (Hudson et al. 1972).

The load-displacement behaviour and measured tensile strength of the gold tailings was found to be highly erratic and very brittle in the pendular regime. However, with an increase in moisture content, results were seen to be significantly more consistent and repeatable and a more ductile behaviour was observed.

For all moisture contents considered, the behaviour of unsaturated gold tailings repeatedly resulted in the following trend: the load displacement curve rose to yield point corresponding to

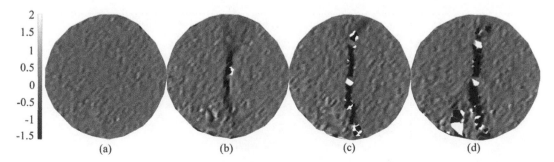

Figure 9. Horizontal strain development.

the mobilisation of tensile strength, then plateaued before rising somewhat further as the crack opened and the sample was crushed.

In terms of the recorded tensile strengths, it was found that, similar to the results of Lu et al. (2009), strengths tended to increase in the pendular regime, achieve maximum strength in the funnicular regime and decrease in the capillary regime.

The present study is ongoing and will need to examine the tensile strength behaviour of other unsaturated fine grained materials and the factors influencing it.

REFERENCES

Consoli, N.C., Cruz, R.C., Consoli, B.S. & Maghous, S. 2012. Failure envelope of artificially cemented sand. *Géotechnique*, 62(6): 543–47.

Erarslan, N. & Liang, Z.Z. 2012. Experimental and Numerical Studies on Determination of Indirect Tensile Strength of Rocks. *Rock Mechanics and Rock Engineering*, 45(5): 739–51.

Heymann, G. & Clayton, C.R.I. 1999. Block sampling of soil: Some practical considerations. In Wardle, G.R., Blight, G.E. & Fourie, A.B., eds. *Geotechnics for Developing Africa*. Rotterdam, 1999. A.A. Balkema.

Hudson, J.A., Brown, E.T. & Rummel, F. 1972. The controlled failure of rock disks and rings loaded in diametral compression. *International Journal of Rock Mechanics and Mining Sciences & Geomechanics Abstracts*, 9(2): 241–248.

Kim, T.H.K. & Changsoo, H. 2003. Modeling of tensile strength on moist granular earth material at low water content. *Engineering Geology*, 69(3): 233–244.

Li, D., Ngai, L. & Wong, N. 2013. The Brazilian Disk Test for Rock Mechanics Applications: *Review and New Insights. Rock Mechanics and Rock Engineering*, 46(62): 269–287.

Lu, N. & Likos, W.J. 2004. *Unsaturated Soil Mechanics*. New Jersey: John Wiley & Sons.

Lu, N., Kim, T.H., Sture, S. & Likos, J. 2009. Tensile Strength of Unsaturated Sand, *Journal of Engineering Mechanics*: 1410–1419.

White, D.J., Take, W.A. & Bolton, M.D. 2003. Soil deformation measurement using particle image velocimetry (PIV) and photogrammetry, *Géotechnique* 53(7): 619–631.

Proceedings of the first Southern African Geotechnical Conference – Jacobsz (Ed.)
© 2016 Taylor & Francis Group, London, ISBN 978-1-138-02971-2

Effect of hydrometer type on particle size distribution of fine grained soil

A. Kaur & G.C. Fanourakis
University of Johannesburg, Johannesburg, South Africa

ABSTRACT: The particle size distribution of soils, including clay content is of utmost importance in the field of Geotechnical Engineering. The hydrometer analysis is the most widely used technique for analyzing the particle size distribution of the fine grained fraction of soil. The various hydrometer test methods (internationally) generally vary mainly in the use of the prescribed dispersing agent. In addition, certain methods vary with the use of the prescribed hydrometer. The purpose of this study was to establish the optimum concentration and volume of the three dispersing agents (calgon, sodium pyrophosphate decahydrate and sodium tetra pyrophosphate) using the ASTM Hydrometer 152H: E100 (instead of the Bouyoucos hydrometer 152H) for three soil classes, selected for their varying activity. The results indicated that the ASTM hydrometer 152H: E100 generally yielded lower results than hydrometer 152H in terms of percentage fines, with the differences varying from 38% to -65% for all the soils and the dispersing agents.

1 INTRODUCTION

Although seemingly simple, the determination of the soil particle size is one of the most problematic areas in the geotechnical engineering. Differences in the laboratory results are still common and substantial, which reinforces the need that the field data should be collected very carefully by the surveyors. The particle size analysis is a method of separation of the mineral part of the soil into various size fractions and the determination of proportions of these fractions. The mechanical or sieve analysis is performed to determine the distribution of the coarser, large-sized particles, and the hydrometer method is used to determine the distribution of the finer particles.

As the clay content of the soil is used to determine the activity of a soil, which in turn is used for design purposes, it is very important to accurately determine the clay content of the soils. Inaccurate clay content determinations have resulted in inappropriate design solutions which have even led to unacceptable damage to the structures.

Sedimentation analysis (Hydrometer analysis) is based on Stoke's law, according to which the velocity of free fall of grains is directly proportional to the square of the particle's diameter. The expression used to determine the particle diameter in the sedimentation analysis is based on this settling velocity. Hence the individual soil particles must be dispersed to enable the determination of particle size distribution accurately. However, the finer grains of soil carry charges on their surface and hence have a tendency to form flocs. Thus, if the

floc formation is not prevented, the grain diameter obtained would be the diameter of flocs and not of the individual grain (Ranjan, 1991). Hence in hydrometer analysis which is a type of sedimentation analysis, deflocculating agents are added to prevent fine soil particles or clay particles in suspension from coalescing to form flocs.

Dispersing agents can either act as a protective colloid on the solid particle or alter the electrical charge on the particle to prevent the formation of flocs (Sridharan et al., 1991). The various hydrometer test methods (internationally) generally vary mainly in the use of the prescribed dispersing agent. In addition certain methods vary with regards to the prescribed hydrometer type. The 152H hydrometer (TMH1 1986, ISRIC 2002), the ASTM Hydrometer 152H: E100 (SANS 3001, ASTM D422–63) and the Specific Gravity hydrometer (IS 2720, 1985 and Lambe, 1951) are some of the most commonly used hydrometers for the sedimentation analysis of the soils. To date, many researchers (Means & Parcher (1963), Herd (1980), Sridharan et al (1991) and Bindu & Ramabhadran (2010)) have researched the effect of different dispersing agents on the particle size analysis. However, the effect of other factors such as the hydrometer type, stirrer type, stirring time, stirring speed, soaking time and the sample size, on which hydrometer analysis depends on, appear to be still unexplored.

The current study focuses on the variation in the results of the sedimentation analysis with the change in the hydrometer type from the 152H to the ASTM 152H: E100 on 3 different soil samples selected with varying activity. The soil samples were treated with the three different dispersing agents. The tests were performed originally with the Bouyoucos hydrometer 152H and the optimum quantity for each concentration (calgon:—35:7, 40:10, 60:10, 70:10 & 80:10, sodium pyrophosphate decahydrate:—3.6%, 5%, 6% and 7% and sodium tetra pyrophosphate:—3.6%, 5%, 6% and 7%) was determined. In this study the results for the optimum quantity and concentration obtained by the Bouyoucos hydrometer 152H were compared with the results obtained when using the other hydrometer i.e. hydrometer 152H:E100 while following the TMH1 guidelines.

2 EXPERIMENTAL DETAILS

2.1 Material used

Three soil samples were collected from various parts of South Africa. The one was a black soil from the town of Brits in the Northwest Province, the second was a light brown sample collected from Linksfield in the Gauteng area and the last one was a red sample from Springfield in the Gauteng area. The Atterberg Limits and Activity of these soils were determined in accordance with the TMH1 (1986) method. The clay content was determined by means of the hydrometer analysis (Method A6 of TMH1–1986) with a deviation in the prescribed dispersing agent type, quantity and adjustment in the readings by subtracting the hydrometer readings obtained on the "blank" companion specimens to account for the effect of the dispersing agent. TMH1 (1986) Method A6 prescribes 5 ml of sodium silicate and 5 ml of sodium oxalate as the dispersing agent. The activities of the soils used for current study were computed by using the clay content obtained from the hydrometer analysis when 125 ml of calgon 35:7 (a solution comprising 35 grams of sodium hexametaphosphate (NaHMP) and 7 grams of sodium carbonate (Na_2CO_3) in 1 litre of distilled water) is used.

The optimum concentration and volume of the three dispersing agents (calgon, sodium pyrophosphate and sodium tetra pyrophosphate)

Table 1. Atterberg limits and activity of the soils sampled.

Properties	Black Soil	Brown Soil	Red Soil
Liquid limit (LL)	56	32.8	28.3
Plastic limit (PL)	22	24.3	15.1
Plasticity index (PI)	34	8.5	13.2
Clay content (%)	32	5.7	29
Activity (A)	1.07	1.5	0.5

was established using the ASTM 152H: E100 hydrometer instead of the 152H hydrometer. Concentrations of 4.2%, 5%, 7%, 8% and 9% solution of calgon, 3.6%, 5%, 6% and 7% solution of sodium pyrophosphate and 3.6%, 5%, 6% and 7% solution of sodium tetra pyrophosphate were prepared by mixing the required quantity in 1000 ml of distilled water. The quantities of chemicals added for the preparation of the stock solutions are given in Table 2.

2.2 Test series description

For each of the three soil types, hydrometer analyses were conducted with both the 152H and 152H: E100 hydrometers at each concentration of the three dispersing agents and for various volumes.

2.3 Testing procedure and calculations

For all the tests performed, 50 grams of the soil sample passing through 425 micron sieve was mixed with the desired quantity of dispersing agent and about 400 ml of distilled water in a canning jar. The soil-water mixture was allowed to stand overnight. After the mixture had been allowed to stand, it was dispersed for 15 minutes with a standard paddle. The paddle was washed with distilled water, allowing the wash water to run into the container with the suspension.

The suspension was then poured into the Bouyoucos Cylinder and the canning jar was rinsed with distilled water from the wash bottle. The cylinder was then filled with distilled water to the 1130 ml mark with the hydrometer (152H:E100) inside. Then, the hydrometer was removed and the cylinder was inverted a few times using the palm of one hand as a stopper over the mouth of the cylinder to ensure that the temperature was uniform throughout. After bringing the cylinder to the vertical position, stop watch timing was initiated. The hydrometer was inserted and the readings were taken at 18 seconds and 40 seconds without removing the hydrometer from the cylinder. The hydrometer was then taken out and rinsed with water and it was again inserted into suspension when the elapsed time was 2 minutes. The reading was noted and the hydrometer was removed and placed in distilled water. This procedure was repeated for taking readings at elapsed times of 5 minutes, 15 minutes, 30 minutes, 1 hour, 4 hours and 24 hours. After taking each hydrometer reading, the temperature was also recorded and used to correct the readings.

Companion hydrometer tests were conducted on blank solutions comprising only distilled water and dispersing agent with no soil in the identical proportions as solutions prepared with

Table 2. Quantity of chemicals added for preparation of Calgon, Sodium pyrophosphate and Sodium tetra pyrophosphate solution.

| Solution of Concentration (%) | Calgon | | NaPP | NaTPP |
	Quantity of NaHMP Added (g)	Quantity of Na$_2$CO$_3$ Added (g)	Quantity of NaPP Added (g)	Quantity of NaTPP Added (g)
3.6	-	-	36	36
4.2	35	7	-	-
5	40	10	50	50
6	-	-	60	60
7	60	10	70	70
8	70	10	-	-
9	80	10		–

those containing soil. In the case of these blank solutions, the dispersing agent and water mixture was also soaked overnight as done for solutions containing soil. The hydrometer readings taken on the solutions containing soils were reduced by subtracting the hydrometer readings obtained on the companion blank solutions. TMH1 does not make any provision for an adjustment to the hydrometer readings to account for the effect of the dispersing agent on these readings. It was observed that the effect of dispersing agent on the hydrometer reading was significant, as in the case of some solutions the hydrometer readings (in grams per litre) exceeded the mass of the soil used in the test (50 grams).

The calibration curve was plotted. The percentage finer than 0.075 mm, 0.05 mm, 0.04 mm, 0.026 mm, 0.015 mm, 0.01 mm, 0.0074 mm, 0.0036 mm and 0.0015 mm were respectively calculated by the readings taken at 18 sec, 40 sec, 2 min, 5 min, 15 min, 30 min, 1 hour, 4 hour and 24 hours, by means by Equation 1.

$$P = \frac{C \times Sf}{Sm} \quad (1)$$

Where, P = Percentage finer than relevant size,
S_m = Mass of soil fines used in analysis (50 grams),
S_f = Percentage soil fines in total sample (<0.425 mm),
C = Corrected hydrometer reading.

The percentage clay content present in each sample was obtained from the particle size distribution curve.

3 RESULTS AND DISCUSSION

3.1 Black soil

The effect of various hydrometers on optimum values of concentration and volume of all three

dispersing agents (calgon, sodium pyrophosphate and sodium tetra pyrophosphate) for the black soil are shown in Figures 1, 2 and 3, respectively. Table 3 shows the clay contents determined by the various concentrations and volumes. 152 H hydrometer test results yielded the maximum clay content of 35.2% with NaPP as dispersing agent while maximum clay content was 28.2% when the tests were performed with 152H:E100 hydrometer with calgon as dispersing agent.

The following is evident in case of the black soil, from Figures 1, 2 and 3 and Table 3:

• The ASTM hydrometer 152H:E100 (H2) yielded lower clay contents than the 152H hydrometer (H) in the case of all the tests. This percentage difference for the two hydrometers ranged from 38% to −2.54% (The negative percentage shows that the hydrometer H2 yielded higher clay contents than the 152H hydrometer). In the case of all three dispersing agents, the percentage difference at the optimum volume for the different concentrations generally decrease with an increase in the concentration. The range in percentage decrease for the calgon, NaPP and NaTPP were 38% to −2.5%, 34.1% to 19.1% and 16.5% to 5.4%, respectively. The average percentage decrease for the three dispersing agents was 24.1%, 24.2% and 11.2% respectively.
• For the hydrometer tests performed with the 152H hydrometer, 80 ml of NaPP (3.6%) yielded the maximum clay content of 35.2% while 200 ml of calgon (4.2%) yielded a maximum clay content of 35%. In the case of tests performed with hydrometer 152H:E100, 50 ml of calgon (9%) yielded 28.2% while 160 ml of NaPP (5%) yielded 27.5%.
• In the case of the 152H hydrometer results, when comparing clay contents at the optimum volumes, an increase in the concentration of dispersing agent generally resulted in a decreased clay content. This trend was most prominent

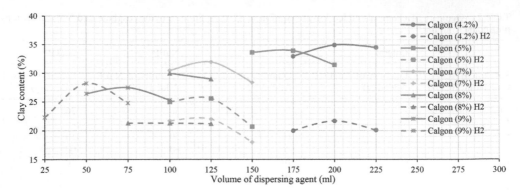

Figure 1. Effect of hydrometers on the optimum values of concentration and volume of calgon in determining the clay content of the black soil.

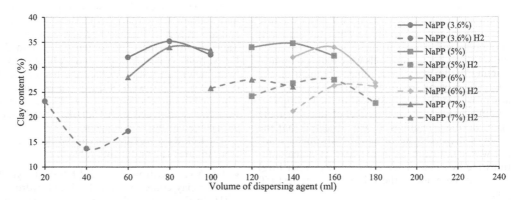

Figure 2. Effect of hydrometers on the optimum values of concentration and volume of sodium pyrophosphate in determining the clay content of the black soil.

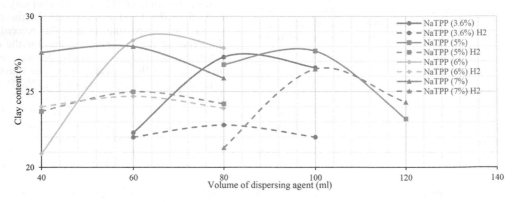

Figure 3. Effect of hydrometers on the optimum values of concentration and volume of sodium tetra pyrophosphate in determining the clay content of the black soil.

in the solutions containing calgon where the optimum volume decreased with an increase in the sodium hexametaphosphate content.

- Hydrometer 152H:E100 is prescribed by two known methods (SANS 3001 and ASTM 1965),

where 125 ml of sodium hexametaphosphate (4%) is the prescribed dispersing agent. So, two additional tests were also conducted with both the hydrometers i.e. 152 H and 152H: E100 with 4% of sodium hexametaphosphate while

Table 3. Variation of clay content depending on hydrometer type on the concentration and volume of dispersing agents added to black soil.

DA type	Volume of DA (ml)	3.6% H	3.6% H2	4.2% H	4.2% H2	5% H	5% H2	6% H	6% H2	7% H	7% H2	8% H	8% H2	9% H	9% H2
	25														22.3
	50													26.4	28.2
	75												21.3	27.5	24.8
Calgon	100					25				30.5	21.7	30	21.3	25.3	
	125					25.6				32	22	29	21.2		
	150					33.7	20.7			28.4	18				
	175			33	20	34									
	200			35	21.7	31.5									
	225			34.6	20.1										
	20	23.2													
	40	13.7													
	60	32	17.2						21.2	28					
	80	35.2							26.3	34					
NaPP	100	32.5							26.2	33.4	25.8				
	120					34	24.2			27.5					
	140					34.8	26.8	32		26.1					
	160					32.3	27.5	34							
	180					22.8	26.9								
	40					23.7	20.9	24							
	60	22.3	22			25		28.4	24.7						
NaTPP	80	27.3	22.8			26.8	24.2	27.9	23.9	27.6	21.3				
	100	26.6	22			27.7				28	26.5				
	120					23.2				25.9	24.3				

following TMH1 guidelines. The 152H and 152H: E100 hydrometers yielded clay contents of 28.8% and 27.3% respectively.

3.2 Light brown soil

The effect of the 152H and 152H:E100 hydrometers on the optimum concentration and volume of all three dispersing agents (calgon, sodium pyrophosphate and sodium tetra pyrophosphate) for the light brown soil is shown in Figures 4, 5 and 6, respectively. The clay contents determined by the different concentration and volume combinations are shown in Table 4. 152H hydrometer test results yielded the maximum clay content of 8.1% with NaTPP as dispersing agent while maximum clay content was 7.5% when the tests were performed with 152H:E100 hydrometer with calgon as dispersing agent.

Figure 4, 5 and 6 and Table 4 show the following.

- ASTM hydrometer 152H:E100 (H2) generally yielded lower clay contents than the 152H Hydrometer (H) in the case of most of the tests. This percentage difference for the two hydrometers ranged from 16.1% to −22.8%. In the case of all three dispersing agents, the percentage difference at the optimum volume for the different concentrations generally decrease with an increase in the concentration. The range in percentage decrease for the calgon, NaPP and NaTPP were 14.5% to −22.8%, 10% to 2.7% and 16.1% to −19.1%, respectively. The average percentage decrease for the three dispersing agents was −1.4%, 5.6% and −1.5% respectively.
- For the hydrometer tests performed with the 152H hydrometer, 80 ml of NaPP (3.6%) yielded the maximum clay content of 35.2% while 200 ml of calgon (4.2%) yielded a maximum clay content of 35%. In the case of tests performed with hydrometer 152H: E100, 50 ml of calgon (9%) yielded 28.2% while 160 ml of NaPP (5%) yielded 27.5%.
- In the case of both the hydrometers i.e. 152H and 152H: E100, with an increase in concentration of dispersing agent beyond optimum concentration, clay content generally decreased.
- Also, when the tests are performed with 125 ml of sodium hexametaphosphate (4%), both

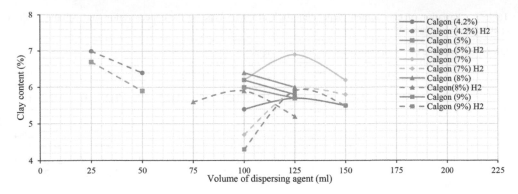

Figure 4. Effect of hydrometers on the optimum values of concentration and volume of calgon in determining the clay content of the light brown soil.

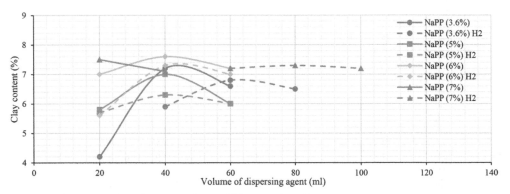

Figure 5. Effect of hydrometers on the optimum values of concentration and volume of sodium pyrophosphate in determining the clay content of the light brown soil.

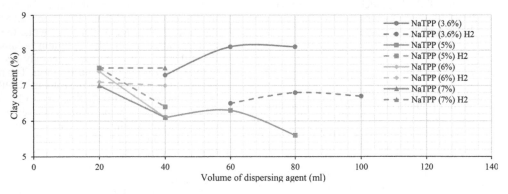

Figure 6. Effect of hydrometers on the optimum values of concentration and volume of sodium tetra pyrophosphate in determining the clay content of the light brown soil.

hydrometers 152H and 152H:E100 yielded the maximum clay content of 8.2%.

3.3 Red soil

The effect of the hydrometers 152H and 152H: E100 on the optimum concentration and the volume of all three dispersing agents (calgon, sodium pyrophosphate and sodium tetra pyrophosphate) for the red soil are shown in Figures 7, 8 and 9, respectively. The clay contents determined by the different concentration and volume combinations are shown in Table 5. 152H hydrometer test results yielded the maximum clay content of 32.3% with

Table 4. Variation of clay content depending on hydrometer type on the concentration and volume of dispersing agents added to light brown soil.

DA Type	Volume of DA used (ml)	Concentration of dispersing agent													
		3.6%		4.2%		5%		6%		7%		8%		9%	
		H	H2	H	H2	H	H2	H	H2	H	H2	H	H2	H	H2
Calgon	25			7		6.7									
	50			6.4		5.9									
	75												5.6		
	100			5.4		6				6.2	4.7	6.4	5.9	6.2	4.3
	125			5.7		5.7				6.9	5.9	6	5.2	5.8	5.9
	150			5.5						6.2	5.8				5.5
	20	4.2				5.8	5.7	7	5.6	7.5					
	40	7.2	5.9			7	6.3	7.6	7.3	7.1					
NaPP	60	6.6	6.8			6	6	7.2	7		7.2				
	80		6.5								7.3				
	100										7.2				
	20					7.5	7.4	7.1	7	7.5					
	40	7.3				6.1	6.4	6.1	7	6.1	7.5				
NaTPP	60	8.1	6.5			6.3									
	80	8.1	6.8			5.6									
	100		6.7												

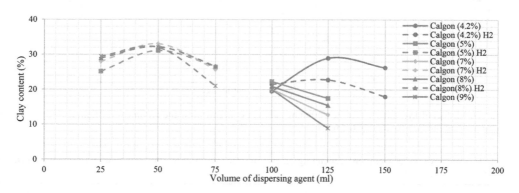

Figure 7. Effect of hydrometers on the optimum values of concentration and volume of calgon in determining the clay content of the red soil.

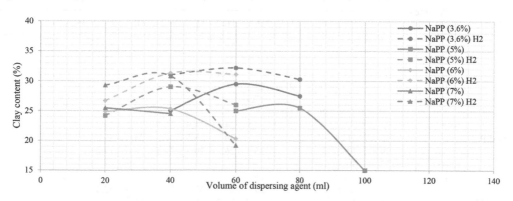

Figure 8. Effect of hydrometers on the optimum values of concentration and volume of sodium pyrophosphate in determining the clay content of the red soil.

NaTPP as dispersing agent while maximum clay content was 33% when the tests were performed with 152H: E100 hydrometer with calgon as dispersing agent.

From Figure 7, 8 and 9 and Table 5, the following is evident.

- In the case of the red soil, ASTM hydrometer 152H: E100 (H2) generally yielded more clay conents than the 152H Hydrometer (H) when calgon and NaTPP are used as dispersing agents, but in the case of NaPP ASTM hydrometer 152H: E100 (H2) yielded lower clay contents than the 152H Hydrometer (H). The percentage difference for the two hydrometers ranged from 30.3% to −65%. In this case the results for the percentage difference at the optimum volume of all three dispersing agents and for all the different concentrations are insignificant. The range in the percentage decrease for calgon, NaPP and NaTPP were 21.4% to −65%, −9.2% to −23.7% and 30.3% to −14%, respectively. The average percentage decrease for the three dispersing agents was −39.4%, −16.9% and 21.3% respectively.

- For the hydrometer tests performed with the 152H hydrometer, the best dispersing agent is 20 ml of NaTPP (6%) which yielded 32.3% of clay content while 60 ml of NaPP (3.6%) yielded a maximum clay content of 29.5%. In the case of tests performed with hydrometer 152H: E100, 50 ml of calgon (7%) yielded 33% while 60 ml of NaPP (3.6%) yielded 32.3%.

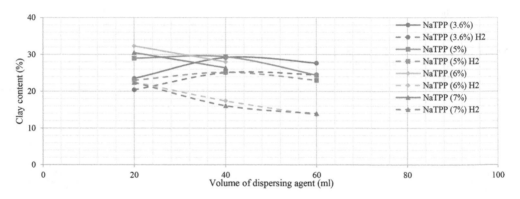

Figure 9.   Effect of hydrometers on the optimum values of concentration and volume of sodium tetra pyrophosphate in determining the clay content of the red soil.

Table 5.   Variation of clay content depending on hydrometer type on the concentration and volume of dispersing agents added to red soil.

DA type	Volume of DA (ml)	3.6%		4.2%		5%		6%		7%		8%		9%	
		H	H2	H	H2	H	H2	H	H2	H	H2	H	H2	H	H2
Calgon	25						25.2				28		29		29.5
	50						31.1				33		32.2		32
	75						26.4				25.7		26.6		21
	100			19.5	21.5	22.3				20		20.9		20	
	125			29	22.8	17.5				12.8		15.5		9	
	150			26.3	18										
NaPP	20					24.2		24.9	26.7	25.5	29.3				
	40	25	31			29		25.3	31.3	24.5	30.8				
	60	29.5	32.2			25		26	20.3	31.1	19.2				
	80	27.5	30.3			25.5									
	100					15									
NaTPP	20	23.5	20.4			29	23	32.3	22.5	30.5	22.4				
	40	29.2	25.1			29.4	25.2	28	17.4	26.4	16.1				
	60	27.7	24.5			24.4	23	13.6		14					

314

- When the tests are performed with 125 ml of sodium hexametaphosphate (4%), hydrometers 152H and 152H: E100 yielded clay content of 20.5% and 21.8% respectively.

From the results discussed above, it has been observed that for the red, black and light brown soil with activities 0.5, 1.07 and 1.5 respectively, the maximum clay content obtained for 152H hydrometer and H2 hydrometer are 32.3%, 35.2% & 8.1% and 33%, 28.2% and 7.5% respectively. When the tests are performed with hydrometer H2, the maximum clay content obtained is higher for more active soil and it decreased for less active soils.

4 CONCLUSIONS

The following conclusions were drawn from the study conducted:

- In the case of the black soil for all three dispersing agents as the concentration increases beyond optimum concentration, the difference in the clay contents yielded by the two hydrometers generally decreases, with H2 hydrometer always yielding lower clay contents. Also, the average of the difference in clay content percentage yielded by the two hydrometers was highest in NaPP solutions (24.2%) and lowest in NaTPP solutions (11.2%).
- When the light brown soil was treated with any of the three dispersing agents, the difference in the clay contents yielded by the two hydrometers generally decreased as the concentration of the dispersing agent increased beyond optimum concentration and the average of the difference in clay content percentage yielded by the two hydrometers was highest in NaPP solutions (5.6%) and lowest in calgon (−1.47%).
- When the red soil was treated with calgon and NaPP, with an increase in the concentration beyond optimum concentration, the difference in the clay contents yielded by the two hydrometers generally increased with H2 hydrometer always yielding greater clay contents. In the case of NaTPP, as the concentration increases beyond optimum concentration, the difference in the clay contents yielded by the two hydrometers generally increased with H2 hydrometer always yielding lower clay contents. Also, the average of the difference in clay content percentage

yielded by the two hydrometers was highest in NaTPP solutions (21.3%) and lowest in calgon (−39.4%).
- The results of this investigation confirm the findings of Means & Parcher (1963) that different dispersing agent types are more effective with certain clay types.
- For tests performed with hydrometer 152H: E100, 125 ml of NaHMP (4%) yielded a lower clay content than the best dispersing agent in case of the black and the red soil and the same clay content in case of the light brown soil.
- When the tests are performed with hydrometer H2, the maximum clay content obtained is greater for more active soil and it decreased for less active soils.
- Finally, the effect of the type of hydrometer on the hydrometer readings was considerable and the clay content obtained by H2 hydrometer is lower for more active soil.

REFERENCES

ASTM. 1965. *Grain size analysis of soils*. ASTM Designation D422–63, ASTM, Philadelphia.
Bindu, J. & Ramabhadran, A. 2010. Effect of concentration of dispersing agent on the grain size distribution of fine grained soil. *Proceedings, Indian Geotechnical Conference*, Mumbai, 275–278.
Head, K.H. 1980. *Manual of Soil Laboratory Testing*, Vol. I, Pentech Press, London.
Indian Standard IS. 1985. *Methods of tests for soils*. IS Designation 2720 Part IV, BIS, New Delhi.
International Soil Reference and Information Centre. 2002. *Procedures for soil analysis. Technical paper 9*, The Netherlands.
Lambe, T.W. 1951. *Soil testing for engineers*, New York: John Wiley & Sons, Inc.
Means, R.E. & Parcher, J.N. 1963. *Physical properties of soils*. Ohio: Charles E. Merrill Book Co.
Ranjan, G. & Rao, A.S.R. 1991. *Basic and applied soil mechanics*, Wiley Eastern Limited, New Delhi.
SANS. 2014. *Civil engineering test methods*. SABS Standard Division Designation 3001 Part GR3, SABS, Pretoria.
Sridharan, A., Jose, B.T. & Abraham, B.M. 1991. Determination of clay size fraction of marine clays. *Geotechnical Testing Journal*, 14(1): 103–107.
TMH1. 1986. *Standard methods of testing road construction materials. Method A6*. Pretoria: National Transport Commission.

Proceedings of the first Southern African Geotechnical Conference – Jacobsz (Ed.)
© *2016 Taylor & Francis Group, London, ISBN 978-1-138-02971-2*

Falling cone test apparatus as liquid limit determination option in South Africa

B.M. Du Plessis
BM Du Plessis Civil Engineering, Pretoria, South Africa

ABSTRACT: Classification of fine soil requires the Atterberg Limits which includes the liquid limit of the material. In the laboratory the plastic limit and linear shrinkage limit can be determined with good repeatability using the standard test methods. However, the liquid limit is the most challenging due to the difficulty of obtaining repeatable results. Two types of liquid limit test apparatus are available in South Africa. Widely known and commonly employed is the Casagrande apparatus. The lesser known test is the falling cone apparatus, currently mainly employed for research, problem soils or other special cases. The author has ten years of experience in commercial soils testing and the falling cone apparatus has been used as the primary liquid limit test. Experience has shown that the apparatus gives reliable and repeatable results and is highly suited for commercial testing. This paper addresses the physical difference between the Casagrande and falling cone apparatuses as well as the differences in the test methods. An overview is given of international and South African specifications for the liquid limit determination. Practical experience with liquid limit determination comparing the Casagrande and the falling cone apparatus results is presented.

1 INTRODUCTION

1.1 Liquid limit

The liquid limit is mainly required in soil classification procedures and in determining Plasticity Index (PI). It is defined as the moisture content, expressed in percentage moist, at which the soil consistency passes from the plastic state to the liquid state, as determined by a liquid limit test apparatus. A fundamental standard method used within the liquid limit evaluation, is the determination of the moisture content of various samples in the procedure. The apparatus used to determine the liquid limit can be the cone penetrometer or the Casagrande apparatus.

1.2 Soil moisture content

The behavior of soil is influenced by the amount of water in the soil and determines the soil condition or phase (solid-semi-solid-plastic-liquid-suspension). Moisture determination forms part of many test methods for soil mechanics and agricultural applications. The measurement of the soil moisture content provides a way of classifying soils and evaluation of soil engineering properties. Soil moisture measurement procedures are simple and basic, but are a critical measurement to be performed correctly. Soil Index tests are relying on the accurate measurement of moisture content of the soils. The oven-drying method for moisture determination is commonly used. Special methods do exist where an oven cannot be used.

Aspects that needs special consideration by the requesting authority, as well as service provider, is the drying temperature appropriate to the material (normally 105–110C), the period of drying (12–24hrs or untill constant weight), representative sample size and the balance or scale capacity, resolution and accuracy.

1.3 Liquid limit test apparatus review

1.3.1 Casagrande apparatus
The classification of soils was first defined by Dr. A. Atterberg for agricultural use in 1911 as reported by Head, (1992). The limits known as Atterberg Limits included the Liquid Limit (LL), Plastic Limit (PL) and Shrinkage Limit (SL) and were originally determined by means of simple hand tests, Casagrande (1932). Further development followed for the application in engineering by Casagrande (1932) by using a percussion apparatus as standard method for the work by Atterberg. Casagrande developed the mechanical device known as the Casagrande apparatus for liquid limit determination. The apparatus became used worldwide for commercial testing and research and has been extensively documented.

In terms of the Casagrande apparatus the liquid limit can be defined as the water content (%), at which a pat of soil in a standard cup and cut by a groove of standard dimensions, will flow together at the base of the groove for a distance of 13 mm when subjected to 25 shocks (blows) from the cup being dropped 10 mm in a standard (Casagrande) liquid limit apparatus operated at a rate of two shocks per second.

This is practically achieved by doing at least 3 determinations (number of blows) at 3 different moisture contents plotting it on a graph with moisture (%) (linear scale) vs. numbers of blows (log scale) followed by a best linear fit through the 3 points and reading the moisture content from the graph correlating to 25 blows.

The linear fit through the semi-logarithmic graph, from the Casagrande results, is referred to as the "Flow Curve". Head (1992) explains that the linear fit is inclined at an angle β to the horizontal axis. Equation 1 below present the flow curve equation expressed as:

$$LL = \omega \left(\frac{N}{25} \right)^{\tan \beta} \tag{1}$$

LL = liquid limit, ω = moisture (%), N = number of blows for the moisture and $\tan \beta$ = slope of the flow curve. Casagrande and Head (1992) notes that various values for the flow curve slope, $\tan \beta$, exists for various types of material.

1.3.2 Falling cone apparatus

The cone penetrometers apparatus has been in use for several decades too. Claveau-Mallet et al. (2012) report that the use of the cone penetrometer method for liquid limit measurement was first proposed by the Geotechnical Commission of the Swedish State Railways in the 1920's, referred to as the Swedish fall cone. Head (1992) reports in the United Kingdom the falling cone method was developed by the United Kingdom's Transport and Road Research Laboratory (TRRL) from various cone tests used in other countries later adopted by British Standard Institute (BSI) in 1975 as a BS standard method. The apparatus is the same as used for bituminous material testing but is fitted with a special cone.

In terms of the cone penetrometer, the liquid limit is defined as the water content (%), at which a pat of soil, placed in a standard cup, is penetrated 20mm in depth with a 30° apex angle cone, weighing 80g, starting the penetration from the surface of the soil placed in the cup.

This is practically achieved by doing at least 4 determinations (penetrations in mm) at 4 different moisture contents, plotting it on graph with moisture (%) vs. the corresponding penetration (mm) followed by a linear fit through the 4 points and reading moisture content from the graph correlating to a 20mm penetration.

1.3.3 Comparison of falling cone and Casagrande methods

Several correlation studies have been performed and research reports by various authors reported by Claveau-Mallet et al. (2012). For liquid limits up 100% there is little difference found between Casagrande and falling cone comparative tests. Head (1992) reports findings by Sherwood & Ryley (1968) that results obtained with the cone method have proved to be more consistent and less liable to experimental and personal errors compared to results obtained by the Casagrande method.

Wood (1990) states the Casagrande apparatus shows poorer repeatability and consistency, is more operator sensitive and in addition cannot provide a direct strength index. A combination of factors influences Casagrande results and Whyte (1970) reports the aspects impacting results most are:

a. Material of base.
b. Base dimension and support on table.
c. Material, dimensions and weight of the cup.
d. Drop height.
e. Operator judgement in groove closing and placing material.

Whyte further reports the proved poor reproducibility by Sherwood (1970) is no surprise considering the variation in operator judgement, machine design and strength and variability soil types.

Jacobsz & Day (2008) presented comparative test results where duplicate samples were sent to South African service providers for soil classification testing in 2006 and again in 2008 and with regards to the liquid limits poor correlations were obtained between service providers using only the Casagrande apparatus. Comparisons between falling cone and Casagrande in South African lacks attention compared to the international trend.

2 SAMPLE PREPARATION AND MOISTURE MIXING

2.1 Sample preparation

The material fraction used for Atterberg Limits testing is specified by all methods, as the material fraction passing the 425 μm sieve. In special cases material fraction passing less than 425 μm could be requested.

The effect of oven drying the material during preparation can cause irreversible changes to physical behavior of soils, especially tropical residual soils. Liquid limits of tropical soils, soils containing organic material, peats and gypsum are severely influenced by oven drying. Best practice would be that any soil used for Atterberg limits tests not be oven dried during preparation (Head, 1992).

Pestle and dry sieving (after oven drying) to prepare liquid limit test material is currently the preparation procedure in the South African National Standard (SANS) test methods. Wet preparation procedures for the removing of material retained on the 425 μm sieve is prescribed by international methods such as BS and ASTM. Wet preparation should be done with distilled water instead of tap water as ion exchange may take place with impurities in the water. Washed material passing the 425 μm sieve should be allowed to settle, then decanted, followed by air-drying to consistency for the first liquid limit determination. Oven drying, even complete drying out by air-drying, should not be permitted (Head, 1992).

2.2 Test sample moisture adjustment

Irrespective of the method used for determining the liquid limit, an important factor affecting success of the test is the thoroughness of the mixing of the soil with the distilled water for the required determination(s), (Head, 1992). Mixing of the water into the soil is a demanding action and requires special care, effort, dedication and skill by the operator. This can be appreciated considering the shape of soil particles, in material less than 425 μm, and its effects on water absorption.

3 TESTING STANDARDS

The Casagrande method is covered in the standards of most countries but mainly the British Standard (BS) and American Society for Testing and Material (ASTM). Other Casagrande method standards are the Indian Standard (IS), German Institute for Standardization (DIN), Canadian standard (CAN), the Swedish Standards (SS) and the South African National Standards (SANS).

The falling cone method is covered by the British, Swedish and Canadian standards.

For the purpose of this paper the BS, ASTM and SANS methods are considered for liquid limit determination.

3.1 British Standard—BS

The British Standards for the liquid limit are given in the BS1377: Part 2:1990 Clause:

- 4.3 Liquid limit (cone penetrometer)
- 4.4 Liquid limit (one-point cone method)
- 4.5 Liquid limit (Casagrande apparatus)
- 4.6 Liquid limit (one-point Casagrande method).

The BS1377:1990 prohibit the drying out of material during preparation. The method emphasizes the importance of (distilled) water mixing on a glass plate till a uniform distribution of water is achieved.

For the Casagrande one-point method the flow curve slope, tan β, is assumed as tan β = 0.092, representing most British soils (Head, 1992). Head reports that in the British Standard the Casagrande method is regarded as a subsidiary method.

The cone method is the BS recognized method for liquid limit testing. The method requires two penetrations to repeat within 0.5 mm for each moisture point before advancing. When this is followed proper mixing is confirmed and results will be accurate without having to repeat the test.

Head (1992) reports the suggested use of the falling cone one-point cone method by Clayton & Jukes (1978) which is based on statistical analysis of experimental data. The one point method is practical in situations where little material is available but less accurate than the multi-point result and should be restricted to special requirements.

3.2 American Society for Testing and Material—ASTM

Head (1992) reports the BS1377: Part 2:1990 Clause 4.5 and Clause 4.6 are similar to the ASTM D4318. Differences are the hardness of the rubber base of the Casagrande apparatus, different grooving tools, four determinations as opposed to three in the ASTM procedure and grooving the material in one stroke as opposed to several strokes in the ASTM procedures.

The ASTM D4318 allows wet and dry preparation of material and specifies wet preparation as default. No oven drying is allowed during wet preparation. It specifies drying out at room temperature together with mixing on a glass plate until the consistency to start the test is reached. The material is then transferred to a "mixing and storage container". Mixing to adjust the moisture during the test procedure is to be done in this container as opposed to a glass plate. Distilled water is recommended for preparation and testing.

The ASTM D4318 makes provision for three point and one point liquid limit determination where the latter is performed on request only. The standard admits that the correlation used for the one point calculation may not be applicable to all soils. The one point liquid limit is calculated using Equation 1 and the average is reported.

The ASTM standard does not present a test method using the falling cone apparatus to determine the liquid limit.

3.3 South African standards

3.3.1 Technical methods for highways—TMH 1
In South Africa the TMH 1 (1986) was historically the testing standard for the testing of road construction materials. Applicable to Atterberg limits test sample preparation is Method A1(a): "The wet preparation and sieve analysis of gravel, sand and soils samples". Although the methods description states "wet preparation" the material is subjected to several oven drying cycles during preparation.

The liquid limit determination is described in Method A2: "The determination of the liquid limit of soils by means of the flow curve method". Three method options are included in Method A2:

a. Flow curve method (multi points on graph).
b. One-point method (similar to ASTM and BS).
c. Two-point method (not found in other standards).

All TMH 1 liquid limits testing methods are based on the Casagrande apparatus.

3.3.2 SANS—South African National Standards
TMH 1 was changed to the South African National Standard (SANS) in 2013, and the methods applicable to liquid limit testing are reflected as the SANS 3001-GR series. During the process of developing the SANS 3001-GR series the liquid limit determination of the TMH 1 Method A2 has effectively been divided into three individual test standards for determining the liquid limit, still using only the Casagrande apparatus. The applicable standards for the SANS 3001 series are:

- GR1: Wet preparation and particle size analysis. As for TMH 1 Method A1(a), oven drying forms part of the preparation.
- GR10: Determination of the one-point liquid limit, plastic limit, plasticity index and linear shrinkage.
- GR11: Determination of the liquid limit with the two-point method.
- GR12: Determination of the flow curve liquid limit.

The two-point method is not applied in the BS or ASTM standards and appears to be a derivative of the one-point method. For the one- and two-point methods the tests specifies ranges for the numbers of blows to fall in, as well a maximum

of 7% between each liquid limit determined and their average. This requires specimen drying and calculation before success is confirmed. GR 10 and GR11 advise to use GR12 for materials with a plasticity index of more than 20. It fails to specify on how this will be determined before the actual testing is done.

A comparative summary of the testing standards are given in Table 1 including notes with regards to the SANS standards for liquid limit.

4 APPARATUS COMPARISON CASE STUDY

4.1 Background

A client requested indicator tests done for a dam project in Ugie, Eastern Cape, South Africa. The liquid limit determinations were initially done using the BS1377: Part 2:1990 Clause 4.3 Liquid limit (cone penetrometer) described by Head (1992). Results of two samples were redone and the liquid limit was determined using the TMH 1 Method A2 (as for SANS 3001-GR12) with the Casagrande apparatus. The tests were repeated on the same samples as initially prepared for the falling cone testing after retrieved from storage. Sample moisture adjustments were done in the same procedure as for the falling cone. Sample 1 was classified as a Sandy Lean Clay (CL) and Sample 2 a Sandy Elastic Silt (MH) according to the ASTM Unified Classification System based on the falling cone results.

Table 1. Liquid limit standards with notes on SANS standards.

SANS 3001	Description	Notes
GR 1	Wet sieve	Involves oven drying
GR 10	One-point LL, PL & LS	Refers to GR 11 & 12 if LL >20; Range 22–28 blows; Repeated once to get 2 LL's; Equation 1, tan $\beta = 0.121$, used for LL's; Average LL's
GR 11	Two-point LL	Refers to GR 12 if LL >20; Range 20–24 and 26–30 blows; Equation 1, tan $\beta = 0.121$, used for LL's; Average the LL's
GR 12	Flow curve LL	Plots 3 points on graph to make linear fit and read LL for 25 blows. Range 15–22; 22–28 and 28–35 blows
GR 20	Moisture	

4.2 *Falling cone test data*

The falling cone test data initially performed for the two samples are reported in Table 2 and Table 3 below.

Figure 1 and Figure 2 represent the falling cone moisture-penetration relation on a graph for the data in Table 2 and 3. The liquid limit is determined at 20mm penetration.

Table 2. Falling cone test data: Sample 1.

Determination number	Cone penetration (mm)	Averaged penetration (mm)	Moisture content (%)
1	12.7	12.65	43.50
	12.6		
2	17.1	16.90	46.06
	16.7		
3	18.3	18.05	47.28
	17.8		
4	20.2	20.45	48.86
	20.7		

Table 3. Falling cone test data: Sample 2.

Determination number	Cone penetration (mm)	Averaged penetration (mm)	Moisture content (%)
1	9.6	9.85	46.96
	10.1		
2	12.8	12.80	50.20
	12.8		
3	18.7	18.65	53.84
	18.6		
4	21.1	21.05	55.48
	21.0		

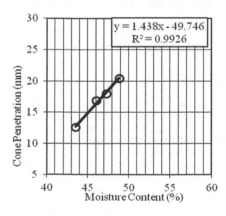

$$y = 1.438x - 49.746$$
$$R^2 = 0.9926$$

Figure 1. Moisture vs. cone penetration Sample 1.

4.3 *Casagrande test data*

The Casagrande test data for Sample 1 and Sample 2 is summarized in Table 4.

The data in Table 4 was used to determine Graph for Sample 1 and Sample 2 in Figure 3 and Figure 4 respectively. On each graph the resulting flow curve has been added fitting an exponential

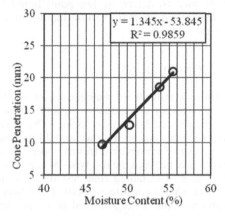

$$y = 1.345x - 53.845$$
$$R^2 = 0.9859$$

Figure 2. Moisture vs. cone penetration Sample 2.

Table 4. Casagrande test data Sample 1 and 2.

Determination number	Sample 1		Sample 2	
	Number of blows	Moisture (%)	Number of blows	Moisture (%)
1	50	40.34	35	50.86
2	41	44.57	31	50.77
3	29*	42.67	39	48.59
4	22*	43.67	20*	52.10
5	29*	44.11	26*	52.42
6	24*	47.76	17	56.89
7	13	78.58	13	56.33

* In SANS two-point range: 20–24 & 26–30.

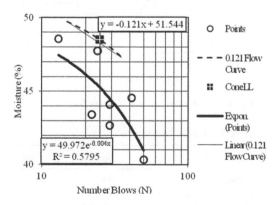

$$y = -0.121x + 51.544$$
$$y = 49.972e^{-0.004x}$$
$$R^2 = 0.5795$$

Legend: ○ Points; - - - 0.121 Flow Curve; ▓ ConeLL; ▬ Expon (Points); — Linear (0.121 FlowCurve)

Figure 3. Moisture vs. log of blows Sample 1.

321

trend line with its formula to calculate the liquid limits corresponding 25 blows.

Figure 3 and Figure 4 in addition shows the related liquid limit determined with the falling cone as ConeLL with the flow curve at slope at tan $\beta = 0.121$ intersecting 25 blows, taking that the falling cone gave the correct liquid limit. Points falling in the ranges to perform the two-point method were used to determine two-point liquid limits. Table 5 summarizes the various liquid limit results for the samples.

5 CONCLUSIONS

Various standards are making use of the falling cone as testing apparatus for liquid limits applied in commercial laboratory testing. The reliability and accuracy has been shown and documented by a number of researchers. In this paper a practical experience between the two types of apparatus was presented. The falling cone test data was found to give reliable consistent results. The falling cone apparatus and testing method is independent of operator interpretation.

With the Casagrande apparatus the results became more comparable with the falling cone after more data points were obtained, although these points do not fall in the prescribed number of blow ranges. In practice, if any of these three points (or two points in the two-point method) would be used, it is evident by inspection that in such a case the results will be more different, with some result even unrealistic.

The Casagrande shows a poor curve fit even with an increased number of data points. The flow curve slope prescribed in the standards does not always align to the specified slope of tan $\beta = 0.121$.

Pinning of the flow curve on the 25 blow vertical line is related to the curve slope of the scattered data points. As a result using the two-point method, with it applied on a flow curve slope position for the data at hand, gives an incorrect intersect with the 25 blows vertical line, but still complies with the quality control measures in the standard. Wrongly, the liquid limit is effectively randomly determined and checked for quality.

The specified ranges of blows for the Casagrande gives a 3% moisture window where the falling cone would permit 15% moisture range determinations.

Comparing the graph plots for the different apparatus it is evident that the Casagrande presents inconsistencies compared to the falling cone. The Casagrande apparatus is very sensitive in the rather small ranges specified to be operated in, and the sensitivity relates to the various variables influencing the Casagrande results. Compared to the falling cone many more repeats is required by the Casagrande method if done correctly.

The falling cone requires a larger sample which slows mixing. However, after four penetrations the result is certain and a test repeat it not likely required.

Considering the known and document challenges the Casagrande apparatus test methods pose, it is recommended that the South African standard be reconsidered for liquid limit testing with regards to the following aspects:

- Liquid limit sample preparation practice.
- Liquid limit moisture adjustment and mixing.
- Apparatus and test methodology incorporating to the falling cone as standard testing apparatus.
- One-point and two-point liquid limit test only allowed for special condition testing.

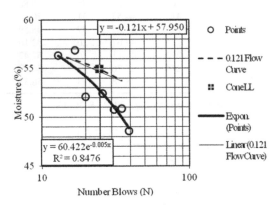

Figure 4. Moisture vs. log of blows Sample 2.

Table 5. Liquid limit results.

Apparatus	Sample 1	Sample 2
Falling cone	49	55
Casagrande flow curve	45	53
Casagrande two-point	43 or 46	52

REFERENCES

ASTM D4318. 2008. *Standard test methods for liquid limit, plastic limit and plasticity index of soils*. ASTM International, West Conshohocken, PA.

BS 1377: Part 2:1990. *Methods of test for soils for civil engineering purpose*.

Casagrande, A. 1932. Research of Atterberg Limits of Soils. *Public Roads*. Vol. 13(8): 121–136. Baltimore.

Claveau-Mallet, D. Duhaime, F. & Chapuis, R.P. 2012. Practical considerations when using the Swedish fall cone. *Geotechnical Testing Journal*. Vol 35(4): 618–628.

Clayton, C.R.I. & Jukes, A.W. 1978. A one point cone penetrometer liquid limit test? *Géotechnique*. Vol. (4): 469–472.

Head, K.H. 1992. *Manual of soil laboratory Testing. Vol. 1: Soil Classification and compaction tests*, 2nd Ed. London.

Jacobs, S.W. & Day, P. 2008. Are we getting what we pay for from geotechnical laboratories? *Civil Engineering*, April 2008:8–11.

SANS 3001-GR10:2013, *Civil engineering test methods—Part GR10: Determination of the one-point liquid limit, plastic limit, plasticity index and linear shrinkage.* Ed. 1.2, Pretoria, SABS.

SANS 3001-GR11:2013, *Civil engineering test methods—Part GR11: Determination of the liquid limit with the two-point method.* Ed. 1.2, Pretoria, SABS.

SANS 3001-GR12:2013, *Civil engineering test methods—Part GR12: Determination of the flow curve liquid limit.* Ed. 1.2, Pretoria, SABS.

SANS 3001-GR1:2013, *Civil engineering test methods—Part GR1: Wet preparation and particle size analysis.* Ed. 1.2, Pretoria, SABS.

SANS 3001-GR20:2013, *Civil engineering test methods—Part GR20: Determination of the moisture content by oven-drying.* Ed. 1.1, Pretoria, SABS.

Sherwood, P.T. & Ryley, D.M. 1968. An examination of the cone penetrometer methods for determining the liquid limits of soils. *RRL report. Vol. 233 of TRRL Laboratory reports*, Transport and Research Laboratory, Crowthorne, Berk.

Sherwood, P.T. 1970. The reproducibility of the results of soil classification and compaction tests. *TRRL Laboratory reports, Report LR 339*, Transport and Research Laboratory, Crowthorne, Berk.

TMH 1:1986, *Standard methods of testing road construction materials.* 2nd ed. Pretoria: National Transport Commission.

Whyte, I.L. 1970. Soil plasticity and strength—a new approach using extrusion. *Ground Engineering*, Vol 15:16–20,

Wood, D.M. 1990. *Soil behaviour and critical state of soil mechanics.* New York, Cambridge University Press.

Soil reinforcement and slopes

Proceedings of the first Southern African Geotechnical Conference – Jacobsz (Ed.)
© *2016 Taylor & Francis Group, London, ISBN 978-1-138-02971-2*

Energy considerations in intense-rainfall triggered shallow landslides

G.W. Waswa
Masinde Muliro University of Science and Technology, Kakamega, Kenya

S.A. Lorentz
SRK Consulting Ltd, Pietermaritzburg, South Africa

ABSTRACT: Previous studies indicate that most rainfall-triggered shallow landslides are initiated by a spiked rainfall-intensity, which typically occurs several hours into a critical rainfall event, in which the slide is triggered. The critical rainfall event also usually occurs after several days of the antecedent rainfall. Rainfall triggers landslides via rapid increase in pore-water pressure, commonly associated with rapid infiltration. However, based on the above timings of landslide occurrences, this paper argues that the rapid increase in pore-water pressure is a result of diffusive and rapid transmission of intense-rainfall induced pressure-head into a tension saturated (or near-saturated) soil profile. This argument is supported by a conceptually new pressure-head diffusion equation. Antecedent and critical rainfalls are significant in creating tension saturated continuous pore-water phase, necessary for the rapid transmission of the induced pressure head to a potential failure plane.

1 INTRODUCTION

1.1 *Landslide triggering by rain infiltration*

Rainfall triggers shallow landslides via rapid increase in pore-water pressure, common in slopes with residual soils and areas that experience intense rainfall (Premchitt et al. 1986, Rahardjo et al. 2008). However, while the empirical relationships between rainfall and shallow landslides have been greatly investigated (Caine 1980, Wilson & Wieczorek 1995, Guzzetti et al. 2008), the involved physical processes are not yet clear. A common explanation is that the intense rainfall influences landslide-triggering pore-water pressure by rapid infiltration into the soil profile. For instance, Campbell (1974) states that high intensity critical rainfall is required so that water infiltrates at a faster rate than it can drain away through the underlying subsoil, thereby causing landslide triggering transient pore-water pressure. Fourie (1996) argues that a prolonged antecedent rainfall and a critical rainfall of particular intensity and duration are required for a wetting front to arrive at a critical depth for failure to occur.

1.2 *Antecedent and critical rainfall*

A critical rainfall event, however, typically occurs after several days of antecedent rainfall (Table 1 & Fig. 1), when antecedent soil moisture levels are high. The critical rainfall event is characterized by spiked intensities that cause landslide-triggering

rapid pore-water pressure responses. For instance, in Hong Kong, a study by Premchitt et al. (1986) concluded that the number of landslides increased dramatically with the hourly intensity of the critical rainfall and the majority of the landslides occurred within four hours of peak rainfall-intensity; some studies have recorded lesser lag-times (Table 1). The pattern of the critical rainfall has also been found to be significant (Tsai 2008), and that hyetographs with a peak at the end of a rainfall event have a stronger destabilizing effect than those with a constant rainfall or with a peak at the beginning of a critical rainfall (D'Odorico et al. 2005).

1.3 *Tension saturation soil profile*

From the above studies, it appears that the rapid pore-water pressure responses that trigger shallow landslides are a result of the action of an intense rainfall on a pre-wetted soil profile. This was evident in the experiments reported by Iverson (2000), where the soil was pre-wetted by application of low-intensity rainfall to raise moisture contents to near saturation levels without producing positive pressure heads; higher-intensity rainfalls were then used to elevate groundwater pressures and trigger slope failure. This report indicates that for the spike rainfall-intensity to rapidly elevate pore-water pressure and trigger landslide, the soil profile should be in a state of tension saturated (or near-saturated). Certainly, Waswa et al. (2013) found that when an intense rainfall acts at the surface

Table 1. Characteristics of intense-rainfall triggered landslides from some selected studies.

| Reference | Location | Date & time of landslide | Antecedent rainfall | | Critical rainfall | | Peak intensity | | Lag-time* |
			Period (days)	Amount (mm)	Period (hours)	Amount (mm)	mm/h	Time	(Hours)
Hurliman et al. (2003)	Switzerland	06.08.2000, 7 am	38	189	20	106	11	4 am	3
Guzzetti et al. (2004)	Italy	23.11.2000, 5 pm	45	600	24		30	4 pm	1
Chen et al. (2005)	Taiwan	01.12.2000, 2 pm	20	580	38	261	35	11 am	3
Cardinali et al. (2006)	Italy	05.12.2004, 5 am	64	360	24	45	10	3 am	2
Dahal et al. (2006)	Japan	20.10.2004, 3 pm			32	450	76	1 pm	2
Kuriakose et al. (2008)	India	22.01.2007, 6 am	25	997	14	147	147	3 am	3

*Lag-time between the peak rainfall intensity and the time of landslide.

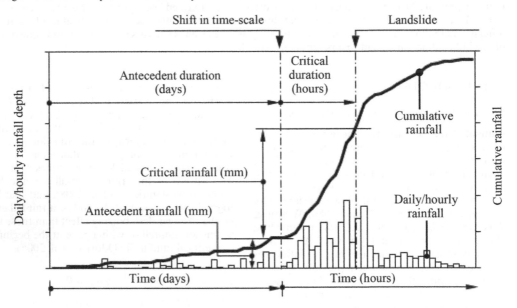

Figure 1. Antecedent and critical rainfall parameters (modified from Aleotti 2004).

of a tension saturated soil profile, an additional energy, proportional to the rainfall intensity, is rapidly induced into the profile, elevating the pressure head at every depth. These findings have been supported by a pressure-head diffusion equation (transient pressure wave model: Waswa & Lorentz 2015a, b) that simulates the diffusive transmission and elevation of pressure head in a pre-wetted soil profile. While a diffusion type equation has been used by Reid (1994), to estimate rainfall induced slope failure, this was still based on an infiltration process, as have many subsequent studies (Godt et al. 2012, Shao et al. 2015, Suradin 2015).

The objective of this paper is to extend the transient pressure wave model (Waswa & Lorentz 2015a, b) to the intense-rainfall triggered shallow landslide phenomenon.

2 TRANSIENT PRESSURE WAVE MODEL

2.1 Differential equation

The diffusive transmission of pressure head from the ground surface through pore-water in a tension saturated soil profile can be described by the equation (Waswa 2013, Waswa & Lorentz 2015a):-

$$\frac{\partial h_w}{\partial t} = d_e \frac{\partial^2 h_w}{\partial y^2}, \tag{1}$$

in which, h_w [L] is the pressure head (potential energy per unit weight), d_e [L²T⁻¹] is pressure head diffusivity coefficient (Waswa & Lorentz 2015b) and, y[L] and t [T] are the space and time coordinates, respectively

Equation 1 is solved for the reduced initial conditions:

$$h_w(y,0) = 0, \text{ for } y \geq 0 \tag{2}$$

and for the boundary conditions:

$$h_w(0,t) = h_w(t), \text{ for } t \geq 0 \tag{3}$$

at the ground surface (upper boundary), and

$$\frac{dh_w(y,t)}{dy} \rightarrow 0 \qquad \text{as } y \rightarrow \infty \tag{4}$$

as a semi-infinite boundary condition.

The solution to Equation 1, which also satisfies the initial and boundary conditions in Equations 2–4, is:

$$h_w(y,t) = h_w(0,t)\,\mathrm{erfc}\left(\frac{y}{\sqrt{4d_e t}}\right), \tag{5}$$

analogous to the heat conduction solution (Carslaw & Jaeger 1959). $h_w(y, t)$ is the pore pressure head at any given depth y and time t after the imposed surface conditions $h_w(0, t)$.

2.2 Evaluation of the model

Equation 5 was evaluated using field pressure-head observations from a wetland zone in the Weatherley research catchment, located in the northern Eastern Cape province of South Africa, and for the rainfall events of the summer season of 2000/2001. The physical parameters of the soil at the observation site were as shown in Table 2. The instrumentation comprised two ceramic-cup tensiometers, installed at 0.2 m and 1m below the ground surface.

Eleven rainfall events caused rapid pore-water pressure responses at the observation site. These events occurred in the middle of the rainfall season, when the water table at the observation site was shallow and, based on the pore air entry pressure head value of the soil (Table 2), the zone of tension saturation extended to the ground surface. For the present purpose of evaluating the model (Eq. 5),

the tensiometric responses at the observation site in two events, namely Event 47 and Event 70 (Table 2), are used. The observed values of pressure head at the shallower tensiometer (0.2 m below ground surface) were used as the upper boundary conditions (Eq. 3) in Equation 5 to predict the pressure head at the deeper tensiometer. The predicted and the observed values agreed well (Fig. 2)

2.3 Application of the transient pressure wave model

Equation 5 can be combined with an Infinite Slope Model (ISM), commonly used to predict the stability of a shallow slope based on a dimensionless Factor of Safety (FS). FS greater than 1 indicates stable conditions and FS less than 1 indicates unstable conditions. The ISM can be expressed as (Iverson 2000):

$$FS = \frac{\tan\varphi}{\tan\alpha} + \frac{c}{\gamma_s y \sin\alpha\cos\alpha} - \frac{h_w(y,t)\gamma_w \tan\varphi}{\gamma_s y \sin\alpha\cos\alpha} \tag{6}$$

Table 2. Physical parameters of the soil at the observation site and the characteristics of rainfall Events 47 and 70 of the summer season 2000/2001.

Parameter and units	Symbol	Value
Hydraulic and energy parameters		
Saturated hydraulic conductivity (m/h)	K_{sat}	0.07
Pore air entry pressure head (m)	h_a	0.45
Density of water (N/m³)	γ_w	9800
Energy diffusivity coefficient (m²/h)	d_e	0.186
Soil physical parameters		
Saturated unit weight	γ_s	22000
Soil composition	Loamy sand	
Coarse sand (%)	CS	13
Medium sand (%)	MS	17
Fine sand (%)	FS	36
Silt and clay (%)	C&S	36
*Mechanical properties		
Slope angle (degrees)	α	35
Friction angle (degrees)	φ	20
Cohesion (N/m2)	c	30000
Rainfall characteristics	Event 47	Event 70
Date of the event	18 Jan 2001	10 Mar 2001
Time of the event	1800–1941	1315–1424
Event duration (minutes)	101	69
Amount (mm)	11.2	47.2
6-min peak intensity (mm/h)	22;26	36;92
Time of peak intensity	1810;1817	1339;1409

*Assumed/hypothetical values.

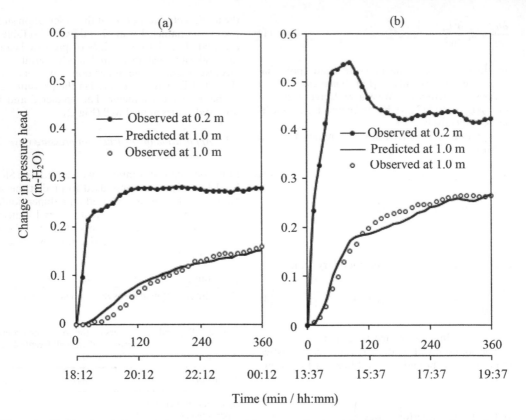

Figure 2. Predicted and observed changes (responses) in pressure head at 1.0 m below ground surface at the observation site in the Weatherley research catchment in South Africa in (a) Event 47 and (b) Event 70 of the 2000/2001 summer season.

where, φ is the friction angle of the soil in degrees, α is the slope of potential failure plane in degrees, c is the cohesion of the soil in N/m^2, γ_s is the saturated unit weight of the soil in N/m^3, γ_w is the unit weight of water in N/m^3, and $h_w(y, t)$, generated by Equation 5, is the space and time dependent pore pressure head in m. Note that the structure of Equation 6 allows the factor of safety to take on negative values, when the third term on the right hand side is larger than the sum of the first two terms.

All the terms in Equation 6 depend on soil water content. However, when the soil profile is in a tension saturated (or near-saturated) state, as it usually appears during the latter period of the critical rainfall but before landslide initiation, soil water content may not vary significantly and rapidly to account for the observed rapid landslide-triggering pore-water pressure (Premchitt et al. 1986). Therefore, Equation 6 can be reduced to be dependent only on depth and time as follows:-

$$FS = A + F_c(y) + F_w(y,t) = A + \frac{B}{y} + \frac{C.h_w(y,t)}{y}, \quad (7)$$

where, $F_c(y)$ is the depth-dependent cohesion factor and $F_w(y,t)$ is the depth-and time-dependent pore-water pressure head factor; and A, B and C are constants: $A = \tan\varphi/\tan\alpha$, $B = c/\gamma_s \sin\alpha\cos\alpha$ and $C = \gamma_w \tan\varphi/\gamma_s \sin\alpha\cos\alpha$.

2.4 Illustration with a hypothetical hillslope

Here, we use a hypothetical hillslope of the mechanical properties indicated in Table 2, but with the soil characteristics and hydrological conditions as the observation site in the Weatherley Research Catchment. The predicted pressure-head distribution down the soil profile in the two events, using Equation 5, is as shown in Figure 3. The predicted pressure head values were input in Equation 7 to generate the factor of safety down the soil profile (Fig. 4).

3 DISCUSSION

Field results of pressure head observations indicated that the extension to the ground surface of

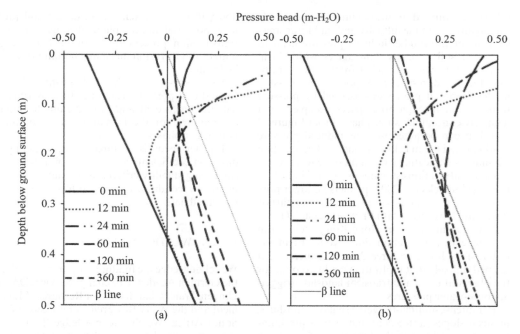

Figure 3. Predicted pressure head in the soil profiles at the observation site in the Weatherley research catchment in South Africa during (a) Event 47 and (b) Event 70 of the 2000/2001 summer season. 0 minutes corresponds to 1812h and 1337h for (a) Event 47 and (b) Event 70, respectively (see Fig. 2).

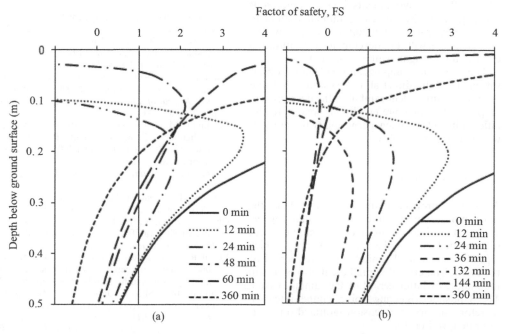

Figure 4. Factor of safety as a result of change in pore-water pressure (Fig. 3) and for a hypothetical slope of soils mechanical properties and slope angle indicated in Table 2). 0 min corresponds to 1812h and 1337h for (a) Event 47 and (b) Event 70, respectively (see Fig. 2).

a tension saturated (or near-saturated) soil profile was significant in the diffusive and rapid transmission of intense-rainfall induced pressure head to the lower soil horizons. This agrees with Marui et al. (1993), who concluded from a field study, that the rapid response of pore-water pressure in the deep soil profile was assisted by pressure transmission through pore spaces occupied by a relatively continuous water phase. From Figures 2 and 3, it can be noted that the induced pressure head reached 0.2 m below ground level in less than 12 minutes, indicating a rapid pressure wave transmission. Infiltration estimates using the HYDRUS model (Šimůnek et al. 2005) predicts far lower pore water pressure responses at 1 hour after the start of the event. This agrees with Premchitt et al.'s (1986) report that landslides in Hong Kong occurred in soils in which infiltration times would be in the order of 14 hours to three days; and therefore it seemed unlikely that infiltration could account for the observed rapid and transient landslide-triggering pore-water pressures.

For intense-rainfall-triggered shallow landslides, in steep slopes, antecedent and critical precipitation are significant in the formation of a tension saturated or near-saturated continuous water phase from the ground surface to a potential failure plane. This requirement of the presence of a continuous water phase distinguishes the significance of the antecedent rainfall and critical rainfall in different environments. Freely drained soils will require significant critical rainfall amount to develop near saturated conditions, while low permeability soils will start developing near saturation condition with significant antecedent rains.

This discussion supports and explains the observations by Rahardjo et al. (2008) who found that the role of antecedent rainfall in the development of pore-water pressure conditions was more significant in soils with low permeability than in soils with higher permeability.

4 CONCLUSION

A review of previous studies has indicated that the rapid pore-water pressure response that triggers shallow landslide during a critical rainfall event is generated by a spike rainfall intensity, which typically occurs at the middle or latter period of the event. The critical rainfall event also occurs in the middle or latter period of the rainfall season. The landslide-triggering spike rainfall-intensity, therefore, occurs on a tension saturated (or near-saturated) soil profile.

The pressure-head diffusion equation presented in this paper indicates that the intense rainfall-induced pressure head is rapidly transmitted down through the tension saturated soil profile to a potential failure plane through an energy transmission process that can be better represented by a diffusion mechanism than infiltration hydraulics.

Since pore-water is the medium through which the induced pressure head is diffusively transmitted, the antecedent and critical rainfall amounts are required for the formation of a continuous pore-water phase.

Further focus is required in determining the diffusion coefficients for various soil types and in simulating stability criteria using rainfall intensity boundary conditions in the diffusion equations presented here.

ACKNOWLEDGEMENTS

We acknowledge the funding support received from the German Academic Exchange Service (DAAD), Umgeni Water Chair for Water Resources Management and the Water Research Commission (WRC) South Africa. We also acknowledge the technical support received from the School of Engineering and Centre for Water Resources Research, University of KwaZulu-Natal.

REFERENCES

Aleotti, P. 2004. A warning system for rainfall-induced shallow failures. *Engineering Geology* 73: 247–265.

Caine, N. 1980. The rainfall intensity: duration control of shallow landslides and debris flows. *Geografiska Annaler. Series A. Physical Geography*: 23–27.

Campbell, R.H. 1974. Debris flows originating from soil slips during rainstorms in Southern California. *Quarterly Journal of Engineering Geology and Hydrogeology* 7(4): 339–349.

Cardinali, M., Galli, M., Guzzetti, F., Ardizzone, F., Reichenbach, P. & Bartoccini, P. 2006. Rainfall induced landslides in December 2004 in south-western Umbria, central Italy: types, extent, damage and risk assessment. *Natural Hazards and Earth System Science* 6(2): 237–260.

Carslaw, H.S. & Jaeger, J.C. 1959. *Conduction of heat in solids*. New York: Oxford University Press.

Chen, C., Chen, T., Yu, F. & Lin, S. 2005. Analysis of time-varying rainfall infiltration induced landslide. *Environmental Geology* 48: 466–479.

Dahal, R.K., Hasegawa, S., Yamanaka, M. & Nishino, K. 2006. Rainfall triggered flow-like landslides: understanding from southern hills of Kathmandu, Nepal and northern Shikoku, Japan. *Proceedings of the 10th International Congress of IAEG, the Geological Society of London, IAEG2006 Paper* (819): 1–14.

D'Odorico, P., Fagherazzi, S. & Rigon, R. 2005. Potential for landsliding: dependence on hyetograph characteristics. *Journal of Geophysical Research* 110: 2003–2012.

Fourie, A.B. 1996. Predicting rainfall-induced slope instability. *Proceedings of the ICE-Geotechnical Engineering* 119(4): 211–218.

Godt, J.W., Şener-Kaya, B., Lu, N. & Baum, R.L. 2012. Stability of infinite slopes under transient partially saturated seepage conditions. *Water Resources Research* 48 W05505 doi:10.1029/2011WR011408:1–14.

Guzzetti, F., Cardinali, M., Reichenbach, P., Cipolla, F., Sebastiani, C., Galli, M. & Salvati, P. 2004. Landslides triggered by the 23 November 2000 rainfall event in the Imperia Province, Western Ligura, Italy. *Engineering Geology* 73: 229–245.

Guzzetti, F., Peruccacci, S., Rossi, M. & Stark, P. 2008. The rainfall intensity-duration control of shallow landslides and debris flows: an update. *Landslides* 5: 3–17.

Hürlimann, M., Rickenmann, D. & Graf, C. 2003. Field and monitoring data of debris-flow events in the Swiss Alps. *Canadian Geotechnical Journal* 40(1): 161–175.

Iverson, R.M. 2000. Landslide triggering by rain infiltration. *Water Resources Research* 36: 1897–1910.

Kuriakose, S.L., Jetten, V.G., Van Westen, C.J., Sankar, G., & Van Beek, L.P.H. 2008. Pore water pressure as a trigger of shallow landslides in the Western Ghats of Kerala, India: some preliminary observations from an experimental catchment. *Physical Geography* 29(4): 374–386.

Marui, A., Yasuhara, M., Kuroda, K. & Takayama, S. 1993. Subsurface water movement and transmission of rainwater pressure through a clay layer. *Proceedings of the Yokohama Symposium, on the Hydrology of Warm Humid Regions*, IAHS Publication No. 216: 463–470.

Premchitt, J., Brand, E.W. & Phillipson, H.B. 1986. Landslides caused by rapid groundwater changes. *Geological Society, London, Engineering Geology Special Publications* 3(1): 87–94.

Rahardjo, H., Leong, E.C. & Rezaur, R.B. 2008. Effect of antecedent rainfall on pore-water pressure distribution characteristics in residual soil slopes under tropical rainfall. *Hydrological Processes* 22(4): 506–523.

Reid, M.E. 1994. A pore pressure diffusion model for estimating landslide-inducing rainfall. *Journal of Geology*, 102:709–717.

Shao, W., Bogaard, T.A., Bakker, M. & Greco, R. 2015. Quantification of the influence of preferential flow on slope stability using a numerical approach. *Earth Syst. Sci.* 19:2197–2015.

Šimůnek, J., van Genuchten, M.Th. & Šejna, M. 2005. The HYDRUS-1D softwarepackage for simulating the one-dimensional movement of water, heat, and multiple solutes in variably-saturated media. Version 3.0. HYDRUS. Softw. Ser. 1. Dep. of Environ. Sci., Univ. of California, Riverside, CA.

Suradi, M. 2015. *Rainfall induced failure of natural slopes in tropical regions*. PhD Dissertation, School of Civil, Environmental and Mining Engineering, University of Western Australia.

Tsai, T. 2008. The influence of rainfall pattern on shallow landslide. *Environmental Geology* 53: 1563–1569.

Waswa, G.W. & Lorentz, S.A. 2015a. Transmission of pressure head through the zone of tension saturation in the Lisse effect phenomenon. *Hydrological Sciences Journal* (just-accepted).

Waswa, G.W. & Lorentz, S.A. 2015b. Energy considerations in groundwater ridging mechanism of streamflow generation. *Hydrological Processes* 29(23): 4932–4946.

Waswa, G.W. 2013 *Transient pressure waves in hillslopes*. PhD Thesis, School of Engineering, KwaZulu-Natal, Durban, South Africa. 120 pages.

Waswa, G.W., Clulow, A.D., Freese, C., Le Roux, P.A.L. & Lorentz, S.A. 2013. Transient pressure waves in the vadose zone and the rapid water table response. *Vadose Zone Journal* 12(1). DOI: 10.2136/vdzj2012.0054.

Wilson, R.C. & Wieczorek, G.F. 1995. Rainfall thresholds for the initiation of debris flow at La Honda, California. *Environmental and Engineering Geoscience* 1(1): 11–27.

Proceedings of the first Southern African Geotechnical Conference – Jacobsz (Ed.)
© *2016 Taylor & Francis Group, London, ISBN 978-1-138-02971-2*

Analysis of a slope failure above critical infrastructure

L. Newby & R. Armstrong
SRK Consulting, Johannesburg, South Africa

G. Maphanga
Nkomati, Machadodorp, South Africa

ABSTRACT: Following the observed instability of a slope immediately above the primary crusher installation at Nkomati mine in Mpumalanga, it was necessary to investigate the cause of the failure and define the remedial measures required to stabilise the slope. The stability condition of the slope was determined by means of a back analysis through the calculation of factors of safety and probabilities of failure, as is customarily done for slope design. Three scenarios were considered in terms of slope idealization: (1) Using partially saturated conditions calculated from the foundation indicator test results; (2) Using the total stress as given by triaxial test results combined with an undrained shear strength and cohesion varying with depth; (3) Using the effective stress as given by the triaxial test results combined with a worst case surface water table. The most critical scenario was further analysed in terms of possible remedial measures.

1 INTRODUCTION

In early 2010, a small toe failure and cracks developed on the slope above the primary crusher at Nkomati and a review the slope design was requested. The analysis indicated that, even at the relatively low design angle of 23°, the slope was sensitive to groundwater and slope angle. It was recommended that the slope remained unchanged with the installation of survey and groundwater monitoring systems.

Following an instability of the slope during December 2012, it was necessary to provide recommendations on managing the failure (Armstrong and Moletsane, 2014), investigate the risk to the crusher installation and operating staff, investigate the cause and redefine the remedial measures required to provide long term stability.

2 SITE INVESTIGATION

2.1 *Site observations*

A large crack, approximately 3 m deep, had developed to the north of the slope and further smaller cracks formed on the slope extending from the 1375 m level to the upper section of the slope. A layout of the slope is presented in Figure 1 and a representative cross section through the slope in Figure 2.

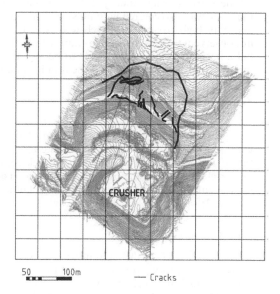

Figure 1.   Cracks, indicated by dotted lines, above the primary crusher which is located on the flat area to the south.

Observations of the slope indicated that degradation of the toe had occurred resulting in undercutting of the slope.

Waste rock was in the process of being placed at the toe in an attempt to buttress the slope and

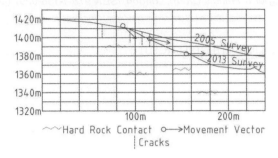

Figure 2.   Observed cracks and hard rock contact level as indicated by boreholes.

prevent further failure. It was assumed that the slope was undercut vertically by approximately 5 m, which resulted in an increase of the overall slope angle from 23° to 25°.

2.2 Groundwater regime

Water was observed in an open drill hole on the 1382 m bench, near the central and highest part of the slope, during January 2013. This part of the slope was also covered with moss, which is evidence for high groundwater levels when seepage is not observed. Further inspections during low rainfall months revealed the drill hole to be dry. Two months after the initial site visit the drill hole had no standing water.

2.3 Borehole logs

Logs corresponding to a total of 236 m of core were considered for the estimation of the properties of the geotechnical units. The hard rock contact level indicated by the boreholes is shown on a section cut through the center of the slope failure (Figure 2).

2.4 Monitoring data

Monitoring data from the radar monitoring system as well as monitoring prisms are used to interpret vertical and horizontal movement vectors along the main sections of the slope, and aid in identifying the failure surface.

2.5 Inclinometer readings

Three inclinometers were installed on the 1382 m bench in July 2013. These readings were analysed in order to ascertain the appropriate depth of the failure surface and indicated that a deep seated failure occurred at approximately 20 mbgl (meters below ground level) and a shallow failure surface

formed at approximately 6 mbgl. Using these readings, together with the analysis from the previous sections, an approximate failure profile and geometry of the slope are determined.

3 MATERIAL CHARACTERISATION

3.1 Laboratory testing program

The laboratory testing program included foundation indicator tests and triaxial compression strength tests, undertaken on a range of undisturbed samples selected from the slope. The triaxial compression test comprised a saturated, consolidated undrained test. Since no drainage is permitted the moisture content remains constant during compression, and the resulting changes in pore pressure are measured at the base of the specimen.

The Atterberg limits from the foundation indicator tests were used to classify the failed material as an ML to CL using the USCS classification.

3.2 Strength parameters

The estimation of the strength parameters was based on site observations, on the results of the laboratory testing program as well as a series of back analyses carried out on the slope.

In order to obtain an accurate representation of the failed slope, three cases were considered for the analysis, namely: (1) effective stress under partial saturation conditions; (2) total stress in undrained conditions; and (3) effective stress in undrained conditions.

The saturated as well as the partially saturated density of the failed material was determined using the triaxial test results. From the triaxial tests, the cell pressure is measured to be the minor principal stress (σ_3) and the major principal stress (σ_1) is taken as the sum of the all-round pressure and the applied axial stress. The pore pressure (μ) is measured by means of an electronic pressure transducer under "no flow" conditions of the triaxial test, and enables the results to be expressed in terms of effective stress (σ'_1 and σ'_3).

The stress path parameters, denoted by s and t (in kPa), are given by:

$$s = \tfrac{1}{2}(\sigma_1 + \sigma_3) \tag{1}$$

$$t = \tfrac{1}{2}(\sigma_1 - \sigma_3) \tag{2}$$

A number of specimens were tested from each sample, each under a different value of all-round pressure. The failure envelope as denoted by s and t, can thus be drawn and the shear strength parameters of the soil determined. The failure envelopes of the triaxial tests for effective and total stress parameters are presented as Figures 3 and 4. A line of best fit is drawn through the plotted points, and the shear strength parameters are calculated and summarised in Tables 1 and 2.

These results are only applicable in situations where the site drainage conditions correspond to the test conditions (in this case undrained). Since the water conditions on site are uncertain, it becomes necessary to also consider the shear strength of the failed material under partially saturated conditions. In the case of normally consolidated clays, the undrained shear strength (C_u) is assumed to increase linearly with an increase in effective vertical stress σ'_v (i.e. with depth). Using the results from the triaxial tests, C_u can be expressed in terms of total stress with the equation:

$$C_u = (\sigma_1 - \sigma_3) \tag{3}$$

In addition to this, Skempton (Craig, 2004) proposed the following correlation between the ratio C_u/σ'_v and the plasticity index (PI) for normally consolidated clays:

$$\frac{C_u}{\sigma'_v} = 0.11 + 0.0037PI \tag{4}$$

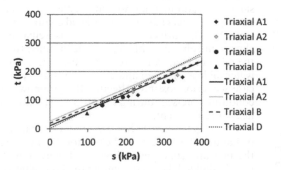

Figure 3. Effective stress parameters.

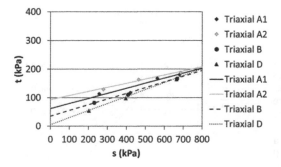

Figure 4. Total stress parameters.

Table 1. Effective stress failure envelopes.

Test	Effective stress failure		Effective stress envelope	
	Slope	Intercept	Φ' (°)	c' kPa
A1	0.49	19.67	29.2	22.5
A2	0.50	22.88	29.8	26.4
B	0.48	17.39	28.7	19.8
D	0.55	1.30	33.1	1.6

Table 2. Total stress failure envelopes.

Test	Total stress failure		Total stress envelope	
	Slope	Intercept	Φ' (°)	c kPa
A1	0.18	60.89	10.1	61.8
A2	0.14	92.00	8.2	93.0
B	0.20	35.15	11.2	35.8
D	0.24	5.53	13.8	5.7

In order to investigate the change in C_u with depth, the equivalent depth (E) representing the effective and total stress from the triaxial test results are determined by:

$$E' = \frac{s'}{\frac{9.81(ysaturated - 1000)}{1000}} \tag{5}$$

$$E = \frac{s}{\frac{9.81(ysaturated - 1000)}{1000}} \tag{6}$$

A plot of C_u versus σ'_v is presented in Figure 5. In order to take the C_u into account as calculated using Skempton's formula, the C_u as calculated in terms of total stress was plotted together with Skempton's C_u against the equivalent depth in Figures 6 and 7. The reference line in Figure 7 indicates the "best fit" line and follows the plot of y = 3.5x + 20, where y represents C_u and × represents depth.

4 SLOPE STABILITY ANALYSIS

4.1 Back analysis

Combinations of strength parameters and pore pressures that give a Factor of Safety (FoS) of 1.0 as well as the observed slip surface are investigated in this section. In summary, the objectives of the back analysis are:

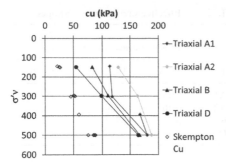

Figure 5. C_u versus σ'_v.

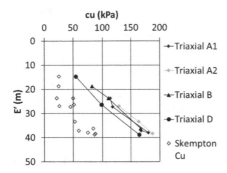

Figure 6. C_u versus E'.

1. Obtain shear strength material properties (co-hesion and friction angle) at failure which will correspond with ML to CL type materials in the USCS classification as well as the laboratory data; and

2. Estimate general water conditions at failure for the design of remedial measures.

The stability of the pit slopes was assessed through limit equilibrium stability analyses utilising SLIDE from RocScience. This section describes the analysis carried out for the evaluation of the stability of the slope.

4.2   Geotechnical section

The slope stability analyses were carried out on a section based on the current topography and the geology derived from the geotechnical boreholes and laboratory testing data. Since the open drill holes installed on the slope indicated transient groundwa-ter conditions, a range of groundwater scenarios are analyzed in order to identify one that replicated the sliding surface geometry at a FoS of 1.0.

The pore pressure is calculated by means of a H_u coefficient ranging from 0–1, where $H_u = 0$ rep-resents a dry soil and $H_u = 1$ indicates hydro-

static conditions. SLIDE automatically estimates the pore pressure using the formula: $H_u = Cos^2\alpha$, where α represents the inclination of the water sur-face to the horizontal. The resulting pore pressure (μ) is then calculated as:

$$\mu = H_\mu y_w h \qquad (7)$$

4.3   Limit equilibrium analysis

The stability analyses were carried out for the three scenarios considered in terms of assumed site con-ditions along the section.

4.3.1   Effective stress under partial saturation conditions

A partially saturated condition was analysed using the effective stress envelope of $c' = 20$ kPa and $\Phi' = 29°$ (Table 1) coupled with a low water table condition. The partially saturated unit weight as calculated was used to represent the unit weight of the failed material. The results of the analysis show that a high FoS of 2.03 is achieved, indicat-ing that either the cohesion and/or friction angle is too high or that a high water table/pore pressure is acting on the slope.

4.3.2   Total stress in undrained conditions

The undrained slope condition was analysed con-sidering a saturated unit weight coupled with total stress parameters based on undrained compression strength. The surface cohesion value as well as the change in cohesion with depth is determined from the reference line in Figure 7. A conservative, high water table is assumed and corresponds to a worst case scenario, possibly reflecting the groundwater conditions pre-sent during the rainy season after a heavy rainfall event.

This resulted in a FoS = 0.93 and a Probability of Failure (PoF) = 22.5%. The relationship between the FoS and cohesion is shown in Figure 8, with the highlighted data representing the FoS achieved in the SLIDE model.

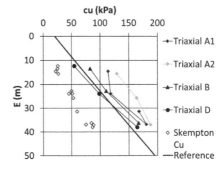

Figure 7. C_u versus E.

4.3.3 *Effective stress in undrained conditions*

The effective stress condition was analysed using the saturated unit weight of the material and the effective stress parameters. The analysis was completed assuming a surface water table and a range of Hu from 0–1, automatically selected by SLIDE. From the analysis, a FoS of less than 1 corresponds to an H_u value ranging from 0.9–1.0. The relationship between the FoS and the Hu value is graphed in Figure 9. From the figure, a FoS smaller than or equal to 1, which represents the conditions on site at the time of the failure, corresponds to an Hu value ranging from 0.9–1.0.

The worst case scenario of $H_u = 1$ was thus used as an input to further compare the value of cohesion with that of friction angle. By specifying a range of realistic values for the Mohr-Coulomb parameters, various combinations of these resulting in a specified FoS can be analysed, as shown in Figure 10. The graph indicates a specified lower and upper bound FoS of 0.85 and 1 respectively, which depicts the site conditions on the slope at the time of the failure.

The final selected combination of strength parameters for the material from Figure 10 are thus c = 20 kPa and $\Phi = 29°$.

The PoF of this model is calculated at 85.8%, which makes it the most critical model compared to those analysed previously. This model will there-fore be used to depict the slope parameters at the time of failure.

5 SLOPE STABILITY REMEDIATION

5.1 *Evaluation of options*

An analysis was conducted to review and design appropriate remedial measures to provide long term stability to the slope. The options considered in the analysis included:

– leaving the existing buttress as is;
– adding to the existing buttress with additional waste rock or stock pile material;
– dewatering;
– backfill of the existing crack;
– with the displacement that has already taken place in the slope, anchorage is not considered to be an option; and
– with the steep incline of the existing topography, a cut back and flattening of the slope is not considered to be an option.

It was assumed that the waste rock material will be compacted in place. The material is therefore expected to have relatively high cohesive strength.

For this analysis, a value of c = 50 kPa and $\Phi = 37°$ has been assumed. It is also necessary to adjust the cohesion value for the failed material to accurately represent the site conditions on the slope post-failure. A cohesion value of 20 kPa was used to represent the site conditions prior to the slope failure, however since we are now dealing with a failed material, it becomes necessary to utilise the residual strength of the soil. A cohesive strength of 0 kPa is thus a more realistic estimate of the current slope condition. The adopted strength parameters of the material as calculated from the back analysis are c = 0 kPa and $\Phi = 29°$.

5.1.1 *Buttressing*

Analysis of the existing buttress resulted in a FoS of 1.06, which correlates to the current site conditions

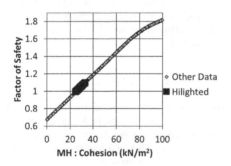

Figure 8. The relationship between FoS and cohesion for undrained conditions.

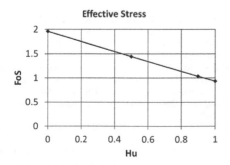

Figure 9. The relationship between FoS and *Hu*.

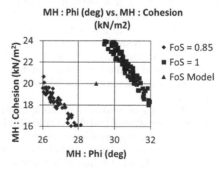

Figure 10. The relationship between cohesion and friction angle.

of experiencing small movements. This value is too low for long term stability, and an additional layer of waste rock was added to the existing buttress. This yielded a FoS of 1.46. The internal stability of the new buttress was analysed and produced a FoS of 1.33. In order to further increase the FoS, slope dewatering was recommended.

5.2 Dewatering

The heavy rainfall close to the escarpment will cause the water table in the slope to rise, increasing hydraulic pressure in the slope. It was recommended that the groundwater be controlled by installation of a parallel array of horizontal drains in order to intercept the groundwater flow and reduce the piezometric buildup. An analysis considering the effect of installing free draining horizontal wells between the 1382 m bench and the crest of the slope was conducted. The result of lowering the groundwater table increases the overall slope stability to a FoS of 1.93 with the PoF approaching 0%.

5.3 Surface water control and backfill of existing tension crack

The existing tension crack and surface water drain should be back filled so as to prevent ponding and additional water ingress into the slope. The crack should be cleared of existing debris then filled and compacted in 500 mm increments. The bottom two layers (1 m) should comprise a clayey material so as to prevent water ingress. The top 500 mm should also consist of a clayey material, while general fill may be used in between these layers. Water should not be allowed to pond on the existing 1364 m bench and run off from the slope should be routed to the stream locations on either side of the cut.

5.4 Monitoring and risk management

The mine currently utilises a robotic system to monitor various prism stations at critical locations on the slope. It is important that the prism locations and movement/vector data be plotted on plans and sections to illustrate movement within both stable and unstable areas of the crusher slope. This monitoring will also provide an indication of the effectiveness of the sub horizontal drains, buttress and backfilled tension crack in stabilising the slope movement.

Observations of rainfall and slope monitoring records in Armstrong and Moletsane (2014) indicated that there is a significant increase in slope movement following large rainfall events (greater than 30 mm). It was recommended that any rainfall event greater than 20 mm should be considered a "trigger" event and monitoring data should be reviewed and the area evacuated if necessary.

6 CONCLUSIONS

In order to maintain long term stability of the slope, various options were analysed with the following conclusions and recommendations.

- Strength parameters for the rock mass have been estimated from the data collected with the drill core and laboratory testing investigations with an adequate reliability for the purpose of assessing the stability conditions of the crusher slope.
- The failure occurred in the silty clayey material (weathered diabase) which is classified as an ML to CL soil with shear strength properties of $\Phi' = 29°$ and $c' = 0$–20 kPa.
- The results of the stability analysis indicate that additional waste rock needs to be added to the existing buttress, and horizontal dewatering holes should be installed to prevent the build-up of groundwater pressure in the slope.
- It is recommended that the buttress be extended with a 5 m step out at the toe of the slope and the surface of the buttress extending 60 m from the 1382 m bench level.
- The existing crack should be cleared, filled and shaped with layers of clay and general fill.
- Constant slope monitoring should continue, mak-ing additional use of the installed inclinometers.
- Significant rainfall events (>20 mm) should be considered as potential trigger events and the monitoring data should be reviewed following each event.

ACKNOWLEDGEMENTS

The authors acknowledge the owners and management of the Nkomati Mine for the permission to publish the results of this study and Peter Terbrugge of SRK Consulting for his ongoing guidance with the project.

REFERENCES

Armstrong, R. & Moletsane, K. 2014. Management of the Nkomati Mine Crusher Slope Failure. *The Journal of The Southern African Institute of Mining and Metallurgy*, 114, pp. 1–4.
BSi Standards. 1990. Part 8: Shear strength tests (effective stress). *British Standard Methods of test for Soils for civil engineering purposes*, BS 1377: Part 8.
Craig, R.F. 2004. *Craig's Soil Mechanics Seventh Edition*. SPON.
Rocscience. 2002. Slide v.5.0–2D Limit Equilibrium Slope Stability Analysis.
Terbrugge, P.J. & Dlokweni, T. 2010. Review of the stability of the Nkomati Primary Crusher Slope. *SRK Consulting May 2010*.

Proceedings of the first Southern African Geotechnical Conference – Jacobsz (Ed.)
© 2016 Taylor & Francis Group, London, ISBN 978-1-138-02971-2

Stability of an undermined quay wall in Port Elizabeth

G. Wojtowitz & J.U.H. Beyers
Aurecon, Pretoria, South Africa

ABSTRACT: This paper presents detailed analyses undertaken to investigate the stability of existing quay wall structures along two berths of the Charl Malan Quay, Port of Port Elizabeth. The quay wall was undermined by ship movement, the extent of which varied along the wall and ranged from 10% to 25% of base area. At first hand calculations were undertaken considering ultimate limit states. Plaxis 2D finite element analyses were undertaken to assess soil-structure interaction in more detail and investigate failure mechanisms that were not anticipated in the initial calculations. The value of a geotechnical investigation, in providing representative design parameters, is shown in the analyses as this was only conducted following an initial assessment for which representative parameters had to be assumed. The analyses provided a deeper understanding of the behaviour of the quay wall subjected to undermining and could be used to evaluate the risks associated with operation of the quay.

1 INTRODUCTION

The existing quay wall structures along berths 102 and 103 of the Charl Malan Quay in the Port of Port Elizabeth were severely undermined due to ship movement. The Transnet National Ports Authority (TNPA) appointed Aurecon to investigate the current stability of the existing quay wall structures. This assessment would provide information to inform a decision on whether to allow further ship mooring at the respective berths and if so, whether to allow the use of the crane.

Port Elizabeth is located on the south eastern coast of South Africa in the Eastern Cape Province. This assessment formed part of a pre-feasibility study for the establishment of additional container capacity in Nelson Mandela Bay, focusing on solutions within the Port of Port Elizabeth to complement the Port of Ngqura.

At first, the stability assessment was undertaken by means of hand calculations considering the ultimate limit states of overturning, sliding and bearing capacity. Very limited geotechnical information was available and the nature and parameters of the founding material, as well as the retained material, was assumed. Plaxis 2D finite element analyses were undertaken in order to assess soil-structure interaction in more detail, and investigate failure mechanisms that may have not been anticipated in the initial calculations. Due to limited available ground information, a parametric study was undertaken to assess the representativeness of assumed parameters. The developed finite element model was used to determine the degree of undermining required to cause failure of the wall under different load cases.

Following this assessment, a geotechnical investigation was undertaken to determine the nature of the founding and retained material. This provided information with regards to the geological profile and representative geotechnical design parameters. The stability assessment of the wall was undertaken using this updated information. This paper shows how the analyses provided a deeper understanding of the behaviour of the quay wall subjected to undermining and could be used to evaluate the risks associated with operation of the quay. In addition, the value a geotechnical investigation adds to a geotechnical design problem is highlighted.

2 QUAY WALL

The quay wall structure comprises a blockwork wall with a stone backing. A schematic drawing of the quay wall is shown in Figure 1. The wall is 17.7 m high with a base width of 9.5 m and a top width of 3.4 m.

The quay wall was undermined by ship movement over a 30 m length along the existing berth. Divers' inspections and survey results indicated that the extent of the undermining below the quay wall ranged from 10% to 25% of base area varying along the extent of the wall. The divers' survey results are summarised in Table 1. The undermined base width is calculated as the distance of the undermined section underneath the wall measured from the toe of the wall. According to the survey, the maximum base width that has been undermined is 2.4 m and the average is 1.2 m. The

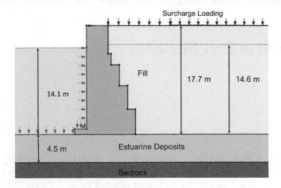

Figure 1. A schematic drawing of the quay wall.

Table 1. Summary of divers' survey results.

Position*	Base width undermined (m)	Percentage of base width undermined (%)
1	0.0	0
2	2.4	25
3	2.4	25
4	1.0	10
5	1.0	10
6	1.8	19
7	0.0	0
Average	1.2	13

* Measurements taken at 6m centres along the length of the wall. The percentage is expressed as the percentage of the full base width of 9.5 m.

percentage of wall that is undermined is expressed as the percentage of the full base width of 9.5 m. Undermining of an average dimension of 1.2 m across the entire undermined section of quay wall relates to approximately 13% of the foundation contact area.

3 INITIAL ASSESSMENT

The initial stability assessment considered the ultimate limit states for overturning, sliding and bearing capacity. Very limited geotechnical information was available and the nature of the ground profile had to be assumed. The founding material was assumed to be a granular type material. This assumption was based on the interpretation that the quay wall could not be founded directly on rock due to the fact that undermining of the wall had occurred. This was supported by information from Transnet reporting a recent telephone conversation with an employee who had been engaged on the scheme from 1933 to 1939. The employee described the quay wall as being "founded on

compacted sand probably with stones in it". The fill material retained behind the wall was assumed to be a sandy material. The geotechnical design parameters assumed for the initial assessment are included in Table 3.

In order to investigate the effect of the extent of undermining on the stability of the wall, an analysis was undertaken for a range of 0%, 10%, 20%, and 38.5% of the base area undermined. The analysis for the ultimate limit states considered the equilibrium of the loads acting on the wall as well as the mass making up the wall. Undermining of the wall was accounted for by a reduction in the lever arm when considering moments causing stability and instability of the wall. A summary of the stability results for the initial assessment is shown in Table 2.

As observed in Table 2, the factor of safety for the overturning limit state approaches 1 at a percentage undermined base width of 38.5%. The factors of safety determined for the other two ultimate limit states do not decrease significantly within the undermined range considered, thus the overturning limit state is the most critical. Working with a factor of safety of 2 for overturning of

Table 2. Summary of stability results for initial assessment.

Percentage of base width undermined (%)	Factor of safety		
	Overturning	Sliding	Bearing Capacity
0	2.1	1.4	16.2
10	1.8	1.4	12.5
20	1.5	1.4	8.7
38.5	1.0	1.4	1.6

Table 3. Comparison of geotechnical design parameters assumed for the initial and Plaxis 2D analyses.

Design Parameter	Initial analyses		Plaxis 2D analyses	
	Fill	Founding material	Fill	Founding material
Unit weight (kN/m³)	19	20	19	20
Cohesion c' (kPa)	1.0	1.0	4.0	4.0
Friction angle φ' (°)	34	30	42	40
Dilatancy angle ψ (°)	0	0	5.0	4.0
Young's Modulus E (MPa)	–	–	40	60
Poisson's ratio v	0.3	0.3	0.3	0.3

a retaining wall shows that any undermining of the wall increases the risk of instability. The results indicate that the failure point of the wall would occur when the wall is undermined by 38.5% or a distance of 3.7 m from the toe of the wall. The divers' survey results shown in Table 1 indicate the maximum distance undermined was 2.4 m. Thus from these initial results the quay wall's stability is a high risk.

4 FINITE ELEMENT ANALYSIS

The above initial assessment, which is governed by the overturning limit state, purely considers the equilibrium of the loads and masses, and does not account for soil-structure interaction. In addition, undermining of the wall cannot be explicitly represented in these calculations. A Plaxis 2D finite element analysis was conducted in order to assess soil-structure interaction in more detail as well as investigate failure mechanisms that may not have been anticipated in the initial calculations. The Plaxis 2D finite element input also allowed one to attempt to explicitly model undermining of the wall.

A Plaxis 2D finite element model was set up based on the geometry of the wall as shown in Figure 1. At first the validity of the parameters assumed for the initial assessment was investigated by means of a parametric study. The "representativeness" was judged by virtue of whether the wall could be made to stand up safely under the design loading conditions.

The wall was unstable using the design parameters assumed for the initial hand calculations. Hence, the need for the parametric study. A comparison of the resultant parameters determined from the sensitivity study to those used in the initial assessment is shown in Table 3.

The initial assessment assumed a founding material comprised of granular material. The results from the parametric study represent a material more in line with "compacted sand containing stone or gravel" or an otherwise densified granular material. In addition, it was originally assumed that the fill material behind the wall comprised a sandy material. A drawing subsequently provided by Transnet portrayed the fill material behind the wall as a rock backfill. The design parameters resulting from the study represent a rock fill material, which permitted raising the expected shear strength parameters of the backfill.

The rock backfill and founding material were modelled using a Mohr-Coulomb material model. The concrete quay wall was modelled as a linear-elastic, non-porous material. Each individual block forming the wall was not modelled, instead the quay wall was modelled as one material. The granular founding material was assumed to be 4.5 m thick below founding level with bedrock underlying this layer. The depth to bedrock was unknown. The rock was assumed to be a linear elastic drained material with a Young's Modulus of 30 MPa. The assumed rock stiffness is high, given that the quartzitic sandstone encountered in Port Elizabeth is known to be folded and fractured. The purpose of this assumption however was to "force" the mode of failure to occur underneath the footing as opposed to occurring into the rock and to limit the contribution of rock settlement in relation to what the quay wall would experience.

The loads applied to the wall consisted of container stack loads of 5 kPa applied a distance of 20 m from behind the wall, 10 kPa from 20 m to 40 m behind the wall and 40 kPa from 40 m to the end of the model as well as a bollard horizontal pull force of 30 kN/m and a vertical crane rail load of 561 kN/m. The resultant bollard load from ships mooring occurs at an angle, and was resolved into a vertical and horizontal component. However, only the destabilizing horizontal component was included in the model. The in-plane load and uplift load due to the angled load were ignored as they were perceived to be stabilizing loads. A tidal lag of 1 m was applied (Hong Kong Civil Engineering Department, 2004) as a vertical distributed load behind the quay wall as well as a triangular horizontal stress distribution on the back of the wall.

The effect of undermining on stability was considered for the operational load case consisting of the container stack loads, tidal lag, crane loading and bollard load. The boundaries of the model were set such that boundary effects did not influence the displacement or stress profile of the modelled structure. The model input in Plaxis 2D is shown in Figure 2. The model was used to determine the degree of undermining required to cause failure of the wall under the applied load case.

Undermined areas were modelled by removing elements from the finite element mesh. This effectively created a void with finite dimension in plane and infinite dimension out of plane. The calculation phasing of the model comprised of setting up the initial stresses; wishing in place the quay wall; placement of the fill material behind the wall in consecutive layers; application of the distributed container loads behind the wall and tidal lag; application of the bollard pull force and crane rail load; application of respective undermining for the model. A strength reduction analysis was performed for the final step to determine the global factor of safety against failure.

The simulated failure mechanism following undermining is shown in Figure 3 (showing the zone of elements reaching the Mohr-Coulomb

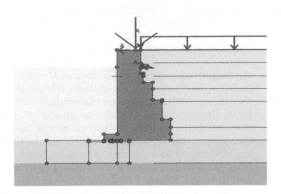

Figure 2.   Model input into Plaxis 2D.

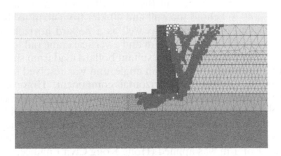

Figure 3.   The failure mechanism from the Plaxis 2D analysis showing the deformed mesh.

failure criterion). The deformed mesh showing a visualization of the quay wall and ground movement response upon failure is shown in Figure 4. The failure mechanism shows an active wedge-like failure zone behind the quay wall that extends below the wall. A passive failure zone occurs at the toe of the quay wall with a mechanism representing sliding occurring along the base of the quay wall. This failure mechanism could not be anticipated during the initial "hand" calculations.

The results of the Plaxis 2D analyses are shown in Table 4. It is evident that any undermining poses an increase in risk on the stability of the quay wall. Failure occurs at 20% undermining of the wall. This is different to the result from the initial assessment where failure occurred at 38.5% undermining. The 13% undermining analysis was conducted as it was found as the average percentage undermining according to the divers' survey.

5   GROUND INVESTIGATION

Following the stability assessments described above, a detailed ground investigation was undertaken comprising the drilling of seven onshore and

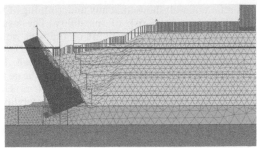

Figure 4.   The deformed mesh showing the failure of the quay wall from the Plaxis 2D analysis.

Table 4.   Summary of Plaxis 2D results.

Percentage of base width undermined (%)	Base width undermined (m)	Global Factor of Safety	Horizontal movement (mm)
0	0.0	1.4	13
10	0.95	1.2	39
13	1.2	1.2	56
20	1.9	1.0	143
25	2.4	Failed	-

seven offshore boreholes and ten Dynamic Probe Super Heavy (DPSH) tests onshore. The fill material encountered behind the quay wall comprised loose to medium dense, silty sand with quartzite gravel, cobbles and boulders. In general the fill is compacted, however less competent zones occur within areas affected by tidal fluctuations. This differed substantially from the initial assumption of a rock fill.

The quay wall concrete footing and fill material are underlain by estuarine deposits comprising silty sand. This in turn is underlain by interlayered residual mudrock and sandstone. Bedrock comprises of completely weathered to highly weathered, closely to medium jointed, very soft rock siltstone or mudrock (Beyers, 2014). Standard Penetration Test (SPT) results indicated the silty sand fill material to have an overall dense consistency. This was confirmed by the Dynamic Probe Super Heavy (DPSH) tests.

Revised geotechnical parameters were determined based on the geotechnical investigation and are shown in Table 5. These can be compared with the parameters used in the initial assessments as shown in (Table 3). The friction angle φ' and cohesion c' are lower than those initially anticipated.

The Plaxis 2D finite element analyses were repeated using these revised parameters. The results of the revised analyses are shown in Tables 6 and 7 which include a comparison with the results from the initial analyses. This comparison shows that the model is extremely sensitive to changes in the Mohr Coulomb strength parameters assumed. This is due to the fact that the Paxis 2D software uses a strength reduction factor to calculate the global factor of safety. This entails a reduction in the $tan\varphi'$ and c' parameters of the materials until failure of the system occurs.

According to the revised analyses, failure is observed at 13% undermining. Table 7 shows the horizontal movement of the top of the wall due to undermining. Accelerated horizontal movements occur from 10% undermining of the wall. Hence, as concluded from the initial assessments, any undermining poses an increase in risk on the stability of the quay wall.

The analyses represent the percentage undermining as a plane strain problem. However, the undermining varied with a maximum of 2.4 m of undermining underneath the foundation of the wall. This represents 25% of the base per 6 metres length of quay wall. The quay wall did not show any clear evidence of significant distress at the time of the analyses and geotechnical investigation. This indicated that secondary effects such as interlocking of the quay wall building blocks,

Table 5. Comparison of geotechnical design parameters assumed for the initial and revised stability analyses.

Designses Parameter	Initial analyses		Revised analyses*	
	Fill	Founding material	Fill	Founding material
Unit weight (kN/m³)	19	20	19	20
Cohesion c' (kPa)	4.0	4.0	2.0	2.0
Friction angle φ' (°)	42	40	34	38
Dilatancy angle ψ (°)	5.0	4.0	5.0	4.0
Young's Modulus E (MPa)	40	60	75	114
Poisson's ratio ν	0.3	0.3	0.3	0.3

Table 6. Comparison of initial and revised stability analysis results: global stability.

Percentage of base width undermined (%)	Base width undermined (m)	Global factor of safety. Initial analysis	Global factor of safety. Revised analysis
0	0.0	1.4	1.1
10	0.95	1.2	1.1
13	1.2	1.2	1.0
20	1.9	1.0	Failed
25	2.4	Failed	Failed

Table 7. Comparison of initial and revised stability analysis results: horizontal movement of wall.

Percentage of base width undermined (%)	Base width undermined (m)	Horizontal movement initial analysis (mm)	Horizontal movement revised analysis (mm)
0	0.0	13	16
10	0.95	39	215
13	1.2	56	243
20	1.9	143	Failed
25	2.4	Failed	–

arching of the founding material, stiffness of the foundation concrete and steel grillage found below the quay wall were possibly providing additional support to the quay wall. These secondary effects could not be effectively modelled within the finite element analyses. However, reliance on such secondary effects would have been optimistic.

6 DISCUSSION

The initial assessment whereby only the overturning, sliding and bearing capacity ultimate limit states were considered, showed the overturning limit state to be the most critical, with failure occurring at 38.5% undermining of the wall. These calculations did not account for soil-structure interaction and the global failure mechanism observed in Figures 3 and 4. The Plaxis 2D finite element analyses accounted for soil-structure interaction and resultant global failure mechanisms could be observed. This allowed a better understanding of the potential failure mechanisms.

The failure mechanism observed from the finite element analyses showed an active wedge-like failure zone behind the quay wall that extended below the wall. In addition, a passive failure zone appeared to occur at the toe of the quay wall, with a potential sliding occurring along the base. This failure mechanism is more complex than pure overturning, as considered in the initial hand calculations. Remediation measures could thus be designed to account for possible failure mechanisms.

7 CONCLUSIONS

Initially, a parametric study was conducted to determine representative parameters for the retained and founding material as no geotechnical data was available. Following the initial assessment, a geotechnical investigation was conducted which resulted in lower strength parameters than those originally assumed. The stability analysis using the revised parameters resulted in failure occurring at 13% undermining of the wall. This was lower than that observed from the initial assessment. Herein the value of a geotechnical investigation is highlighted as initially a lack of information resulted in unrepresentative parameters being used. The value a geotechnical investigation adds to the representativeness of the applied data and parameters required for a particular design problem as well as contributing to the understanding of the ground profile cannot be under-estimated and should be a requirement for every project.

One limitation of the Plaxis 2D analyses was that it represented a plane strain model effectively modelling an infinitely long quay wall. Thus the percentage undermining for each analysis was applied for this long wall. In reality the undermining varied along the wall and this variation along the length of the wall could not be applied in the plane strain case. However, withstanding this limitation, the analysis allowed one to truly investigate the effect of undermining on the mechanics and global stability of the wall.

An additional limitation of the model, was its inability to account for load transfer from one section of the wall to the next which would improve stability. This possibly explains why the wall is standing with the current undermining profile observed below the wall. In addition, the varied profile of the undermining below the wall would also further explain why the wall is standing for a percentage undermining that is shown to cause failure for the 2D plane strain modelled case. This could also signify that the wall may be at imminent failure as the maximum percentage undermining along the wall of 25% is greater than the undermining of 13% shown to cause failure in the 2D plane strain case.

Withstanding these limitations, the Plaxis 2D analyses assisted in providing a deeper understanding of the behaviour of the quay wall subjected to undermining. It allowed an observation anticipated global failure mechanisms that could not be predicted with simple equilibrium hand calculations. The finite element models also allowed one to input explicitly the undermining of the wall by removing elements below the wall. The analyses were used to evaluate the risks associated with operation of the quay and indicated that the stability of the wall posed a high risk. In addition, the analyses showed that any future remedial work to be done to the undermining needed to account for the possibility that the wall may indeed be at imminent failure in certain sections. As such, remediation measures need to consider actively stabilising the wall as opposed to merely filling the voids.

The analyses showed that the quay was operating at a much reduced design standard to what was initially intended. Based on the results, it was recommended that the berthing of ships along the undermined wall should be discontinued until remediation of the wall occurred.

REFERENCES

Beyers, J.U.H. 2014. Stability analysis of the Charl Malan quay wall in the port of Port Elizabeth. In, *Cultivating the future of geotechnics; Proc. 8th South African Young Geotechnical Engineers Conference, Stellenbosch, South Africa, 17–19 September 2014*.

Civil Engineering Office, Civil Engineering Department. 2004. *Port Works Design Manual part 2, Guide to Design of Piers and Dolphins*. Hong Kong.

Geotechnical Control Office, Civil Engineering Services Department. 1982. *Guide to Retaining wall Design*. Hong Kong.

Proceedings of the first Southern African Geotechnical Conference – Jacobsz (Ed.)
© *2016 Taylor & Francis Group, London, ISBN 978-1-138-02971-2*

Investigating the effect of reinforcing fine grained soils using shredded waste plastic

A.S. Chikasha & D. Kalumba
University of Cape Town, Cape Town, South Africa

F. Okonta
University of Johannesburg, Johannesburg, South Africa

ABSTRACT: The main objective of this study was to investigate the effect of plastic inclusions on the shear strength of fine grained soils. Soil classification tests were performed prior to the direct shear test to determine physical properties. Relationships between angle of friction and the varying parameters; and cohesion and the varying parameters were deduced. It was graphically confirmed that at optimum conditions, the cohesion of fine grained soils increased by up to 100% while angle of friction showed a continuously decreasing trend by up to 50%. Based on the results obtained from the laboratory tests, it was concluded that the structural integrity of the fine grained soil, is ultimately reduced by inclusion of shredded plastics, as indicated by the decreasing shear strength and deteriorating slope stability. The minimum acceptable plastic contamination in fine grained soil backfill that would not induce greater than 10% reduction in properties (shear strength) was found to be 0.5%.

1 INTRODUCTION

1.1 *Background to study*

Soil forms a fundamental basis for several geotechnical engineering and civil engineering construction work. In its natural form, however, may not exhibit sufficient strength characteristics. Most soil types are naturally well compacted and rigid, while others are not well cemented. Such, problem soils, incorporate fine grained soils. It is therefore a challenge to improve the strength of weaker soil types so that purpose may be found for the composites in developments such as slope stabilization and cut off trenches.

Fine grained soils are characterised by relatively high cohesion and low shear strength. This increased particle interaction is a result of the friction caused between soil particles and the reinforcement elements. Due to the resulting friction, forces are transmitted between the soil particles and the reinforcing elements which then cause increased cohesion to the whole mass. Previous studies have also shown that reinforced soils reinforced using dispersive reinforcement elements exhibit cohesion in all directions which therefore allows the construction of reinforced earth structures in any desired shape. The stresses developed in the reinforcement of such structures depend on the total contact actions between the soil particles. Consequently, shearing and sliding can be avoided

by designing and placing the reinforcement properly.

1.2 *Objectives of study*

The main objectives of this study were to investigate the effect of including shredded waste shopping plastic as a reinforcement element on the shear strength of fine grained soils. Other specific objectives were to investigate the effects of varying parameters such as length, width and concentration of plastic strips on cohesion and angle of friction; and, finally performing a slope stability analysis to determine the impact if waste plastic on slope stability.

2 SOIL REINFORCEMENT MECHANISM

Mandal (1987) describes reinforced soil as soil whose performance has been improved by the inclusion of elements in form of solid plates, fibres or fibrous membranes to interact with soil through the effect of friction and adhesion in order to resist tensile forces.

Reinforcement materials are generally classified into two categories which are extensible and inextensible reinforcement materials. Extensible reinforcement is such that the tensile strain exhibited by the reinforcement elements is equal to or larger than the horizontal extension required in developing an active plastic state in the soil. A good example is geosynthetics (McGown 1977). Inextensible reinforcement is such that the tensile strain in the reinforcement is significantly less than the horizontal extension required to develop an active plastic state in the soil. Steel is a good example. (McGown 1977).

The mechanism of combining two materials of different strengths characteristics to form a composite material of greater strength is a familiar practice in civil engineering and has been in use for a long time. Low shear strength soils can also be strengthened by the inclusion of materials with high tensile strength. This mobilisation of tensile strength is obtained by surface interaction between the soil and the reinforcement elements through friction. The theory of soil reinforcement is classified in regards to the predominant forces that are generated in the inclusions (McGown 1977). For reinforcement elements, the primary resisting forces are the tensile stresses generated within the elements; hence the interaction mechanism developed in soil reinforced by tensile inclusions is characterized by the mobilization of frictional forces (shear stresses τ) along the surface of the inclusion.

2.1 Effects of reinforcing materials on soils

In a study based on reinforcement of a slopes and embankments, it was found that the effects and functions of reinforcing materials are based not only on the deformation conditions of the soil, but also the direction of reinforcing materials (Tokue & Umetsu 1982). The following conclusion was made: effects of reinforcement are visible when reinforcement operates as tensile material and contrarily, not visible when reinforcement operates as a compressive material, as sometimes the strength of reinforced soil reduces.

Reinforced soils are considered a cost effective technique when compared to other construction techniques and they bring the following benefits: shear resistance capability of soil is increased thereby allowing problem soils to be used as structural founding components, land acquisition can be kept at a minimum as reinforced structures can be made steeper than normally possible, and construction time can also ultimately be reduced.

Bergado (1993) undertook a study to investigate the effect of reinforcing clayey soil with steel grids using pullout tests. It was concluded that the pullout failure mechanism was a function of both the soil and the reinforcement grid parameters and subsequently the inclusion of steel grids improved the pullout resistance in clayey soil.

In another study by Sobhee (2010) to investigate the effect of reinforcing sand with plastic waste on two sands of different particle shapes it was concluded that the effect of the plastic strips as a soil reinforcement element varied from sand to sand. The shape of the sand particles influenced the outcome as sands with particles of angular shapes exhibited higher strengths than those with round shaped particles.

In further tests by Chegenizadeh and Nikraz (2012), it was concluded that inclusion of natural fibres in clayey sands increases the shear strength as increasing either the concentration or length of fibres at fixed normal stress, resulted in increasing shear strength.

3 MATERIALS AND METHODOLOGY

3.1 Kaolin clay

Kaolin was selected because, being an industrially manufactured product, it was readily available and consistent hence repeatability was guaranteed. Prior to commencing the test procedures, mechanical properties of the soil were determined as presented in Table 1 below.

3.2 Polymeric materials

The material used as a source of plastic strips for this study was low density polythene waste shopping bags since this study was based on the use of plastic waste as reinforcement element. A process of randomly collecting used shopping bags from the student residences surrounding the University of Cape Town was undertaken. However, for consistency and repeatability purposes, one type of shopping bag, with respect to shop brand name, had to be selected for use in the investigation. Tensile strength tests were performed on shopping bags from four retail supermarkets in South Africa

Table 1. Summary of Kaolin properties.

Property	Unit	Magnitude
Specific gravity, Gs	Mg/m³	1.78
Natural moisture content	%	0.36
Average dry density	kg/m³	1580
Moisture content	%	22.0
Maximum dry density	kg/m³	1825
Plastic limit	%	25.39
Liquid limit	%	36.87
Shrinkage limit	%	3.67
Plastic index	%	11.49
Angle of friction, φ'	Degrees	17.76
Cohesion, c'	kN/m²	9.22

to determine the most suitable type of plastic bag to use as reinforcement element.

Since a large amount of plastic strips were required for the experiments and a high level of accuracy had to be employed in cutting the strips for consistency and repeatability, a laser cutter was used with a bed size of 960 mm x 610 mm and a laser cartridge of 40W. The dimensions of strips required were drawn in AutoCAD and fed to the machine where each type of strip was cut at a time.

3.3 Test method

The direct shear test was used to perform the investigation of shear strength behaviour and the apparatus used was the Wykeham Farrance SBI Constant strain rate direct shear apparatus. Identical experiments were performed at three different normal pressures of 25 kPa, 50 kPa and 100 kPa. Length of plastic strips was varied as follows: 0, 7.5 mm, 15 mm, 22.5 mm, 30 mm and 45 mm; while concentration of plastic strips was varied as follows: 0, 0.05 g/m^3, 0.1 g/m^3, 0.2 g/m^3 and 0.3 g/m^3; and, width of plastic strips was varied as follows: 0, 3 mm, 6 mm, 12 mm and 18 mm. Only a single parameter was varied at a particular time but three identical experiments were performed at normal pressures of 25 kPa, 50 kPa and 100 kPa for the same parameter varied.

To carry out slope stability analyses, the relationship between shear stress (the component of stress that operates in the down-slope direction) and shear strength (the properties that resist shear stress, i.e., cohesion + normal stress) is often used. As such, the ratio of shear strength to shear stress is one possible derivation of the factor of safety. When the ratio of shear strength to shear stress is greater than 1, shear strength is greater than shear stress and the slope is considered stable. However, when this ratio is equal to 1, shear strength is equal to shear stress. And, when this ratio is less than 1, shear strength is less than shear stress and the slope is considered unstable.

Incorporating the properties of Kaolin clay soil and various physical properties of a slope, the mathematical formula below shows the derivation of factor of safety for analysis. Figure 1 illustrates the distribution of forces on a typical slope.

$$FS = \frac{c + (\rho g H \cos\theta - \rho_w g\, W)\tan\phi}{\rho g H \sin\theta}$$

c = cohesion
ρ = density of soil mass
H = regolith thickness
θ = slope angle

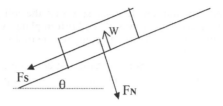

F_S = shear stress
F_N = normal stress
θ = slope angle
W = height of water table

Figure 1.   Force distribution on a slope.

g = acceleration due to gravity
ρ_w = density of water
φ = angle of internal friction
W = height of water

3.4 Precautions during testing

A mass of 1200 g of Kaolin was used to prepare test specimens for three tests at different normal stresses. This mass was obtained and computed based on the required densities after the proctor test and also based on the dimensions of the shear box (Petersen 2009). The required amount of plastic to mix with the sand was obtained by calculating the mass which corresponded to the concentration of plastic required for the experiment being done. Concentration of the reinforcement in a composite was measured in terms of a percentage of the weight of the sand and not in terms of the number of plastic strips.

Having weighed both the plastic strips and the clay, the plastic strips were added to the clay. 363 ml of distilled water was added to achieve the desired moisture content of 33% then random mixing was done using an electric mixer in order to ensure an even composite, since the mix was wet and hence less workable. The moisture content of 33% was chosen instead of the optimum moisture content of 22%. At 22% moisture content the mix was unevenly crumbled; hence moisture content within the liquid and plastic limit boundaries had to be selected for better workability through eye judgement and physical inspection.

The composite was then divided into three portions and was ready to be placed in the direct shear box. Assembling of the direct shear box was done according to BS 1377-7:1990. In order to allow drainage during the shearing process, perforated plates were included in the shear box since the composite was wet. The shear box was filled in two layers. A first layer of composite was placed in the shear box and a spatula was used to level the clay and also to make sure that the soil particles filled the grooves of the bottom plate. A hand tamper was then used to compact the first layer of composite.

In order to ensure that the test conditions were reproducible for each test, the following precautions were taken:

- New plastic strips were used for each test.
- A particular soil composite was strictly used for only one test.
- The first test performed was repeated three times to ensure the results were repeatable.
- All research equipment and apparatus was properly calibrated.
- All masses were recorded using the same balance which was accurate to 3 decimal places.
- All tests were conducted as per the British Standards for global comparability.

4 RESULTS AND DISSCUSSIONS

The build-up of shear stress resistance in the kaolin composites increased with an increase in normal stress. As a result, an increase in the resisting interparticle shear stresses was developed as normal loads increased from 25 kPa through 50 kPa to 100 kPa.

The introduction of plastic strips generally decreased the angle of friction of Kaolin and this could possibly be explained by the effect of shapes of the Kaolin particles. Clay particles in the unreinforced state all maintain a uniform angularity in shape but the introduction of plastic strips may disturb the structural orientation of the soil with respect to creating a line of separation between particles and hence a discontinuity.

As the contact surface area for the plastics and soil particles increased, it increased randomly throughout the soil mass; hence introducing disturbances in several intact regions of the soil mass by affecting the packing of particles. This ultimately resulted in the decreasing angle of friction.

Increasing length and width of plastic strips resulted in an initial gradual increase in cohesion until a peak value of 9.22 kN/m² was reached, after which further increases resulted in decreasing cohesion. This is explained by the fact that as the dimension of inclusions increased, the surface area on which soil particles press also increased hence more soil particles were in contact with the plastic strips. Further increase in plastic strip dimensions, however results in soil particles being surrounded by longer or wider strips which will cause sliding of particles against the plastics; hence lowered cohesion.

The inclusion of plastics in the lattice of Kaolin clay soil ultimately results in a decrease in the factor of safety and hence reduced slope stability. This is a result of the gradual decrease in shear strength with increasing length or concentration of the plastic inclusions. The films of plastic introduce discontinuities within the soil regolith hence creating lines of weakness which then lead to a less stable soil structure.

5 CONCLUSIONS AND RECOMMENDATIONS

- Increasing the length of plastic strips within a Kaolin-plastic composite initially results in a decrease in friction angle then further increase results in gradually increasing the friction angle.
- Increasing the length of plastic strips in a Kaolin-plastic composite increases cohesion. However, further increase in length beyond the length of 15 mm decreases cohesion.
- Introducing plastic strips into Kaolin clay soil reduces the shear strength of the composite giving a safety factor of less than 1. It was therefore concluded, in the combination studied, that plastic inclusions result in soil instability.
- The introduction of plastic strips into Kaolin ultimately reduced the mobilization of shear resistance due to the progressive reduction of the soil's particle-to-particle interaction by the introduction of discontinuities within the soil structure brought about by the plastic inclusions.
- Running tests at different moisture contents ranging from dry to wet is recommended in order to consider the effect of water tables in environmental application.
- The general conclusion is that the implementation of shredded low density polythene plastic strips as reinforcement to cohesive soils (clay soils), does not yield positive engineering value and hence cannot be applied to land rehabilitation needs such as reinforcement of embankments.

REFERENCES

Bergado, D. 1993. Performance of a Welded Wire Wall with Poor Quality Backfills on Soft Clay. *Proceedings of ASCE 1991 Geotechnical Congress* (pp. 908–922). Boulder: ASCE.
Chegenizadeh, A., & Hamid, N. 2012. *Numerical Shear Test on Reinforced Clayey Sand.* Perth: EJGE.
McGown, A. (1977). *Performance of Reinforced Soil Structures.* Britain: British Geotechnical Society.
Mandal, J. 1987. Geotextiles in India. *Geotextiles and Geomembranes*, 6 (4): 253–274.
Sobhee, L. 2010. *Soil Reinforcement Using Perforated Plastic (Polythene) Waste. BSc Civil Engineering Undergraduate Thesis*: Cape Town: University of Cape Town.
Tokue, T. & Umetsu, K. 1982. *Influence of Reinforcement Direction on Slope Stability.* Japan: SMFE.

Proceedings of the first Southern African Geotechnical Conference – Jacobsz (Ed.)
© 2016 Taylor & Francis Group, London, ISBN 978-1-138-02971-2

Design and construction of a mining facility over weak soil using modern geogrid technology

J. Klompmaker
BBG Bauberatung Geokunststoffe GmbH & Co. KG, Espelkamp-Fiestel, Germany

A. Post & S. Westhus
NAUE GmbH & Co. KG, Espelkamp-Fiestel, Germany

ABSTRACT: Geogrids used in combination with good quality fill material allow construction equipment access to sites where the soils are normally too weak to support the initial construction work. They also enable compaction of initial lifts on sites where the use of ordinary compaction equipment is very difficult or even impossible. Geogrids reduce the extent of stress on the subgrade and prevent base aggregate from penetrating into the subgrade. Geogrid/geotextile composites help on prevent subgrade fines from pumping or otherwise migrating up into the base. Experience has shown that geosynthetics increase subgrade strength over time. All these proven benefits were part of the design considerations in the presented Australian mining facility project, which had challenging subgrade conditions, bearing in mind the loads to be expected.

1 INTRODUCTION

1.1 *General*

Due to the globalization of the world economy existing ports are being developed and new ports are being built in order to cope with the increasing volume of goods in transit. New container terminals, stockyards or port extensions are mostly built on reclaimed land from the sea. As dredged fill with its low or medium density provides insufficient bearing capacity to take up the final loads of the container terminals, the long-term stability and trafficability of the gained land must be improved. An economic measure to improve the bearing capacity is the use of geogrid/nonwoven composite material as reinforcement and separation layers.

As the geogrid can absorb greater tensile stresses than the base course itself, the tension in the reinforced granular material is reduced. This paper will give an overview using geogrid/nonwoven composite materials to increase the bearing capacity of base courses, from various projects. Harbour storage areas carry large traffic volumes and typically have concrete or paved surfacing over a base layer of aggregate. The combined surface and base layers act together to support and distribute traffic loading to the subgrade. Problems are usually encountered when the subgrade consists of soft clays, silts and organic soils. These types of soils are often water sensitive and, when wet, unable to adequately support traffic loads. If unimproved,

the subgrade will mix with the road base aggregate, which leads to a reduction in strength, stiffness and drainage characteristics, promoting distress and early failure of the roadway. Contamination with fines makes the base course more susceptible to frost heaving.

1.2 *Geosynthetic separation function*

The separation function of the geotextile is defined by prevention of intermixing of soils with different gradation characteristics, where mixing is caused

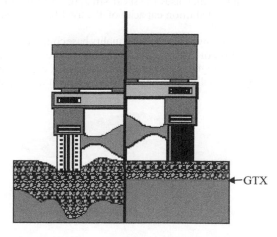

Figure 1. Illustration of geotextile separation function.

by mechanical actions. The mechanical actions generally arise from physical forces imposed by construction or operating traffic and may cause the aggregate to be pushed down into the soft subgrade and/or the subgrade being squeezed up into the base aggregate.

A properly designed geotextile separator allows the base aggregate to remain "clean", which preserves its strength and drainage characteristics.

The use of geotextile separators ensures that the base course layer in its entirety will contribute and continue to contribute its structural support of vehicular loads. The separator itself does not contribute structural support to the aggregate layer.

Yoder and Witczak (1975) state that as little as 20% by weight of the subgrade mixed in with the base aggregate will reduce the bearing capacity of the aggregate to that of the subgrade. This highlights the importance of a geotextile separator with regard to the performance of base aggregate layers on fine-grained subgrades.

1.3 Geosynthetic reinforcement function

Vehicular loads applied to the surface of trafficked areas create a lateral spreading motion of the unbound aggregate layers. Tensile lateral strains are created at the interface subgrade/geogrid as the aggregate moves down and sideways due to the applied load. Through shear interaction of the base aggregate with the geogrid, which is known as inter-locking, (see Figure 2), the aggregate is laterally restrained or confined and tensile forces are transmitted from the aggregate to the geogrid.

As the geogrid is much stiffer in tension as compared to the aggregate, the lateral stress is reduced in the reinforced base aggregate and less vertical deformation at the road surface can be expected. This interaction between geogrid and base course material increases the shear strength and thus the load distribution capacity of the used base course material.

In many projects, good quality base course aggregate is not available on site or close to the site. As a result, high transport costs of imported, expensive good quality base aggregate have a great influence on the total project costs. In such cases geosynthetic reinforcement and separation products can help to save money by reducing the amount of imported fill material, needed to achieve the specified bearing capacity, for the expected loads on the base course.

To combine the function of reinforcement and separation in one product, so called Geocomposites have been developed. Geocomposites, such as Combigrid®, allow faster construction rates compared to separately installed geogrid and geotextile components.

2 UTAH POINT BERTH, PORT HEDLAND, AUSTRALIA

2.1 Project history

Australia is the world's largest iron ore exporter and as a producer, ranks third after China and Brazil. Port Hedland is one of the largest ore export facilities in Australia. The facilities have been expanding due to the growing global demand for base metals from Australia. In 2008 the Port Hedland Port Authority commenced construction of a multiuser bulk iron, manganese and chromite ore export facility comprising roads, stockyards (comprising of thirteen stockpiles), conveyors and a wharf. An aerial photograph of the site and drawing showing the arrangement of the stockyard and the perimeter road are shown in Figure 3. Figure 4 shows the stockpile and perimeter arrangements. The method of geotechnical analysis and the earthwork construction process are described in the following sections.

2.2 Geological and geotechnical information

2.2.1 Subsurface condition

The geotechnical investigation comprised of the excavation of test pits, hand augered boreholes, hand held Dynamic Cone Penetrometer (DCP) tests and shear vane tests. The laboratory investigation comprised of classification tests and consolidation tests on the Mangrove mud. The subsurface

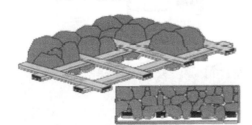

Figure 2.   Interaction of aggregate with geogrid.

Figure 3.   Aerial view of construction site.

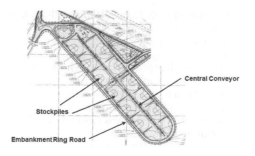

Figure 4. Stockpile and perimeter road arrangements.

profile of the stockyard and associated facilities consist of 0.5 m to greater than 1.8 m of mangrove mud followed by a deep layer of calcarenite.

Geotechnical parameters adopted for the mangrove mud layer were derived from the results of field vane shear tests and laboratory consolidation tests. An undrained shear strength (s_u) of 10 kPa was adopted

2.2.2 Subsurface condition

The weak subsurface condition and loading from embankment and stockpiles warranted use of high strength geogrid reinforcements for slope stability and bearing capacity purposes. As basal reinforcement underneath the embankments laid and welded geogrid types Secugrid® 400/40 R6 and Secugrid® 120/40 R6 were used.

2.2.3 Design loads adopted

The following design loads were adopted in the stability and deformation analysis. Average vehicle traffic loads on the perimeter embankment road of 20 kPa; and iron ore stockpiles up to 16.5 m in height with an assumed iron ore stockpile dry unit weight of 22 kN/m³. The stockyard floor will be subject to a stockpile load of up to about 300 kPa.

3 ANALYSIS AND RESULTS

3.1 Method of analysis

Limit equilibrium method of stability analysis as well as finite element technique was undertaken with the following conditions and assumptions:

– Use of soil reinforcement to achieve end of construction and long term global stability;
– Embankment global slope stability Factor Of Safety (FOS) ≥ 1.5 for static loading, FOS ≥ 1.1 for dynamic (seismic) loading and FOS ≥ 1.4 for a flood event.
– The horizontal acceleration coefficient or hazard factor a_h of 0.12 g (1 in 500 years annual

probability of exceedance) and a_h of 0.09 g (1 in 250 years annual probability of exceedance) were considered for quasi-static earthquake analysis.
– The analysis was based on a design life of 20 years.

3.2 Strength of mangrove mud layer

The end of construction stability of the perimeter road embankment and stockyard floor were analyzed taking into consideration the gain in strength of the mangrove mud layer due to consolidation during the construction of the embankment fill. This was estimated and taken into account in the analysis, as follows:

– The undrained shear strength (s_u) of the Mangrove mud layer was assumed to be approximated by Equation 1 (Ladd, 1986).

$$\frac{s_u}{\sigma'_v} = 0.2 \cdot (OCR)^{0.8} \qquad (1)$$

Where: σ'_v = Effective vertical stress and OCR = Overconsolidation Ratio.

– An s_u of 10 kPa was adopted for the mangrove mud. The stress history of the mud was assessed to vary from an OCR of 3.6 at shallow depth, decreasing to 2.6 at about 1.8 m depth.
– With the full embankment height, the resultant OCR of the mangrove mud layer was estimated. Then based on the resultant OCR and using the above equation the undrained shear strength of the mangrove mud layer underneath the embankment was calculated.

3.3 Construction procedure

The following construction sequence was adopted.

– Chain saw the mangrove trees to ground level without disturbing the root zone below the

Figure 5. Clearing of mangrove wetlands.

ground surface thus retaining some of the root system as a reinforcement zone.

- Lay out the geotextile drainage separator layer on the mud with a minimum overlap of 1.0 m.
- Lay out the basal geogrid reinforcement layer and hand tension to remove any folds or slack in the geogrid. Overlap of rolls to be 600 mm with cable ties at 300 mm intervals in both directions to maintain tension over joints. The reinforcement layer was a laid and welded geogrid Secugrid® 400/40 as shown on Figure 6.
- Spread the first layer of fill material over the geotextile layers in an initial 1.0 m thick bridging layer using light tracked construction equipment with low ground pressures. Ensuring that all equipment only operates over at least 1.0 m of fill. The bridging layer is anticipated to be compacted by loading from the construction equipment. The fill in this initial layer was a free draining sandy or sandy gravelly material, having a fines content (grain sizes finer than 75 μm) of not more than 8% by weight.
- Install geotechnical instrumentation (settlement cells, inclinometers, piezometers) and associated survey datum at selected locations.
- In some areas of the stockyard floor (where required), lay out the geogrid reinforcement layer Secugrid 120/40 R6 and hand tension to remove any folding or slack in the geogrid.

Figure 6. Installation of geotextile and Secugrid geogrid.

Figure 7. Installation of fill on top of geogrid.

Overlap of rolls to be 600 mm with cable ties at 300 mm intervals in both directions to maintain tension over joints.

- For succeeding fill layers to achieve the required elevation, dredged materials, with a soaked CBR of not less than 12%, shall be used and compacted in a loose layer lift not exceeding 250 mm. These layers were compacted to achieve a dry density ratio of not less than 95% of Modified Maximum Dry Density.

4 CONCLUSIONS

The increase of global trade and transport of goods creates growing demands to handle cargo and bulk goods. To accommodate growing volumes, existing ports are extended and new ports are being built. Soft subgrades are often the basis for the foundation design of the port pavement system. As economic construction method geogrids are often used in this case to improve the low bearing capacity, for the expected traffic and storage loads.

The technology of geosynthetic reinforced soil provides an economic construction method for the development of new ports. With the improved structural load-bearing capacity of geogrid reinforced fill, stress concentrations on soft subgrades can be reduced, which minimizes differential settlements at the pavement surface and automatically improves the transport safety on the installed fill layers.

In the case of the presented project the awarded contractors successfully installed approx. 500,000 m² of geogrid over the work areas. In 2011 Port Hedland Port Authority and PINC Group (Project Management) won the Australian Engineering Excellence Award in the "Resource Development" Category for the Utah Point Multi-User Bulk Export Facility project.

REFERENCES

Giroud, J.P. & Noiray, L. 1981. *Design of Geotextile Reinforced Unpaved Roads*, J. Geotechnical Eng. Div., ASCE, Vol. 107, No. GT9, 1981, pp. 1233–1254.

Klompmaker, J., Heerten G. & Lesny, C. 2009. *Improvement of the long-term trafficability of container storage areas in harbours with composite geogrid reinforcement*, Geosynthetics 2009, Salt Lake City, USA.

Ladd, C.C. 1986. *Stability Evaluation during Staged Construction*. The Twenty-Second Terzaghi Lecture, Journal of Geotechnical Engineering, 117 (4), 537–615.

Woodroof, P. & Chakrabarti, S. 2012. *Use of high strength geogrid reinforcement for embankments on soft soil in Western Australia*. ANZ 2012, Melbourne, Australia.

Yoder, E.J. & Witcak, M.W. 1975. *Principles of Pavement Design*. 2nd Edition, John Wiley and Sons, 711.

Proceedings of the first Southern African Geotechnical Conference – Jacobsz (Ed.)
© *2016 Taylor & Francis Group, London, ISBN 978-1-138-02971-2*

Optimal geogrid reinforcement of clay liners

D.H. Marx & S.W. Jacobsz
University of Pretoria, Pretoria, South Africa

ABSTRACT: Finite element analyses of a clay liner subject to differential settlement, reinforced with geogrids at four positions, were conducted. The analysis allowed the definition of a function to predict the maximum tensile strain in the liner for any combination of stiffness of the four geogrids. Consequently, the function was used to determine the optimum placement of reinforcement within the clay liner. This was done by minimising the maximum tensile strain in the clay liner induced by differential settlement, as well as cost. The resulting relationship between the maximum tensile strain generated as a function of reinforcement, a Pareto front, is presented. The Pareto front allows for informed decision making on the optimal placement and stiffness requirement of geogrids, for solutions of different costs. Preliminary analysis indicate that for solutions of the lowest cost, the optimal position of the geogrid reinforcing is at the bottom of the liner. However, for the greatest reduction in tensile strain, the liner should be reinforced both at the top and at the bottom.

1 BACKGROUND

Current waste disposal facilities are strained by continued population growth. This, combined with a reduction in available air space and sites suitable for landfills, necessitates the exploration of vertical extension of current landfill sites, i.e. a piggyback landfill, as an alternative. However, most landfills in South Africa were constructed before the establishment of the Minimum Requirements for Waste Disposal by Landfill (Department of Water Affairs and Forestry 1998). Consequently, vertical expansions of old landfills should be designed to the same specifications as a new landfill on a greenfield site.

General municipal waste is a highly heterogeneous material with potential for differential and local settlement throughout the waste body. Despite its ductility, a clay liner will eventually fissure and crack at certain strains when subjected to these settlements. Thus, only limited distortion of the clay liner can be permitted before critical cracking (and thus an increase in permeability) occurs. The so-called distortion level a/l, where a is the maximum central settlement and l the width of the settlement trough, can be used to quantify the differential settlement (Viswanadham & Rajesh 2009).

Geogrid reinforcement can be used to mitigate the effect of differential settlement on the integrity of clay liners. Due to practical considerations usually more than one layer of geogrid is required to bear the load of the overlaying waste. By investigating the reinforcement mechanisms of geogrids

in clays this paper aims to determine the optimal placement and spacing of the geogrids.

Based on the work of Kuo & Hsu (2003) on pavements, Rajesh & Viswanadham (2015) suggested that the optimal placement of geogrid reinforcement is in the top half of the liner. However, significant tensile strain occurs at both the top and the bottom. Accordingly, reinforcement might be required at both positions. Furthermore, a clay liner reinforced at the top only has greater potential to split apart (delaminate) along the plane of the geogrid. Accordingly, there is room to investigate alternative reinforcement positions for the geogrids.

Van Eekelen & Bezuijen (2014) investigated basal reinforced piled embankments founded in a weak foundation layer. They found that: 1) the behaviour of soil reinforced with two identical geogrids is equivalent to soil reinforced with a single geogrid with double the stiffness and 2) soil reinforced with two grids placed directly on top of one another behaves the same as soil reinforced with two grids separated by some material. However, the performance of the system was evaluated in terms of the load generated on top of the piles and the deflection of the geogrid. For clay liners the performance of the system depends on the integrity of the liner. The integrity is a function of the fracture and tensile strain in the clay and not only total deflection. Accordingly, given the difference in performance criteria, there is reason to investigate whether the second finding by Van Eekelen & Bezuijen (2014) holds for the current application.

2 SURROGATE SURFACE GENERATION

In order to optimise the reinforcement solution to the problem, the problem must first be defined as a mathematical function. This function will act as a substitute (surrogate) to the actual problem. Subsequently, an optimisation algorithm can be used to find the valleys (or peaks) of the function value by varying the dependent variables. These minima (or maxima) are the optimal solutions to the problem. This section discusses the process to define the current problem (maximum tensile strain in a clay liner as a function of the stiffness of the four geogrids) as a mathematical function.

For some physical phenomena, e.g. 1D stress and strain, the relationship between the variables can be defined as a simple mathematical relationship, e.g. $\sigma = Ec$. For other phenomena the relationship is too complex to be defined as a simple function. If a number of function values are known, an arbitrary curve, or surface (for multiple variables), can be fitted to the known data points to model the relationship. This allows for interpolation between data points. This arbitrary curve/surface usually has fitting parameters (for example the slope and intercept in the case of a straight line) that can be changed to improve the fit of the function to the actual problem.

The current problem defines maximum tensile strain as a function of the stiffness of the four geogrids (the vector J). It is a complex problem, thus no simple functional relationship exists. Accordingly, an arbitrary surface has to be fitted to the numerical data (results of a FE analysis for this problem) to define the relationship. The specific surrogate surface used is known as a Radial Basis Function (RBF) (Forrester & Keane 2009). The RBF ($s(J)$) consists of a series of bell-shaped (radial) functions ($\varphi(J)$) that are added together to form the final surface and fitted through the available data points:

$$s(J) = \sum_{k=1}^{n} \lambda_k \varphi_k(J) \qquad 1$$

The Gaussian function, $\varphi_k = e^{(E\,J - J^k)^2}$, was used to as basis. The maximum tensile strain in a liner can be estimated for different designs (combinations of geogrid stiffnesses) where no experimental data is available, by fitting the RBF to the data points obtained from the finite element analysis conducted.

3 FINITE ELEMENT MODELLING

To generate the data points of known behaviour required to fit the RBF, 900 Finite Element (FE) analyses were conducted. For each analysis the combination of stiffness of the four geogrids differed. The stiffness combinations (design vectors) were randomly generated using Latin hypercube sampling. This type of sampling generates stratified random vectors and is discussed by Ross (2013), e.g. for two-dimensional data in a square grid, each row and column will be sampled only once. The random vectors were scaled to consider stiffness between 0 kN/m and 2500 kN/m.

3.1 Modelling approach

Plane strain FE analyses were conducted with ABAQUS 6.13–3. The models were built to be comparable to the physical and numerical modelling done by Rajesh & Viswanadham (2011, 2012, 2015) on geogrid reinforced liners. Those tests consisted of a 14 m long clay liner, 1.2 m thick. Their models, however, considered only one layer of reinforcement (with an axial stiffness (J) of 10000 kN/m) placed in the top quarter of the liner. Central settlement of 1 m was modelled, resulting in a settlement trough 16 m wide, equivalent to a distortion level (a/l) of 0.0625.

The numerical model for the present study consisted of a clay liner reinforced at four levels with geogrids, and an underlying waste body (see Figure 1). The grids were spaced vertically at 0.3 m intervals to allow for the position of the top geogrid (1) to coincide with the models in literature. The thickness of the clay liner (1.2 m) and total length (14 m) was chosen to allow for validation to the experimental data from literature. The clay was modelled using 2992 continuum, plane strain, eight-node elements. The material model used for the clay liner is discussed in the following section. The geogrids were modelled as 187 two-node truss elements. The geogrid material was linear elastic with a Poisson's ratio of 0.3 and no compressive strength. The axial stiffness (J) of the geogrids was varied depending on the analysis conducted.

The left-hand boundary of the model was fixed against movement. The right-hand boundary was fixed against movement in the horizontal direction only, modelling symmetry. A 25 kPa overburden pressure was applied to the top of the clay liner. Differential settlement was modelled by forcing the waste surface to deform in a Gaussian shaped trough (Equation 2), similar to the surface settlement trough due at a distance x away from the centre of the trough is a function of the maximum settlement (Δy_{max}) and the distance to the point of inflection (i). For a trough width (w) of 16 m, assuming $w = 5i$ (New & O'Reilly 1991), i is equal to 3.2 m.

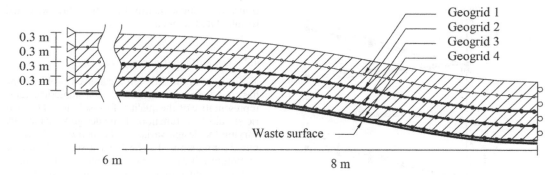

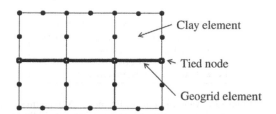

Figure 1. Diagram of the FE model analysed.

$$\Delta y = \Delta y_{max} \cdot e^{-x^2 /(2i)^2} \qquad (2)$$

The clay liner and underlying waste surface (only), was modelled as two separate parts. This was done to allow the clay to separate from the waste surface should the tensioned membrane effect of Giroud et al. (1990) develop. The interaction between the two parts was modelled as a hard contact (preventing both the penetration of the waste into the clay and the transfer of tensile stress between the two). Friction between the waste surface and the bottom of the clay layer was not investigated (i.e. frictionless tangential behaviour was assumed).

Geogrids reinforce soil by 1) disrupting the failure planes and tensile strain fields (Jones 1985) and 2) by sharing the applied load. For the analyses presented fracture induced failure planes were not considered (following Rajesh & Viswanadham (2015)) and thus, the first mechanism was not modelled.

Subsequently, the reinforcement provided by the geogrid was modelled by tying the geogrid nodes to the coinciding clay elements (see Figure 2). Thus, the displacement of the clay node is not only a function of the material properties of the clay, but also of that of the connected, stiffer, geogrid. Should the tensioned membrane effect occur, failure planes will develop and the aforementioned approach will not be valid. However, this did not occur for a maximum central settlement of 1 m.

3.2 Validation

The FE model was validated against the centrifuge test data of Rajesh & Viswanadham (2011, 2012), and was compared to the numerical analysis of Rajesh & Viswanadham (2015). Three different material models were considered for the clay:

1. a simple linear elastic model ($v = 0.3$, $E = 2620$ kPa (Rajesh & Viswanadham 2015)),

Figure 2. Modelling of geogrid reinforcement by tying the geogrid nodes to the clay elements (seen in elevation).

since compacted clay liners are generally overconsolidated;

2. the linear elastic model of (1) combined with undrained Mohr-Coulomb plasticity ($\varphi = \psi = 0°$, $cI = 19$ kPa (Rajesh & Viswanadham 2015)). Undrained behaviour was assumed due to the relatively fast rate of settlement imposed in the centrifuge models;

3. a linear elastic, undrained Mohr-Coulomb plasticity model (model 2), with a tension cut-off. Thusyanthan et al. (2007) developed a relationship to determine the tensile stress (σ_T) required for a clay to crack (Equation 3). Assuming the initial consolidation stress (σ_0) to be 630 kPa, equivalent to a saturated Proctor density of 95% (Jessberger & Stone 1991), and the initial effective stress (σ_0) to be 25 kPa, a conservative tension cut-off would be -37.65 kPa.

$$\sigma_T \approx -(0.45 \pm 0.15) \cdot {}^j \sigma^I \qquad (3)$$

All validation analyses were done with all four geogrids included in the model. For the validation of the unreinforced clay liner the stiffness of the four geogrids was set to zero. For the validation of the liner reinforced at the top only, the stiffness of the top grid was set to 10000 kN/m and that of the lower three grids to zero.

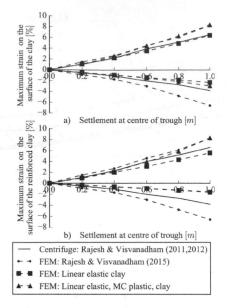

a) Settlement at centre of trough [m]

b) Settlement at centre of trough [m]

— Centrifuge: Rajesh & Visvanadham (2011,2012)
← → FEM: Rajesh & Visvanadham (2015)
■ ■ FEM: Linear elastic clay
▲ ▲ FEM: Linear elastic, MC plastic, clay

Figure 3. Validation of the FE model. Maximum strain at the surface of a) the unreinforced liner, and b) the reinforced liner. Compressive strain is positive and tensile strain is negative.

The maximum tensile and compressive strains at the surface of the unreinforced clay liner, for a given central settlement, for both the centrifuge tests and numerical analyses, are shown in Figure 3a. A similar plot is shown for the reinforced clay liner in Figure 3b. Due to the relatively high tension cut-off modelled there was no difference in the behaviour of clay models (2) and (3).

Maximum surface tensile and compressive strains of both the reinforced and unreinforced liners, as calculated by the FE analyses, were compared to the centrifuge models. Clay model 2 (linear elastic with Mohr-Coulomb plasticity), was selected for all further modelling. For each of the four data sets the coefficient of determination (R^2) between the modelled and experimental strain data was greater than 0.995, illustrating the validity of the model.

4 OPTIMISATION

4.1 Fitting of RBF

The fitting parameter (c) was optimised to fit the RBF to the calculated response for a given design (i.e. a combination of different stiffnesses for the four geogrids). Firstly, all but one of the known designs (and corresponding maximum tensile strain) were used to generate the surface of func-

tion values. Subsequently, the new surface was used to calculate the maximum.

4.2 Generation of pareto front

To find the optimal design for a given problem one has to determine a metric, that is a numerical value representative of the quality of the solution. The metric should be a function of the design variables. By varying the design variables, the metric can be minimised (or maximised). The design associated with the minimum value of the metric is considered optimal.

For the current problem the maximum tensile strain in the clay liner is such a metric. Tensile strains result in cracking of the liner and cracks are assumed to represent the permeability, and thus integrity, of the liner. The strain is related to the design variables (the geogrid stiffnesses, J) through the RBF, as discussed in Section 2. Accordingly the scalar metric (or function value) of maximum tensile strain (f_1) can be defined in terms of the design variables as:

$$f_1(J) = s(J), J = [J_1, J_2, J_3, J4] \tag{4}$$

Another important metric of the optimality of the design would be the total cost of the reinforcement. For the current problem it was assumed that the cost of each of the 900 design options analysed is directly related to the sum of the stiffness of the four geogrids, i.e. a direct relationship between stiffness and cost was assumed. In reality the relationship may be more complex. This scalar metric of cost (f_2) can be defined in terms of the design variables as:

$$f_2(J) = \sum_{i=1}^{4} J_i \tag{5}$$

These two metrics are, however, in conflict. Reducing the cost of the reinforcement would result in increased strain and vice versa. Consequently, each design is a trade-off between the cost and the integrity of the liner. Accordingly, the perceived optimality of the design is a combination of both the maximum tensile strain and the cost. This combined metric (f_3) can be defined as a function of the first metric (f_1) and the second (f_2). By using a weighted sum approach, the relative influence of the first two metrics (cost and strain) on the value of the third metric can be varied: more weight assigned to the strain metric results in a stiffer, more expensive optimal design; more weight assigned to the cost metric results in a cheaper, less stiff optimal design. The combined metric (f_3) can be defined as:

$$f_3(J) = (1-\alpha)\frac{f_1(J) - f_1^{min}}{f\,max - f\,min} + \alpha\frac{I_2(J) - I_2^{min}}{fmax - fmin} \tag{6}$$

where changing α between 0 and 1 varies the relative importance of the two metrics, and $f_{1/2} = max/min$ is used to scale the respective metric between 0 and 1.

This combined metric can be optimised for different weights of f_1 and f_2. As the weighting of f_1 and f_2 changes, the value the two individual metrics (strain and cost) varies at the optimum for the system. This can be presented graphically by plotting the strain value on the x-axis and the cost value on the y-value, at the system optimum, for all the different weights. The line that emerges is called a Pareto front (Pareto 1906). Since the points are generated by optimising f_3, all points on the front are optimal designs.

To optimise f_3 and generate the Pareto front for the current problem, the so-called weighted min-max method (Arora 2004) was implemented. The SciPy (Jones et al. 2014) implementation of the differential evolution algorithm (Storn & Price 1997) was found best suited for the optimisation. Again the stiffness of any of the four geogrids was limited between 0—2500 kN/m to investigate the reinforcement mechanism as the grids progressively increased in stiffness.

The cost-strain-Pareto front ensuing from the optimisation is presented in the first graph of Figure 4. This Pareto front presents the minimum

attainable tensile strain in the liner for a given cost spent e.g. if geogrids with a summed stiffness of 2 MN/m is installed (indicative of the cost) the minimum attainable tensile strain in the liner is 2.72%. The remaining four graphs of Figure 4 show the optimal stiffness of each of the four geogrids for the corresponding point on the Pareto front. For example, for a summed geogrid stiffness of 2 MN/m the optimal design will entail a geogrid with a stiffness of 1.05 MN/m at position 1, and a stiffness of 0.94 MN/m for Geogrid 4. The Pareto front is, however, specific to the current problem with tensile strain for the remaining design. This estimated strain was compared to the actual value from the FE analysis to determine the error in the fitted surface (so called leave-out-one-cross-validation). The process was repeated until each of the 900 designs were used for validation. For the current problem, the surface generated makes an average prediction error of 0.03% strain over a range of 1.21% strain.

5 DISCUSSION

The horizontal strain distribution through the liner, after imposing a Gaussian shaped settlement with a central displacement of 1 m, is presented in Figure 5 for: a) an unreinforced clay liner; b) a clay liner reinforced at position 4 (bottom) with a 2.5 MN/m geogrid; and c) a liner reinforced with a similar geogrid at position 1 (top) only. Refer to Figure 1 for positions. This strain distribution, as well as the Pareto front and corresponding optimal designs, deliver some insights into the behaviour of a clay liner reinforced with geogrids at multiple levels, subject to the imposed differential settlement.

Firstly, a significant initial decrease in maximum tensile strain (>1%) can be achieved by installing a fairly low stiffness geogrid (0.64 MN/m) at the base of the liner. In comparison, an additional summed stiffness of more than 5.4 MN/m is required to reduce the strain by a further percentage point. The region of highest curvature of the liner is at

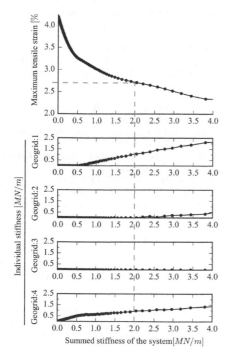

Figure 4. Cost-strain-Pareto front and corresponding optimum designs of geogrid stiffness for a clay liner subject to a central Gaussian settlement of 1 m.

Figure 5. Horizontal strains in a liner subject to a Gaussian displacement profile from the FE analysis: a) unreinforced liner; b) reinforced at the bottom only (J_4 = 2500 kN/m); and c) reinforced at the top quarter only(J_1 = 2500 kN/m).

the bottom centre (see Figure 5a). Consequently, installing a geogrid there significantly disrupts the tensile strain field and reduces the maximum strain in the liner (see Figure 5b). When a geogrid improves soil by the second mechanism (that is by sharing the applied load), only the soil above it is strengthened. Therefore, reinforcing at position 1 (at the top) has minimal effect since significant tensile strains occur elsewhere, i.e. at the bottom of the liner.

Secondly, the design that results in the lowest strain in the liner consists of, for all practical purposes, only geogrids at the base (Geogrid 4) and in the top of the liner (Geogrid 1). These regions contains the outer fibres of the clay liner where the most significant tensile strains develop. Geogrid 1 improves the liner as a whole by absorbing the surcharge load and disrupting the tensile strain field at the surface of the liner. Geogrid 4 improves the bottom of the liner by disrupting the associated tensile strain field but adsorbs surcharge to a lesser degree. The system is more efficient if the bulk of the surcharge is absorbed by Geogrid 1. If not, more load is transferred through the clay to the lower geogrids.

The greater load in the clay induces more strain. Accordingly, the stiffness required for Geogrid 4 plateaus while the optimum stiffness for Geogrid 1 continues to rise up to the maximum allowed.

6 CONCLUSIONS

The results of various finite element analyses were used to investigate the optimal position of geogrid reinforcement in a clay liner. A Pareto optimal front was generated expressing maximum tensile liner strain as a function of the combined stiffnesses of the geogrids used. The Pareto front can be used to determine the reduction in maximum tensile strain in the liner for a given cost. The analysis showed that, for the current problem, reinforcement of the clay liner in the top only is not optimal. Furthermore, the analysis showed that clay should be reinforced with two grids separated by some material, rather than a single grid of the same combined stiffness. It was also found that an initial significant reduction in maximum tensile strain in the clay resulted from the placement of a relatively low stiffness geogrid at the bottom of the liner. For the maximum reduction in tensile strain the liner should be reinforced both at the bottom and at the top quarter, with the stiffer grid placed in the top quarter.

ACKNOWLEDGEMENTS

The first author would like to acknowledge the Geosynthetics Interest Group of South Africa for the funding provided.

REFERENCES

Arora, J. 2004. *Introduction to Optimum Design.* Academic Press.

Department of Water Affairs and Forestry 1998. *Waste management series, second edition: Minimum Requirements for Waste Disposal by Landfill.*

Forrester, A. & Keane, A. 2009. Recent advances in surrogate-based optimization. *Progress in Aerospace Sciences* 45(1–3): 50–79.

Giroud, J.-P., Bonaparte, R., Beech, J., & Gross, B. 1990. Design of soil layer-geosynthetic systems overlying voids. *Geotextiles and Geomembranes* 9(1): 11–50.

Jessberger, H. & Stone, K. 1991. Subsidence effects on clay barriers. *Géotechnique* 41(2): 185–194.

Jones, C. 1985. *Earth Reinforcement and Soil Structures.* Butterworth and Co. Ltd.

Jones, E., Oliphant, T., & Peterson, P. 2014. {*SciPy*}: Open source scientific tools for {Python}.

Kuo, C. & Hsu, T. 2003. Traffic induced reflective cracking on pavements with geogrid-reinforced asphalt concrete overlay. In *Proceedings of the 82th Annual Meeting at the Transportation Research Board (CD-ROM).*

New, B. & O'Reilly, M. 1991. Tunnelling induced ground movements; predicting their magnitude and effects. In *Proceedings of the 4th International Conference on Ground Movements and Structures, invited review paper, Cardiff, Pentech Press, London*: 671–697.

Pareto, V. 1906. *Manuale di Economica Politica.*

Rajesh, S. & Viswanadham, B. 2011. Hydro-mechanical behavior of geogrid reinforced soil barriers of landfill cover systems. *Geotextiles and Geomembranes* 29(1): 51–64.

Rajesh, S. & Viswanadham, B. 2012. Centrifuge Modeling and Instrumentation of Geogrid-Reinforced Soil Barriers of Landfill Covers. *Journal of Geotechnical and Geoenvironmental Engineering* 138(1): 26–37.

Rajesh, S. & Viswanadham, B. 2015. Numerical Simulation of Geogrid-Reinforced Soil Barriers Subjected to Differential Settlements. *International Journal of Geomechanics* 15(4): 1–15.

Ross, S. 2013. *Simulation (5th Edition) 10.3 Latin Hypercube Sampling* (5th ed.). Elsevier Ltd.

Storn, R. & Price, K. 1997. Differential evolutiona simple and efficient heuristic for global optimization over continuous spaces. *Journal of global optimization* 11(4): 341–359.

Thusyanthan, N., Take, W., Madabhushi, S., & Bolton, M. 2007. Crack initiation in clay observed in beam bending. *Géotechnique* 57(7): 581–594.

Van Eekelen, S. & Bezuijen, A. 2014. Is 1+1 = 2? Results of 3D model experiments on piled embankments. In *10th International Conference on Geosynthetics,* Berlin.

Viswanadham, B. & Rajesh, S. 2009. Centrifuge model tests on clay based engineered barriers subjected to differential settlements. *Applied Clay Science* 42(3–4): 460–472.

Proceedings of the first Southern African Geotechnical Conference – Jacobsz (Ed.)
© 2016 Taylor & Francis Group, London, ISBN 978-1-138-02971-2

New results of high-tensile steel meshes tested in first large scale field test application

P. Baraniak & M. Stolz
Bern University of Applied Sciences, Burgdorf, Switzerland

B. Schoevaerts
Geobrugg Southern Africa (Pty) Ltd., Gauteng, South Africa

C. Wendeler
Geobrugg AG, Romanshorn, Switzerland

ABSTRACT: The stability of newly cut or natural slopes is an important issue of geotechnical engineering. Nowadays, one of the most frequently chosen methods for slope stabilization is soil nailing in combination with flexible facing. In this configuration, the soil nailing is designed to stabilize deep-seated instabilities, while the instabilities near the surface have to be stabilized by a flexible facing. One option is the use of steel wire mesh. In order to ensure proper slope stabilization, the soil nails and the flexible facing have to act as one integrated system. Such system has been lately tested in large scale within a research project. For this reason, the large scale setup, consisting of an inclinable large box (12 × 10 × 1.2 m), was established. This setup allowed for testing various flexible facing systems in the conditions simulating the real slope conditions in best possible way. The present paper is a documentation of the setup and the results of the large scale tests.

1 INTRODUCTION

The slope stabilization method based on flexible facing systems has become one of the most commonly used solutions for unstable soil and rock slopes. There are many reasons why this method became so popular, for instance: high resistance allows installation in almost every soil and rock

Figure 1. Example of the railway slope stabilized with flexible facing system and soil nailing (Source: Geobrugg Polska sp. z o.o.).

condition on steep slopes and even on overhangs; flexibility allows to install the mesh on slope surface; homogenous strength of the flexible facing allows for installation using flexible nail pattern. Figure 1 presents an example of the slope stabilized with flexible facing system and soil nailing.

Lately, several different flexible facing systems, including some new developments, have been tested in large scale within the frames of research and development project. The entire project was supported by Swiss Commission for Technology and Innovation and elaborated by Bern University of Applied Sciences in cooperation with Geobrugg AG and AGH University of Science and Technology.

2 OBJECTIVES

The objective of the research project was to investigate the behaviour of various flexible facing systems in series of large scale field tests in changing soil conditions and different nail patterns. This investigation would allow improving old and developing new system elements, comparing different flexible facings. The gathered data and the results presented in this paper will allow an engineer to compare the performance of different flexible systems and will help to choose the most suitable solution for particular slope stabilization projects.

3 LARGE SCALE FIELD TESTS

3.1 Background

All the tests were conducted between 2012 and 2014 in a small private quarry in Winterthur, Switzerland. The aim of the tests was to check whether the objectives of the research project were correct and to ensure the settings would imitate the real slope conditions in the best possible way.

Initial experiments of the test period allowed for observation and optimization of the test setup, testing procedure and data acquisition method. After optimization of the test setup the following experiments were conducted in a repetitive way to guarantee reliable and comparable experimental results. All individual experiments were conducted until the ultimate inclination of the setup was reached (85 degrees) or until the whole flexible system got ruptured.

3.2 Test setup and conditions

The test setup was an artificial slope represented by a large scale inclinable steel frame ($12 \times 10 \times 1.2$ m), which could be raised on one side by a crane in order to simulate the slope angle. In order to keep the soil material inside the frame its inner side was faced with wooden planks. To create certain friction between wooden lining and the soil, small wooden planks (60×30 mm in distance of every 500 mm) were installed across the floor and the walls of the setup (Figure 2).

The anchors used in tests were represented by threaded steel bars (ø 32 mm), encased in corrugated PVC tube (ø 100 mm) and cemented to simulate the construction conditions as real as possible. The anchors were installed in the box in regular triangular patterns of 2.5×2.5 m, 3.0×3.0 m and 3.5×3.5 m. Some of the anchors were equipped with strain gauges in order to measure stresses occurring in the anchors during testing procedure. All the anchors were rigidly fixed to the steel frame by means of steel foot plates.

In order to analyze the behaviour of flexible facing systems in different soil conditions, two materials with different strength parameters were chosen (Table 1). Both materials were classified according to USCS Soil Classification System. The first soil is classified as poorly Graded Gravel (GP). It's a mixture of 16 mm and 32 mm diameter grains with internal friction angle of 33 degrees. Most of its grains are rounded and therefore the soil is

Figure 2. The inner side of test setup: steel frame and anchors installed in triangular pattern.

Table 1. Properties of the soil types used in the tests.

USCS definition and symbol	Paper definition	Friction angle [°]	Cohesion [kPa]	Bulk density [kg/m³]
Poorly graded gravel GP	Round gravel	33	0	1670
Poorly graded gravel with silt GP-GM	Gravel with silt	38	0	1800

called round gravel. The second soil is classified as poorly graded gravel with silt (GP-GM) with grain size between 0 and 63 mm, and internal friction angle of 38 degrees. The soil filling up the box was slightly compacted while its distribution across the box was done by a small digger which is visible in Figure 2. The internal friction angle of both soils was determined as an average from two different methods: angle of repose and large triaxial test (vacuum triax).

3.3 Instrumentation and measurements

The inclination of the test setup was measured constantly in order to have one reference scale for comparison of different flexible facing systems tested in varying conditions. The inclination was measured with an electrical inclinometer with measuring resolution ≤0.14 degrees and possible measuring range 0–360 degrees. The inclination sensor had to be calibrated directly before the experiment, each time it was meant to be used.

The deformations of the flexible facing systems were measured at every inclination step (0, 30–85 degrees) by pulse laser scanner. In general, the laser scanner measures distance by illuminating a target with a laser and analyzing the reflected light.

In this case, the values of deformations measured at certain inclination were later deducted from the values of the deformations at the initial state. The values of the maximum deformations were calculated from the whole area of the box. In order to avoid errors in the measurements the maximum deformation value was defined as the maximum deformation of an area of at least 0.25m². The horizontal and vertical angular resolution of the laser scanner was set to 0.02 degrees, resulting in a density of 2×10^4 points/m² at 20 m distance and 1×10^4 points/m² at 30 m distance and the accuracy of measurement of 7 and 10 mm, respectively.

3.4 Testing procedure

In order to ensure the same testing conditions for all flexible facing systems, all tests were conducted according to the same, simple test procedure. An effort was also made to ensure that the mesh is laid out, connected and pre-tensioned in the same way each time. The pre-tensioning of the mesh was applied by the torque wrench with a scale. The testing procedure consisted of three phases: preparation, execution and after-tests discussion.

The preparation phase consisted of checking of the test setup and instrumentation, pre-tensioning of the mesh, preparation of data acquisition system and the calibration of the inclination sensor. At that time the reference scan of the flexible facing surface was taken. During the testing phase, the test setup was lifted up from horizontal position up to 30 degrees inclination and that continued in 5-degree steps up to final inclination of 85 degrees or to the point where the tested flexible facing got ruptured. At every inclination step, the lifting procedure was stopped for a few minutes to take a scan of the surface. Within that phase the data acquisition system was also switched on at all times, recording changes in the measuring devices. The last phase was a quick meeting with conclusions regarding the behaviour of test setup and flexible facing.

Each test was recorded with high resolution cameras and documented by on-site notes and observations. Figure 3 presents the experimental setup for testing flexible facing systems in large scale during one of the tests. The test setup and its procedure are more precisely described in Cała et al. (2013).

3.5 Elements of flexible facing systems

The main components of flexible facing systems is the steel wire mesh. The mesh used within the research project was produced of 2, 3 and 4 mm high-tensile steel wires (tensile strength of steel ≥ 1770 N/mm²). The used mesh is a diamond-shaped

Figure 3. Experimental setup for testing flexible facing systems in large scale.

Figure 4. High-tensile steel wire mesh with newly developed large spike plate.

chain-link mesh with the dimensions of 83 × 143 mm of a single mesh and its aperture of 65 mm. This combination of wire diameter, opening shape and the steel parameters allows creating meshes with a longitudinal strength in a range of 65–250 kN/m.

Another primary component of flexible facing systems are the system spike plates which are steel plates connecting the mesh with the nailing system. The steel spike plates are produced in two types. The regular plate is produced with dimensions 330 × 205 × 7 mm, while the large plate is produced with dimensions 667 × 300 × 7 mm (Fig. 4). The plates are designed to transfer the stress from the subsoil through the mesh to the nailing system.

The most important strength parameters of flexible facing systems are their puncturing resistance on the upper edge of spike plate and shearing resistance on the contact between mesh and plate. Both of these parameters depend directly on the wire diameter of the mesh, the geometry, the size of mesh openings and the size of the spike

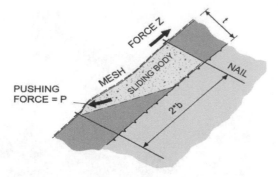

Figure 5. The sliding soil body with schematically marked puncturing force *P* and shearing-off force *Z*.

Table 2. Properties of flexible facing systems examined in large scale field tests.

Mesh aperture	Wire diameter	Plate size	Puncturing resistance	Shearing-off resistance
[mm]	[mm]		[kN]	[kN]
65	2	Regular	40	10
65	3	Regular	90	30
65	4	Regular	140	50
65	4	Large	185	75

Table 3. The variables and results of conducted tests.

Test no.	Punctuing resistance	Soil frition angle	Anchor pattern	Defor- mation at 60°	Failure inclintion Test	Failure inclintion RUV
	[kN]	[°]	[m]	[m]	[°]	[°]
5	90	38	2.5	0.24	>85	>85
9	90	33	3.5	0.64	72	56
10	140	33	3.5	0.55	83	66
11	90	38	3.5	0.45	80	62
12	185	33	3.5	0.48	>85	76
13	140	38	3.5	0.43	83	74
17	40	33	3.5	0.62*	56	47
18	90	38	3.0	0.41	84	73
19	40	38	3.0	0.46	63	58
20	90	33	3.0	0.47	73	65
21	40	33	3.0	0.54	57	52
22	185	38	3.5	0.34	>85	85
23	185	38	3.5	0.36	>85	85
27	90	38	2.5	0.28	>85	>85
29	90	38	3.5	0.52	78	62

* Deformation at 55 degrees inclination.

plate. Figure 5 presents the sliding soil body with schematically marked puncturing resistance force P and the shearing-off force Z. Table 2 presents the puncturing and the shearing-off resistance of specific meshes and plates combined in flexible facing systems.

4 TEST RESULTS

In this paper the results of fifteen out of total thirty one tests are presented. It is because the missing tests were either invalid tests due to imperfect performance of the test setup or test with other type of meshes which are beyond the consideration of this paper. Table 3 presents variables and test results of performed tests. The variables are represented by puncturing resistance of flexible facing, friction angle of the soil and the anchor triangular distance, while the test results are represented by maximum outward deformation of the surface at 60 degree inclination of the setup and failure inclination of the flexible facing recorded during the large scale tests. The last column of Table 3 presents the values of failure inclination, calculated with dimensioning concept, designed

for calculating of the slope layers near the surface of the slope.

4.1 *Deformation of flexible facing*

The analyses of maximum deformation of various flexible facing systems are presented in Figure 6. The graph presents dependence of deformation of flexible facing on its puncturing resistance in various soil conditions and anchor pattern conditions. The two lower lines represent results received in 3.0 × 3.0 m anchor distance, while the two upper lines represent the results from the tests with anchor distance of 3.5 × 3.5 m. On the other hand, the dashed lines represent tests conducted in conditions of round gravel, while the solid lines represent tests in gravel with silt conditions. Since the puncturing resistance of the mesh increase linearly, it was also assumed, that the deformations would be linear to the puncturing resistance of flexible facings, what seems to be correct according to the graph presented in Figure 5.

Further analysis has shown that, while changing the anchor pattern from 3.5 × 3.5 m to 3.0 × 3.0 m, one can expect 17 to 30% smaller deformation in conditions of round gravel and 11 to 18% smaller deformations in conditions of gravel with silt. It is worth noticing that in round gravel conditions, bigger difference in deformations occurred in facing with higher puncturing resistance compared to the facing with lower puncturing resistance. In gravel

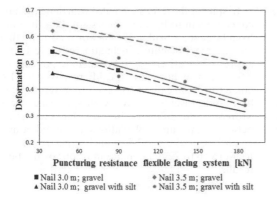

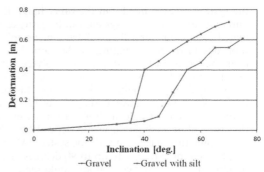

Figure 6. Dependence of deformation of flexible facing on its puncturing resistance in various soil conditions and anchor pattern conditions.

Figure 7. Comparison of deformations of the same flexible facing depending on soil conditions (3 mm diameter wire mesh, regular size spike plate and 3.5 × 3.5 m anchor pattern).

with silt conditions it appeared to be opposite—bigger difference in the deformations occurred in lower resistance facing compared to higher resistance facing.

On the other hand, keeping the anchor pattern constant, and changing only the soil conditions from round gravel to gravel with silt, one can expect 9 to 14% smaller deformations in flexible facing with 3.0 × 3.0 m anchor pattern, and 14 to 27% smaller deformations with 3.5 × 3.5 m anchor pattern. Here the situation is similar to previous example. In the anchor pattern of 3.0 × 3.0 m, the difference in the deformations decrease together with increasing puncturing resistance of the facing, while in the anchor pattern of 3.5 × 3.5 m, the difference of the deformations increases together with increasing puncturing resistance of the flexible facing.

Figure 7 presents comparison of deformations of the same flexible facing system (3 mm diameter wire mesh, regular size spike plate and 3.5 × 3.5 m anchor pattern) tested in different soil conditions.

Moreover, installation of flexible facing in gravel with silt conditions with anchor pattern 3.5 × 3.5 m, would result in only 4% bigger deformation compared to installation of the same facing in round gravel conditions with a 3.0 × 3.0 m anchor spacing.

4.2 Failure inclination of flexible facing

All presented tests were conducted either up to 85 degrees inclination (safety reasons) or to the failure inclination of flexible facing system. The failure of flexible facing means, that at certain point the whole system or one of its elements breaks and testing procedure is no longer possible. Within each experiment, the failure inclination of

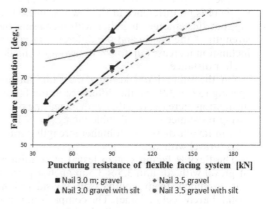

Figure 8. ailure inclination of various flexible facing systems recorded within large scale tests.

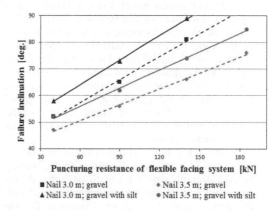

Figure 9. Failure inclination of various flexible facing systems estimated with dimensioning concept. The analysis was done with all partial and safety factors set to 1.0.

the system was recorded. On the other hand, the failure inclination of the system can be easily estimated using commercially available dimensioning tool, designed for calculating of the slope layers near the surface of the slope. The analyses with dimensioning concept were done with all partial and safety factors (cohesion, friction angle, factors from positive and negative loading) set to 1.0. Similarly to analysis of deformations, it was assumed that the failure inclination would depend linearly on the puncturing resistance of the system. All the graphs were prepared only up to 90 degrees inclination what in natural conditions would represent perfectly vertical slope.

The observations made on the failure inclination from the large scale field tests (Fig. 8), show that the anchor pattern has smaller influence on the failure inclination compared to the soil strength. While keeping the soil type constant and changing the nail pattern from 3.0×3.0 m to 3.5×3.5 m, there is no difference in the low resistance facing, but this changes together with increasing the strength of the facing and the difference in failure inclination increases up to 5% in favour of smaller anchor distance.

On the other hand keeping the same anchor spacing (3.0×3.0 m), the difference in the failure inclination increase, together with increasing puncturing resistance of the flexible facing, from 10 to 13% in favour of soil with higher strength parameters (gravel with silt).

The flexible facing system tested in conditions of gravel with silt and 3.5×3.5 m anchor pattern was excluded from comparison considerations (solid, barely inclined line). The comparison is not possible due to unexpectedly low failure inclination (83 degrees) of relatively strong system (puncturing resistance of 140 kN). Deducting from the behaviour the same system in the same anchor distance, but round gravel soil conditions, much higher performance of this system was expected.

In case the results estimated with the dimensioning concept (Fig. 9), reduction of the anchor distance from 3.5×3.5 m to 3.0×3.0 m, the difference in failure inclination increases from 10 to 18% together with increasing puncturing resistance of the system (round gravel conditions). In gravel with silt soil conditions, the situation is almost the same, but the increase is 10 to 17% in favour of smaller anchor pattern.

On the other hand, comparison of the failure inclination in constant nail pattern, but changing the soil strength would result in the same difference of 10% in favour of stronger soil (gravel with silt) along the whole range of strength of flexible facing systems.

In spite of the results estimated with dimensioning concept seem to be more consistent than the ones from large scale field tests, it is worth noticing that the difference of failure inclination of corresponding tests is relatively high. For instance, in case of 3.0×3.0 m anchor patter the difference is 9 to 11% and 8 to 13% in case of soil conditions represented by round gravel and gravel with silt, respectively. Moreover, the difference in the failure inclination between tests and calculations in round gravel conditions and 3.5×3.5 m anchor pattern would result in 16 to 22% difference in favour of the field results.

5 CONCLUSIONS

The present paper is a documentation of large scale field tests conducted on flexible facing systems in the frame of research and development project. In the consideration the flexible facing systems consisting of 2, 3 and 4 mm diameter high-tensile strength meshes and two types of spike plates were included. These systems were tested under changing soil and anchor pattern conditions.

The presented results show that the deformations of the surface as well as failure inclination of the slopes protected with flexible facing systems strongly depend on bearing resistance of flexible facing itself, the soil shear strength and the distance between anchors. Moreover, even though the dimensioning concept is based on simplified model approach, the results acquired from the calculations agree well with the test results and the experience gathered within over the last 15 years.

REFERENCES

Baraniak, P., Schwarz-Platzer, K., Stolz, M., Shevlin, T. & Roduner, A. 2014. Large scale field tests for slope stabilizations made with flexible facings. Kingston, GEOHAZARDS 6.

Cała, M., Flum D., Roduner A., Rüegger, R. & Wartmann, S. 2012. *TECCO Slope Stabilization System and RUVOLUM. Dimensioning Method.* Bad Langesalza: Beltz Bad Langesalza GmbH.

Cała, M., Stolz, M., Baraniak, P., Rist, A. & Roduner A. 2013. Large scale field tests for slope stabilizations made with flexible facings. *Proceedings of EUROCK 2013 Wrocław*, 23–26 September 2013. Rotterdam: Balkema.

Geobrugg AG. 2011. TECCO *Slope Stabilization System. Summary of Published Technical Papers* 1998–2011.

Proceedings of the first Southern African Geotechnical Conference – Jacobsz (Ed.)
© 2016 Taylor & Francis Group, London, ISBN 978-1-138-02971-2

Effect of water on the strength behaviour of fiber reinforced soils

V. Oderah & D. Kalumba

University of Cape Town, Cape Town, South Africa

ABSTRACT: This paper investigated the effect of 12 cycles of wetting and drying as well as soaking in water of sugarcane bagasse fiber included randomly in a soil matrix. It had been established from a previous study that inclusion of sugarcane bagasse improves the shear strength behavior of typical South African soils. Thus, this study was aimed at expanding the research by comparing the shear strength behavior obtained from submerged composites reinforced with 1.0% bagasse fiber with those of the dry composites. Wetting and drying cycles showed insignificant changes in the shear strength although smoothening of the fibres was noted. Soaking results on the other hand showed a 15% reduction in the shear strength after 2 days of soaking, and an insignificant reduction afterwards. This implies that for the bagasse-soil composite to retain its strength coating mechanisms must be constituted as well as a proper drainage system.

1 INTRODUCTION

The use of natural fibers has been shown to be feasible in improving the engineering properties of weak soils. Natural fibers such as coir (Sivakumar & Vasudevan 2008, Maliakal & Thiyyakkandi 2013), palm (Marandi et al. 2008; Estabragh et al. 2013; Sarbaz et al. 2014), sisal (Prabakar & Sridhar 2002) and wheat straws (Qu et al. 2013) have been studied to improve the bearing capacity of soft soils, reduce the formation of cracks, and in the repair of failed slopes.

Furthermore, a simple review by Hejazi et al. (2012) underscored the benefits of using natural fibers by highlighting that inclusion of 0.2–2.0% by dry mass of soil of the natural fibers increases the shear strength of soil and reduces the loss in the residual strengths. The study concluded that the interaction matrix formed between the fiber and soil contributes to the improvement in the shear strength of fiber-reinforced soils. The interaction from fibers in this case is dependent on their morphology, which consists of cellulose fibrils embedded in a lignin matrix—lignocellulosic. The prevailing theory is that as fiber content is increased, the shear strength behavior of the soil is enhanced up to a maximum fiber content beyond which an increase in fiber content reduces the shear strength.

A further study was constituted by Oderah (2015) to investigate the possible use of sugarcane bagasse fiber. This was due to the lack of similar studies conducted on bagasse fiber as a standalone soil reinforcing material. The large direct shear test revealed that using bagasse fiber increases the

shear strength behavior of a typical South African sandy soil up to 30% at a fiber content of 1.4%. Results also showed a reduction in the loss of the post peak shear strengths with the addition of fibers, which depended on the normal pressures applied. The research ultimately concluded the possibility of using bagasse fiber in geotechnical applications as a way of minimizing the effects posed by its disposal—for the sugar industries who have not explored bagasse alternative uses. The main objective of this subsequent study was to improve the findings on the use of sugarcane by investigating the effects of prolonged exposure of soils reinforced with bagasse fiber in a moist environment—the prolonged exposure in-terms of wetting and drying as well as soaking in water. Wetting in this context is defined as addition of water to a non-compacted composite until all the individual particles become moist while soaking entails the complete submergence of the compacted composite over time. It was anticipated that understanding the loss of strength in bagasse-reinforced soils would aid in exploring the possible mitigating measures.

2 RESEARCH MATERIALS AND LABORATORY PROCEDURES

2.1 Soil material

The soil type used in this investigation was Klipheuwel sand sourced in Cape Town, South Africa. Klipheuwel sand is a well-graded medium dense, reddish brown sand with sub-angular particles.

The sand was clean and consistent which ensured repeatability of results since identical samples could be easily reproduced. Table 1 shows its mechanical properties.

2.2 Sugarcane bagasse materials

The sugarcane bagasse fiber used for the experiment was obtained from TSB Sugar Company, found in Malalane, South Africa, at a moisture content of approximately 50%. The sample was therefore air dried for to eliminate the effect of moisture and preserve its morphological composition prior to testing.

The fiber bagasse, which is mainly the outer rind and hard fibrous material, was of 50–80 mm in length, 2–4.75 mm in diameter and had a tensile strength of 40–80 MPa. Figure 1 shows the fiber bagasse.

2.3 Sample preparation

The sample preparation and direct shear tests were conducted in accordance with ASTM 3080.

Table 1. Physical properties of Klipheuwel sand.

Soil property	Units	Value
Specific Gravity, Gs	-	2.65
Dry density	kg/m³	1833
Average dry density (Loose)	kg/m³	1606
Average dry density (dense)	kg/m³	1912
Particle range	mm	0.063–1.18
Mean grain size, D_{50}	mm	0.28
Coefficient of uniformity, Cu	-	1.8
Coefficient of curvature, C_c	-	0.95
Angle of internal friction, ϕ	degrees	32.8

Figure 1. Bagasse fiber.

Fiber bagasse in 1.0% mass of dry soil was mixed randomly with Klipheuwel sand before carrying out the durability tests. The obtained composite was then mixed using a 20 liters rotary base mechanical mixer for 3 minutes to obtain a homogenous mix. Mixing for more than 3 minutes caused floatation of fibres and hindered homogeneity. In addition, care was taken while compacting the sample in the shear box to avoid further segregation.

2.4 Experimental procedures

2.4.1 Wetting and drying procedure

Wetting of the composite was done for 24 hours then dried in an oven set at 35°C for another 24 hours. This was repeated 12 times before conducting shear tests. For the wetting cycles, a predetermined 10 liters of water was added to the already mixed sample in a tray then kept sealed for 24 hours. This quantity of water was arrived at after several trials, the guiding principle being ensuring that all the particles get moist homogenously. The procedure simulated the effect of wetting and drying conditions experienced in-situ due to the drawdowns in the ground-water levels. The temperature of 35°C ensured complete drying of the composite with no damage to the fibers. It should be noted that the choice of this temperature was informed by a similar study conducted by Azwa et al (2013). At the end of the 12th wetting cycle, the sample was directly compacted in a 305 mm by 305 mm direct shear box in 3 layers to attain 55% relative density.

2.4.2 Soaking tests procedure

The bagasse-soil mixture was compacted dry in the uniquely designed durability molds then submerged in water for a duration of between 6 hours and 14 days. The compaction was done in 3 layers of 25 blow each using a 2.5 kg tamping rod raised at a height of 300 mm to attain a relative density of 55%. Figure 2 shows a randomly mixed sample, soaked for 14 days, ready to be placed in the shear box for testing. The soaking on the other hand was done by completely submerging the mold in water and keeping a 20 mm level of water beyond the height of the mold. Maintaining this level necessitated daily monitoring and addition of water whenever there was a drop in the level, since fibers absorbed part of the water. The most critical period was between 6 and 24 hours.

After the set period, shear tests were then performed at three different vertical pressures of 50 kPa, 100 kPa and 200 kPa at a shearing rate of 1.0 mm/min. A maximum displacement of 60 mm was attained corresponding to 20% axial strain.

Figure 2.   14 days soaked composite ready for testing.

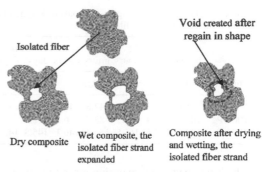

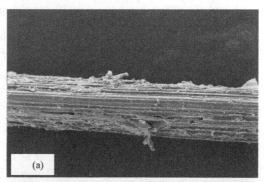

Figure 3.   The mechanism of wetting and drying.

3   RESULTS, ANALYSIS AND DISCUSSIONS

3.1   *12 cycles of wetting and drying*

The 12 cycles of wetting and drying of Klipheuwel sand reinforced with 1.0% fiber reduced the angle of internal friction from 40.6° to 38.8°, which corresponds to a 4.4% reduction. This is an insignificant change attributed to the smoothening of the fiber surface due to the addition of water. It however increased the cohesion from 18.7 kPa to 20.6 kPa.

As water was added to the composite, bagasse absorbed most of it. Upon drying, the fibers resumed their original structure, except for the smoothened surfaces. This regain in structure offered the interaction matrix but at a reduced magnitude compared to the dry fiber-soil composite not subjected to wetting and drying cycles.

Another possible explanation could be the voids created by the fibers expanding during wetting and keeping their original shape when dry. These voids weakened the interaction matrix and consequently the shear strengths obtained, especially at low normal pressures. Figure 3 explains this mechanism.

A dry single strand of fiber was viewed under X10,000 magnification using an Scanning Electron Microscope (SEM). The results are shown in the micrograph figures 4(a) and (b), note the smooth surface obtained after wetting and drying.

3.2   *Effects of soaking on the shear strength parameters*

The soaking/submerging of the composite between fiber and Klipheuwel sand in water for a minimum

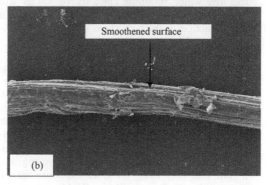

Figure 4.   Micrograph of an isolated dry fiber strand (a) before and (b) after 12 cycles of wetting and drying (X10,000).

of 6 hours and a maximum of 14 days produced the results summarized in table 2.

From the results, it is evident that submerging the fiber-Klipheuwel sand composite in water for 6 hours insignificantly affected the peak shear strength of reinforced Klipheuwel sand. This could be attributed to the restriction of water by allowing it to access the mold in one direction (from the top) reducing the surface area for the composite wetting. In addition, compaction of the composite produced little "pockets" of uncompact regions,

Table 2. Peak shear strengths of soaked Klipheuwel sand reinforced with 1.0% fiber bagasse.

	0 hours	6 hours	12 hours	24 hours	48 hours	7 days	14 days
Duration	Peak shear strengths (kPa)						
50 kPa	61.9	58.3	52.8	47.8	42.3	43.5	43.4
100 kPa	104.1	102.4	87.5	83.7	76.8	72.7	72.6
200 kPa	190.4	184.5	166.5	153.4	146.1	145.3	145.5

which retained water requiring more time to infiltrate the whole composite. This meant that not all the bagasse and soil particle got moist within the 6 hours of soaking, hence the insignificant change in peak strengths.

As hours of soaking progressed to 48 hours, the ingress of water increased causing a significant reduction in the peak shear strengths as shown in table 2. Increased water absorption dissolved part of the fiber cellulose and lignin weakening the interaction matrix by making the outer covering (rind) smooth. This combined with the lubrication of the soil particles, limited the resistance within the shear plane. Nevertheless, this limitation was prevented during consolidation and shear as flow of water out of the shear box was evident due to the low water retaining capacity of sands with increase in normal pressure. An overall percentage reduction in the peak shear strength of 31.6%, 30.2% and 23.7% was observed after 7 days for 50 kPa, 100 kPa and 200 kPa, respectively.

After 2 days of soaking, the angle of internal friction reduced from 40.6° to 34.7° representing a 15% reduction as shown in figure 5. From 7 to 14 days no change was realized on the angle of internal friction. This behavior may be attributed to the optimum water absorption achieved by the fibers. Fiber was fully saturated after about 2 days limiting further smoothening of the outer covering. This meant that increasing the fiber exposure time in water did not affect the fiber cells. However, some breakage of the fibers was observed as shown in figures 7(a). Furthermore, the configuration of the mold used in the submergence limited full anaerobic condition that would have facilitated decaying, thereby maintaining the tensile strength of the fibers. This observation was confirmed through the visual inspection of the exhumed fibers shown in figure 7(b). Similar results were obtained by Sarbaz et al (2013) who concluded that submerging fibers in water insignificantly affected the CBR of reinforced sand after 5 days of soaking.

Similar reduction in cohesion was observed with increased water exposure time of the reinforced Klipheuwel sand as shown in figure 6. This was attributed to the increased lubrication of the soil

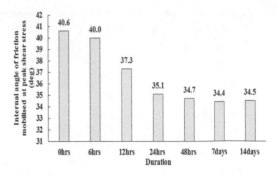

Figure 5. Effect of soaking on the angle of internal friction of Klipheuwel sand.

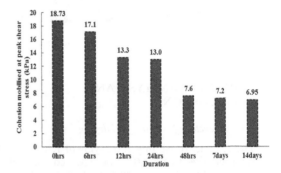

Figure 6. Effect of submerging on the cohesion of reinforced Klipheuwel sand.

Figure 7. Exhumed fibers after soaking (a) visual (b) micrograph (X10,000 magnification).

particles, reduced interaction and the consequent breakages of fiber with water ingress over time. In theory, apparent cohesion of soils increases with the increase moisture up to an optimum value due the apparent cohesion induced by water within the particles and build-up of negative pore pressure during shear. Introducing biodegradable fibers offers an opposing behavior by absorbing most of the water added to the composite. This coupled with the flow of water out of the shear box during consolidation stage could explain the reduction in cohesive strength observed in the study.

4 CONCLUSIONS

The study investigated the effect of prolonged water exposure of a sandy soil reinforced with 1.0% bagasse by conducting a direct shear test on a soaked composite as well as repeated cycles of wetting and drying. The following observations can therefore be concluded.

1. The 12 cycles of wetting and drying insignificantly affects the angle of internal friction of Klipheuwel sand. However, it smoothens the fiber surface and creates voids, especially at low pressures, in the composite matrix during the drying period that lowers the fiber friction.
2. On the other hand, soaking of bagasse-soil composite in water for 14 days decreases the peak shear strengths as well as the angle of internal friction of Klipheuwel sand by about 15% after 2 days. This implies that, in general, for the bagasse-soil composite to retain its strength proper drainage systems must be in place. Of course, an elaborate drainage system could be unfeasible, but in long-term the reduction in strength becomes minimal thus justifying any conventional drainage system. In addition, the fibers could be coated or treated with polymers to reduce their hydrophilic properties.

Owing to these results, bagasse reinforced soils can be used as an embankment fill in low load applications, temporary applications or in projects where settlement after the consolidation stage is non-critical. This is based on the insignificant reductions observed after the 2 days of soaking, indicating a long-term non-critical trend. A further study is however recommended over longer durations to investigate the effect of water on the bagasse fiber reinforced soil using a wide range of soils.

ACKNOWLEDGEMENTS

The authors appreciates the financial support received from ARISE Consortium, and TSB sugar for providing the materials used in the study. Profound gratitude also goes to Mr Charles Nicholas, of the Civil engineering workshop, University of Cape Town for his expertise applied in manufacturing the molds used for the soaking tests.

REFERENCES

Azwa, Z.N. Yousif, B.F., Manalo, A.C. & Karunasena, W. 2013. A review on the degradability of polymeric composites based on natural fibres. *Materials & Design.* 47(0):424–442.

Estabragh, A., Bordbar, A. & Javadi, A. 2013. A Study on the Mechanical Behavior of a Fiber-Clay Composite with Natural Fiber. *Geotechnical and Geological Engineering.* 31(2):501–510.

Hejazi, S.M., Sheikhzadeh, M., Abtahi, S.M. & Zadhoush, A. 2012. A simple review of soil reinforcement by using natural and synthetic fibers. *Construction and Building Materials.* 30100–116.

Lovisa, J., Shukla, S.K. & Sivakugan, N. 2010. Shear strength of randomly distributed moist fiber-reinforced sand. *Geosynthetics International.* 17(2), pp. 100–106.

Maliakal, T. & Thiyyakkandi, S. 2013. Influence of randomly distributed coir fibers on shear strength of clay. *Geotechnical and Geological Engineering.* 31(2):425–433.

Marandi, M., Bagheripour, H., Rahgozar, R. & Zare, H. 2008. Strength and ductility of randomly distributed palm fibers reinforced silty-sand soils. *Am J Appl Sci;* 5:209–220.

Oderah, V., 2015. Shear strength behaviour of sugarcane bagasse reinforced soils. *Masters dissertation, University of Cape Town, Cape Town.*

Prabakar, J. & Sridhar, R., 2002. Effect of random inclusion of sisal fibre on strength behaviour of soil. *Construction and Building Materials,* 16(2), pp. 123–131.

Qu, J., Li C., Liu B., Chen X., Li M. & Yao Z., 2013. Effect of Random Inclusion of Wheat Straw Fibers on Shear Strength Characteristics of Shanghai Cohesive Soil. *Geotechnical and Geological Engineering,* 31(2), pp. 511–518.

Sarbaz, H., Ghiassian, H. & Heshmati, A.A. 2014. CBR strength of reinforced soil with natural fibres and considering environmental conditions. *International Journal of Pavement Engineering.* 15(7):577–583.

Sivakumar Babu, G. & Vasudevan, A. 2008. Strength and stiffness response of coir fiber-reinforced tropical soil. *Journal of Materials in Civil Engineering.* 20(9):571–577.

Proceedings of the first Southern African Geotechnical Conference – Jacobsz (Ed.)
© *2016 Taylor & Francis Group, London, ISBN 978-1-138-02971-2*

Utilizing geotextile tubes to extend the life of a Tailings Storage Facility

P.J. Assinder, M. Breytenbach & J. Wiemers
HUESKER Synthetic GmbH, Gescher, Germany

F. Hörtkorn
SRK Consulting, Johannesburg, South Africa

ABSTRACT: As environmental legislation becomes more stringent, the required capital and operational investment for the construction and operation of mine Tailings Storage Facilities (TSF) is increasing significantly. The paper describes the system of geotextile tubes filled with tailings to create structural elements to enable the use of filled tubes to create extensions/rises for new and existing TSF's. Tubes can be placed as a new dam foundation or stacked on top of existing tailings dams, resulting in dam embankments comprising filled geotextile tubes, which leads to increased storage capacity for the mine operator and a subsequent expansion of the life of TSF's. Additionally, it is possible to use a broad range of tailings within the geotextile tube due to the potential to dewater the fills with flocculation agents, if required, during the filling process. The construction process is described and the required geotechnical stability aspects explored.

1 INTRODUCTION

1.1 Environmental legislation

Legislation (Waste Act 2008, National Norms and Standards 2013) in South Africa relating to the environmental requirements of mine waste storage has been updated (Waste Amendment Act 2014) with a more stringent requirement required from the mine owners and operators with respect to the treatment of their waste, including tailings. Consequently, higher amounts of capital and operational investment is likely to be required to deal with the deposition of mine tailings. A possible solution to reduce the mine investment is to incorporate encapsulated geotextile tubes to form the main structural body of the TSF dam/embankments. The financial benefits of including geotextile tubes is that even the finer fraction of the tailings could be used within the construction of the tailings dam which leads to a cost-effective use of a traditional waste product. There is also potential sustainability benefits due to the re-use of a former waste product, which substitute's coarser grained aggregates traditionally used to construct the tailing dam structures.

1.2 Geosynthetics

For several decades geosynthetics have been used successfully for mine related applications including the capping of sludge lagoons, geogrid reinforcing of retaining structures, including dam walls, reinforcement of access roads, and lining of TSF's and other impoundment facilities. In fact, almost all types of geosynthetics for the purposes of separation, filtration, drainage, reinforcement, protection, sealing and erosion protection can be applied within the wide application field of tailings dams.

This paper presents the first concepts of dewatering tailings using geotextile tubes. The dewatering processes, tube dimensioning and methodologies for stability analysis are introduced.

2 DEWATERING TUBE COMPONENTS

2.1 Geotextile tubes

Geosynthetic dewatering tubes were originally developed for the purpose of dewatering of sewage and various sludge sediments. The standard dimensions vary from small tubes with 30 m³ storage volume up to 65 m long tubes with a containment capacity of approximately 1,600 m³ per tube. In combination with carefully selected flocculation agents almost every type of tailings/sludge can be dewatered within the geotextile tubes, including both organic and inorganic substances. Once filled, the dewatering tubes are elliptically shaped, stable long geotextile containment elements designed with a dewatering and storage function.

Normally the dewatering tubes are furnished with inlets, distributed along the longitudinal axis of the tube. The tube filling is undertaken through these nozzle inlets with the processed tailings slurry. Tubes can be installed in a single layer or stacked with multiple layers (Figure 1) to form a pyramidal type geometry (see Wilke et al 2015 for case study on stacked tubes).

2.2 *Flocculation*

The majority of the tailings is likely to be at the lower end of the grading curve i.e. clay and silt sized particles. Such particle sizes would take a long time to dewater within the geotextile tube, therefore additional flocculation agents are added to the tailings before they are pumped into the geotextile tubes to increase the speed of dewatering (the exception would be coarser tailings similar to a medium grained sand size which would have sufficient inter grain pore space to dewater under pumping pressure and self-weight).

There are several ways to agglomerate the finely suspended solids in order to increase the water release capacity and enhance the dewatering performance of the tailings (Wilke & Breytenbach 2015). Two basic bonding or agglomeration principles exist: coagulation and flocculation. Flocculation, which to date has been most commonly used in conjunction with geotextile dewatering tubes, will be briefly explained.

The flocculation is the step where destabilized colloidal particles are assembled into aggregates which can then be efficiently separated from water. Flocculants clarify water by combining with suspended solids, in such a way as to enable these particles to be quickly and easily separated from the water.

Flocculation agents can be produced from different raw materials including polyacrylamides, starch, chitin and minerals. Depending on the particle characteristics (size, charge, etc.) and the sludge properties (pH, concentration of suspended solids) an appropriate flocculation agent can be selected. This type of dewatering methodology extends the range of different dewatering possibilities and a potential high efficiency relating to the final dried solid content, the containment function and the overall safety factor.

3 DIMENSIONING OF GEOTEXTILE TUBES

The correct dimensioning of the geotextile dewatering tubes combines two different design aspects: the estimation of the required tensile strength of geotextile and selection based also on filter criteria.

3.1 *Required strength of tube*

The required tensile strength of the geotextile tube is estimated using the linear membrane theory and depends on the circumference, maximum filling height and maximum pumping pressure (Leshchinsky *et al.* 1995). To simply the calculation procedure and to add speed, computerized software including GeoCoPs (Adama) and SOFFTWIN (Palmerton) can be used for the estimation of the tensile strength in the tube fabric. Additionally, specific reduction factors for the geotextile tube can be applied to estimate the design/allowed strength. The reduction factors represent the loss of strength due to creep (RF_{CR}), installation damage (RF_{ID}), weathering (RF_W) and chemical and biological effects (RF_{CH}) related to the short term tensile strength of the fabric. Further information on the derivation of the required reduction factors are described in detail in ISO TR 20432:2007.

Special attention should be paid to the fact that the required tensile design strength increases exponentially with increasing filling height. During the tube dimensioning, a proper optimization of the theoretical diameter, maximum filling height, storage capacity of the tube and the required tensile design strength of the geotextile tube all have to be performed to ensure an optimal safe and stable design for the filling and dewatering operation.

3.2 *Required filter criteria*

With regard to the ideal filter behavior of the geotextile tube several classic design criteria can be applied. A detailed overview of filtration of natural filers and geotextile filters can be found in Giroud (2008).

Figure 1. Stacked dewatered geotextile tubes (5 tube layers high) after Wilke et al 2015.

4 CONSTRUCTION OF TAILINGS DAMS

4.1 *Traditional construction*

The construction of tailings dams is closely linked to mining operations and the characteristics of the tailings highly depends on the type and composition of the mined mineral. The tailings dam is raised in line with the mine waste storage requirements over the life of mine. The basic construction concepts for tailings dams are the upstream, downstream, or centre-line methods. The selection of the most favourable construction method is based on the tailings, embankment fill or discharge requirements, water storage suitability, seismic resistance, raising rate and relative costs (U.S. EPA 1994). All construction techniques are linked by the construction of a starter dam/dyke. This generally is built before the mining process starts and therefore is built similar to conventional water storage dams. If necessary, the starter dam has to be equipped with draining and sealing functions. During the mine operation tailings are discharged and the tailings dam height is increased successively by using local or imported fill materials. More detailed information is given in ICOLD Bulletin 74 (1994). The use of fine tailings in the dam body is not accepted (sand sized tailings are allowed) due to their poor mechanical and hydraulic characteristics especially in seismic areas, although their use naturally represents a high potential for cost reduction and extension of storage capacity.

4.2 *Construction methodologies using geosynthetic tubes*

When considering construction of a typical downstream Tailings Dam embankment using geotextile dewatered tubes, the following practical aspects need to be considered:

4.2.1 *Dam embankment footprint (dewatering area)*

Traditionally geotextile dewatering tubes are placed on a prepared area capable of bearing loads expected to be imposed by the filled dewatering tubes. The footprint should be capable of bearing the expected loads. Additionally, the dewatering area has to allow for sufficient drainage capacity of the dispersed water. Normally the dewatering pad consists of a containment bund, a flexible membrane liner and a gravel drainage layer.

The set-up of the dewatering field can be adapted to specific project requirements. Nevertheless, some points always have to be taken into account:

- The lining system design has to be adjusted to the degree of contamination of the sludge.
- The area on which the dewatering tubes will be placed has to be erosion-resistant. Otherwise the effluent water may erode the surface. This is particularly relevant for downstream dams of existing TSF's and also for the upstream face.
- The area has to be horizontally levelled (slope perpendicular to the longitudinal axis of the tubes $\leq 0.1\%$; slope in direction of the longitudinal axis of the tube $\leq 1.0\%$).

4.2.2 *Filling of geotextile dewatering tubes*

Tailings are transported into the dewatering tubes on the dewatering area by a manifold system. The inclusion of the flocculation agent is introduced via a mobile dosing unit (Figure 2) which can easily be positioned at a convenient location on or adjacent to the TSF. The dosing unit has an inlet pipe and an oulet pipe and the flow of tailings can be controlled via buffer tanks if necessary. A computerized dosing unit accurately controls the required inflow of the flocculent agent into the tailings stream. Additionally, the flow can be diverted in a controlled manner to every tube by use of several valves and tube inlets. Careful consideration has to be given to the tailings water retention characteristics and grading.

4.2.3 *Staged filling of dewatering tubes*

Practically, dewatering by means of geotextile tubes comprises a cyclical process, schematically shown in Figure 3.

During the first filling cycle the dewatering tube is filled to the given maximum initial design height (as determined by the design), and the filling is stopped. The static drainage of the sludge by gravitation commences as soon as the filling process is halted and following a degree of dewatering the tube can be re-filled again. During this cycle the water within the sludge is extracted, therefore

Figure 2. Example of mobile dosing unit for addition of flocculation agent into tailings flow (Courtesy of Clariant).

the volume is reduced and the solids concentration of the residual dewatered material increases. The principal process is repeated until the tube is completely filled. Subsequent consolidation and further desiccation occurs. Wilke & Cantré (2016) provide further information on the geotechnical characteristics of the dewatered soils within geotextile tubes following a consolidation period of up to 6 months. In order to maintain efficient staged construction of the tailings dam embankments and to allow for optimum utilization of tailings storage facilities, the dewatering tubes have to be placed and filled in such a manner so as to mimic regular downstream or upstream construction methodologies. Starter dam construction using dewatering tubes requires the tubes to be placed in the direction of the tailings dam embankment. By doing so this will enable the relatively fast installation of the subsequent geomembrane barrier and geomembrane protection layers. As future tailings dam raises become required, the dewatering tubes have to be installed in the same directional manner (Figure 4).

When considering the equal loading of the base of the tailings dam footprint it can be considered to place the geotextile dewatering tubes which do not form part of the starter walls perpendicular to the embankment direction, this will also provide greater embankment stability.

5 STABILITY AND SAFETY CONSIDERATIONS

The failure of tailings dams can result in huge environmental damage as well as loss of life. Therefore the long-term functionality and safety of TSFs are of the highest concern and aim of the tailings dam design. As detailed within ICOLD Bulletin 74 (1989), a careful theoretical and experimental investigation of new design concepts and/or unconventional construction methods and materials are greatly recommended. Thus the first safety observations are presented.

The knowledge of all potential failure mechanisms is the base of all safety considerations. The innovative construction methodology presented has to fulfil all safety provisions compulsory for conventionally designed and constructed TSFs. In accordance with U.S. EPA technical report EPA 530-R-94-038 (1994) checks are required including slope failure from rotational sliding, foundation failure, erosion, piping and overtopping.

A guideline for structures built with single geotextile-encapsulated sand elements including geotextile bags, mattresses, tubes or containers was developed by Bezuijnen & Vastenburg (2013). The fault tree shown in Figure 5 comprises all potential failure mechanisms for those systems described.

5.1 Inadequate stability

5.1.1 Waves and currents
The stability safety related to 'waves' and 'currents' (Figure 5) can be neglected for the tailings dam application.

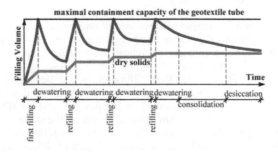

Figure 3. Dewatering cycle by use of geosynthetic tubes (adapted and modified from Lawson (2008).

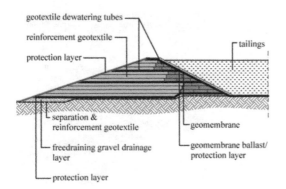

Figure 4. Concept layout of a downstream starter dam using geotextile tubes and geotextile reinforcement (if required to ensure stability).

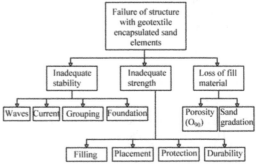

Figure 5. Fault tree for a structure with geotextile-encapsulated sand elements (Bezuijnen & Vastenburg 2013).

5.1.2 Grouping

The analysis of the 'grouping' stability should be performed in test fields (Figure 1) and/or large scale laboratory tests, in order that the understanding of interaction behaviour between the individual tailings tubes can be expanded.

5.1.3 Foundation/overall stability

The geotextile of the dewatering tube can be considered as a reinforcing element within the tailings dam embankment. Consequently, the tailings dam constructed with geotextile tubes can also be classified as geotechnical structure that is reinforced with geosynthetics. A large number of national guidelines for reinforced soil structures are available worldwide. In Germany, EBGEO (2011) deals with different geotechnical applications. However, regulations for structures built with geotextile tubes are not mentioned. Nevertheless, the provisions for geosynthetic reinforcement can be transferred to the construction methodology described above. In accordance with EBGEO, geosynthetic reinforcements applied to a slope or an embankment have to be analysed with regard to the following failure mechanisms:

- Rupture of geotextiles reinforcement
- Pull out of geotextile reinforcement
- Slope Stability
- Sliding on the embankment base
- Sliding along reinforcement layers
- Bearing failure
- Squeezing out of subsoil.

Recommended design procedures to check the above failure mechanisms are included in greater detail in EBGEO or other National Guidelines (e.g. BS8006, SANS: 207). The required geotechnical ultimate limit state analysis determining the safety against failure of the structure conventionally are proven using analytical methods. The possibilities of considering geosynthetic tubes in a conventional analytical slope stability software are limited. The improving/contributing effect of geotubes can be implemented via apparent shear parameters and/ or using their design tensile strength. Xu & Sun (2008) introduced an apparent cohesion arising in small geotextile bags using Eq. 1. This formula was developed and numerical proven for bags measuring several decimetres.

$$C = \frac{T}{\sqrt{K_p}} \cdot \left(\frac{K_p}{H} - \frac{1}{B} \right) \quad (1)$$

where c is apparent cohesion [kN/m2], T the tensile design strength of soil bags [kN/m2], K_p the passive earth pressure coefficient ($Kp = (1 + sin\,\phi)/(1 - sin\,\phi)$), H the height of the

single geotextile bag [m] and B the width of a single geotextile bag [m].

The possibility of transferring the apparent cohesion approach to large dewatering tubes has not been proven yet. The apparent cohesion for three different tensile strengths was estimated (Figure 6) with the width of geotextile tubes varying between B = 1.0 m and B = 15.0 m. The results show shows that there is only a small increase in apparent cohesion between the narrow and wide width of the geotextile tube.

Another method adopts slope stability analysis using commercial software (GGU Stability). An embankment of H = 10 m and slope inclination of 1:2 was modelled with a dam crest width of L_w = 5.0 m. and a friction angle of 20° was used for the tailings. A tensile strength of 40 kN/m was assumed for the dewatering tubes (H = 2.0 m, B = 10.0). This relates in accordance with Eq.1 and Figure 6 to c' = 26 kN/m².

The stability analysis is based on the global safety concept without taking into account partial safety factors. The resulting Factor Of Safety (FOS) is presented in Table 1. The apparent cohesion approach considers the friction angle and results in a higher factor of safety. In comparison the results using the tensile design strength approach appears more conservative.

In further analysis the use of these approaches shall be determined with regard to their applicability for tailings dams. In this comparison drained shear parameters of the tailings have been taken into consideration. Another important point is the stability during the filling of the tubes. With regards to this, test results showing the development of the undrained (vane) shear strength are presented in Wilke (2016). The applicability of the undrained shear strength in the slope stability analysis has to be analysed in further detail as it depends on the filling and construction schedule. Due to the large period of construction time

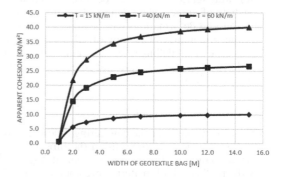

Figure 6. Apparent cohesion v width of geotextile container calculated using Eq. 1, H = 2,0 m. φ' = 20°.

377

Table 1. Results of comparison between different design approaches.

Approach	Angle of friction	Cohesion c	Design strength (single layer) T	Design strength (double layer) T	Factor of safety
	[°]	[kN/m²]	[kN/m]	[kN/m]	[–]
Apparent cohesion	20.0	26.0	–	–	2.38
Tensile design strength	20.0	0.0	40	80	1.49

the undrained conditions may only dominate in the top dewatering tubes. The occurrence of pore water pressures in the tailings dam has to be known over the entire life span of the tailings dam structure.

In addition to the slope stability, failures due to sliding on the dam base and along individual geotextile tube surfaces have to be considered. The possible slip surfaces can be in between geotextile tube fabric and tailings or in between two adjacent geotextile dewatering tubes. In this regard the friction angle Φ should be estimated for example from shear box tests. It is likely that the friction between two dewatering geotextile tubes is the critical slip surface within a tailings dam. Figure 4 shows the addition of horizontal layers of geotextile reinforcement and these may be required if the safety against internal slope failure is not sufficient with the dewatered geotextile bags alone. The additional wrapping of the horizontal reinforcement around the geotextile tubes at the upstream and downstream edges also increases the anchorage. A higher strength horizontal geosynthetic reinforcement (high strength geogrid or geotextile) may be required at the base of the dam structure to control overall stability, sliding stability and/or foundation extrusion. The requirement of horizontal reinforcement is determined at the design stability stage. In addition to the ultimate limit state analysis discussed an estimation of the serviceability during both construction and service phase can be investigated with the help of numerical modelling. The stress distribution and deformations can be analysed as well as the interaction behaviour of all components. This type of modelling research is currently being undertaken and the results will soon be available and published.

5.2 Inadequate strength

The dimensioning of the dewatering tube in terms of both 'inadequate strength' in the tube geotextile in relation to 'filling' and 'placement' together with the 'loss of fill material' are components of the dewatering operation described/referenced in Section 4.

The 'protection' and 'durability' of the geotextiles tubes once they have been filled and are in place within the tailings dam are primarily related to issues of U.V. protection, vandalism and accidental impact forces. Typically, a thin veneer of cover soils will protect the tubes.

6 SUMMARY

In this paper concepts about the utilization of geotextile dewatering tubes to form the embankments of tailings dams have been presented. This innovative construction methodology enables the increase of the storage capacity in the pond by using the tailings as embankment fill material within the tailing dams. The use of dewatering geotextile tubes has a large potential to lower the transportation and installation of imported granular fill materials and the associated capital expenditure and carbon footprint. The principles of the dewatering process and the dimensioning of dewatering tubes with regard to the required tensile strength and filter stability of the fabric was demonstrated. In recent dewatering projects experience in the high stacking of dewatered geotextile tubes has successfully been undertaken (Wilke 2015). With regard to practical aspects the main components of a tailings dam built with geosynthetic dewatering tubes as well as the downstream construction method were also introduced. In order to avoid any performance failures of a tailings dam embankment constructed with dewatering geotextile tubes, a precise design considering the main failure modes is required. The current lack of design guidance can be compensated with a combination of existing guidelines dealing with tailings dams, encapsulated sand containers and geosynthetics in retaining structures. It is considered that methods of apparent cohesion and/or tensile strength design approaches, coupled with appropriate numerical modelling, will provide sufficient design robustness to prove the safety of such structures. The undertaking of further research, comprising further stability analysis (analytical and numerical), test fields and laboratory tests, will improve our knowledge of the behaviour and interaction of stacked dewatered geotextile tubes to form dam embankment structures. The knowledge of these subjects remain essential for the stability analysis of tailings dams built with dewatering geotextile tubes and new research will continue to be published through 2016.

REFERENCES

Adama Engineering, 1996. GeoCoPS Computer Program. ADAMA Engineering, Newark, Delaware.

Bezuijen, A. and Vastenburg. E.W. 2013. Geosystems—Design Rules and Applications, 2013, Leiden: Balkelma.

EBGEO. 2011. Recommendations for Design and Analysis of Earth Structures using Geosynthetic Reinforcements. DGGT, Ernst & Sohn, Essen-Berlin.

Giroud, J.P. 2010. Criteria for geotextile and granular filters, Proceedings of the *9th International conference on geosynthetics*, Brazil: 45–64.

ICOLD, 1989. Tailings Dams safety, ICOLD Bulletin 74, ICOLD, Paris.

ISO TR 20432: 2007. Guidelines for the determination of the long-term strength of geosynthetics for soil reinforcement, ISO. Geneva, Switzerland.

Lawson, C.R. (2008) Geotextile containment for hydraulic and environmental engineering, *Geosynthetics International*, 15 No. 6: pp. 384–427, IGS.

Leshchinsky, D. Leshchinsky, O. Ling, H.I. and Gilbert, P.A. 1995: Geosynthetic Tubes for confining pressurized Slurry: Some Design Aspects. *Journal of Geotechnical Engineering*, Vol. 122, No. 8, pp. 682–690.

National Environmental Management: Waste Act 2008 (Act 59 of 2008). Department of Environmental Affairs, Republic of South Africa.

National Norms and Standards for disposal of waste to landfill. National Environmental Management: Waste Act 2008 (Act 59 of 2008). 7 Notice R 636 August 2013. Department of Environmental Affairs, Republic of South Africa.

National Environmental Management: Waste Amendment Act 2014 (Act 26) Government Gazette, June 2014. Deptartment of Environmental Affairs, Republic of South Africa.

Palmerton, J.B. 1998. SOFFTWIN—Simulation of Fluid Filled Tubes for WINdows. Computer Program. 1998.

Syllwasschy, O. Wilke, M. 2014. Sludge Treatment and Tailings Pond Capping by the Use of Geosynthetics, *7th International Congress on Environmental Geotechnics*, Melbourne, 10. November 2014.

U.S. EPA, 1994. Design and Evaluation of Tailings Dams, Technical Report EPA 530-R-94–038.

Van Kessel, M-T. Breytenbach, M. 2016. Common practice and innovations in tailings dams using geosynthetic tubes. (ed.), 84TH *ICOLD general meeting, Johannesburg, 15–20 May 2016.* (submitted and accepted, to be published).

Wilke, M. Breytenbach, M. Reunanen, J. & Hilla, V-M. 2015. Efficient environmentally sustainable tailings treatment and storage by geosynthetic dewatering tubes, Vancouver 25–28 October. *Proc. intern. symp Tailings and Mine Waste*

Wilke, M. & Cantré, S. 2016. Harbour maintenance dredging operations—Residual characteristics after treatment by geosynthetic dewatering tubes, Miami, 10–13 April, 2016. *GeoAmericas 2016—The 3rd Pan-American Conference* (submitted and accepted, to be published).

Xu Y. Huang J and Sun D. 2008. *Earth reinforcement using soilbags*, Geotextiles and Geomembranes, Volume 26, Issue 6, June 2008, 279–289.

Proceedings of the first Southern African Geotechnical Conference – Jacobsz (Ed.)
© 2016 Taylor & Francis Group, London, ISBN 978-1-138-02971-2

Geogrid reinforced load transfer platforms on piles—design, practice and measurement results

C. Psiorz & L. Vollmert
BBG Bauberatung Geokunststoffe GmbH & Co. KG, Espelkamp-Fiestel, Germany

S. Westhus & A. Post
NAUE GmbH & Co. KG, Espelkamp-Fiestel, Germany

ABSTRACT: Weak soils often provide the risk of settlements which can have massive effects on the project design. In many cases settlements have to be limited or avoided. Measures like piling foundation provide sufficiently investigated technical solutions to meet project requirements. Such solutions result in large construction efforts and can lead to significant additional costs for a project. To deal with available project budgets geosynthetic solutions can be considered to decrease the required number piles and lead to a more economic construction method. Using geosynthetic reinforcement products to bridge larger distances between piles and furthermore to limit the amount of (differential) settlements within the so called load transfer platform, has become state of the art. Geogrids are best suited for such applications. Several standards present design approaches (e. g. EBGEO, BS 8006) which have been modified during the last years based on new research results and large-scaled field trials. Verification is provided by individual projects, for example the Hamburg HafenCity. Measurement techniques have been used in other projects as well to determine resulting stresses and strain in the reinforcement and to get a better understanding for load transfer between the piles.

1 INTRODUCTION

The beginning of the construction method "Geogrid reinforced load transfer platform on piles" starts in the mid of the 1990s. Detailed information on the first experience can be found at Rogner & Stelter (2002) for road constructions and for railway applications at Gartung et al. (1996) and Vogel (2006). Main idea of the construction is the use of geosynthetics, spanning the pile caps, for improved load transfer of life and dead loads. The second version of EBGEO from 2010 defines design rules which have been introduced the first time in Germany for piled embankments in combination with geosynthetic reinforcement, spanning the pile caps.

The geosynthetic works as a membrane, spanning the pile caps, as well as reinforcement to reduce the risk and deformations of punching effects. The vertical loads are redistributed and concentrated on the stiff pile caps due to arching effects within the fill. A drawing illustrating the principle of this effect is shown in Figure 1.

The main design approaches used worldwide are EBGEO and BS 8006. Both design approaches consider utilization of the geosynthetic reinforcement as tensioned membrane, and resulting chance

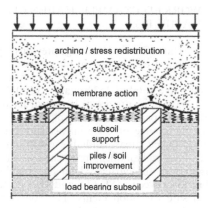

Figure 1. Principle of soil-geosynthetic interaction of load transfer platforms (LTP) (Heitz, 2005).

to increase the pile spacing for a more economic construction.

The compound system leads to a concentrated loading on the pile caps, while the geosynthetic reinforcement is absorbing the reduced loading between the piles (compare Figure 1). The stresses acting on the reinforcement can be reduced due to consideration of the available Young's modulus of the weak soil layers. In contrast to the loading on the weak soil layer can be reduced by increase of tensile stiffness of the geosynthetic reinforcement layer(s). In special cases (e.g. the drop of the groundwater table and progressive subsoil subsidence) a complete load absorption by the geosynthetic has to be considered. Usually a multi-layered system leads to a higher ductility of the system is preferred. Most applications require the installation of two reinforcement layers anyway due to required biaxial load transfer.

Furthermore the reinforcement can absorb lateral forces from embankment structures which can be considered in the design.

Apart from the Ultimate Limit State (ULS) the Serviceability Limit State (SLS) has to be analyzed as well. While the ultimate limit state considers the creep rupture strength of the reinforcement, the SLS is focusing on limited strain levels (e.g. 3% according to BS 8006). Therefore the long-term stress strain behavior of the reinforcement product needs to be considered (pay attention to product specific characteristics) and requirements on maximum creep strain are defined.

The following case studies will focus on the design and construction to provide an overview on the capability of geosynthetic reinforced load transfer platforms.

2 WORKED EXAMPLE: HAFEN CITY HAMBURG

At the HafenCity in Hamburg, previously part of the harbor area, the current inner-city area south of the historic warehouse district, is being extended by 40% to a total of 157 hectares. 5,500 dwellings for 12,000 residents are to be constructed, along with office space for 40,000 employees.

In the course of these infrastructure projects, the trafficked areas—with the exception of the quay and embankment promenades—are being raised from the current Mean Sea Level (MSL) +5 m to between MSL +7.5 and +8.0 m, to make them safe for flood events. The soil conditions around the present Honkongstrasse (formerly Magdeburger Strasse) are typical for the HafenCity Hamburg. Fill material of low bearing capacity overlies soft organic layers of clay and peat which in turn overlie firm sands. Raising the level of the road embankment by approximately 3.0 m would have resulted in long-term settlements of between 300 and 400 mm, and significant differential settlements would have been expected.

The system is characterized by vertical columns (lime-cement treated gravel, unreinforced) and an overlying sand layer horizontally reinforced with geogrids. The placement grid of the supporting elements should be designed to transfer the geogrid loads in an orthogonal manner. For Hongkongstrasse, this resulted in a rectangular grid with a spacing of 2.3 m normal to the embankment axis and 2.5 m in the axial direction, the diameter of the elements was 0.6 m. Reinforced concrete columns with continuous steel reinforcement were used at the edge of the structure to cope with a bending moment (e.g. should any excavation be required at a later date) as a result of lateral pressure. The geogrid reinforcement is installed 150 mm above the columns in order to guarantee adequate safety against shear during the construction phase, and in case of large settlements. The design required a short-term tensile strength of 400 kN/m.

To avoid the risk of subsequent construction activity of investors endangering or destroying sections which have already been built, an area of 1.0 m of the traffic section must be able to be removed. Fill which intrudes into investor areas must also be removable. In order to prevent any damage to the embankment support system, the geogrid-reinforced fill layer was built with sufficient overlap. The outer section can thus be removed in the course of normal earthworks.

The geogrid-reinforced layer was designed according to a verification concept which has already been used and proven itself several times at HafenCity Hamburg. The limited bearing capacity was first verified in accordance with a suggestion from Kempfert et al. (1997). A conservative value for subgrade reaction was used. This verification procedure does not enable any deformations to be inferred. However, verification of serviceability and of deformation limitation is compulsory for all construction projects in the HafenCity Hamburg. A complementary design procedure was therefore adopted, using a method developed and extended from EBGEO, which at the time of planning was only available in its 2004 draft stage. This extended design method was verified for similar subsoil and loading conditions (Vollmert et al. 2006). The anticipated further settlements at the level of the reinforcement were estimated at less than 50 mm after termination of construction. Comparisons with the current EBGEO (2010), available in its final form now that construction is complete, show that the design and verification of the system is sufficiently robust to cater even for the special case "Loss of Subgrade Reaction" in Load Case 3.

Extensive monitoring using horizontal inclinometer, strain gauges and stress transducers has been done for this project, reported by Weihrauch et al. (2013) and Schäfer et al. (2014). The measurement results documents the ductility of the system and robustness against subsoil influences gained by an excavation for the new Greenpeace building, right to the construction.

Figure 2 gives and overview of the planned cross-section and the installation of the system on site.

3 WORKED EXAMPLE: SCHLEUSE NEUER HAFEN, BREMERHAVEN

An essential part of the project New Harbor in Bremerhaven, Germany, in the years 2003 to 2005 was the new development of a sluice as connection between the New Harbor and the Weser River.

For the crossover of a road across the outer head of the sluice, fillings with thicknesses of up to 5 m were required in the ramp area. Due to deep clay layers with a low bearing capacity different measures for the reduction of the expected large settlements were required. Amongst others, due to the earlier installed sheet piles a dam filling on cement stabilized columns was planned. Above the columns a geogrid reinforced load distribution platform was installed to support the bearing arch and to transfer the dam loads into the pile group. The sequence of strata is particularly affected by highly compressible, holocene organic soft layers in depths of up to approx. −5 m below sea level or in a total thickness of between 15 m and 20 m. The sea silt is characterized by a predominant weak consistency, a medium oedometric modulus of $E_S = 1.5 \text{ MN/m}^2$ and an undrained shear strength of $c_u \geq 15 \text{ kN/m}^2$.

On an approximately 0.3 m thick sand layer a double-layered biaxial geogrid reinforcement together with an intermediate 0.3 m thick sand layer is installed, followed by the embankment fill (compare Figure 3). The geogrids with a biaxial short-term strength of 80 kN/m (Secugrid® 80/80/Q6) are characterized by a very high stiffness of $J_{2\%} \geq 2000 \text{ kN/m}$.

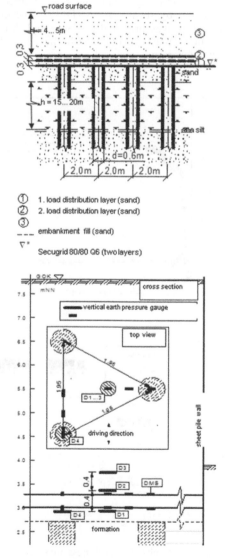

① 1. load distribution layer (sand)
② 2. load distribution layer (sand)
③
- - - embankment fill (sand)
▽* Secugrid 80/80 Q6 (two layers)

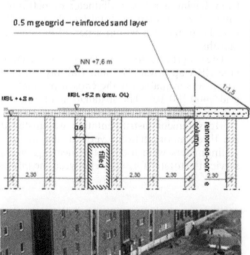

0.5 m geogrid−reinforced sand layer

NN +7,6 m

UBL +4,2 m   UBL +5,2 m (preu. OL)

Figure 2. Cross section—piled sand layer, Hamburg HafenCity.

Figure 3. Schematic drawing of cross section and layout of monitoring devices.

To prove the serviceability of the load distribution platform a comprehensive instrumentation of the geogrids with strain gauges is carried out. The instrumentation has been complemented by vertical pressure gauges above the cement stabilized piles and between the piles as well as by settlement measurements. To register the arching effect earth pressure cells have additionally been installed in the middle of the field between the piles (compare Figure 3) and on one pile cap. The earth pressure cells in the middle of the field are thereby ranged in the altitude in such a way that reductions of vertical stress due to arching effects can be comprehended, Figure 4. The strains occurring within the reinforcement are registered by strain gauges (DMS). The instrumentation was prepared in-house and thus was installed in-situ together with the geogrids, in only a few hours.

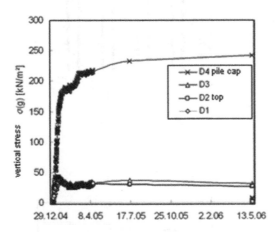

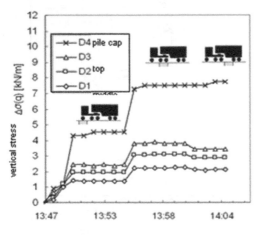

Figure 4.   Results of vertical stress transducers illustrating the load concentration on top of the piles from dead load (top) and life load (bottom).

4   WORKED EXAMPLE: REHABILITATION OF RAILWAY LINE NASSENHEIDE-LÖWENBERG

The existing railway track between Rostock and Berlin was planned to be rehabilitated for an increase in the maximum speed from 120 km/h (max. axle load 22.5 t) up to now 160 km/h (max. axle load 25 t). The railway track is crossing four moor areas with three of them having a length of ~800 m and weak soil layers up to 10 m thickness. For the planned track embankment heights of up to 10 m have to be considered which are resulting in massive permanent loading acting on the foundation system.

The design was carried out considering the design approach according to EBGEO resulting in two reinforcement layers in each direction (longitudinal and transverse) with a tensile strength of 400 kN/m.

During a complete closure for a period of 10 months the piled foundation was installed (precast concrete piles) and fill-up of the embankment structure was realized, see Figure 5.

The pile spacing (s) was considered with s = 1.80 m (Moor 1) and s = 2.00 m (Moor 2 + 3) respectively having a pile cap diameter of 600 mm. In total 7,000 piles with a total length of 76,300 m have been installed within a period of only three months.

The geogrid reinforcement was requested to provide EBA approval (German Railway Authority) which was available for the chosen product Secugrid® 400/100 R6. The installation of the multilayered system is shown in Figure 6.

Measurement instrumentation was installed in the three moor sections, vertical and horizontal inclinometers were used. After nearly 2 years in service the total deformation between the piles is in

Figure 5.   Installation of piled foundation in wintertime 2012 in moor section 3. (Source: Centrum Pfähle GmbH.).

Figure 6. Installation of load transfer platform with multi-layered geogrid reinforcement.

the range of approximately 1.5 cm which is resulting in sufficient serviceability of the structure. At trial level no deformation has been documented. Further measurements with earth pressure cells confirm the load concentration on the piles as expected.

5 CONCLUDING REMARKS

Piled embankments with lateral geosynthetic reinforcement as soil improvement techniques were widely used in the last 20 years and certain progress in understanding of the structures could be made. Exemplarily, three projects are presented here. The prime objective of the soil improvement is, in all cases, the reduction of subsequent settlements in order to guarantee the serviceability of the trafficked areas under operating conditions. Depending on the construction process of the columns, this method can be used even close to existing structures. Under the conditions applying at the HafenCity, the pile-supported, geogrid-reinforced method is usually somewhat cheaper to construct than an expanded-clay fill layer.

A realistic estimation of bedding modulus and deep foundation system play the key role to realize safe and still economic designs for geosynthetic reinforced load transfer platforms. Among the available design methods, the calculations following the current EBGEO and BS 8006 design codes lead to satisfactory results.

REFERENCES

BS 8006–1 (2010): *Code of practice for strengthened/ reinforced soils and other fills*. BSI Standard Publication, 2010.

Deutsche Gesellschaft für Geotechnik e. V., Arbeitskreis 5.2. 2010. *Empfehlungen und die Berechnung für den Entwurf und die Berechnung von Erdkörpern mit Bewehrungseinlagen aus Geokunststoffen* (EBGEO). Berlin: Wilhelm Ernst & Sohn.

Kempfert, H.G., Stadel, M. & Zaeske, D. 1997. *„Berechnung von geokunststoffbewehrten Tragschichten über Pfahlelementen"*, Bautechnik, Nr. 74(12), S. 818–825, 1997.

Rogner, J. & Stelter, J. 2002. *Bauverfahren beim Straßenbau auf wenig tragfähigem Untergrund—Aufgeständerte Gründungspolster*. Berichte der Bundesanstalt für Straßenwesen, Heft S 26, Bergisch-Gladbach.

Tost, S. 2015. *Weicher Untergrund trifft steife Geogitter— zweigleisiger Ausbau der DB-Strecke im Bereich zwischen Nassenheide und Löwenberg mit aufgeständertem Fahrweg*, NAUE-Kolloquium, Montabaur.

Vogel, W. 2006. *Erfahrungen über den Einsatz geogitterbewehrter Bodenkörper auf Säulen für Eisenbahnfahrwege bei Neubau- und Ausbaumaßnahmen der DB*. Symposium Geotechnik—Verkehrswegebau und Tiefgründungen, 26. September 2005. In: Schriftenreihe Geotechnik, Universität Kassel, Heft 18, September 2005.

Vollmert, L. 2014. *Piled embankments in soft estuary clay—Experience from design and field measurements for redevelopment of harbor areas in Northern Germany*. XV Danube—European Conference on Geotechnical Engineering (DECGE 2014). Paper No 132.

Vollmert, L., Kahl, M., Giegerich, G. & Meyer, N. 2006. Schleuse *Neuer Hafen, Bremerhaven—In-situ Verifizierung eines erweiterten Berechnungsverfahrens für geogitterbewehrte Gründungspolster über vertikalen Traggliedern*. Beitrag zur Baugrundtagung. Bremen: DGGT.

Weihrauch, S., Oehrlein, S. & Vollmert, L. 2013. *Subgrade improvement measures for the main rescue roads in the urban redevelopment area HafenCity Hamburg*. Proceedings of the 18th International Conference on Soil Mechanics and Geotechnical Engineering, Paris.

Zaeske, D. 2001. *„Zur Wirkungsweise von unbewehrten und bewehrten mineralischen Tragschichten über pfahlartigen Gründungselementen"*, Dissertation, Schriftenreihe Geotechnik Universität Kassel, Heft 10, 2001.

Proceedings of the first Southern African Geotechnical Conference – Jacobsz (Ed.)
© *2016 Taylor & Francis Group, London, ISBN 978-1-138-02971-2*

Long-term experience with a geogrid-reinforced landslide stabilization

D. Alexiew & P. Assinder
Huesker Synthetic GmbH, Gescher, Germany

A. Plankel
3P Geotechnik, Bregenz, Austria

ABSTRACT: In summer 1994 a landslide occurred just below a ski-lift station in the Austrian Alps. It blocked the road at the toe of slope and endangered the stability of the lift station. A quick solution for the slope stabilization and reconstruction had to be developed and executed. The solution had to meet a wide range of requirements, some of them controversial. Finally, a geogrid-reinforced full-height slope was designed and constructed reusing the local soils and reconstructing approximately the former natural slope shape. The system was successfully built in less than two months and is still stable after twenty years of service. This is most probably the first geogrid landslide stabilization at least in Europe. Problems, boundary conditions, philosophy and design from 1994 are described together with the unknown factors and specific solutions, and the construction technology and experience as well. The current state is described and commented.

1 INTRODUCTION

The region of Lech is one of the most famous ski regions in the Austrian Alps (Figure 1). A huge net of ski trails and lifts is available. They have to be integrated in an optimal way into the natural landscape.

In summer 1994 a landslide occurred in a natural slope just below the lower ski-lift station at the so called Steinmähder Wand (Figure 1).

2 OVERVIEW AND SOME FACTS

In Lech snow fall usually starts in October and ends in April. The landslide occurred (surprisingly) in July 1994. There were some days at that time with up to 30 mm rain per day, but such values are believed not to be critical. Average precipitation over the year is in the range of 30 to 40 mm/day without a significant scattering. The most probable trigger was may be a "global" over-wetting due to massive "delayed" snow melting on the slopes above the lift station (Figure 1).

The slid soil mass blocked an unpaved but important alpine road at the toe of slope. The upper part of the sliding surface reached the foundation of the lift station on top of slope destroying the earth platform in front of the station and endangering the entire building (Figure 2) despite the micro-piles below the front part of foundation. The width of landslide amounted to about 50 m, the height varied from 23 to 25 m.

Twenty years ago spontaneous quick landslides occurred relatively rare. As known, in the meantime the problem of "surprising" landslides and avalanches has increased worldwide, Alexiew (2005), Alexiew & Bruhier (2013). Thus, the case described herein is even of greater importance.

Approximately in the same position a former smaller lift station had existed for 14 years; the new one was built in 1993, i.e. one year before the landslide occurred.

3 GEOTECHNICAL CIRCUMSTANCES

Long-term experience with the "older" smaller station did not indicate possible instabilities or

Figure 1. The ski region of Lech in the Austrian Alps (the Steinmähder Wand is marked).

geotechnical problems. However, because of some scattering in the geotechnical and hydrological data from the reports in 1993 (for the new station), intuitive slope stability doubts and the more or less permanent risk of oversaturation, in 1993 a proper drainage around the new station was recommended and—on the safe side—the installation of micro-piles below the foundation near the slope.

After landslide neither a specific soil layer as possibly pre-defined sliding plane (e.g. clay), nor concentrated wells (despite some extremely wet areas) could be visually identified. The failure surface was three-dimensional, non-planar and quite irregular (Figure 2).

The typical data of the local slope soil were:

talus material, sandy gravelly silt, coefficient of uniformity ca. CU = 50, gravel 22%, sand 27%, fines 51% (silt 42%, clay 9%), unit weight ≈ 18 kN/m^3, Proctor density ≈ ρd = 2.07 g/cm^3, w_{opt} = 11.3%. The soil fraction < 2 mm (fines and sand) possesses an angle of internal friction φ'peak = φ'post-peak = 36.5° (say no indication of "progressive" shear failure tendency) and a cohesion c' ≈ 0 to 5 kPa (a bit surprisingly modest). The coefficient of permeability had not been tested, but based on granulometry, soil classification and local experience it was obvious, that the soil is not "free draining".

Figure 2. Part of landslide with endangered lift station (Photos from 01 Aug 1994).

4 WHAT HAD TO BE DONE AND HOW?

The slope had to be "repaired" as quickly as possible due to both the endangered lift station and the blocked road. The frame conditions for an "ideal solution" from the point of view of the owner "Skilifte Lech" were:

– guarantee in the long term the stability and serviceability of the lift station (i.e. a permanent solution was required),
– reconstruct the platform in front of the station,
– protect the entire slope against similar problems in future,
– reconstruct to the greatest possible extent the natural shape of terrain as prior to failure,
– reuse the local soil (slid masses) so far as possible,
– minimize the transport (import) of any construction materials in terms of weight and volume,
– put the lift station in operation latest end of September (beginning of the ski season), say the time to find an optimal solution, to design it and to build was about two months,
– use a simple and adaptive technology avoiding a special contractor's qualification and skills,
– "green" solution (vegetated surface) to fit the green natural surroundings,
– keep the total costs as low as possible (!).

These requirements were complex, to some extent controversial and not easy to meet.

5 LEVEL OF KNOWLEDGE AND DIFFICULTIES IN 1994

The solution finding process was in the same time very intensive and quick due to the stringent time limitations mentioned above. A geosynthetic reinforced soil block stabilizing the slope and regenerating the landscape was identified as generally the best solution. All the thoughts and discussions from this time cannot be explained herein due to brevity.

Note, that at that time the experience with high geosynthetic reinforced slopes—being today a routine—was quite modest, and no geosynthetic-based system or solution for landslide stabilization was known (even to our best knowledge today it had never been built at least in Europe before 1994).

Some additional difficulties were faced:

– scattering of soil parameters was not known, and there was no time for additional sufficient soil testing. There was also no time for analysis of the failure and its reason(s). Thus some hypotheses and assumptions had to be made by engineering judgement and on the safe side.

- the hydrological situation was not clear enough. Consequently, an extensive drainage had to be provided.
- there were no standardized or codified calculation and design procedures for such systems, not to mention corresponding software.
- difficult access to the site: unpaved narrow mountain roads.
- there was no experienced contractor for such a work.

6 SUGGESTED SOLUTION

A typical simplified cross-section is depicted in Figure 3.

The solution as depicted is may be the third or fourth (and final) version. Some previous drafts became step by step modified.

Some short comments:

It was decided to reuse totally the local soil (slid mass) as fill (compare Chapters 4 and 5). The authors were optimistic due to its good compactibility (wide gradation) and its sufficient shear resistance (Chapter 3) despite the relatively low permeability expected. In 1994 such a decision seemed risky: state-of-the-art was the exclusive use of high-quality non-cohesive completely free-draining fills.

However, due to the modest permeability intensive drainage had to be foreseen.

Geosynthetic drainage mats were preferred to a gravel layer in the interface to the sliding surface: they are easier to transport and install. The mats were connected to a drainage pipe (Figure 3). They had to be installed upslope vertically in a "zebra" pattern: 1 m mat, 1 m spacing, i.e. after each mat

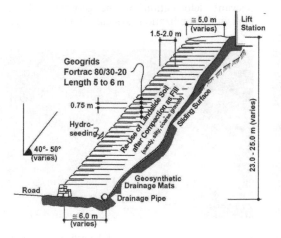

Figure 3. Typical cross-section of the landslide stabilization solution (Note: geometry is generally varying).

there was 1 m of direct contact of the compacted fill to the local soil. The reason was that usually the interface "mat to soil" has a lower shear resistance (bond) then the internal shear resistance of the contacting soils itself; due to the extremely tight schedule there was no time for testing. Thus, it was believed that the risk of creating an artificial surface prone to sliding in the case of continuous full-area drainage mat installation is too high.

As geosynthetic reinforcement providing the required stability high-tenacity low-creep flexible polyester geogrids from the Fortrac® family were chosen being easy to transport and install due to their low weight and flexibility.

Due to logistic reasons, the extremely tight time schedule and the not precisely known total final amount of reinforcement it was decided to use a unified single type of geogrid for the entire system, although a differentiation could result in a more cost-efficient solution. The geogrids were 5 m wide and exhibited based on previous tests for other projects a very high coefficient of interaction (bond) to a wide range of soils. There lengths varied typically from 6 m to 5 m. They had to be installed in the so called "wrap-back" manner using removable formwork without e.g. supporting steel meshes in front, thus creating a "soft" nature-alike reconstituted slope surface to be vegetated by hydro-seeding.

The owner decided that no additional anti-erosion geotextile should be installed at the inner side of the geogrids at the facing (front) as it was suggested by the authors.

For stability analysis a design software under MS DOS was used based on a simplified version of a polygonal block-sliding method in combination with some additional checks using Bishop's method. In 1994 there was no complete specialized software available in the market dealing specifically with geosynthetic reinforcement. Finally, some decisions had to be made by engineering judgment.

For the choice of the unified geogrid (see above) two options seemed possible: a lighter geogrid with an Ultimate Tensile Strength (UTS) of 55 kN/m at 0.5 m vertical spacing and a stronger one with an UTS of 80 kN/m at 0.75 m. The latter option was chosen (Figure 3) to make installation, fill compaction and construction quicker, although at that time no experience with such a big vertical spacing was available.

Due to brevity no further details can be explained herein.

A fill compaction of at least $D_{pr} = 98\%$ was prescribed.

The design and technology concept avoided any export or import of soils, steel, concrete etc.: only the light-weight, low-volume geosynthetic rolls had to be transported.

7 EXECUTION AND EXPERIENCE

Construction started mid of August 1994 and was completed just in time end of September. A removable formwork was used for the front. Some construction stages are shown in Figures 4 and 5.

Figure 4 shows at a glance all components of the system and the compaction procedure. Note the use of heavy compaction for the 0.75 m thick fill layers and the installation of drainage mats with a space in between (Chapter 6). One can see (Figure 5) that due to the lack of experience (the contractor had never built such a system) and the extreme time pressure the geometry at the front of the first geogrid-soil layers was not really precise. However, because this was not critical for stability and due to the anyway intended nature-alike final shape, this was tolerated.

Quality of compaction was controlled at several points at every fill layer.

All the time the structure geometry had to be adapted to reconstruct so far as possible the shape of the previous natural slope (Chapter 4) keeping always the design basics as per Figure 3. This procedure was quite unique: strictly speaking complete precise execution drawings had never existed; instead, engineering judgement and ad hoc decisions were applied. It was a demonstration of the flexibility and adaption capability of such systems.

Figures 6 & 7 show the geogrid-reinforced stabilization just after completion and one year later.

It should be noted that considering all the problematic factors and circumstances such a short construction time was a real achievement.

It is recognizable how the adapted soft, to some extent irregular 3D-geometry fits successfully the natural alpine landscape.

In summer 1995, after almost a year after construction and after a winter-spring cycle there were no indications of stability or serviceability problems of any type. (Note that a year ago at that time the landslide failure occurred). The hydro-seeding seemed to be quite successful although not perfect.

Then for a very long time the authors had no contact to the project. However, in such cases the motto applies: no news, good news...

8 AUGUST 2014: TWENTY YEARS LATER

In August 2014 the authors had the opportunity to visit the structure again together with the project leader of the owner in 1994 being still active.

In Figures 8, 9 and 10 the present situation is depicted.

By the way: it is interesting to compare Figures 6 and 7 from 1994 with Figure 8 from 2014.

During the site inspection no indications of any relevant deformation of the geogrid-reinforced landslide stabilization system from 1994 were

Figure 4. Construction: first layers, fill compaction by heavy equipment, formwork, geogrids, drainage mats.

Figure 5. Construction: partial view of the geogrid-reinforced slope. Note the "bellies" at front of the first layers.

Figure 6. Geogrid reinforced landslide stabilization after completion end of September 1994 (Photo from 20 Oct 1994).

found: neither vertical, nor horizontal. Although no precise measurements had been done over the twenty years passed, the systematic inspections of the owner over this period had attested no problems from the point of view of stability or serviceability of the reinforced slope or of the platform and station foundation on top.

The facing was to more than 90% vegetated, although only marginal maintenance had been done. In the meantime the typical local vegetation dominated the picture. The only spots with meager vegetation were the almost vertical parts of some "belly" layers (Figure 10). Some erosion of fines was visible (Figure 10) due to the missing anti-erosion geotextile behind the geogrids (decision of the owner in 1994, see Chapter 6). However, this had no further negative consequences. Note, that the completely uncovered geogrids at this spot are still intact despite twenty summer-winter cycles and UV impact.

Figure 7. The system in summer 1995, almost a year after completion and after a winter-spring cycle.

Figure 8. August 2014: left up: overview, reinforced slope is marked; right down: persons as a scale for the structure.

Figure 9. August 2014: view upslope.

Figure 10. August 2014: left: zoom of a vertical layer of the geogrid facing with a "belly"; right: same section in September 1994.

The inspection of the runout of the drainage system (Figure 3) showed a small but steady flow of clear water indicating its effective operation as well.

In summary: after twenty years in operation the system is completely in a good condition without any additional measures or corrections in that period.

9   FINAL REMARKS

It seems that the geogrid-reinforced landslide stabilization structure designed and built in less than two months in 1994 proves to be generally a successful, efficient and durable solution despite the unfavorable climatic, hydrological and geotechnical conditions and a number of specific disadvantageous circumstances in 1994 resulting in ad hoc solutions based to some extent on engineering judgment. Even the design procedures and tools were in 1994 quite modest.

The solution included a number of aspects being non-common in 1994 (Chapters 5 and 6).

In our opinion this demonstrates the flexibility, adaptiveness and robustness of such structures.

It is may be the first geogrid application for rehabilitation and stabilization of a landslide.

In our understanding the positive experience gained should encourage the intensive application of such solutions nowadays especially in consideration of the globally increasing landslide problems due to climate change.

ACKNOWLEDGEMENTS

The authors appreciate very much the competence, open mind and consistent support of Mr. Manhart from "Skilifte Lech".

REFERENCES

Alexiew, D. 2005. Design and construction of geosynthetic-reinforced "slopes" and "walls": commentary and selected project examples. *12th Darmstadt Geotechnical Conference. Darmstadt Geotechnics No. 13, TU Darmstadt, Institute and Laboratory of Geotechnics*: 167–186. Darmstadt.

Alexiew D., Bruhier, J. 2013. Some case studies of geosynthetic solutions while dealing with landslides and unstable slopes. *Proc. ICLR2013 (Int. Conf. on Landslide Rehabilitation)*: on CD, no pages. Draham, Tunesia.

Proceedings of the first Southern African Geotechnical Conference – Jacobsz (Ed.)
© 2016 Taylor & Francis Group, London, ISBN 978-1-138-02971-2

Author index